"十三五"国家重点图书出版规划项目

"十三五"国家重点图书出版规划项目
2018年度输出版优秀图书

中外物理学精品书系

高瞻系列·13

Progress in Nanoscale Characterization and Manipulation

纳米表征与调控研究进展

王荣明 主编
王 琛 张洪洲 陶 靖 白雪冬 等 编著

图书在版编目(CIP)数据

纳米表征与调控研究进展 = Progress in Nanoscale Characterization and Manipulation / 王荣明主编；王琛等编著. — 北京: 北京大学出版社, 2017.5
（中外物理学精品书系）
ISBN 978-7-301-28306-6

Ⅰ.①纳… Ⅱ.①王… ②王… Ⅲ.①纳米材料—研究 Ⅳ.① TB383

中国版本图书馆 CIP 数据核字 (2017) 第 102687 号

书　　　名	Progress in Nanoscale Characterization and Manipulation （纳米表征与调控研究进展） Nami Biaozheng yu Tiaokong Yanjiu Jinzhan
著作责任者	王荣明　主编　王琛　张洪洲　陶靖　白雪冬　等　编著
责任编辑	尹照原
标准书号	ISBN 978-7-301-28306-6
出版发行	北京大学出版社
地　　　址	北京市海淀区成府路 205 号　100871
网　　　址	http://www.pup.cn
新浪微博	@北京大学出版社
电子信箱	zpup@pup.cn
电　　　话	邮购部 62752015　发行部 62750672　编辑部 62752021
印　刷　者	北京中科印刷有限公司
经　销　者	新华书店
	730 毫米 ×980 毫米　16 开本　42.25 印张　插页 6　724 千字 2017 年 5 月第 1 版　2022 年 10 月第 2 次印刷
定　　　价	250.00 元

未经许可，不得以任何方式复制或抄袭本书之部分或全部内容。
版权所有，侵权必究
举报电话：010-62752024　电子信箱：fd@pup.pku.edu.cn
图书如有印装质量问题，请与出版部联系，电话：010-62756370

"中外物理学精品书系"
编委会

主　任：王恩哥
副主任：夏建白
编　委：(按姓氏笔画排序，标 * 号者为执行编委)

王力军　　王孝群　　王　牧　　王鼎盛　　石　兢
田光善　　冯世平　　邢定钰　　朱邦芬　　朱　星
向　涛　　刘　川*　　许宁生　　许京军　　张　酣*
张富春　　陈志坚*　　林海青　　欧阳钟灿　　周月梅*
郑春开*　　赵光达　　聂玉昕　　徐仁新*　　郭　卫*
资　剑　　龚旗煌　　崔　田　　阎守胜　　谢心澄
解士杰　　解思深　　潘建伟

秘　书：陈小红

序　言

物理学是研究物质、能量以及它们之间相互作用的科学。她不仅是化学、生命、材料、信息、能源和环境等相关学科的基础,同时还是许多新兴学科和交叉学科的前沿。在科技发展日新月异和国际竞争日趋激烈的今天,物理学不仅囿于基础科学和技术应用研究的范畴,而且在社会发展与人类进步的历史进程中发挥着越来越关键的作用。

我们欣喜地看到,改革开放三十多年来,随着中国政治、经济、教育、文化等领域各项事业的持续稳定发展,我国物理学取得了跨越式的进步,做出了很多为世界瞩目的研究成果。今日的中国物理正在经历一个历史上少有的黄金时代。

在我国物理学科快速发展的背景下,近年来物理学相关书籍也呈现百花齐放的良好态势,在知识传承、学术交流、人才培养等方面发挥着无可替代的作用。从另一方面看,尽管国内各出版社相继推出了一些质量很高的物理教材和图书,但系统总结物理学各门类知识和发展,深入浅出地介绍其与现代科学技术之间的渊源,并针对不同层次的读者提供有价值的教材和研究参考,仍是我国科学传播与出版界面临的一个极富挑战性的课题。

为有力推动我国物理学研究、加快相关学科的建设与发展,特别是展现近年来中国物理学者的研究水平和成果,北京大学出版社在国家出版基金的支持下推出了"中外物理学精品书系",试图对以上难题进行大胆的尝试和探索。该书系编委会集结了数十位来自内地和香港顶尖高校及科研院所的知名专家学者。他们都是目前该领域十分活跃的专家,确保了整套丛书的权威性和前瞻性。

这套书系内容丰富,涵盖面广,可读性强,其中既有对我国传统物理学发展的梳理和总结,也有对正在蓬勃发展的物理学前沿的全面展示;既引进和介绍了世界物理学研究的发展动态,也面向国际主流领域传播中国物理的优秀专著。可以说,"中外物理学精品书系"力图完整呈现近现代世界和中国物理科学发展的全貌,是一部目前国内为数不多的兼具学术价值和阅读乐趣的经典物理丛书。

"中外物理学精品书系"另一个突出特点是,在把西方物理的精华要义"请进来"的同时,也将我国近现代物理的优秀成果"送出去"。物理学科在世界范围内的重要性不言而喻,引进和翻译世界物理的经典著作和前沿动态,可以满足当前国内物理教学和科研工作的迫切需求。另一方面,改革开放几十年来,我国的物理学研究取得了长足发展,一大批具有较高学术价值的著作相继问世。这套丛书首次将一些中国物理学者的优秀论著以英文版的形式直接推向国际相关研究的主流领域,使世界对中国物理学的过去和现状有更多的深入了解,不仅充分展示出中国物理学研究和积累的"硬实力",也向世界主动传播我国科技文化领域不断创新的"软实力",对全面提升中国科学、教育和文化领域的国际形象起到重要的促进作用。

值得一提的是,"中外物理学精品书系"还对中国近现代物理学科的经典著作进行了全面收录。20世纪以来,中国物理界诞生了很多经典作品,但当时大都分散出版,如今很多代表性的作品已经淹没在浩瀚的图书海洋中,读者们对这些论著也都是"只闻其声,未见其真"。该书系的编者们在这方面下了很大工夫,对中国物理学科不同时期、不同分支的经典著作进行了系统的整理和收录。这项工作具有非常重要的学术意义和社会价值,不仅可以很好地保护和传承我国物理学的经典文献,充分发挥其应有的传世育人的作用,更能使广大物理学人和青年学子切身体会我国物理学研究的发展脉络和优良传统,真正领悟到老一辈科学家严谨求实、追求卓越、博大精深的治学之美。

温家宝总理在2006年中国科学技术大会上指出,"加强基础研究是提升国家创新能力、积累智力资本的重要途径,是我国跻身世界科技强国的必要条件"。中国的发展在于创新,而基础研究正是一切创新的根本和源泉。我相信,这套"中外物理学精品书系"的出版,不仅可以使所有热爱和研究物理学的人们从中获取思维的启迪、智力的挑战和阅读的乐趣,也将进一步推动其他相关基础科学更好更快地发展,为我国今后的科技创新和社会进步做出应有的贡献。

"中外物理学精品书系"编委会 主任
中国科学院院士,北京大学教授
王恩哥
2010年5月于燕园

Preface

Nanoscale characterization has enabled the discovery of many novel functional materials which started from understanding important relationships between material properties and morphologies. Therefore, nanoscale characterization has become an important research topic in nanoscience. It fosters the foundation for the design of functional nanodevices and applications of these nanomaterials.

The book "Progress in Nanoscale Characterization and Manipulation" is focused on charged-particle optics and microscopy as well as their applications in materials sciences. Prof. Rongming Wang acts as editor-in-chief of this volume. This book involves many cutting-edge theoretical and methodological advances in electron microscopy and microanalysis, testifying their crucial roles in modern materials research. It will be of primary importance to all researcher who work on ultramicroscopy and/or materials research.

While nanomaterials find wider and more significant applications in almost every aspect of modern science and technology, researchers have been trying to gain detailed knowledge of novel materials with atomic (even sub-A) scale resolution that are responsible for their unique properties, including chemical composition, atomic organization, coordinates, valence states, etc. This has been driving the development of ultramicroscopy. This book addresses the growing opportunities in this field and introduces the state-of-the-art charged-particle microscopy techniques. It showcases the recent progress in scanning electron microscopy, transmission electron microscopy and helium ion microscopy including the advanced spectroscopy, spherical-corrected microscopy, focused-ion imaging and in-situ microscopy. To appreciate the synergies of the above-mentioned charged-particle methods, the common features of their optical systems are summarized in the first chapter.

Our authors are active international researchers working at the forefront of the field, while we have received direction and assistance from several senior Chinese scientists (Prof. Hengqiang Ye, Prof. Fanghua Li, Prof. Ze Zhang, Prof. Junen

Yao and Prof. Xiaofeng Duan, etc.). Based on their extensive expertise in ultramicroscopy, our authors have provided many their cutting-edge research outputs and demonstrated the indispensable roles of charged-beam microscopy in the development of modern materials research. This defines the unique style of the book: an excellent integration of fundamental theories and practical applications. Therefore, it can meet the needs of a range of readers who are either working on those microscopy techniques or applying them to the investigation of advanced materials. While the development of Cs-corrected microscopy, in-situ microscopy and high-resolution spectroscopy is stepping into a golden age, it is clearly imperative to gain a big picture of the development. The book covers many timely topics and it can serve as a good reference for researchers or students working in many fields such as materials sciences, physics, chemistry, electronics, the semiconductor industry and biology.

We have received numerous constructive suggestions and comments from many colleagues. On behalf of all the editors, I would like to offer our sincere gratitude to those who have contributed to the book.

On behalf of the editors, I would like to thank all our authors. They have been working diligently. I would like to thank my fellow editors for their hard work. I hope this book can facilitate the development of microscopy techniques, inspire young researchers, and make due contributions to the field.

Prof. Rongming Wang
April, 2017

Contents

1 Electron/Ion Optics · 1
   1.1 General ray diagram of TEM · · · · · · · · · · · · · · 2
   1.2 Electron sources · 4
   1.3 Optics · 13
   1.4 Detectors · 27
   1.5 Ion optics · 32

2 Scanning Electron Microscopy · 44
   2.1 Introduction · 44
   2.2 Fundamentals of the SEM · · · · · · · · · · · · · · · · · 45
   2.3 Analytical capabilities of the SEM · · · · · · · · · · · · · 56

3 Transmission Electron Microscopy · · · · · · · · · · · · · · · · · · · 87
   3.1 Introduction · 87
   3.2 High-resolution transmission electron microscopy imaging · · · · · · 89
   3.3 A new approach to image analysis in HRTEM · · · · · · · · · · · 101
   3.4 Focal series reconstruction · · · · · · · · · · · · · · · · · 140
   3.5 Convergent beam electron diffraction · · · · · · · · · · · · 163
   3.6 Lorentz electron microscopy · · · · · · · · · · · · · · · · 177
   3.7 Electron holography · 209

4 Scanning Transmission Electron Microscopy (STEM) · · · · · · · · · · 270
   4.1 Introduction · 270
   4.2 The Principle of reciprocity · · · · · · · · · · · · · · · · 273
   4.3 Principle of STEM imaging · · · · · · · · · · · · · · · · 277
   4.4 HAADF imaging · 288
   4.5 ABF imaging · 311
   4.6 Scanning Moiré fringe imaging · · · · · · · · · · · · · · · 322
   4.7 Application on micro-area analysis · · · · · · · · · · · · · 326
   4.8 Discussion and conclusion · · · · · · · · · · · · · · · · · 327

5 Spectroscopy · 334
   5.1 Introduction · 334
   5.2 Principle of EDS and EELS · · · · · · · · · · · · · · · · · · 336
   5.3 EDS+TEM and EDS+STEM · · · · · · · · · · · · · · · · · 348
   5.4 EELS-TEM · 359
   5.5 EELS-STEM and applications · · · · · · · · · · · · · · · · · 369
   5.6 Spectrum imaging · 381

6 Aberration Corrected Transmission Electron Microscopy and Its Applications · 392
   6.1 Basics of aberration correction · · · · · · · · · · · · · · · · · 392
   6.2 Aberration corrected electron microscopy · · · · · · · · · · · 414
   6.3 Applications of aberration corrected electron microscopy · · · · 441

7 In Situ TEM: Theory and Applications · · · · · · · · · · · · · · · 499
   7.1 In situ TEM observation of deformation-induced structural evolution at atomic resolution for strained materials · · · · · · · · · · · · 499
   7.2 In situ TEM investigations on Ga/In filled nanotubes · · · · · · 515
   7.3 In situ TEM electrical measurements · · · · · · · · · · · · · 547
   7.4 Several advanced electron microscopy methods and their applications on materials science · · · · · · · · · · · · · · · · · 581

8 Helium Ion Microscopy · 629
   8.1 Introduction · 629
   8.2 Principles · 632
   8.3 Imaging techniques · 646
   8.4 Applications · 650
   8.5 Current/Future developments · · · · · · · · · · · · · · · · · 660
   8.6 Conclusion · 662

1
Electron/Ion Optics

Jing Tao, Rongming Wang and Hongzhou Zhang

Transmission electron microscopes (TEM) have very delicate optical systems in order to fulfill their research purpose, i.e., the characterization of material's structures in a versatile manner in modern science. High-energy electrons are generated, forming an electron probe with variable sizes to bombard the sample materials. The scattered electrons, which carry details of material's structures, are reflected to produce a set of data including diffraction patterns, images and spectra. The processes of the electron generation, interaction with the materials and data acquisition are all accomplished in high-level vacuum inside the microscopes. Unlike optical microscopes, most of the lenses in the TEM use electric currents in a particular geometry to generate magnetic fields, which are used to bend the path of the electron beam. Therefore, the experimental operation of TEM and interpretation of the TEM results require background in electromagnetism, vacuum physics, scattering theory and specimen-related solid state physics, among which understanding the optical structure of the microscopes should be considered as the first step of the TEM-based study to but not limited to those who plan and perform experiments using TEM.

Although this chapter attempts to provide a comprehensive overview of the optical system of the TEM and cover the main components inside the electron microscopes, we note that this is a developing research area, in which new designs of the electron paths and new generations of lenses/detectors are emerging very quickly in academic institutes and industries all over the world. Some of the forefront technology in high-end TEM, such as aberration correctors both for the probe forming and imaging, will be elaborated in detail in following Chapters.

1.1 General ray diagram of TEM

The optical system of a conventional TEM consists of three main components: illumination system, imaging system and projection system. As shown in Fig. 1.1.1, those three systems, each with their own apertures, lenses and other necessary pieces, are arranged following the sequence of the electron beam path. In particular, 1) high-energy electrons are released from an electron gun and such an elec-

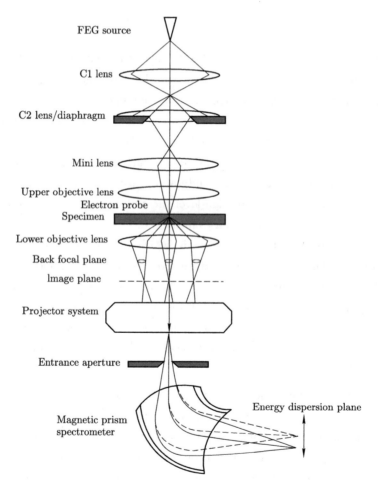

Fig. 1.1.1 A general ray diagram of conventional transmission electron microscopes.

tron beam is confined by a multiple-aperture-lens system before the electron beam reaches the specimen. Sometimes a mini lens or a combination of lenses is added to the microscope, together with the upper objective lens, for further confinement of the beam size and shape. Controllable parallel illumination is thus formed on the specimen by the lenses and apertures described above. 2) The objective lens and objective apertures constitute the imaging system. After the interaction between the electron beam and the specimen, scattered electrons interfere at the back focal plane of the objective lens to form electron diffraction patterns and at the imaging plane of the objective lens to form images of the specimen. Usually objective apertures are placed at or very close to the back focal plane so that the direct beam or a particular diffracted beam can be selected to form bright-field (BF) or dark-field (DF) images, respectively. 3) However, both the diffraction patterns and the images formed by the imaging system might still be too small to view, and thus a projection system, which is composed of several lenses, is required to significantly magnify the patterns and images for data detection and acquisition.

In addition, electrons that undergo inelastic scattering with the specimen are analyzed by a magnetic prism spectrometer to reveal their energy distribution in the electron energy-loss spectra formed at the energy dispersion plane of the spectrometer system. Most of the spectrometer systems are attached at the end of the electron paths, however some commercial TEMs have the spectrometer system incorporated in the design of the column.

In the following sections of this chapter, details of the illumination system, imaging system and projection system will be introduced in this sequence. Due to the importance and the uniqueness of the electron sources to the microscopes their description including classification and characterization of the electron sources are discussed in the next section. Moreover, the quality of the TEM data is critically dependent on the detectors. Thus the detectors with recent updates and applications will be described after the optics section. Different types of detectors and their functions in a TEM will be discussed in Section 1.4. For example, energy-dispersive X-ray (EDX; the X-ray is generated from specimen by the electron beam) spectroscopy, which is not shown in Fig. 1.1.1, plays a key role in charactering the heavy metal elements in materials with catalytic applications. Finally, a different but similar microscope using ion beams, as well as the details of their unique optic system, will be discussed in this Chapter.

1.2 Electron sources

The type and quality of electron sources are essential to the performance of a microscope and to the final results coming out from the TEM. Before using a TEM, a researcher should know the type of the electron source and the basic characteristics of the electron beam, such as how bright it is and the energy spread of the beam, in order to achieve their particular research purposes. Usually users of a TEM do not need to measure by themselves the characteristics of the electron source, which could be provided by the manufacturer or the instrument maintainer. For the above reasons, this section will focus on the types of the electron sources and the characterizations of the electron beam that is emitted from the gun assembly. The optics of a TEM is specifically designed for certain types of the electron sources. Therefore, the type of the electron source in one TEM is not interchangeable between thermionic and field emission kinds.

1.2.1 Types of electron sources

The types of the electron sources are determined by the physical mechanisms of electron emission from the gun tip materials. There are basically three kinds of electron emission in most of the TEM guns: thermionic emission, field emission and Schottky emission, where Schottky emission can be considered as a mixture of thermionic and field emission (Williams and Carter, 2009; Richardson, 2003; Murphy and Good, 1956).

Thermionic emission: Thermionic sources have often been called filaments and sometimes cathodes. If the material of the source is heated and electrons gain sufficient thermal energy to overcome the potential barrier at the material's surface, i.e., the material's work function ϕ, the electrons can be emitted from the surface of the source. The electron flow is described by current density J and has the following relationship with the heating temperature T in the Richardson equation

(Richardson, 2003):

$$J = AT^2 e^{-\frac{\phi}{kT}}$$

where k is the Boltzmann constant. The proportional coefficient, A, is called Richardson's constant, determined by the source material. The work functions of most metal materials are about a few eV. In order to have a large current of electron emission and a long lifetime of the electron source, a low material work function is required and the source should not melt or evaporate at the operating temperature. Williams and Carter's book shows a good comparison of the work function, operating temperature and lifetime etc., between different source materials. Tungsten and lanthanum hexaboride (LaB_6) crystals are the most common materials selected for thermionic TEM sources, mainly for their low work functions and high melting temperatures. On the other hand, thermionic guns are measured to have a saturate emission current and the operating temperature/heating current is set at the optimum value below the saturation condition in order to maximize the emission current and the source lifetime (Williams and Carter, 2009).

Field emission: A TEM with a field emission gun (FEG) costs much more than a TEM with a thermionic source. The mechanism of field emission is shown in Fig. 1.2.1. Electron emission of this kind is induced by an electrostatic field applied at the source (Fowler and Nordheim, 1928). There is a potential gradient such that

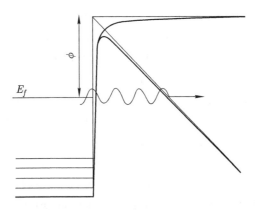

Fig.1.2.1 Principle of field electron emission by quantum tunneling of electrons from surface into vacuum induced by an electrostatic field.

the energy level at vacuum is tremendously lowered. Thus the energy barrier at the surface is narrow enough so that the electrons inside the source material have a certain probability to tunnel out and generate emission current.

Field emission takes place only when the electrostatic field is very high at the source. To produce high electrostatic fields, the source material is manufactured to be a needle tip because the sharper the tip is the higher the electrostatic field will be. Without help from thermal heating the emission is called cold field emission. Usually the cold field-emission source is made of single-crystal tungsten that is a sharp needle tip. If the electrons are heated and an electrostatic field is still applied at the tip, thermal energy can assist the electron escape over the reduced barrier and the emission current is significantly enhanced. This is known as Schottky emission or field-enhanced thermionic emission. Schottky emitters are the predominant source in modern TEMs. They are made of tungsten tips coated with a layer of zirconium oxide (ZrO_2) to improve the emission stability. The emission current from a Schottky emitter can be determined by Richardson equation with modification of the work function term.

The vacuum in the surrounding space of the cold field-emission gun should reach ultra-high vacuum (UHV; $< 10^{-9}$ Pa) level to maintain the gun tip in pristine condition without contamination or oxidation. However, as time goes by, contamination of the residual gas occurs at the tip surface of a cold FEG. In order to remove the contaminants accumulated at the tip, we need to flash the tip once or twice a day in practice by heating the tip in a very short time to evaporate the contaminants. No flashing is needed for Schottky FEG and thermionic sources.

Without regard to the cost, each of the three types of the electron sources has its pros and cons in terms of brightness, coherency and stability, which will be introduced in Section 1.2.2. The development of TEM technology includes finding new materials and establishing new types of the electron source, always pursuing better performance of the electron source with versatile functions. A new generation of ultra-bright electron sources has been studied and developed recently based on advanced materials. For example, carbon nanotubes have shown excellent field-emission properties to be good candidates for this purpose (de Jonge and van Druten, 2003). Furthermore, polarized electron source (PES) has been developed for many years, especially for low-energy (< 40 keV) electron microscopy. Recently such a technique was established in TEM with operating voltage ~ 20 kV at Nagoya University in Japan (Kuwahara et al., 2011). By modifying the setup of the illumination system of a conventional TEM, the instrument with PES has achieved a high level of electron spin polarization, quantum efficiency and large electron dose on the specimen. With PES, it is possible that the materials' magnetic structures that remain undetectable by conventional TEM will be unveiled by the spin-polarized TEM in the future.

1.2.2 Characteristics of electron beams

The electron sources introduced above have clear definitions that are specifically referred to the electron emitter. However, electron beams are more complicated than simply the electron flow out from the emitter. Indeed, a TEM needs a gun assembly to have the electrons accelerated and focused at a crossover. Once an electron beam is formed, we are able to characterize the beam in terms of its brightness, coherency and stability. The outcome results, including high resolution TEM images, electron diffraction, electron energy-loss spectroscopy, etc., by using a TEM critically depend on these characteristics of the electron beam.

Gun assembly: Fig. 1.2.2 shows a schematic diagram of a gun assembly of a thermionic source. Modern TEM instruments with thermionic sources use LaB_6 crystal as the filament. The high tension is applied between the filament and the anode to accelerate the electrons with the anode at the earth potential. Wehnelt cylinder acts as the first lens in a TEM along the electron path. Unlike the magnetic lens, it is indeed an electrostatic lens. The role of the Wehnelt cylinder is to focus the emitted electrons at a crossover before the anode by applying an optimum bias at the Wehnelt cylinder. The purpose of the Wehnelt bias is to maximize the beam brightness. The beam's diameter at the crossover is d_0. The emission beam that goes through the anode has a divergent semi-angle α_0 and the current i_e. The beam from this point can be considered as the object of the first lens in the illumination system of TEM optics.

On the other hand, the gun assembly of an FEG has two sets of anodes. The first anode has a positive potential of a few kV with respect to the tip, which enhances

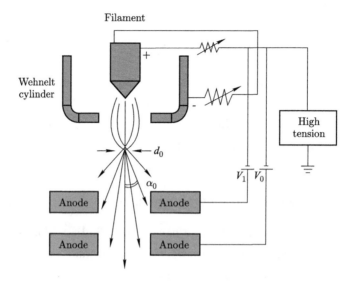

Fig.1.2.2 A schematic drawing of the gun assembly of a thermionic source. A high tension is placed between the filament and the anode. The Wehnelt cylinder is an electrostatic lens which creates a crossover between the Wehnelt cylinder and the anode.

the electron emission from the tip. The second anode accelerates the electron to the working energy of the beam. In both the thermionic gun and FEG the filaments and tips are susceptible to thermal shock so they should be treated carefully and thermal processes should be processed slowly such as heating or cooling and/or turning on the gun voltages.

Brightness: With the parameters defined in the gun assembly section, now we are able to express the brightness β in the following equation:

$$\beta = \frac{4i_e}{(\pi d_0 \alpha_0)^2}$$

Firstly, we should note that the brightness is defined as the current density per unit solid angle of the beam going into the illumination system of the TEM. The unit of the brightness is $A/m^2 \cdot sr$. What is measured here is different from the term "intensity" of the beam or the term "electron dose" on the specimen that are often used in TEM operation. The latter two terms not only relate to the brightness of the TEM gun, but also are a function of the set-up of illumination system. For instance, the "electron dose" on the specimen is measured by counting how many electrons hit on the specimen per unit area in a second, which is a function of the beam brightness and can be varied by changing the condenser apertures and the beam convergent angle. Secondly, illumination with high brightness significantly increases signal-noise ratio of TEM results from a specimen and thus improve the resolution of the data. The equation here suggests a number of ways to raise the emission brightness. Because the emission current is linearly increasing with increased operating voltage, TEM instruments with higher operating voltage (300 kV– 3 MV) have been developed to provide brighter beams with improved resolution due to shorter wavelength of the emitted electrons and less severe C_c effect and better approaches for thicker specimens (Takaoka et al., 1997). Alternatively, FEGs provide much higher beam brightness than the thermionic guns operating at the same voltage. However, electron illumination with high voltage and/or high brightness on the specimen is likely to produce more sample damage. The possibility and mechanisms of beam damage will be discussed next to this section.

Coherency: In physics, coherency is an ideal property of waves that enables stationary (temporally and spatially constant) interference, i.e., a term to describe how well the waves are "in-step" when oscillating with one another. Basically two parameters determine whether two waves are coherent or not: the frequency and the phase. According to quantum physics, electrons can be treated as waves as well as particles. Therefore, the coherency of electron sources is referred to two aspects of emitted electron waves. One is how consistent are the frequencies of all the electron waves and this description is called temporal coherency. The other is whether all the electrons emitted from different places on the filament or the tip have their wave packets "in-step". The latter is related to the spatial distribution of the filament/tip, and it's called spatial coherency.

As the frequency of a wave is directly related to the energy of the particle representation of the wave, the coherence length, as a measure of the temporal coherency can be determined by $\lambda_c = \dfrac{vh}{\Delta E}$, where ΔE is the energy spread of the beam, v is the electron velocity, and h is Planck's constant. With stable power supplies and high tension supplies, typical energy spread of thermionic sources is greater than 1.5 eV at an accelerating voltage of 100 kV, while the cold FEG is about 0.3 eV and a Schottky FEG is about 0.7 eV at the same condition. The energy spread of the beam can be further reduced by energy-selecting systems called monochromators, however a sacrifice of some intensity of the beam occurs. Electron beams that contain electrons with small energy dispersion are reflected through multi-mirror system in monochromators and the emitted beam has selectable frequency/energy with a much narrower band than the incident beam. Today we can achieve high energy electron beams (200 keV) with energy spread of $\sim$0.1 eV or even lower values with monochromators (Krivanek et al., 2009). Increasing the temporal coherence has obvious merit in TEM studies as it may reveal finer electronic structures illustrated by the electron energy-loss spectroscopy (EELS) and other methods with the energy value of the electron beam involved in the data analysis.

Spatial coherency is determined by the size of the emitter. High spatial coherency can be provided by: 1) A sharp source tip. 2) A small subtended angle by the source at the specimen (which can be controlled by adding small condenser apertures in the illumination system). 3) A low accelerating voltage. A simple demonstration of spatial coherency of a beam is the Fresnel fringes at the inside edge of the TEM image of an aperture. Fig. 1.2.3 shows a TEM image of a condenser aperture $\sim$10 µm in diameter with a few clear Fresnel fringes, using a Schottky FEG at 200 kV. Similar to the Fresnel biprism measurements in light optics, using carbon nanotubes as a nanobiprism has recently enabled precise quantification of the

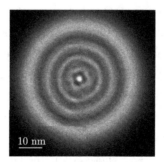

Fig.1.2.3 A TEM image of a parallel electron beam with diameter $\sim$50 nm in the NBD mode of JEOL 2100F. The condenser aperture size is $\sim$10 µm. The Fresnel fringes arise from good coherency of the electron beam.

spatial coherency of an electron beam (Cho et al., 2004). Spatial coherency is of great significance for the quality of TEM data, especially high resolution TEM images. For this aspect, field emission sources are better than thermionic sources.

Stability: The stability of electron beams is a combination of a number of parameters, including the stability of the power supply and high tension supply and the stability of the emission current from the source. Regarding the stability, thermionic and Schottky FEG sources are better than the cold FEG sources. Practically, the emission currents of thermionic and Schottky sources are reasonably stable in an operating day, while a cold FEG could show observable variation in emission current in a few hours. Moreover, the lifetime of the different types of sources varies very much. With good maintenance of the instrument, Schottky FEG and cold FEG sources have much longer lifetime than thermionic sources.

With the comparison of different types of electron sources, it is difficult to select the best one regarding all the characteristics of the electron beam. Schottky FEG is the most popular one in modern TEM instruments for their excellent properties in small source size with good spatial coherency, relatively low electron energy spread, high brightness, good stability, low short-term noise, low vacuum requirements and low cost for operation/maintenance. In recent years, there are increasing demands on the TEM instruments using cold FEG sources for significant improvements in brightness, energy spread and spatial coherency. Although the stability of the electron current from a cold FEG is not as good as that of a Schottky FEG, it has been improved by increasing the vacuum level of the UHV chambers inside the TEM. Despite the high cost of the source, cold FEG displays many advantages in forefront research areas studying the cutting-edge scientific projects.

1.2.3 Beam damage and voltage effect

There is another factor that the user of a TEM should consider besides the above characteristics of the electron sources, i.e., radiation damage of the electron beam. Electron radiation damage is an effect generated by both the incident electron beam and the nature of the specimen mainly through inelastic scattering. Generally all the materials under high-energy electron beam illumination could be more or less damaged and it changes case by case whether the beam damage affects the scientific information extracted from TEM data. On the beam side, the radiation damage is usually a function of beam energy, which is directly dependent on the accelerating voltage.

There are three major radiation damage forms. 1) Radiolysis results from electron-electron interactions and could change the chemical bonding in polymers and covalent and ionic crystals. Thus the material's chemical and crystalline structures may vary under the electron beam irradiation. The way to lower the cross section of electron-electron interaction is to increase the accelerating voltage. 2) Knock-on damage is a phenomenon often observed in TEM with accelerating voltages higher than 100 kV. High-energy electrons transfer sufficient energy to the material atoms to break the bonding between the atoms and their neighbors. Knock-on damage could create defects such as vacancies or sputter atoms. A few things affect the strength of knock-on damage. The electron energy depends on the operating voltage. The higher the operating voltage is, the more the knock-on damage will be. In addition, if the specimen is very thin, the observation of the knock-on damage would be more obvious than that for thick specimens. Therefore, it is common to lower the operating voltage for TEM observations of thin specimens with light elements in order to minimize the knock-on damage. For example, 60 kV or 80 kV of operating voltage has often been used to obtain high resolution TEM images from graphene and other carbon based molecules. 3) Heating effect arises from the electron-phonon interactions and this form of damage converts a certain amount of the beam's energy to heat in materials. This effect is particularly important when the specimen is a poor thermal conductor. Then heat induced by incident electrons could accumulate in the material bulk and raise the temperature very high, even to the point of melting or evaporating the material. Polymers, biological materials and insulating crystals are very susceptible to heating damage. Higher electron voltage lowers the cross section of the electron-phonon interaction and therefore is recommended in this aspect. Other suggestions in order to prevent severe heating damage on the specimen are cooling the specimen to low temperatures such as liquid N_2 temperature, illuminating the specimen using low electron dose and coating the specimen with a thermal conducting film. The cooling process has been commonly accepted as a method to minimize the heating damage. One exception was reported in carbon nanotubes that more knock-on defects were observed at low temperatures than at high temperatures (Urita et al., 2005) due to the thermal process after the damage as follows. Once an atom is kicked off slightly from its equilibrium position, thermal energy would help the atom to oscillate back, while at low temperatures the atoms lose such capability.

To summarize the voltage-dependent beam damage effect, high operating voltage is not necessarily good or bad, which indeed is a choice for optimum condition of different materials. Besides the influence of operating voltage on radiation damage, as stated above, low accelerating voltage increases the temporal coherency, while high accelerating voltage increases the brightness of the electron emission. Before starting a TEM, researchers should consider those facts and select the optimum voltage for their projects because the alignments of the whole optics of the TEM will be varied with different operating voltages.

1.3 Optics

This section will be an introduction of all essential components in modern TEM optics except the gun assembly part. Carrying charge with low mass and the wave-like nature are the most important features of electrons in TEM, which dominate the beam interaction with materials and form the base for lens design. The first part of the optics section will demonstrate the basic physics of electron motion in magnetic lenses. A brief introduction of the structure of the magnetic lenses and the optical principles are included as well.

Electron optics was born in the 1920s and Hans Busch was the first to describe the possibility to focus electrons in the language of geometric optics in 1927 (Hawkes, 1972). The first TEM with condenser, objective and projector lenses was built by Max Knoll and Ernst Ruska in 1933, and Ernst Ruska was awarded the Noble Prize in physics for this in 1986. After Ernst Ruska moved to Siemens, he was involved in developing the first commercial electron microscope which was produced in 1939. Today, most of the modern TEM instruments installed in many laboratories across the world are commercially made by manufacturers, such as

JEOL, Philips/FEI, Hitachi, LEO and Nion.

In this section, the description of the optics along the TEM "column" will follow the traditional way to divide the "column" into three systems. The first system that the electrons go into after emission from the gun assembly is the *illumination system*. The main purpose of the illumination system is to form the electron beam into a certain condition, condensed and aligned with respect to the specimen. The electron beam then interacts with the specimen and enters the *imaging system*. The most important lens in TEM, the objective lens, is in this system, which forms electron diffraction and TEM images at its back focal plane and image plane respectively. The electron diffraction patterns and TEM images that formed in the imaging system are magnified significantly by the *projection system*. After the projection system, electrons meet the viewing and recording components. In addition to the fundamental optics in a TEM, extra optics for the purpose of performing secondary electron microscopy (SEM), electron energy-dispersive X-ray spectroscopy (EDX), backscattered electron (BSE) and electron energy-loss spectroscopy (EELS) are sometimes installed in the microscope system. A brief description of the optics for EELS will be introduced at the end of this section. Moreover, this book is not a handbook for TEM operation. Therefore procedures to operate the instruments such as alignments and specifications of the lenses/apertures/pumps/holders in detail are not included here.

1.3.1 Electrons in magnetic lenses

Since Louis-Victor de Broglie made his famous hypothesis of matter-waves which was confirmed later, this theory has contributed greatly in modern physics and materials science. Based on the theory, every particle of matter with certain momentum p can be considered as a wave with its wavelength λ and particularly for electrons with relativistic effects (Williams, Carter, 2009):

$$\lambda = \frac{h}{p} = \frac{h}{\left[2m_0 eV\left(1 + \frac{eV}{2m_0 c^2}\right)\right]^{1/2}}$$

In the expression, h is Planck's constant, m_0 is the electron rest mass, e is electron charge, V is the electron accelerating voltage and c is the speed of light in vacuum.

At an operating voltage of 200 kV, the wavelength of the electron beam is about 2.5 pm. Although most of what we do with electron microscopes uses the wave properties of electrons in order to interpret the scattering from electron-material interactions, we need to understand how a magnetic field changes the trajectories of the electron beam, where electrons are treated as particles described by the equation of motion: $\boldsymbol{F} = -e\boldsymbol{v} \times \boldsymbol{B}$, where $\boldsymbol{F}$ is the Lorentz force, e is the electron charge, $\boldsymbol{v}$ is the electron velocity and $\boldsymbol{B}$ is the magnetic flux density or magnetic induction. The direction of the Lorentz force is perpendicular to the velocity and the magnetic field. The magnetic field at the center of the magnetic lenses is parallel to the optic axis. Therefore, electrons passing right along the optic axis do not feel the magnetic fields. Only in the case that the velocity of electrons has perpendicular component the magnetic field gives the lens effect. In a uniform magnetic field, electrons with perpendicular velocity component has a helical trajectory through the optic axis, which give rise to image rotation under different magnifications that involve different lens settings. As we will discuss in this section, magnetic lenses show very common functions as those of optical lenses, such as focusing capabilities and optical principles.

Apertures: An aperture in electron microscopes is a hole in a metal plate or disk that is usually made of Pt or Mo. The role of apertures is to block a certain amount of electrons and to allow electrons pass through within the collection angle. Apertures can affect the resolution of the lens, the depth of field and the depth of focus, image contrast and change the collection angle of EELS, electron diffraction, etc. The real size of the hole for an aperture varies from one micron to hundreds of microns in a TEM.

Magnetic lens: Before a description of structures and principles of magnetic lenses, the concept of the resolution of a single lens should be introduced. The resolution of a lens is the finest distance between two objects that a lens is able to resolve. Assuming an object's image is blurred by the lens to have Gaussian-like distribution for every ideal point on the object, resolution of the lens, defined by the smallest distance δ to resolve two points, in vacuum is commonly given by the classic Rayleigh's criterion: $\delta = \dfrac{0.61\lambda}{\sin \beta}$, where λ is the wavelength of the incident beam and β is the semi-angle of the collection for the lens. From this expression, it can be inferred that the resolution of a lens is limited by the aperture size and the wavelength of the beam. The above expression is the theoretical resolution of a lens without imperfection and is called the diffraction limited resolution.

Fig. 1.3.1(a) shows the simplest form of a magnetic lens, which has the cylindrical symmetry with its center axis as the optical axis. Soft iron parts near the slit of the iron shell are called pole pieces. Most of the magnetic lenses have upper and lower pole pieces. An electric current runs through the copper coils, generating a homogeneous magnetic field. The permittivity of the magnetic loop is high and the gap is sufficiently small so that the change of the magnetic flux density due to the gap is little. The magnetic field is the weakest right at the axis and strengthened off the axis. Hence the electrons away from the axis are deflected stronger than the electrons close to the axis. By this manner, electron beam can be focused on the optic axis.

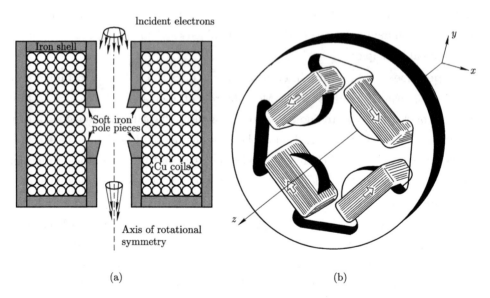

Fig.1.3.1 (a) A schematic of a magnetic lens in a TEM. Electric currents through the Cu coils generate a magnetic field with cylindrical symmetry. A gap at the center consists of soft iron pole pieces. (b) A schematic of a magnetic lens with a quadrupole lens to correct spherical aberration of other lenses.

Magnetic lenses are very imperfect. In order to correct the imperfection of the magnetic lenses and to increase the resolution of the electron microscope, magnetic lenses with quadrupole, sextupole, or octupole pieces are used in special settings of lenses such as aberration correctors or in electron energy-loss spectrometers. A schematic drawing of a quadrupole lens is shown in Fig. 1.3.1(b). In those lenses, adjacent pole pieces have opposite polarity and the magnetic field as a function of x, y and z axes is very complicated to create a desired magnetic field at the center (Hawkes and Kasper, 1996).

The electron path through an ideal magnetic lens follows the optical principles. There are a number of important points and planes noted on Fig. 1.3.2 and discussed as follows. H_1 and H_2 are two principle planes for which the magnification is 1. F_1 and F_2 are front focal plane and back focal plane, respectively. N_1 and N_2 are nodal points on the axis which means a ray passing through N_1 is conjugated to a ray passing through N_2, i.e., $\theta_o = \theta_i$. To form an image of an object, a beam of light (or electrons) needs to obey three basic principles of the lens:

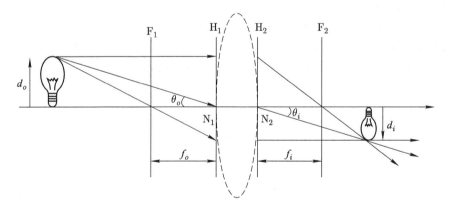

Fig.1.3.2   Optical principles of rays passing through a lens and image forming. For magnetic and optical lenses that the optical indices are the same for both object side and image side $f_i = f_o = f$.

1) A ray parallel to the optical axis will be refracted and pass through the back focal point (cross point of the F_2 plane and the optical axis). A beam of parallel rays, no matter whether it is parallel to the optical axis of the lens, will be focused

to one point on the back focal plane. Particularly, rays parallel to the optical axis can be converged on the back focal point. An alternative description is that F_2 plane is the conjugated plane of $-\infty$.

2) A ray passing through front focal point (cross point of the F_1 plane and the optical axis) will be refracted to be parallel to the optical axis. Alternatively, F_1 is the conjugate plane of $+\infty$.

3) A ray passing through nodal point N_1 will continue from N_2 with the same incident angle ($\theta_o = \theta_i$).

For example, following the three principles, three representative rays from the same point on an object will be converged to one point after they pass through the lens. The plane that contains this point and is perpendicular to the optical axis is called the imaging plane conjugate to the object plane where the object is placed. The lateral magnification (M) of the lens is defined by the ratio of the dimension of the image to that of the object, i.e., $M = d_i/d_o$.

For magnetic lenses and optical lenses the optical index are the same for both object side and image side, i.e., the distance from F_1 plane to H_1 plane (f_o) is equal to the distance from H_2 plane to F_2 plane (f_i) which is generally noted as f. The distance from the object to the H_1 plane (U), the distance from H_2 plane to the image (V) and the focal distance f have the well-known relationship:

$$\frac{1}{U} + \frac{1}{V} = \frac{1}{f}$$

There are two more important characteristics of a lens for the quality of images, the depth of field (D) and depth of focus (D_f). The depth of field is a distance on the object side so that the image is still practically on focus by moving the object within that distance. On the other hand, the depth of focus is a distance on the image side that how much the image can be moved along the axis without noticeable defocus. These two terms are certainly related to one another ($D_f = M^2 D$), and are dependent on the aperture size, the dimension of the object and the lateral magnification of the lens. In practice for a TEM instrument, the value of the depth of field can be tens or hundreds of nanometers, while the final depth of focus has very large values from a few meters to even kilometers. Therefore, imaging detectors including different CCD cameras and films/imaging plates can be inserted at various places with the recorded images all in focus.

Imperfections of magnetic lenses: All the lenses are not perfect in terms of how well the basic optical principles can govern all the rays passing through the lens, especially for magnetic lenses. In the book of "Transmission electron microscopy: A textbook for materials science" by David B. Williams and C. Barry Carter (2009), the qualities of magnetic lenses are compared to optical lenses that magnetic lenses perform like "the bottom of a Coca-Cola^{TM} bottle" in optical microscopy. The major imperfections of magnetic lenses in TEM are discussed as follows (Reimer, 1989).

1) Spherical aberration C_s. This effect affects the resolution of TEM greatly especially in phase imaging and the shape of fine electron probes. The effects of spherical aberrations are schematically shown in Fig. 1.3.3. The further the ray passes away from the optical axis, the larger the deviation of the deflected ray from zero C_s. Therefore, instead of forming a perfect point image of a point object at the theoretical imaging plane, the point image spreads into a blurred disc with different diameters at featured planes shown in Fig. 1.3.3. Electron microscopists have put tremendous effort into the area of correcting the effects of spherical aberrations of magnetic lenses. Special components called C_s correctors have been developed for both the objective lens that forms high-resolution images in TEM mode and the condenser lenses that form a fine electron probe in STEM mode.

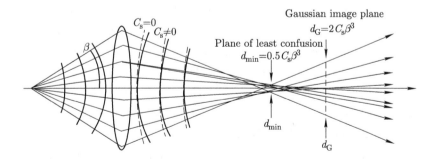

Fig.1.3.3 A schematic drawing shows the distortion on the wavefronts and ray bending caused by spherical aberration of the lens. Consequently, the image of a point object is blurred and the resolution of the lens is lowered.

2) Chromatic aberration C_c. The term "chromatic" is often referring to the effects related to frequency or energy of waves, and chromatic aberration C_c arises from the energy spread of the electron beam. The focusing power of the magnetic lenses to rays with varied energy is different. Therefore, the rays emitted from an ideal point object forms a blurred image, limiting the resolution of a lens. Although the effect of C_c is generally much smaller than the effect of C_s, the effect of C_c can be corrected by recent developments.

3) Astigmatism. This effect is due to the unavoidably inhomogeneous magnetic field generated inside the magnetic lens, which comes from the imperfection in the geometry/symmetry of the magnetic pole pieces or the lens-aperture settings. One of the examples for astigmatism of lenses is that the fast Fourier transform of a defocused image of amorphous materials shows elliptical shapes or other asymmetric shapes instead of circles. This effect can be corrected by stigmators to achieve high-resolution TEM images at a nearly astigmatism-free condition.

Besides these there are other imperfections of the lenses such as coma i.e., the imaging point depends on both the angle and the distance of electrons traveling with respect to the optical axis, and distortions that could take place at low magnifications.

1.3.2 Illumination system

The purpose of the illumination system of a TEM is to form a beam on the specimen. The essential parts to accomplish that purpose are condenser lenses and condenser apertures. The illumination system of a TEM can have lenses at different excitations and adjustable apertures and is fully capable of generating electron beams with great flexibility. The electron beam can be focused down to tens of picometers with a probe forming aberration corrector, or the illumination area of the beam can be as large as a few hundred microns. Such flexibility makes TEM a versatile and powerful tool in materials research.

There are multiple condenser lenses in the illumination system. Shown in the schematic Fig. 1.3.4, the first condenser lens is called C_1 lens with its task to form an image of the gun crossover, i.e., C_1 crossover. The C_1 crossover could be the magnified or demagnified image of the gun crossover, which depends on the type of the gun. For example, FEG has a very small gun crossover and C_1 lens forms a magnified image for FEG. Namely C_1 lens has certain power in controlling the illumination spot size. The second condenser lens is called C_2 lens with a C_2 aperture to confine the beam and to adjust the convergent semi-angle (α) of the electron probe on the specimen.

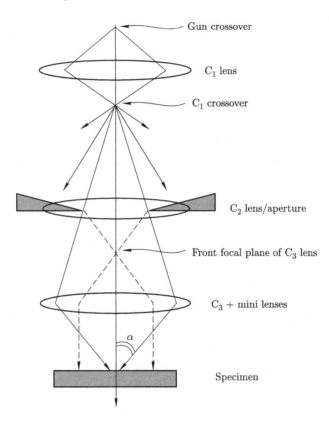

Fig.1.3.4 A ray diagram showing the basic components in the illumination system of a TEM with parallel or convergent beam.

When the C_2 lens is on, it can bend the beam in different ways, i.e., to be focused, overfocused or underfocused, and the beam on the speimen could be parallel or nearly parallel with a small convergent angle with the help of the C_3 and the mini lens. Such an operation mode is noted as a parallel beam mode or TEM mode. The C_3 and mini lenses in the illumination system are a combination of several lenses. Sometimes the lens set is called the condenser/objective lens (c/o lens) because it could be part of the objective lens, i.e., upper objective lens. This lens set still plays a role in the beam illumination and could be stronger than the lower objective lens which solely plays a role in imaging. To create a parallel beam for the purposes such as obtaining coherent nano-area electron diffraction (NAED), C_2 lens generates a crossover right at the front focal point of the C_3+ mini lens set and the size of the parallel beam is controlled by the C_2 aperture (Zuo et al., 2004). Using JEOL-manufactured electron microscopes as an example, normally TEM images and electron diffraction patterns are obtained using the parallel beam mode such as TEM mode or NBD mode (Zuo et al., 2004).

When the C_2 lens is off, electron beam goes through and forms a probe with the help of the C_3 and the mini lens. In this case, the electron probe has a convergent angle that is much larger than that in the TEM mode, and this operation mode is noted as probe mode. Again using JEOL-manufactured electron microscopes as an example, the convergent-beam electron diffraction (CBED) (Spence and Zuo, 1992), STEM results and sometimes EELS and EDX are obtained using probe forming mode such as STEM mode or CBED mode. In CBED mode with a nearly-perfect point probe, the convergence angle is determined by the C_2 aperture.

There are multiple sets of scan coils (not shown in Fig. 1.3.4) in the illumination system of a TEM, which enable us to scan the electron probe over the specimen and obtain particular results including high-angle annular dark-field (HAADF) images, spectrum imaging and other scanning TEM data. Scanning techniques in TEM, combining the local information and the mapping capability, have become very important in modern research due to their uniqueness in chemical and structural/electronic characterization.

1.3.3 Imaging system and holders

Objective lens and apertures: After the interaction between the incident beam and the specimen material, electrons are scattered to different angles with different energy loss by elastic and inelastic scatterings. Two typical sets of patterns are formed directly by the objective lens in TEM: diffraction patterns and images. It should be emphasized here that forming diffraction patterns does not necessarily require an objective lens, i.e., they form at infinity distance without a lens under parallel illumination. This is why sometimes electron diffraction technique is called a lensless technique compared to other imaging techniques. The role of the objective lens in forming electron diffraction patterns is to significantly shorten the distance from infinity to the back focal plane of the lens. There are a few ways to form a diffraction pattern, including selected-area electron diffraction (SAED) and nanoarea or nanobeam electron diffraction (NAED). The former one will be introduced in the next section and the latter one is obtained using a nanometer-sized parallel beam formed in the illumination system. Using a convergent beam, CBED patterns contain more information from electron beams in many directions and are very sensitive to strain, charge density and other detailed electronic structures. The principles of the CBED technique will be introduced in Section 3.5.

On the other hand, an important task of the objective lens is to form a focused image of the specimen area at the image plane. Under parallel illumination, the wave function at the back focal plane is a Fourier transform of the object with both real and imaginary parts under ideal lens and kinematic approximation, i.e., a complex diffraction pattern. An image is formed by the interference of point sources at the diffraction plane. As discussed in Section 1.3.1, a magnetic lens is far from being perfect and the image quality is significantly affected by the defects of the lens such as spherical aberration. A spherical aberration corrector can be installed in the imaging system to improve the TEM image quality.

Since the image is formed by point sources at the diffraction plane, an objective aperture placed at the back focal plane can select different point sources in the diffraction pattern and form images with different contrast. Fig. 1.3.5 shows the principle of so-called bright-field (BF) and dark-field (DF) images. BF images are formed using direct beam or more generally a number of beams including the direct beam, while DF images are formed using a particular diffracted beam or more generally a number of diffracted beams without the direct beam. In principle, any TEM images should be called BF images even without an objective aperture inserted because the direct beam is included when forming the image. Knowing what beams are selected for imaging is important for understanding the image contrast. Other imaging conditions, such as the defocus, beam convergence angle and spherical aberration, are all essential to explicitly interpret the contrast of TEM images. More details of high-resolution TEM (HRTEM) images will be described in Section 3.2. In addition, the interaction of the beam and the specimen generates a range of secondary signals, such as X-ray, secondary electrons, auger electrons, and backscattered electrons. Useful information from them can be collected by inserting the proper detectors in the TEM column.

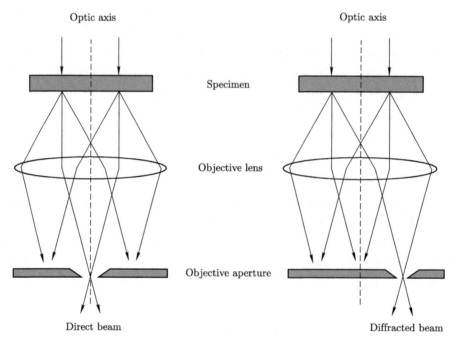

Fig.1.3.5 Principle of forming BF (left panel) and DF (right panel) images.

Holders: The TEM specimen is mounted in a TEM holder which is fitted with and controlled by the TEM stage. The TEM stage together with the holders provide a lot of versatility in TEM observations. Not only is the specimen able to be tilted in x and y directions or rotated in the specimen plane, but also thermal/electric/magnetic/mechanical forces can be applied to the specimen to provide in situ TEM observations in order to study materials' properties under various conditions. In particular, the tilting capability of the TEM holders allows us to explore materials' morphology/structures from different angles or even achieve electron tomography for 3D structural study; cooling and heating TEM holders enable us to perform experiments in temperatures ranged from liquid helium temperature $\sim$4 K to high temperature up to 1000 K; the electrical and/or magnetic properties of material could be explored by using TEM holders where applying electric and/or magnetic fields is possible. In addition, TEM technical developments are always driven by the requests from materials research, so that new TEM holders are being developed for more research purposes. For example, a TEM holder that can guide a laser beam onto the specimen and receive the potential optical response from the specimen has been recently designed and manufactured in order to study materials' optical properties (Zhu et al., 2012).

1.3.4 Projection system and post-column systems

The electron diffraction patterns and the images formed respectively at the back focal plane and the image plane of the objective lens are further magnified by the projection system with multiple lenses (i.e. intermediate and projector lenses). The first intermediate lens can be focused on the image plane or the back focal plane of the objective lens, which depends on whether the user wishes to see images or electron diffraction patterns on the monitor or the camera by adjusting the excitation strength of the intermediate lens. The final images recorded by the camera are usually calibrated with scale bars or denoted by the total magnification. On the other hand, the scale of the electron diffraction patterns recorded on the camera is determined by the camera length and the diffraction patterns should be calibrated at the beginning by using standard specimens such as silicon single crystals with known lattice spacings. The quality of the electron diffraction pattern, no matter if it's the NAED, SAED or CBED, is downgraded by the astigmatism of the intermediate lenses, i.e. having distortions in the diffraction patterns. Hence the astigmatism of the intermediate lenses needs to corrected by the intermediate lens stigmator before obtaining electron diffraction data in TEM. In addition, a biprism can be installed in the projection system for electron holography work. With applied voltage on the biprism, the phase change of the electron beam due to the potential variation on the specimen can be measured. Details of the electron holography technique will be introduced in Section 3.7.

One important technique, SAED, should be introduced as a part of the imaging system. However, the SAED aperture can be, in principle, placed into any plane conjugate with the object, which could be the image plane of the objective lens or other image planes in the projection system. Usually it is placed at the image plane of the objective lens. As shown in Fig. 1.3.6, the SAED aperture blocks the electrons that are outside the area on the specimen which is corresponding to the area in the image selected by the SAED, equivalent to a virtual aperture above the specimen. Then the electron diffraction patterns are formed using the electrons passing through the virtual aperture. SAED patterns have sharp reflections by using parallel illumination under focused conditions of the projection system.

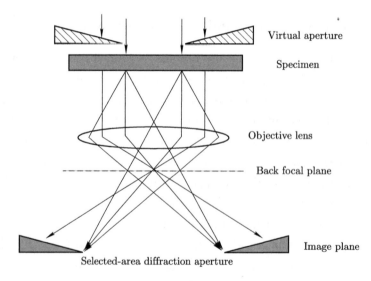

Fig.1.3.6 Principle of SAED patterns that placing a SAED aperture in the imaging plane equivalent to a virtual aperture above the specimen.

The conventional TEM column consists of the illumination system, imaging system and projection system with detectors. It has now become standard configuration that an energy loss spectrometer is attached to the TEM column shown in Fig. 1.1.1. This part has lenses, apertures and magnetic prisms to disperse electrons and form EELS spectra. An EELS system may be attached as a post-column system or as a specially designed piece that is integrated into the column and placed in the imaging/projection system, so-called Omega filter that is manufactured by Leo-Zeiss and JEOL. However, post-column systems made by Gatan are more popular and sometimes have more convenience and flexibility in operation. The principle of the EELS technique will be introduced in Chapter 5. One critical characteristic of EELS is the energy resolution, which is a different concept but related to the energy spread of the electron source. Normally the energy resolution using EELS is referred to the full width at half maximum of the zero-loss peak. Note that the energy resolution will determine, not exclusively, the capability in revealing fine electronic structures in EELS. In TEMs with a cold FEG, the resolution is ~ 0.3 eV and could be less than 0.1 eV when a monochroma-tor is used. Generally, electron scattering techniques have lower energy resolution compared to the other major scattering techniques such as Synchrotron X-ray and neutron scattering (Yakimenko et al., 2012; Meyer et al., 2003). However, electron beam can be focused into a very small probe and the capability of energy resolution can be combined to the real-space resolution, which is a great uniqueness of TEM techniques.

1.4 Detectors

All the TEM results are formed by electron beams or X-ray at various planes in the TEM optics. Those results cannot be directly seen by our naked eyes and they become useful only when they are recorded by the detectors. As introduced in former sections, TEM is a versatile tool. Therefore, detectors inside TEM have different types and structures based on the nature of the results. The distribution of the detectors in a TEM is roughly described in Fig. 1.4.1. The secondary electron microscopy (SEM) detector, the backscattered electron (BSE) detector and the EDX detector are placed above the specimen. Below the specimen, bright-field (BF) and dark-field (DF) detectors for images and diffraction patterns, high-angle annular dark-field (HAADF) detectors and detectors for EELS recording are located through the optics, near or away from the optic axis. Each type of those detectors could have multiple settings in one TEM, for example, multiple CCD

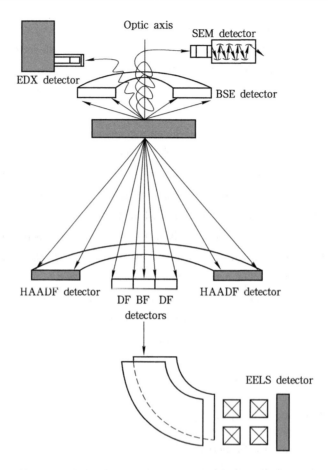

Fig.1.4.1 The distribution of the detectors in a TEM.

cameras for BF/DF images at various places. On the other hand, a TEM instrument may not be equipped with all those detectors. In this Section, details of a few popular detectors will be discussed. Similar to the development of other parts in the optics, detectors for TEM are a rapidly developing area as well in order to speed up data acquisition and improve image qualities. In particular, shortening the necessary data acquisition time has great merit because the shorter the beam dwells on the specimen the less contamination or beam damage would be. Moreover, signal-to-noise ratio is a key factor to measure the quality of a detector, so improving the sensitivity and dynamic range (i.e. the ratio between the largest and the smallest signals that can be authentically recorded) of the detectors are essential for TEM results.

1.4.1 Viewing screen and cameras

A viewing screen is normally coated uniformly with ZnS particles that are ~50 μm in size and it emits green light when being hit by the electrons. Therefore, the BF/DF images and diffraction patterns can be directly viewed through a transparent window in front of the users. Although nowadays some high-end TEM instruments prefer to have remote control that users can view the results and operate the microscope from a distance or even in a different room, directly seeing the results from the viewing screen still has advantages. In particular it brings convenience in TEM operation. For example, in situ TEM study requires real-time observation of the specimen when the user is performing the experiments and needs quick monitor's response that is beyond the capability of other cameras. In addition, a viewing screen has a long life-time and low maintenance cost compared with other detectors.

Today the most popularly-installed camera in TEM is the charge-coupled device (CCD) camera (Boyle and Smith, 1970), used for recording BF/DF/EELS data. Fig. 1.4.2 shows the structure of the device. CCD cameras are built by metal-oxide-semiconductor array which can store charge generated by electrons or lights. The top layer of individual metal-oxide-semiconductor capacitors is made in polycrystalline silicon, or called polysilicon, facing the incident electrons. The next layer is dielectric SiO_2. The amount of charge generated by incident electrons is proportional to their energy and number and the charge is stored in the potential well at the bottom p-Si layer. Each capacitor is electrically isolated from the surrounding cells during the charge-creating process. With applied voltage at the gate, the stored charge can be transferred into its neighbor, shifted one by one and read out by a electric circuit finally. By this way, all the elements of the array can be read out one at a time and the CCD camera is ready for the next image recording. One CCD camera may have millions of capacitors and each cell can be as small as 6 μm. For example, the image recorded by a 2 k×2 k CCD camera has more than 4 million pixels. CCD cameras have a number of advantages. Firstly, they are manufactured very well such that the responses from all the cells under same conditions are uniform. Secondly, the signal output and input ratio is high, i.e., the camera is efficient with low noise level in data recording when the camera is cooled. Thirdly, the output signal is ready for digital analysis, which is convenient for users. However, the dynamic range of today's CCD cameras is not as high as imaging plates. Dynamic range of a detector is important in recording diffraction patterns, especially when a user is trying to obtain the intensities of strong and weak reflections in an electron diffraction pattern. Development of so-called diffraction cameras prevents the leaking of charges from one cell to its neighbors, but doesn't improve the camera's dynamic range.

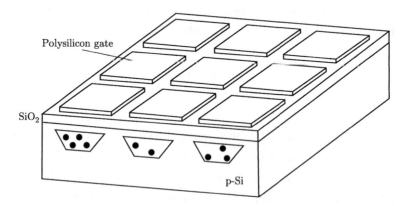

Fig.1.4.2 A schematic drawing shows the structure of CCD detectors that consists of array of metal-insulator-silicon capacitors. The charge packets in the p-Si can be transferred to the neighbor cells by applied voltage at the gate and are finally read out by the connected electric circuit.

In TEM/STEM microscopes, HAADF detectors are installed to collect electrons scattered to high angles. The HAADF detectors are semiconductor detectors of the surface-barrier type and are cut to have a circular shape with a hole at the center. The inside and outside detection angles can be controlled by the magnification of the diffraction pattern. The HAADF detectors are fabricated to have p-n junctions through the inside Si layers and could have a metal coating on the surface as the electrical contact and acting as a depletion layer. The incident electrons create electron-hole pairs and the charges are quickly separated and converted into current or voltage that can be measured by the connected circuit. Semiconductor detectors are highly efficient and sensitive to low electron dose. However, the noise level of semiconductor detectors is high and the response time for intensity change is long.

1.4.2 Films and imaging plates

Film negatives are old-fashioned recording medium that had been widely used in the old generation TEMs. Incident electrons reduce the silver halide (such as AgCl and AgI) emulsion coated on the polymer films or glass plates to silver. In practice, the negatives need to be developed and read out by an optical scanner or a digitizer. So the resolution is limited by the emulsion grain size, the data condition and the scanner parameters. Film negatives cannot be re-exposed or reused.

An alternative product of film negatives is imaging plates as reusable (up to ~ 10000 times) photographic medium that are usually placed at the same place of film negatives. Fig. 1.4.3 shows two photos of the scanners for imaging plates made by Fuji and Ditabis. Data recorded by imaging plates are read out digitally, which is convenient for users. Imaging plates have high real-space resolution (3760×3000 for Fuji system and 5744×5066 for Ditabis system) and very high dynamic range (16 bit dynamic range for Fuji and 20 bit dynamic range for Ditabis) compared with today's CCD cameras; hence it's ideal for recording electron diffrac-

Fig.1.4.3 Photo pictures of the scanners for imaging plates made by Fuji (left panel; 3760×3000 pixels and 16 bit dynamic range) and Ditabis (right panel; 5744×5066 pixels and 20 bit dynamic range).

tion patterns. However, using film negatives and imaging plates requires users to install them in a TEM and take them out from the TEM frequently, which eventually introduces a considerable amount of contamination into the microscope and degrades the high-level vacuum in TEM.

1.4.3 EDX and other detectors

High-energy electrons generate X-rays and secondary electrons by electron bombardment on the specimen. The X-rays can be detected by the silicon drift EDX detectors as fingerprints of the chemical elements in the specimen. The EDX detector needs to be calibrated using standard samples and has high resolution and stability at low temperatures. The EDX detectors are thermally connected to a liquid nitrogen tank. The EDX detector is placed at an angle with respect to the optic axis, the specimen is therefore usually tilted by $15° \sim 20°$ to maximize the signal collection. Secondary electrons are ejected from the conduction or valence bands of the atoms in specimens. Normally secondary electrons have low energies of about tens of eV and cannot pass through many atomic layers. Therefore, the detected secondary electrons are from the specimen's surface, typically one or two monolayers but can vary by the material of the specimen. Recently, a newly developed SEM detector installed inside a Hitachi HD-2700, a dedicated STEM, enables atomic resolution of the SEM micrographs, i.e., atomic structures on a specimen's surface can be resolved simultaneously with the bulk structure by obtaining the SEM images and HAADF-STEM images at the same time (Zhu et al., 2009). Both the BSE and SEM detectors are located in the TEM stage space. Due to different energy levels, these two types of electrons can be easily separated by different electrical bias on the detectors.

1.5 Ion optics

In TEMs, electrons are exploited as the primary beam to interact with the specimen and informative signals are generated and collected. Similarly, we can characterize materials by using the signals induced by ion irradiation, for example, energetic Ga+ beam. That is to say ions can serve as the primary beam and it is expected that ion microscopes can be alternative imaging and analytical tools. Generally speaking, the use of a nuclear microprobe (ions are indeed nuclear) to characterize materials involves many fields and has been extensively investigated (Breese et al., 1996). Two particular ion-beam tools are most relevant to addressing the challenges for nanotechnology: the focused ion beam (FIB) instruments (Bassim et al., 2014) and the helium-ion microscope (HIM) (Economou et al., 2012). The FIB has been widely used for materials science, biomaterials, and semiconductor industry. Since the ions have much larger mass than the electrons and they can cause significant changes to the materials, the FIB is used predominantly for rapid prototyping of individual nanodevices, such as device modification and mask repair. The HIM is a relatively new development, which shows potential in surface imaging as well as nanoscale soft modification (for details, see Chapter 8 HIM). In this section, we outline the optics of ion microscopes with a focus of the generation and manipulation of ions. There are many excellent books and review articles containing more details on the subject (Orloff, 2009; Liu and Orloff, 2005; Gianuzzi and Stevie, 2005).

1.5.1 Ion sources

The ion source technology also plays a crucial role in the development of ion microscopes. For the FIB instruments, the metrics of ion source performance involve available ion species, emission stability, beam current, brightness, size of the probe, energy spread etc. Ga^+ liquid metal ion source (LMIS) is the current industry standard for high-resolution FIB work, and new ion source technologies, such as the gas field ionization source (GFIS), inductively coupled plasma (ICP) ion source, and low-temperature ion sources (LoTIS), have emerged recently to deliver superior performance compared with the Ga^+ LMIS.

1. Liquid metal ion sources (LMIS)

Fig. 1.5.1(a) is a schematic drawing of the Ga^+ liquid metal ion source. The basic functional blocks of the source consists of a gallium reservoir, a coil heater, a tungsten needle, and an extractor electrode. The tungsten needle and the extractor electrode are essentially a blunt field emitter with a typical end radius of ~ 5 μm, which is similar to the electron gun (see Fig 1.2.2). To form the gallium source, the coil heater first increases the temperature at the gallium reservoir to its melting point. The gallium will then wet the tungsten needle and form the capillary flow. It means that a thin film of liquid gallium, i.e. the source, is supported by the needle. When a voltage ($\sim$6 kV) is applied across the extractor electrode and the source, the high electric field ($\sim 10^{10}$ V/m) stresses the liquid. The surface tension of the liquid applies an opposite stress on the liquid. At the very apex of the tip, the liquid takes a conical shape due to the two stresses. The cone is the so-called Taylor cone. The end radius of the liquid Taylor cone is less than 5 nm.

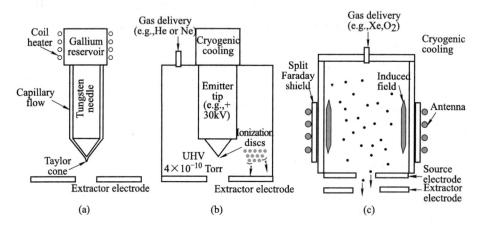

Fig.1.5.1 Schematic drawings show the structure of (a) the liquid metal ion source, (b) the gas-field ionization source, and (c) the inductively-coupled plasma ion source.

Atoms at the apex of the Taylor cone will field evaporated because the potential barrier of evaporation is lowered by the electric field. The evaporated atoms are then ionized by the very high electric field. The ion emitting area appears the same size as the end radius of the Taylor cone. This results in a large ion current density ($\sim 10^{11}$ A/m^2 at $\sim$ 2 µA emission current). The interaction between the ions will change their trajectories (i.e. the space charge effects) and results in a larger optical source size of 30–50 nm. The capillary flow allows gallium atoms migrate from the reservoir to the apex of the Taylor cone so that the continuous emission of Ga$^+$ ions is enabled.

The technology of liquid metal ion sources is principally for gallium. This is mainly because gallium has a low melting point (29.8 °C) and low volatility. Other ions are possible by using corresponding low-melting-point alloys (Bischoff and Akhmadaliev, 2008; Gamo et al., 1983). The reduced brightness of a LMIS is approximately 10^6 A · sr^{-1} · m^{-2} · V^{-1} (Note: the reduced brightness is defined as the brightness per unit acceleration voltage, see section 1.2.2). The energy spread of the LMIS is $\geqslant 5$ eV, which depends strongly on the ion current, mass, and charge. The noise spectrum of the LMIS current shows a flat profile, i.e. white noise, and the typical lift span of a Ga$^+$ LMIS emitter is $\sim$1500 μA-hrs. It should be noted that the Ga$^+$ LMIS is a mature technology. The characteristics of a LMIS emitter set the metrics for any new ion source technologies.

2. Gas-field ionization sources (GFIS)

The gas field ionization source is based on the operating principles of a field ion microscope (FIM) (Economou et al., 2012). FIMs are capable of atomic resolution imaging, but are limited to viewing the tip of very sharp needles ($<\sim 100$ nm radius of curvature). The GFIS uses the emission from a very sharp needle, or emitter, to produce an atomically fine beam source. The most competitive emitter design is the pyramidal W(111)-oriented apex ending in three atoms, which can be electroformed in situ by heating with a high electric field applied ($\sim 5.7 \times 10^{10}$ V/m). Once three atoms are formed at the apex, the field strength is reduced to $\sim 3 \times 10^{10}$ V/m, which is unable to further evaporate the tungsten atoma. The emitter therefore retains its atomically-sharp geometry during operation A noble gas, e.g., He, Ne or H$_2$ is introduced into the emitter chamber and the pressure in the chamber is maintained at ultra-high-vacuum level. With the electric field, neutral gas atoms in the vicinity of the sharpest corners of the apex (i.e. ionization discs) will be ionized via electron tunneling. The ionization discs have a diamete of a few angstroms and a thickness of ~ 25 pm. This means that the energy spread of the helium ion beam is ~ 0.42 eV (at 80 K) and the virtual source size can be ~ 0.3 nm. The reduced brightness of a GFIS emitter can be as high ad 10^9 A·Sr^{-1}·m^{-2}·V^{-1}. Such a high brightness source with such a low energy spread indicates that the He$^+$ beam can be focused to a small probe size on the sample (see Chapter 8-HIM in Vol.1). The high brightness and low energy spread are advantages of the GFIS over the widely adopted LMIS, a limitation of the GFIS is its small beam current range (see table 1.5.1).

3. Inductively coupled plasma sources (ICPS)

As shown in Fig. 1.5.1(c), the inductively coupled plasma source consists of a cylindrical plasma chamber, a solenoid antenna, a faraday shield, and electrodes (Smith et al., 2014). An azimuthal induction field is induced inside the plasma chamber, while a radio frequency (RF) current is passing through the antenna. The RF varying field can heat the electrons, but the ions remain close to room temperature. This is because the RF frequency is below the plasma's electron resonant frequency and above the ion resonant frequency. The electrons in the outer skin (i.e. the yellow colored regions in Fig. 1.5.1(c)) will then gain a sufficient kinetic energy to cause ionization of the resident gas. The ions are then extracted by the source electrodes and the extractor electrodes.

The advantages of this source technology include: 1) a broad range of plasma gases can be generated; 2) a high plasma density can be achieved (10^{13} cm^{-3}); 3) the life time of the emitter can exceed two years. The reduced brightness (Xe$^+$) of the ICP sources ($\sim 10^4$ A · m^{-2} · sr^{-1} · V^{-1}) is lower than the other two techniques discussed previously. The energy spread is about 3.5–5 eV, similar to the standard LMIS. The characteristics of the ion sources discussed are summarized in Table 1.5.1.

Table 1.5.1 Characteristics of typical ion beam sources.

Source technology	Ion species	Acceleration	Beam current	Probe size (nm)	Reduced brightness A/(m^2·sr·V)	Energy spread eV	Life time
LMIS	Ga$^+$	~ 500V -50 kV	~ 1 pA– tens of nA	5	10^6	5	~ 1500 uA–hrs
GFIS	He$^+$, Ne$^+$	$5 - 30$ kV	0.1–10 pA	0.35, 1.9	10^9	1	6 months
ICPS	Xe$^+$, O$_2^+$, O$^-$, He$^+$	15 kV	0.1 pA–μA	50	$10^3 - 10^4$	$3.5 - 5$	2 yrs

1.5.2 Ion optics

In the FIB and HIM, the ion beams are accelerated through a potential down the ion column. The typical acceleration voltage in the FIB ranges from 5–50 kV and 5–30 kV in the HIM. As shown in Fig 1.5.2, the optical column of the ion instruments are similar to that of a conventional scanning electron microscope, which consists of a condenser lens, an objective lens (quadrupoles), apertures, stigmators and scanning coils (octopoles). The condenser lens (Lens 1 in Fig. 1.5.2) is the probe forming lens and the objective lens (Lens 2 in Fig. 1.5.2) is used to focus the beam of ions at the sample plane. In this section, we outline the theoretical framework for dealing with electrostatic lenses. Interested readers can find useful information in dedicated textbooks (Dahl, 1973).

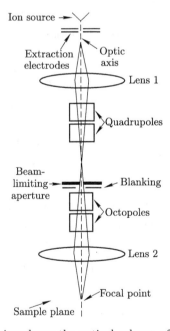

Fig.1.5.2 Schematic drawing shows the optical column of a typical FIB instrument.

Unlike electron microscopes, which use electromagnetic lenses for manipulating the electron trajectory, electrostatic lenses are used for ion beams. The reason for this is that the force experienced by a charged particle when travelling through an electromagnetic lens is directly proportional to its velocity, $\boldsymbol{F} = q\boldsymbol{v} \times \boldsymbol{B}$. Since a helium ion is about 7,295 times heavier than an electron, an electron will gain a velocity 85 times greater than that of a helium ion accelerated through the same potential. It is less effective to use the same set of electromagnetic lenses to focus ions. Using an electromagnetic lens that can manipulate ions effectively is cumbersome. In electrostatic lenses the force experienced by a charged particle is independent of its velocity, $\boldsymbol{F} = q\boldsymbol{E}$. Electrostatic lenses are therefore more effective at focusing the relatively slow ions.

The theory of electrostatic lenses has been developed from 1930 to 1955. Matrix formalism of ion optics with electrostatic lenses can be found in Hinterberger's review article (Hinterberger, 2005). Here, we briefly introduce the ion optics with rotationally symmetric electrostatic lenses. We assume the field involved a is time-independent static field and there is no free-space charge. The electrostatic field in the free space thus follows the Poisson equation, using a polar coordinate:

$$\frac{\partial^2 \varphi}{\partial z^2} + \frac{1}{r}\frac{\partial \varphi}{\partial r} + \frac{\partial^2 \varphi}{\partial r^2} = 0$$

The solution can be expressed in terms of the potential along the symmetry axis $V(z)$

$$\varphi(z,r) = V(z) - \frac{1}{4}V''(z)r^2 + \frac{1}{64}V''''(z)r^4 + \cdots$$

The equipotential surfaces refer to the surfaces of constant scalar potential, which plays important roles in the optical properties of the electrostatic lens. The geometry of the equipotential surface is given by $\varphi(z,r) = const$ and it exhibits hyperbolic profiles in the rotationally symmetric field.

The trajectory of the ions is determined by Newton's equation of motion

$$\frac{d\boldsymbol{p}}{dt} = q\boldsymbol{E} = -q\nabla\varphi$$

The relativistic paraxial trajectory of ions can then be given by

$$x'' = \frac{q}{pv}\left(x'\varphi' + \frac{x}{2}\varphi''\right)$$
$$y'' = \frac{q}{pv}\left(y'\varphi' + \frac{y}{2}\varphi''\right)$$

where x and y are Cartesian coordinates in the plane perpendicular to the symmetry axis all the derivatives are with respect to z, and p is the momentum.

A typical element of electrostatic ion optics is the accelerating tube lens (immersion lens). As shown in Fig. 1.5.3, it consists of two metal tubes with different potentials. We assume the potential along the symmetry axis rises linearly from V_1 to V_2 in the space between the two tubes. By using the paraxial trajectory given in the above equations, we can arrive at the matrix

$$R_x = \begin{pmatrix} 1 & 0 \\ \dfrac{E_2 - E_1}{2p_2 v_2}\dfrac{1}{L} & 1 \end{pmatrix} \begin{pmatrix} 1 & L_{\text{eff}}\dfrac{p_1}{p_2} \\ 0 & \dfrac{p_1}{p_2} \end{pmatrix} \begin{pmatrix} 1 & 0 \\ -\dfrac{E_2 - E_1}{2p_1 v_1}\dfrac{1}{L} & 1 \end{pmatrix}$$

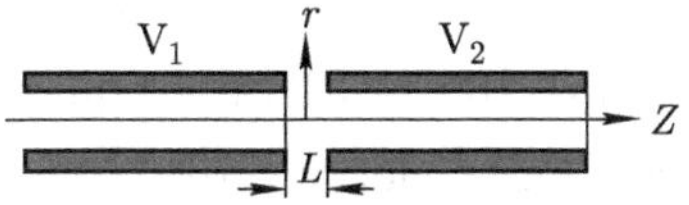

Fig.1.5.3   Schematic drawing shows the accelerating tube lens.

where E, p, v are potential energy, momentum, and velocity respectively. The subscripts represent the areas of the two tubes and

$$L_{\text{eff}} = L\frac{p_1 c}{E_2 - E_1}\ln\frac{p_2 c + E_2}{p_1 c + E_1}$$

where c is the light velocity in vacuum. The transformation of the ray coordinates by the tube lens is given by

$$\begin{pmatrix} x(z) \\ x'(z) \end{pmatrix} = R_x \begin{pmatrix} x(0) \\ x'(0) \end{pmatrix}$$

Using an aperture (a small circular hole in a conduting metal plate separating two regions with different eletric fields) as an example, the trajectory is changed by

$$\begin{pmatrix} x(z) \\ x'(z) \end{pmatrix} = \begin{pmatrix} 1 & 0 \\ -\dfrac{E_2 - E_1}{2p_1 v_1}\dfrac{1}{L} & 1 \end{pmatrix} \begin{pmatrix} x(0) \\ x'(0) \end{pmatrix}$$

This is the matrix of a thin lens with the focusing power

$$\frac{1}{f} = \frac{E_2 - E_1}{2p_1 v_1}\frac{1}{L}$$

It can be seen that the geometry optics discussed in section 1.3 can also be applied in the ion optics.

Electrostatic lenses introduce more severe aberrations than electromagnetic lenses. The focal length is highly dependent on the distance of the particle from the optic axis of the lens, and its angle to the axis. The effects of the strong spherical aberration can be minimized by introducing a small beam limiting aperture to reject off axis particles. This leads to a trade-off between beam current and probe size. For example, very small beam limiting apertures (5 μm diameter) are commonly used in the helium ion microscope for the highest resolution imaging. Such small apertures could cause a problem in the electron microscope due to the effects of diffraction. Helium ions have a larger mass and a shorter de Broglie wavelength, the diffraction effect due to apertures of this size is negligible.

Chromatic aberration is a serious limitation to the probe sizes achievable in conventional ion beam systems. In the FIB the liquid metal ion source generates gallium ions with a relatively large energy spread of 5–10 eV. As these ions travel at different velocities through the lens, the faster ions spend less time in the electrostatic field and experience less deflection, slower ions are more strongly focused due to their extended interaction time with the field of the lens. Ions of different energies are therefore focused on different planes. This results in the ions being focused to a disc area on the sample, instead of a single point. This chromatic aberration currently limits the resolution of the FIB to 3–5 nm at best. However, the gas-field source in the HIM can produce an ion beam with an energy spread of just 0.5 eV, leading to a significant reduction in chromatic aberration over the FIB.

In this section, we briefly discussed several popular ion source technologies and compared their characteristics. The theoretical framework to deal with the electrostatic lens has also been outlined, while the detailed ion optics is a mature field and sophisticated ion systems have been designed by applying the theories.

References

Bassim, N., Scott, K. and Giannuzzi, L. A. (2014) Recent advances in focused ion beam technology and applications. MRS Bull., **39** (4), 317–325.

Bischoff, L. and Akhmadaliev, C. (2008) An alloy liquid metal ion source for lithium. J. Phys. D: Appl. Phys, **41** (5), 052001.

Boyle, W. S. and Smith, G. E. (1970) Charge coupled semiconductor devices. Bell Sys. Tech. J., **49** (4), 587–593.

Breese, M. B. H., Jamieson, D. N. and King, P. J. C. (1996) Materials analysis using a nuclear microprobe. John Wilehy& Sons, INC.

Cho, B., Ichimura, T., Shimizu, R., and Oshima, C. (2004) Quantitative evaluation of spatial coherence of the electron beam from low temperature field emitters. Phys. Rev. Lett., **92**, 246103.

Dahl, P. (1973) Introduction to electron and ion optics. Academic Press.

de Jonge, N., van Druten, N. J. (2003) Field emission from individual multiwalled carbon nanotubes. Ultramicroscopy, **95**, 85–91.

Economou, N. P., Notte, J. A., and Thompson, W. B. (2012) The history and development of the helium ion microscope. Scanning, **34**, 83–89.

Fowler, R. H., Nordheim, L. (1928) Electron emission in intense electric fields. P. Roy. Soc. Lond., 119 (781), 173–181.

Gamo, K., Matsui, T., and Namba, S. (1983) Characteristics of Be-Si-Au ternary alloy liquid metal ion sources. Jpn. j. Appl. Phys, **22** (11), L692–L694.

Giannuzzi, L. and Stevie, F. (2005) Introduction To Focused Ion Beams: Instrumentation Theory, Techniques And Practice. Springer US.

Hawkes, P. W. (1972) Electron Optics And Electron Microscopy. United Kingdom: Taylor & Francis Ltd.

Hawkes, P. W. and Kasper, E. (1996) Principles Of Electron Optics. Elsevier Inc., Part VIII.

Hinterberger, F. (2005) Ion optics with electrostatic lenses. CERN Acelerator School and KVI: Specialized CAS Course on Small Accelerators, 27–44.

Krivanek, O. L., Ursin, J. P., Bacon, N. J., Corbin, G. J., Dellby, N., Hrncirik, P., Murfitt, M. F., Own, C.S., and Szilagyi, Z. S. (2009) High-energy-resolution monochromator for aberration-corrected scanning transmission electron microscopy/electron energy-loss spectroscopy. Phil. Trans. R. Soc. A, **367**, 3683–3697.

Kuwahara, M., Takeda, Y., Saitoh, K., Ujihara, T., Asano, H., Nakanishi, T., Tanaka, N. (2011) Development of spin-polarized transmission electron microscope. J. Phys.: Conf. Ser., 298, 012016.

Life through a Lens. The Nobel Prize in Physics 1986, Perspectives, nobelprize.org.

Liu, X. and Orloff, J. (2005) A study of optical properties of gas phase field ionization sources. Adv. Imag. Elect. Phys., **138**, 147–175.

Meyer, A., Dimeo, R. M., Gehring, P. M., Neumann, D. A. (2003) The high-flux backscattering spectrometer at the NIST center for neutron research. Rev. Sci. Instrum., **74** (5), 2759-2777.

Murphy, E. L., Good, G. H. (1956) Thermionic emission, field emission, and the transition region. Phys. Rev., **102** (6), 1464–1473.

Orloff, J. (2009) Handbook Of Charged Particle Optics. CRC Press, 2nd edition.

Reimer, L. (1989) Transmission Electron Microscopy. Springer Berlin Heidelberg, 2nd edition.

Richardson, O. W. (2003) Thermionic Emission from Hot Bodies. Watchmaker Publishing.

Smith, N.S., Notte, J. A. and Steele, A. V. (2014) Advances in source technology for focused ion beam instruments. MRS Bull., **39** (4), 329–335.

Takaoka, A., Ura, K., Mori, H., Katsuta, T., Matsui, I., & Hayashi, S. (1997) Development of a new 3 MV ultra-high voltage electron microscope at Osaka Universiy. J. Electron Microsc., **46** (6), 447–456.

Urita, K., Suenaga, K., Sugai, T., Shinohara, H., and Iijima, S. (2005) In situ observation of thermal relaxation of interstitial-vacancy pair defects in a graphite gap. Phys. Rev. Lett., 94 (15), 155502.

Williama, D. B. and Carter, C. B. (2009) Transmission Electron Microscopy: A Textbook For Materials Science. Springer US, 2nd edition.

Yakimenko, V., Fedurin, M., Litvinenko, V., Fedotov, A., Kayran, D., and Muggli, P. (2012) Experimental observation of suppression of coherent synchrotron radiation induced beam energy spread with shielding plates. Phys. Rev. Lett., 109, 164802.

Zhu, Y., Inada, H., Nakamura, K. and Wall, J. (2009) Imaging single atoms using secondary electrons with an aberration-corrected electron microscope. Nat. Mater., 8, 808–812.

Zhu, Y., Milas, M., Han, M. G., Rameau, J. D. and Sfeir, M. (2012) Multimodal optical nanoprobe for advanced in-situ electron microscopy. Microsc. Today, 20 (6), 32–37.

Spence, J. C. H. and Zuo, J. M. (1992) Electron Microdiffraction. Springer US.

Zuo, J. M., Gao, M., Tao, J., Li, B. Q., Twesten, R., and Petrov, I. (2004) Coherent nano-area electron diffraction. Microsc. Res. Techniq., 64, 347–355.

2
Scanning Electron Microscopy

Wei Han, Huisheng Jiao and Daniel Fox

2.1 Introduction

The scanning electron microscope (SEM) is the most widely used tool for characterizing and analyzing the surface of solid samples. It is utilized in many research areas as well as in various industry sectors. Using the SEM, material scientists can study nanoscale features of their samples, which enables them to gain knowledge of the formation and properties of the sample. The SEM is also an indispensable tool to investigate biological samples and it facilitates the observation of bacteria, fungi and viruses.

The application history of SEM can be traced back to 1937 when Ardenne (Ardenne, 1938) added scanning coils to a transmission electron microscope (TEM) and acquired the first scanning image. Zworykin et al. (1942) continued to develop the instrument and achieved a resolution of 50 nm. In the 1950s Oatley and his group in Cambridge manufactured the first commercial SEM-"Stereoscan" which was then released by Cambridge Scientific Instrument Company in 1965. There are more than seven SEM manufacturers in Europe, the US and Asia (e.g., FEI, Zeiss, JEOL and Hitachi), and more than 20,000 SEMs have been installed in the world (Amelinckx et al., 1997). Sub-nanometer resolution has been recently demonstrated in the state-of-the-art SEMs, for example, the FEI Verios (0.7 nm at 1kV, released in 2012) (FEI, 2012) and the Hitachi SU9000 (0.4 nm at 30 kV, released in 2011) (Hitachi, 2011).

In this chapter, the principles and applications of SEM will be discussed with an emphasis on new progress of SEMs in nanoscale applications.

2.2 Fundamentals of the SEM

A modern scanning electron microscope consists of an electron optical system, a vacuum system, electronics system, computer and software. The electron optical system involves the formation of the electron probe, which includes the electron gun, the demagnification system (i.e. the condenser lens), the scanning unit, and the focusing system (i.e. the objective lens as sketched in Fig. 2.2.1). The electron-optical system produces a highly focused electron probe ($\sim$1 nm) which is scanned in a raster over a region of the specimen surface. The interaction between the beam electron and the specimen generates a range of signals that can be collected by proper detectors equipped in the SEM to form images or spectra. The image can be displayed on a PC monitor simultaneously, while the SEM is scanning the electron probe on the sample surface. Usually we call the actual scan range horizontal field of view (HFW). If the width of monitor is L, then the magnification $M = L/\text{HFW}$. Since L is a constant, a higher magnification will be obtained if the scan area HFW is decreased.

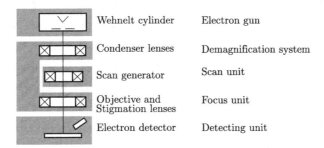

Fig.2.2.1 Main structure of the optical column of the SEM.

The electron-optical system is crucial to the performance of the SEM, for example, the ultimate resolution relies on the quality of the electron gun and the probe. In this section, the components of the electron-optical system will be discussed in detail.

2.2.1 Electron emitters

SEMs can be classified into thermionic and field-emission SEMs according to the methods used for electron emission. The tungsten cathode is now the dominant thermionic gun, while the LaB$_6$ filament is still being used by some SEM manufacturers. Field-emission SEMs (FE-SEMs) have two variants: Schottky FE-SEM and cold FE-SEM, although strictly speaking Schottky field emission is indeed a field-assisted thermionic emission.

The filament of the tungsten SEM (W-SEM) is a hairpin-shaped tungsten wire with an apex diameter about 100 µm (see Fig. 2.2.2(a)). At high temperatures, the electrons in the tungsten wire gain enough kinetic energy and overcome the surface barrier so that an emission current is achieved. The emission current follows the Richardson-Dushman equation:

$$J = AT^2 e^{-B/T} \tag{2.2.1}$$

where A and B are constants. The typical resolution for tungsten thermionic gun SEMs is ~3.0 nm at 30 keV for the standard specimen (gold on carbon). The W-SEM is cheap and reliable, since it does not require ultra-high vacuum and the filament itself is inexpensive.

Fig. 2.2.2(b) illustrates the typical morphology of the cold field emission emitter. A small piece of tungsten single crystal with an apex diameter of 10–100 nm is welded on the top of a hairpin-shaped tungsten wire. There are two anodes in the FE-SEM gun, i.e. extraction and acceleration anodes. A positive bias is applied on the extraction anode, which produces a strong electrical field at the adjacent area of the emitter tip due to the geometric field enhancement effect. The strong electrical field depresses the surface barriers and allows the electron to escape into the vacuum via the tunneling effect. The acceleration anode then accelerates the electron to its working energy, which may vary from a few tens of eV to tens of keV. The current density (A/m^2) is given by the following equation:

$$J = 1.54 \times \frac{10^{-10} E^2}{\varphi} \cdot \exp[-6.83 \times 10^9 \cdot \varphi^{3/2} \cdot k/E] \tag{2.2.2}$$

where E is the electric field at the emitter on the order of 10^9 V/m, φ is the work function and k is a constant. The current densities can reach 10^{12} A/m^2. The Schottky emitter is a tungsten filament coated with ZrO$_2$. The ZrO$_2$ coating lowers the work function and in comparison to the cold FE emitter, the Schottky emitter is heated while it is working.

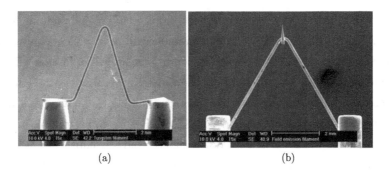

Fig.2.2.2   SEM image of filaments of tungsten SEM (a) and FE-SEM (b).

Table 2.2.1 lists the relevant characteristics of the electron emitters which can be used to evaluate their performance. Gun brightness is defined as the current density per solid angle. Since the brightness is invariant in the electron optics, the higher the brightness the larger the beam current can be acquired when a small probe is formed, which in turn regulates the signal to noise level and limits the demagnification of the probe. It can be seen that cold FE emitters offer the highest brightness. Energy spread in the electron probe can affect the probe size through the chromatic effect, so a small energy spread is in favor of small probe size. The energy spread of the field emission emitters is an order of magnitude smaller than that of the thermionic gun. The advantages of W-SEM (thermionic emitters) are however large probe current, cost efficiency, and less-demanding maintenance.

Table 2.2.1   Comparison of three different emitters.

Characteristics	W	Cold FE	Schottky FE
Brightness(A/cm^2·sr)	10^6	10^9	5×10^8
Energy spread (eV)	2	0.25–0.5	0.35–1.0
Cathode temperature (K)	2800	300	1800
Vacuum (Pa)	10^{-3}	10^{-8}	10^{-7}
Current stability	1%/1 h	5%/15 min	0.4%/1 h
Probe current	10 pA–2000 nA	1 pA–20 nA	1 pA–200 nA
Lifetime	100 hr	> 1 year	> 1 year

The operation of the cold FE emitter requires ultrahigh vacuum in the gun chamber. The residual gases in the gun chamber will nevertheless contaminate the filament tip gradually under the working conditions, and the emission will fluctuate from time to time due to the contamination. After several hours the emitter must be flashed at a high temperature to burn out the contaminant. This instability has limited the application of cold FE SEMs. For example, on-line instruments in semiconductor industry usually don't choose cold FE emitter. The probe current of the cold FE emitter is also lower than that of the Schottky FE emitter, so the cold FE SEM is less suitable for many applications which need higher and more stable currents, such as electron back scatter diffraction (EBSD), wavelength dispersive X-ray analysis (WDX), low vacuum or in situ dynamic applications. Schottky emitters can however meet the requirements of these techniques. Moreover, recent developments have improved the performance of the Schottky emitter to a level close to that of the cold FE emitter in terms of resolution and energy spread. A resolution of 0.7 nm and an energy spread less than 0.2 eV at 1 kV was achieved in the FEI Verios XHR SEM (Schottky Gun) which is equipped with a monochromator.

2.2.2 Electron-optical column in SEM and probe formation

The electron passes through the column of the SEM and the electron beam is demagnified by the condenser lens and focused by the objective lens to form a fine electron probe on the specimen surface. The finer the probe, the better the spatial resolution. The probe size is determined by the quality of the lenses and limited by the wave nature of the electron beam as well as the brightness of the gun. For accelerating voltages from 5 kV to 30 kV, the probe size can be calculated by the following equation (Smith, 1972):

$$d = (C_s^{1/4} \lambda^{3/4}) \left[1 + \frac{I_b}{\beta \lambda^2}\right]^{3/8} \qquad (2.2.3)$$

where C_s is the spherical aberration coefficient of the objective lens, λ the wavelength of electrons, β the brightness of the electron gun and I_b the beam current. For a tungsten cathode thermionic gun, the brightness is about 10^5 A·cm^{-2}·sr^{-1}, and the image performance is not determined by the lens but by the electron gun.

For FE-SEM gun the brightness is about 10^8 A·cm^{-2}·sr^{-1} (Goldstein et al., 1992), the probe size is limited by the performance of the lens (Amelinckx et al., 1997). For normal FE-SEM 1–2 nm probe can be acquired, while the energy spread must be considered at low accelerating voltages (< 5 kV) because of severe chromatic aberration. The size of the probe also increases with increasing beam current, since the statistical coulomb interactions at the beam crossover can modify the lateral velocities of the electrons known as the Boersch effect.

The lenses used in SEM are electromagnetic lenses, except the gun lens which is an electrostatic lens. To meet the requirements of nanoscale applications the modern lens system is designed in combination with the detector system. For the objective lens system there are mainly two types of lenses: the out-of-lens (field-free) type and the in-lens (immersion) type objective lens. For a field-free lens, the sample is placed below the pole piece of the objective lens. When decreasing the working distance the detector efficiency will be decreased. Most high resolution SEMs use this type of objective lens. In the immersion objective lens, the sample is placed in the gap between the upper and lower pole pieces of the lens. The advantage is that a very short working distance can be applied, while the detector efficiency is still quite high with an in-lens detector. This is an effective way to increase resolution in the low kV range due to the decrease in spherical and chromatic aberration. In most ultra-high-resolution SEMs immersion objective lenses are equipped as the standard. However, the magnetic field of the lens could be influenced by magnetic samples, producing stronger astigmatism. Another problem is that it may influence the trace of BSE and induce bending Kikuchi band in EBSD analysis. In commercial SEMs some suppliers combine these two lens types into one system as illustrated in Fig. 2.2.3. Two operation modes can be switched freely, alllowing these problems to be overcome. More details on this subject can be found in dedicated book (Zhou and Wang, 2007).

2.2.3 Signal generation and detection

During the interaction of the electron beam with the specimen, the beam will penetrate into the specimen ranging from several nanometers to several micrometers

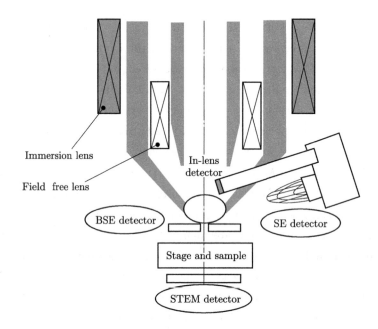

Fig.2.2.3 Illustration of objective lens and detectors in modern SEM system.

depending on the energy of the incident electron and the materials. In this process the incident electrons are randomly scattered. Both elastic and inelastic scattering will happen. The scattered electrons form an interaction volume with a waterdrop shape inside the specimen. Various signals can be generated in the interaction volume and some of them will escape from the sample and be collected by the detectors. This process is illustrated in Fig. 2.2.4. These signals include secondary electrons (SE), backscattered electrons (BSE), Auger electrons, X-rays (including characteristic X-rays and Bremstrahlung X-rays), and cathodoluminescence (CL). The signals carry some information of the specimen. Among them SEs and BSEs are the most widely utilized signals because SEs can offer very high resolution images and BSEs can provide compositional information. In this section, we will discuss SEs and BSEs, while the X-ray and the analytical capabilities of SEM will be introduced in the next section.

The Interaction volume: The resolution of SEM is not only dependent on the primary beam size, but also the interaction volume. The size of the interaction

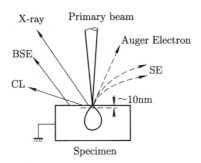

Fig.2.2.4 Signals from the interaction of electron beam and specimen.

volume is determined by the beam energy, the specimen, and the incidence angle. According to Monte Carlo simulation with the same incident beam energy (15 kV), the beam will penetrate into bulk WSi_2 for 1.2 μm in depth, while it will penetrate into bulk KCl for 4.8 μm in depth. The interaction volume has a strong Z effect. Samples with higher atomic number Z have smaller interaction volumes. Another influencing factor on the interaction volume is the beam energy. Smaller incident beam energy corresponds to reduced interaction volume and penetration depth. In Fig. 2.2.5 two images with different accelerating voltages are obtained on a fractured aluminum specimen at 5 keV and 20 keV, respectively. The 5 keV image shows surface details quite clearly, while some information is hidden in the 20 keV image.

In nanoscale research the small interaction volume will be more important. That means low kV performance is very important for nanoscale applications. However, low kV imaging typically has larger beam sizes and higher energy spread. Fortunately technical progress in SEM has overcome these problems in the low kV range. Another advantage comes from the fact that many nano-materials are non-conductive, at low kV the interaction volume is small therefore the yield of SEs will be increased accordingly. This effect can be exploited to balance the rate of electrons entering and exiting the sample, thereby mitigating the charging effect.

Signal generation: As a result of the elastic interaction, the beam electron may be backscattered into the vacuum with a maximum energy equal to the primary

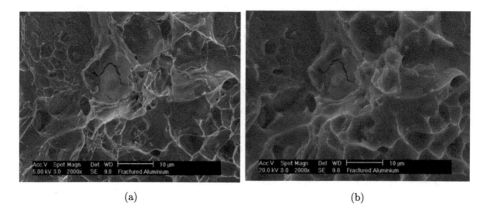

(a) (b)

Fig.2.2.5 A sample of fractured aluminum that has inclusions of copper, SE images at 5 keV (a) and 20 keV (b).

electron energy. Inelastic interaction between the beam electron and the specimen can result in ionization of the specimen atoms, which produces secondary electrons throughout the total interaction volume. These secondary electrons have an average energy of 2 eV to 6 eV and may escape from the specimen from a small depth of about 1 nm for metals, and of the order of 10 nm for carbon. The spectrum of all electrons coming out of a specimen when it is irradiated with an electron beam of energy E_{PE} is shown in Fig. 2.2.6. By convention the electrons with an energy below 50 eV are called secondary electrons (SE) and the others are the backscattered electrons (BSE).

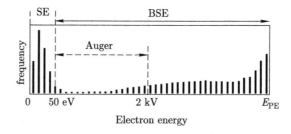

Fig.2.2.6 Energy distribution of electron signals.

It should be noted that a backscattered electron generated deep in the material is energetic enough to produce secondary electrons on its way back to the surface. This type of SE is called SE2, while the secondary electrons generated by the primary beam are called SE1. If a backscattered electron hits the chamber or the pole piece of the electron microscope, some secondary electrons are produced and called SE3. Usually SE1 carries high resolution information, while SE2 and SE3 generate low resolution background. In terms of the detection, it is however not possible to separate SE1, SE2 and SE3, while specific detector configuration may depress the contributions of SE2 and SE3, for example the in-lens (or through lens) detectors.

SE images can generally be easily interpreted as the topographical illustration of the sample surface without any special knowledge. As the beam scans along the surface of the sample, the angle of incidence changes due to the variation in the local roughness of the sample surface. The number of electrons leaving (and therefore being detected) depends on the angle of incidence, which is fundamental for the formation of the topographical contrast. Edges, which have a high angle relative to the incident beam, will produce many SEs compared to the flatter surrounding areas. This bright edge area is referred to as the "edge effect". As shown in Fig. 2.2.7, secondary electron images also exhibit a large depth of focus and the image has stereoscopic effects due to shadow effect. The resolution is higher in comparison to the BSE image.

The most widely used SE detector in SEM is the Everhart-Thornley detector (Everhart and Thornley, 1960), which is a scintillator-photomultiplier system. About 10 kV positive potential is used on the scintillator to accelerate secondary electrons. A Faraday cage made of copper mesh is used to cover the scintillator. The bias on the cage can be varied from −100 V to +250 V. If +250 V is used on the cage secondary electrons will be attracted to the cage. If −100 V is used secondary electrons will be repelled from the Faraday cage, in which case only BSE can be detected. Usually the SE detector is placed inside the specimen chamber above the final lens plane. In high resolution SEM another in-lens SE detector will be installed inside the objective lens or above it.

In Fig. 2.2.6, the large peak around the primary beam energy results from the Rutherford scattering and this process increases with increasing atomic number Z.

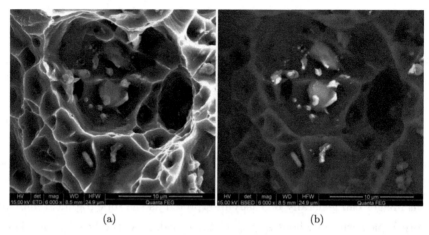

Fig.2.2.7 Comparison of SE image (a) and BSE image (b) of Al fracture in the same field of view.

Therefore, the number of BSEs coming out of the specimen reflects the average Z value of the material. This is the most important contrast mechanism for the backscattered electrons. The right image in Fig.2.2.7 illustrates some features of BSE images. The resolution of the BSE image is not as good as that of the SE image. The image also looks somewhat transparent because the energy of BSE is large enough to escape from up to one third of the total interaction depth. Another feature is the Z-contrast, i.e., the material in the bright areas has higher atomic number than that of other areas.

The yield of BSE increases from 0.05 for carbon to 0.5 for gold. During the scattering process BSE can go any direction. It is difficult to collect BSE with high efficiency because the energy of the BSE is too high to be attracted by high positive potential. BSE detectors are typically placed under the pole piece of the objective lens to increase the collection angle. BSE outside this angle cannot be utilized for signal. Although the yield of BSE is higher than that of SE, the efficiency of BSE detection is much less than that of the SE detection. Another limitation is that the resolution of the BSE image is worse than that of the SE image because of the larger escape depth. No matter how small the actual diameter of the primary beam the interaction volume will degrade the resolution of the BSE image.

The solid state BSE detector (SS-BSD) is the most common BSE detector on the market. It is basically a diode that amplifies the high energy BSE striking the detector, as illustrated in Fig. 2.2.8. When a BSE strikes the detector, electrons in the material move from the valence to the conduction band leaving holes in the valence band. If a potential is applied, the e^- and h^+ can be separated and collected, and the current can be measured. The current is proportional to the number of BSEs that hits the detector.

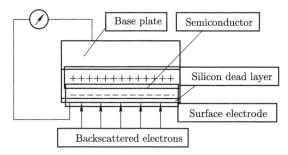

Fig.2.2.8 Structure of solid state BSE detector.

The SS-BSD is usually divided into two detection areas. This will allow for either topographical or elemental imaging of BSE by subtraction or addition signals from two parts respectively. Some BSD systems even separate the SS-BSD into 4 or 5 segments. More combination modes have been developed to meet the requirement of different applications.

In addition to the material contrast, more information about the specimen is necessary to interpret BSE images, especially for crystalline materials where the contrast of BSE image is related to crystal orientation (so called the channeling effect) as well, as shown in Fig. 2.2.9. This channeling effect has been utilized for characterization of crystalline materials. Commercially available detectors have been developed for SEM. This instrument is called EBSD. Applications of EBSD will be discussed in detail in the next section.

BSE signal is indispensable due to its material contrast, for example in the steel industry BSE signal is used for finding inclusions in steels. In forensic science BSE signal is an essential detector for the investigation of gunshot residue (GSR) (ASTM E1588-07; Andrasko and Maehly, 1977). In the natural resources area the

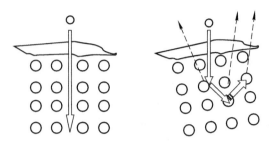

Fig.2.2.9 Schematic illustration of electron channeling effect, in the left figure channeling happens at a particular incidence direction to the crystal. In the right figure there is no channeling effect at random incidence direction.

BSE signal is used for investigating valuable minerals. In the above applications BSE signals can supply a rough distribution of sample composition. In biological samples BSE imaging is now a unique way to analyze resin embedded slices or cross-sections milled by focused ion beam (Li et al., 2013; Knott et al., 2008).

2.3 Analytical capabilities of the SEM

SEM provides a platform to characterize the microstructure of materials. However, with only high resolution images and no analytical capabilities it is difficult to comprehensively characterize the microstructures. Since many different signals are generated when a beam of electrons bombard the surface of a specimen, chemical and crystal information can be acquired through the use of the appropriate detectors. The energy dispersive X-ray spectrometer (EDS) is one of the most useful systems for micro-chemical analysis in SEM. An electron backscatter diffraction (EBSD) system is often used for crystal and orientation measurements in SEM. In the following chapters these techniques will be presented in detail.

2.3.1 X-ray microanalysis and nanoanalysis

1. Generation of X-rays in SEM

In a SEM chamber, a beam of high energy electrons bombards the surface of samples. When an electron with high energy interacts with an atom in samples it may result in the ejection of an electron from an inner electron shell. This will leave a vacancy in this shell and the atom is in an excited state. The excited atom is unstable and an electron from an outer shell fills the vacancy. Electrons in different shells have different energy and the minimum energy required to remove an electron from a particular energy level is known as the critical ionization energy E_c or X-ray absorption edge energy. The critical energy has a specific value for any given energy level and is typically referred to as the K, L or M absorption edge. The change in energy during the de-excitation or "characteristic energy" is determined by the electronic structure of the atom which is unique to the element.

There are two approaches to release this "characteristic" energy. One is the emission of an X-ray photon (fluorescence yield = w) with a characteristic energy related to the energy of different electron shells of an atom. The second way is by releasing so called Auger electrons (Auger yield = a).

The probabilities of energy release by these two ways can be written as:

$$a + w = 1 \qquad (2.3.1)$$

For any given shell, the probability of X-ray emission is relative small for light elements. For example, the value of w for the Si K shell is 0.047, Co K shell is 0.381 and 0.764 for the Mo K shell.

If a vacancy is in the K shell of an atom, X-ray emission from de-excitation is called the K line. Similarly, X-ray emission from de-excitation due to a vacancy in the L or M shell is called the L line or M line. Furthermore, if an electron from the L shell fills the K shell vacancy, the X-ray emitted in this transition is termed the K_α line. If an electron from the M shell fills this vacancy the X-ray emitted in this transition is termed the K_β line. The K shell is the closest shell to the nucleus, it therefore requires the most energy to remove electrons from this shell. The K line has the highest energy if an atom has K, L and M lines.

The intensity of a given line primarily depends on the probability of X-ray generation as a result of a given transition. The relative probability of generating X-rays at the various ionization energies for a given element depends on the value of the incident energy and the excitation cross section for the relevant shell. The cross section itself can be expressed in terms of the overvoltage, U ($U = 2 \sim 3$, gives the highest probability for X-ray generation), which is simply the incident beam energy divided by the critical ionization energy for a particular shell.

2. EDS system for microanalysis

EDS systems are commonly used on SEM for micro chemical analysis. An EDS system is composed of three basic components: 1) X-ray detector, which detects and converts X-rays into electronic signals. 2) pulse processor, which measures the electronic signals to determine the energy of each X-ray detected and 3) analyzer, which displays and interprets the X-ray data.

Fig. 2.3.1 shows a diagram of an old Si (Li) EDS detector. This is a typical Si(Li) detector and a modern EDS detector still uses this configuration except the liquid nitrogen Dewar for cooling the crystal. In an EDS detector, the core component is crystal, which is a semiconductor device, often made of Si, converting X-ray photons into electric charges. These charges are transferred to a field effect transistor (FET) just behind the crystal and converted to voltage output. In front

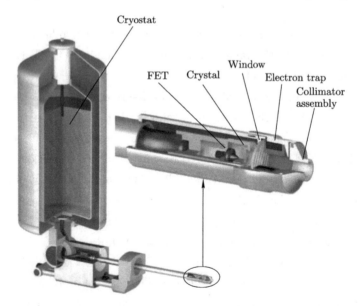

Fig.2.3.1   Diagram showing components in a Si (Li) EDS detector (Oxford Instrument, 2002).

of the crystal, a thin window made of Be or polymer seals the crystal from outside. A Be window is robust but heavily absorbs low energy X-rays. That means only X-rays from elements above Na can be detected. Nowadays EDS detectors on SEMs all use ultra-thin polymer windows. This enables the detection of X-rays down to less than 100 eV (Be). These ultra-thin polymer films are supported on a silicon grid to withstand the pressure difference between the detector vacuum and a vented microscope chamber at atmospheric pressure. An electron trap made of permanent magnets is placed in front of the window to keep electrons away from the crystal. Electrons could cause serious background artifacts. At the beginning of the detector there is a collimator which ensures that only X-rays from the area being excited by the electron beam are detected and stray X-rays from other parts of the microscope chamber are not included in the analysis.

During the last few decades the crystal in the EDS detector has experienced a revolution. Si (Li) was the first material used in EDS detectors and remained the most common choice until the first decade of 21st century. The most common crystal is Si, into which is drifted lithium (Li) to compensate for small levels of impurity. The modern EDS system uses the Silicon drift detector (SDD) which was first manufactured in the 1980s for radiation physics. The development of this detector, and advances in its fabrication methods, have led to a liquid nitrogen free EDS detector to replace the Si(Li) detector with better performance and productivity in SEM. Fig. 2.3.2 shows the difference between these two types of detectors.

Normally a Si (Li) crystal is about a 3 mm thick disc and an SDD crystal is only about 500 µm in thickness. Also, the anode in an SDD crystal is reduced to about 50 µm in diameter for charge collection. In an SDD crystal a negative bias is applied to both sides of the disc. But on one side of the planar structure the bias is graduated across the device by means of a series of 'drift rings' such that a strong transverse electric field component is developed within the structure. This is used to direct electrons produced as a result of X-ray interactions towards a small anode. On the opposite side of the device (X-ray entrance side) there is a uniform shallow implanted junction contact to allow good low-energy X-ray sensitivity and minimum charge loss. This drift detector structure has very low capacitance which provides excellent energy resolution at relatively short electronic processing times

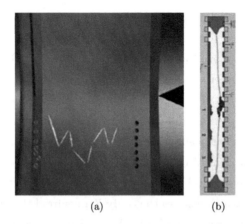

Fig.2.3.2 Cross section of X-ray detectors. Diagram showing generation of electron-vacancy pairs in a traditional Si (Li) crystal (a) and in a SDD crystal(b) (Ketek GmbH, 2013).

and also allows operation at very high count rates (Oxford Instruments, 2012). SDD detectors clearly possess an important advantage because they maintain best resolution performance at much higher count rates than Si (Li). Higher productivity is achieved by collecting data faster with no loss of analytical performance.

3. Qualitative and quantitative EDS analysis

EDS analysis in SEM is one of the most convenient techniques for obtaining chemical information of interesting features. It gets not only elemental information but also the weight and atomic percent of ingredients and it is one of the most accurate methods for composition analysis on the micron or nano level. In general, the detection limits of SEM EDS reach 0.1 wt.% level, and even hundreds of ppm for heavy elements (Goldstein et al., 1992). It is well known that electron probe microanalysis (EPMA) is considered as the most accurate technique for quantitative composition analysis. However with the progress of the EDS technique both in hardware and software, recent experiments have revealed that the SDD EDS could give very high quality data, which is comparable with EPMA.

Micro chemical characterization of SEM EDS analysis falls into two categories:

(a) qualitatively to determine the element species and their distribution, and (b) quantitatively obtain the concentrations of each element. For most systems, as shown in Fig. 2.3.3, elements with obvious peaks in an EDS X-ray spectrum can be identified correctly. However, if two or more element peaks are close to each other or a minor peak is not obvious, it will be difficult to identify the peak correctly. An important specification for characterizing performance of an EDS system is its ability to resolve peak overlaps, termed as energy resolution. Resolution is quoted as the full width at half maximum (FWHM), normally using FWHM of the X-rays of the manganese (Mn) K_α line (Mn K_α). This is convenient for manufacturers of EDS detectors, because they can use a Mn K_α emitting ^{55}Fe radioactive source to monitor individual sensor and detector performance without the need of an SEM. The lower the number the better the resolution a detector has and the better it will be at resolving peaks due to closely spaced X-ray lines. Currently, the energy resolution of a modern SEM SDD is in the range of 120–130 eV for Mn K_α.

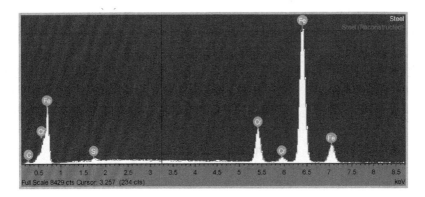

Fig.2.3.3 Comparison between simulated spectrum (red line) and acquired spectrum.

If there are peak overlaps or minor peaks in an EDS spectrum, manual peak identification is always needed for element confirmation. Commercial software always provides special tools, such as spectrum simulation based on identified elements, to check for any possible overlaps or minor elements. For minor peaks, in order to identify the element without doubt, an effective approach is to acquire more counts by elongating the acquisition time to make these peaks more obvious.

EDS analysis therefore benefits from SDD detectors due to their enhanced energy resolution and 10 times higher count rate than Si (Li) detectors.

After the elements have been identified, quantitative analysis may be needed to give the concentration of each element. This requires the accurate measurement of the peak intensities. As there is a contribution from a non-linear background over the entire energy range, a 'top-hat' filtering method is one of the best algorithms to suppress the background. The peak intensities are then obtained using a least squares fitting routine. Once these intensities have been determined matrix corrections (i.e. XPP correction) are applied to determine the concentration of each element. XPP has favorable performance for situations of severe absorption such as the analysis of light elements in a heavy matrix and for samples that are tilted with respect to the incident electron beam. Table 2.3.1 gives the quantitative analysis of two samples with certified concentration showing the accuracy of modern EDS analysis.

Table 2.3.1 Quantitative analysis of two EDS standards.

Anorthoclase			Benitoite		
Element	wt. %	Certificate	Element	wt. %	Certificate
O	46.84	46.54	O	34.49	34.83
Na	2.69	2.75	Si	19.78	20.37
Al	10.07	10.53	Ti	12.02	11.57
Si	30.65	30.06	Ba	33.71	33.21
K	9.57	9.46			
Ca	0.17	0.17			

From table 2.3.1 it is seen that the EDS gives very accurate measurements in comparison with the certified concentration, even for the light element oxygen and minor elements with concentration as low as 0.17%.

4. Spectrum mapping and spatial resolution

For an unknown sample spectrum mapping is the most efficient way to learn about the sample. A field on the sample is defined and then the X-ray spectrum map performs the simultaneous acquisition of X-ray data for all possible elements from each pixel on this area. Beam dwelling time per pixel and the number of frames can be set for spectrum map acquisition in SEM. In a spectrum map each pixel is associated with a single spectrum, it is therefore possible to display a map of any element from stored data without the need to define the element list before acquisition. It is also easy to reconstruct spectra, linescans or maps after acquisition. Element maps can be used to quickly highlight compositional variations in a sample.

The spatial resolution of an element map depends on the interaction volume between the electron beam and the sample. The overall dimensions of the interaction volume depend strongly on the incident beam energy, since the cross section of elastic scattering follows an inverse square dependence with the electron beam energy. As the beam energy of the incident electrons increases, they are able to travel further into the sample. An additional factor which affects the overall dimensions of the interaction volume is the rate at which electrons lose energy. The rate of energy loss is inversely proportional to the energy of the electron. This means that with an increase in electron beam energy, the rate at which these electrons lose energy decreases, such that they are able to travel further into the sample. The shape of the interaction volume depends strongly on the atomic number of the material. The cross section for elastic scattering is proportional to the square of the atomic number of the material. This means that for a fixed beam energy, electrons entering a high atomic number material will be scattered away from their original directions, giving the volume "width" and reducing penetration into the material. However, in a low atomic number material, electrons will penetrate into the sample much deeper, losing energy as they undergo inelastic scattering events, until the energy of the electrons is such that the probability of elastic scattering begins to dominate. This gives rise to the shape of the so called "pear shaped" volume. Eventually, the electrons do not have sufficient energy to scatter further into the sample, which corresponds to the "boundaries" of the electron interaction volume.

$$Z_m = 0.033 \cdot (V_0^{1.7} - V_k^{1.7}) \cdot A/\rho Z \tag{2.3.2}$$

where Z is the atomic number, V_0 is the accelerating voltage (kV), V_k is the excitation energy (kV), A is the atomic weight and ρ is the density of the sample (g/cm^3).

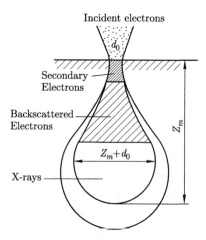

Fig.2.3.4 Diagram showing the interactive volume of electron beam and a sample (Reimer, 1979).

The interaction of the electron beam with the sample is complex, where a whole host of interaction and scattering events are possible. The Monte Carlo method is a mathematical technique, which attempts to model the shape of the interaction volume by simulating a large number of electron trajectories through the solid materials (Joy, 1995). Fig. 2.3.5 shows the interaction volume in bulk Ni at different beam energies.

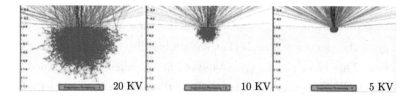

Fig.2.3.5 Monte Carlo simulation of the interaction volume in Ni with different beam energy.

It clearly shows that the interaction volume decreases when the beam energy decreases. The spatial resolution of element maps will be significantly improved when beam energy is changed from 20 kV to 5 kV due to the smaller interaction volume. As the accelerating voltage increases, the interaction volume excited within the sample increases. It means that both X-rays and backscattered electrons can be generated much deeper, affecting both imaging and analysis. Fig. 2.3.6 shows Ni, Cr and Nb overlapped map under 20 kV and 5 kV. Obviously, low kV will give an element map with better resolution because of the shallow penetration of electron beam and small interaction volume.

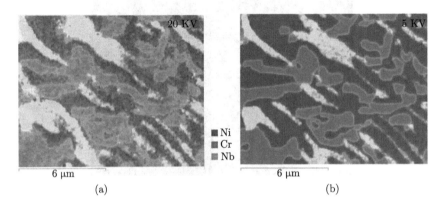

Fig.2.3.6 Overlapped Ni, Cr and Nb map showing the better spatial resolution at low kV.

To achieve better spatial resolution, low kV should be used for elemental mapping. However, low kV and small interaction volume mean less X-ray signals generated. In order to collect X-ray maps with moderate statistics, spectra collection may take tens of minutes or even hours. Recently, large size SDD detectors (up to 150 mm^2 active area) were fabricated which can collect more X-rays under the same SEM operating conditions. For comparison, the active area of a crystal in commonly used EDS detectors is in the range of 10–30 mm^2. With these large area SDD detectors, X-ray maps with a resolution of tens of nanometers can be achieved, as shown in Fig. 2.3.7. For nanoscale element X-ray mapping, the accelerating voltage is low. The performance of SDD detectors at low energy, such as energy resolution and sensitivity, becomes more important. When small features < 1 μm in size are being analyzed, the beam voltage needs to be reduced to avoid generation of X-rays from surrounding material. However, at low kV only low energy lines are available for analysis and X-ray lines are closer together at low energy. Fortunately, excellent low energy resolution is already achieved with very large sensor sizes to maximize count rate and peak sensitivity under low kV operating conditions or when analyzing beam sensitive samples.

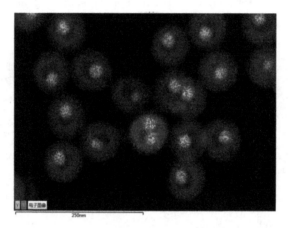

Fig.2.3.7 High spatial resolution element mapping shows the distribution of Si and Y in core/shell nanoballs. The diameter of the balls is about 80 nm. Image courtesy of Wu. W.

2.3.2 Wavelength dispersive X-ray spectrometry in SEM

As well as an EDS system, a wavelength dispersive X-ray spectrometer (WDS) can also be fitted onto an SEM chamber for chemical analysis by collecting X-rays generated by the electron beam. Data collection and analysis with EDS is a relatively quick and simple process because the complete spectrum of energies is acquired simultaneously. However, using a WDS, the spectrum is acquired sequentially as the full wavelength range is scanned. Although it takes longer to acquire a full spectrum, the WDS technique has much improved energy resolution compared to EDS. Typical energy resolution of an EDS detector is 70 eV to 130 eV (depending on the element), whereas peak widths in WDS are 2 eV to 20 eV. The combination of better resolution and the ability to deal with higher count rates typically allows WDS to detect elements at an order of magnitude lower concentration than EDS.

The development of WD spectrometers goes back long before EDS detectors. The first electron probe microanalyzer (EPMA) was developed during the 1940s and used an optical microscope to observe the position and focus the electron beam on the sample. Later, WDS spectrometers were fitted to SEMs, which allowed the specimen to be positioned more precisely under the electron beam and also made possible a visual picture of the distribution of a chosen element—the X-ray map.

In addition to an EDS detector, the WDS spectrometer can also be fitted on a port of the SEM, usually at an angle inclined to the horizontal so that it provides an identical X-ray take off angle to the EDS detector. Although the WDS technique often requires a higher SEM beam current than that typically used for EDS, the X-ray data is usually acquired from EDS and WDS simultaneously. If the EDS detector is fitted with a variable collimator there is no compromise in performance for either technique.

1. Components in WDS and working principle

Inside the spectrometer, analyzing crystals of specific lattice spacing are used to diffract the characteristic X-rays from the sample into the detector (Fig. 2.3.8). The wavelength of the X-rays diffracted into the detector may be selected by varying the position of the analyzing crystal with respect to the sample, according to Bragg's law ($n\lambda = 2d\sin\theta$), where n is an integer referring to the order of the reflection, λ is the wavelength of the characteristic X-ray, d is the lattice spacing of the diffracting material, and θ is the angle between the X-ray and the diffractor's surface. A diffracted beam occurs only when this condition is met and therefore interference from peaks of other elements in the sample is inherently reduced. However, X-rays from only one element at a time may be measured on the spectrometer and the position of the crystal must be changed to tune to another element.

The wavelength dispersive spectrometer consists of two major components-the analyzing crystal and the proportional X-ray detector. The spectrometer shown in Fig. 2.3.9 is of the fully focusing, or Johansson type, where the crystal, the X-ray

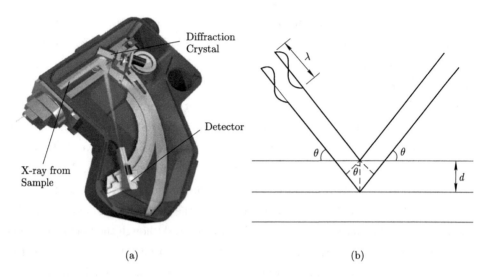

Fig.2.3.8 Components inside a WDS spectrometer (a) and diagram of Bragg's law (b) (Oxford Instruments, 2004).

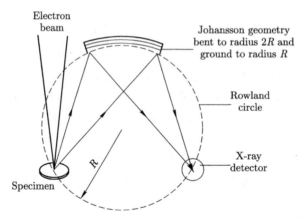

Fig.2.3.9 Diagram of a fully focusing geometry for a WDS (Oxford Instruments, 2004).

source, the sample, and the detector all remain on a circle of constant radius. This circle is known as the Rowland circle. The crystal planes are bent to twice the radius of the Rowland circle, and the crystal itself is ground to the radius of the circle. The crystal moves along a screwed rod in a linear fashion, simultaneously rotating through an angle θ. To maintain the fully focusing geometry, the detector moves through an angle of 2θ.

The spectrometer in Fig. 2.3.9 is known as a fully focusing linear type. The second type of crystal geometry is not fully focusing, known as Johann geometry, in which the diffracting crystal is bent to a radius of $2R$ and only part of the X-rays can be diffracted into the detector. In either crystal geometry the output of the detector is connected to an amplifier where it is converted to a voltage pulse which is then counted or displayed on a rate meter.

In a commercial WDS mechanical limitations make it impractical for one analyzing crystal to cover the entire elemental range. To cover the range of elements that need to be detected, a range of crystals is offered in a wavelength spectrometer. Crystals of larger d spacing are used to diffract the longer wavelengths from the lighter elements. Pseudo crystals have been developed for these light elements, and are known collectively as layered synthetic microstructure (LSM) crystals. The LSM crystals are available in a range of different d spacings (e.g. 6, 8 and 20 nm), optimized for different elements. Other crystals commonly used for X-rays with short wavelengths from heavy elements are LiF (Lithium Fluoride), cleaved along either (200) or (220) plane, TAP (Thallium acid phthalate) and PET (Pentaerythritol).

Detectors used in WDS are usually of the gas proportional counter type. Generally, X-ray photons are diffracted into the detector through a collimator (receiving slit), entering the counter through a thin window. They are then absorbed by atoms of the counter gas. A photoelectron is ejected from each atom absorbing an X-ray. The photoelectrons are accelerated to the central wire causing further ionization events in the gas, so that an "avalanche" of electrons drawn to the wire produces an electrical pulse. The detector potential is set so that the amplitude of this pulse is proportional to the energy of the X-ray photon that started the process. Electrical pulse height analysis is subsequently performed on the pulses to filter out noise. There are two types of gas proportional counters: sealed counter (SPC) (usually xenon or a xenon-CO_2 mixture) and gas flow counter (FPC) (usually P-10: argon with 10% methane). Generally, SPCs are used for high-energy X-ray lines, while FPCs are used for low energy X-ray lines.

To maintain the correct geometrical relationship between specimen, crystal and detector for the full range of diffracted angles, it is necessary to maintain all three on the Rowland circle. To analyze a particular element it is important that the crystal and detector are positioned accurately and associated counting electronics are set up correctly. In the past this was a tedious and complex procedure, but automation and PC control have made WDS operation very straightforward, routine and reliable.

Using EDS, all of the energies of the characteristic X-rays incident on the detector are measured simultaneously and data acquisition is therefore very rapid across the entire spectrum. However, the resolution of an EDS detector is considerably worse than that of a WDS spectrometer. The WDS spectrometer can acquire the high count rate of X-rays produced at high beam currents, because it measures a single wavelength at a time. This is important for trace element analysis. For EDS detector, situations may arise in which overlap of adjacent peaks becomes a problem. Many of the overlaps can be handled through deconvolution of the peaks. Others, however, are more difficult, particularly if there is only a small amount of one of the overlapped elements. An example of the difficult overlap situation is shown in Fig. 2.3.10. In this figure, Si K lines are overlapped with W M lines in

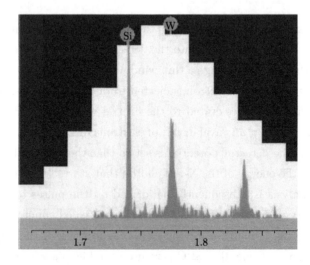

Fig.2.3.10 Overlapped Si and W X-ray peak in EDS but separated in WDS scan.

the EDS spectrum, but with WDS, different lines from two elements are clearly separated. In practice it is advantageous to use the speed of EDS for an initial survey of an unknown sample because major elements will be rapidly identified. However, if trace elements are present they will not be identified, and it may be difficult to interpret complex overlaps. Following the initial EDS survey, WDS can be used to check for overlaps and to increase sensitivity for trace elements.

2. New parallel beam WDS

In addition to the WDS described in previous section, a new type of WDS, called parallel beam WDS, was developed recently, as shown in Fig. 2.3.11. Conventional WDS uses curved crystals to disperse and focus the X-rays onto a detector. Parallel beam WDS collects a large solid angle of X-rays diverging from the sample and re-directs them into a parallel beam incident on various flat diffracting crystal and then into a detector. Normally, parallel beam WDS systems are optimized for only one or two elements for each diffractor and a single WDS is optimized either for light elements or heavy elements. In comparison with traditional WDS

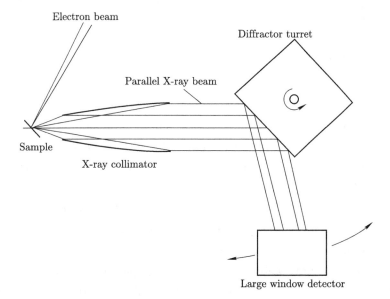

Fig.2.3.11 Diagram of parallel beam WDS optics (Parallax Research, 2013).

using the Rowland circle, parallel beam WDS systems can collect more X-ray counts under the same SEM conditions, but exhibit lower energy resolution than traditional WDS.

2.3.3 Electron backscatter diffraction and applications

Electron backscattered diffraction (EBSD) is a technique which allows crystallographic information to be obtained from samples in SEM. In EBSD a stationary electron beam strikes a tilted crystalline sample and the diffracted electrons form a pattern on a fluorescent screen. This pattern is characteristic of the crystal structure and orientation of the sample region from which it was generated. The diffraction pattern can be used to measure the crystal orientation, measure grain boundary misorientation, discriminate crystal structures between different materials, and provide information about local crystalline imperfections. When the electron beam is scanned in a grid across a polycrystalline sample, the resulting map will reveal the constituent grain morphology, orientations, and boundaries. This data can also be used to show the preferred crystal orientations (texture) present in the material. A complete and quantitative representation of the sample microstructure can be established with EBSD.

1. History in brief

EBSD like patterns were first observed by Nishikawa and Kikuchi in TEM in 1928 (Schwarzer et al., 2009). Kikuchi found that electron diffraction patterns from thin films of mica contained the expected diffraction spots on a background of linear structures, which consisted of pairs of parallel excess and defect lines, now known as Kikuchi lines. Venables and Harland (Schwarzer et al., 2009) observed EBSD in SEM by using a 30 mm diameter fluorescent imaging screen and a closed circuit television camera which gave an angular range of ∼60°. This method allowed on-line examination of specimens and the measurement of crystal orientation at high spatial resolution. Later, Dingley developed the combination of phosphor and television camera, and combined them with a graphical overlay. This allowed an operator to indicate the positions of zones in an EBSD and the software to display a simulation. Using this method, several hundreds of grain orientations could be measured in a day.

Software improvements lead to more intelligent analysis packages, e.g. CHANNEL that could analyze an EBSD from knowledge of the possible phases present. Application of the Hough transform allowed EBSD orientation measurements to be automated. Through the 1990s, automation of EBSD systems became increasingly common. Although initially slow, by 1996 automated systems could index at rates of 4–5 patterns per second, although this was generally applied to higher symmetry phases. EBSD systems could by this stage take control of the SEM beam and, sometimes, the stage as well. This allowed the routine mapping of large areas on the surface of samples, but this often took many hours.

By the end of the 1990s the reliability of indexing and camera sensitivity had improved enough to make orientation mapping of more complex materials, such as geological samples, a reality. However, the speed of such systems was still limited by hardware and rarely exceeded 10 patterns per second.

The development of faster frame grabbers, computers and then the launch of fully digital CCD cameras in early 21st century has had a dramatic effect on the speed of the EBSD technique. At the end of first decade the maximum speed of data acquisition leapt from 100 patterns per second to more than 600 patterns per second. At the time of writing this chapter, it is reported that the acquisition speed reaches 1000 patterns per second. Orientation maps could now be acquired in a short time, typically 10 to 30 minutes for a statistically representative dataset. The reason for this dramatic speed increase is mostly due to the improved performance of the new digital detectors-these have an order of magnitude more sensitivity and also the capability of pixel binning for increased speed.

2. EBSD basics

The principal components of an EBSD system are shown in Fig. 2.3.12.

In a SEM chamber, a sample is tilted 70° from the horizontal towards a phosphor screen. The phosphor screen is fluoresced by electrons from the sample to form the diffraction pattern. A sensitive charge coupled device (CCD) video camera is placed behind the phosphor screen to view the diffraction pattern on it. A vacuum interface is needed for mounting the phosphor screen and camera in an SEM port.

The camera monitors the phosphor through a lead glass screen in the interface, and the phosphor can be retracted to the edge of the SEM chamber when not in use. Electronic hardware is required to control the SEM, including the beam position, stage, focus, and magnification. A computer is used to control EBSD experiments, analyze the diffraction pattern and process and display the results. An optional electron detector mounted below the phosphor screen for electrons scattered in the forward direction from the sample can also be used.

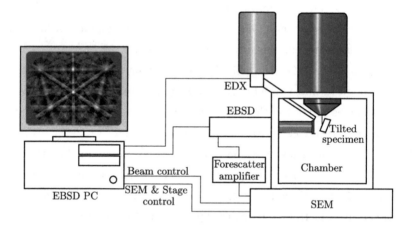

Fig.2.3.12 Diagram showing the components of an EBSD system on SEM.

For EBSD, a beam of electrons directly bombard a point of interest on a tilted crystalline sample in the SEM (Fig. 2.3.13). The mechanism by which the diffraction patterns are formed is complex, but the following model describes the principal features. The atoms in the material inelastically scatter a fraction of the electrons, with a small loss of energy, to form a divergent source of electrons close to the surface of the sample. These electrons are diffracted to form a set of paired large angle cones corresponding to each diffracting plane. When used to form an image on the fluorescent screen, the regions of enhanced electron intensity between the cones produce the characteristic Kikuchi bands of the electron back scattered diffraction pattern.

The mechanisms giving rise to the Kikuchi band intensities and profile shapes are complex. As an approximation, the intensity of a Kikuchi band I_{hkl} for the

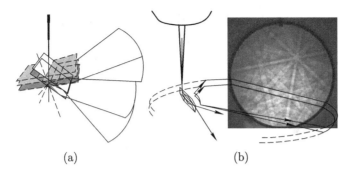

Fig.2.3.13 The interaction of an electron beam with a crystal (a) and the formation of an EBSD pattern on phosphor screen (b).

(hkl) plane is given by

$$I_{\text{hkl}} = \left[\sum_i f_i(\theta) \cos 2\pi(hx_i + ky_i + lz_i)\right]^2 + \left[\sum_i f_i(\theta) \sin 2\pi(hx_i + ky_i + lz_i)\right]^2 \quad (2.3.3)$$

where $f_i(\theta)$ is the atomic scattering factor for electrons and x_i, y_i, z_i are the fractional coordinates in the unit cell for atom i. An observed diffraction pattern should be compared with a simulation calculated using Eq. (2.3.3), to ensure only planes that produce visible Kikuchi bands are used when solving the diffraction pattern. This is especially important when working with materials with more than one atom type.

In indexing EBSD patterns, the Kikuchi band must firstly be positioned. The central lines of the Kikuchi bands correspond to the intersection of the diffracting planes with the phosphor screen. Hence, each Kikuchi band can be indexed by the Miller indices of the diffracting crystal plane which formed it. The intersections of the Kikuchi bands correspond to zone axes in the crystal and can be labeled by zone axis symbols. The crystal orientation is calculated from the Kikuchi band positions by the computer processing the digitized diffraction pattern collected by the CCD camera. The Kikuchi band positions are found using the Hough transform (Hough, 1962). The transform between the coordinates (x, y) of the diffraction pattern and the coordinates (ρ, θ) of Hough space is given by Eq. (2.3.4):

$$\rho = x \cos \theta + y \sin \theta \quad (2.3.4)$$

A straight line is characterized by ρ, the perpendicular distance from the origin and θ, the angle made with the x-axis and, so is represented by a single point (ρ, θ) in Hough space. The transformation is shown in Fig. 2.3.14.

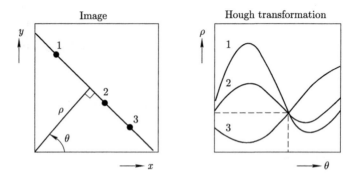

Fig.2.3.14 The Hough transformation coverts lines into points in Hough space. The set of all straight lines going through a point in the (x, y) plane corresponds to a sinusoidal curve in the (ρ, θ) plane, Thus, a straight line formed by a set of points could be characterized by the intersection point of related sinusoids.

Hough transformation makes Kikuchi band detection fast and reliable. Kikuchi bands transform to bright regions in Hough space which can be detected and used to calculate the original positions of the bands. The angles between the planes producing the detected Kikuchi bands can be calculated. These are compared with a list of inter-planar angles for the analyzed crystal structure to allocate Miller indices to each plane. The final step is to calculate the orientation of the crystal lattice with respect to coordinates fixed in the sample. This whole process takes less than a few milliseconds with modern computers.

3. Applications on materials research

In EBSD point analysis the beam is positioned at a point of interest on the sample, a diffraction pattern is collected and then the crystal orientation is calculated. In crystal orientation mapping, the electron beam is scanned over the sample on a grid of points and at each point a diffraction pattern is obtained and the crystal orientation is measured. The resulting data can be displayed as a crystal orientation map and processed to provide a wide variety of information about the sample microstructure.

The diffuseness or quality of the diffraction pattern is influenced by a number of factors including local crystalline perfection, surface contamination and the phase and orientation being analyzed. Pattern quality maps will often reveal features invisible in the electron image such as grains, grain boundaries and surface damage. The typical applications of EBSD in materials research are summarized in the following paragraphs.

Phase discrimination: EBSD can be used to distinguish crystallographically different phases and to show their location, abundance and preferred orientation relationships. EBSD can discriminate crystallographically dissimilar phases by comparing the interplanar angles measured from the diffraction pattern, with calculated angles from a set of candidate phases, and selecting the best fit. For phase identification, EDS will be firstly used to investigate the chemical information, then a rapid search for possible phases in databases based on the constituent elements is performed. Finally, EBSD patterns are used to search the best fit from candidate phases. For phases with the same crystal structure, EDS will be used to help phase discrimination. Fig. 2.3.15 shows the inclusions in a tool steel. To investigate the carbide types in this tool steel, EBSD mapping is performed with 4 phases selected-ferrite (Fe-BCC), austenite (Fe-FCC), M_7C_3 and $M_{23}C_6$ structures. Simultaneously EDS counts for 3 elements (Cr, Mo, V) are taken. After EBSD mapping, two phases were identified in this sample: austenite matrix with $M_7C_3(Cr_7C_3)$ inclusions. However, from the element maps of Mo, Cr, and V, it was realized that the Mo and V rich areas were something else other than austenite. By investigating the composition of the Mo and V rich area using EDS, these areas were re-indexed as $C_{45}Mo_{18}V_{37}$ with the same FCC structure as austenite. Finally, the phases in this sample were all identified correctly as shown in Fig. 2.3.15(f).

Grain size and grain boundary analysis: Unlike an optical or scanning electron micrograph, the crystal orientation map must reveal the positions of all grains and grain boundaries in the sample microstructure. In crystal orientation maps a grain is defined by the collection of neighboring pixels in the map, which

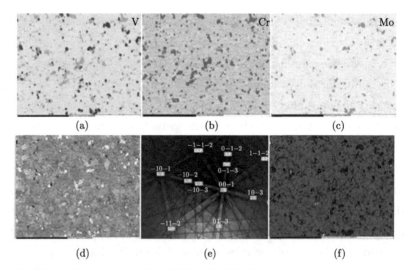

Fig.2.3.15 Phase discrimination by EBSD. (a)–(d) Elemental mapping as recorded with EDS. (e) EBSD pattern of $C_{45}Mo_{18}V_{37}$ phase. (f) EBSD map revealing the phase distribution of ferrite (red), $C_{45}Mo_{18}V_{37}$ (blue) and Cr_7C_3 (green).

have a misorientation less than a certain threshold angle. The distribution of grain sizes can be measured from the data collected for the map. In addition, the distribution of grain boundary misorientation angles can also be shown. Fig. 2.3.16 shows the orientation map, grain boundary angle distribution and grain size measured from a Nickel alloy.

Grain boundaries are characterized by the misorientation axis and angle and the boundary plane. Some boundaries satisfy certain geometrical criteria and their presence in a material may confer particular properties. When crystal lattices share a fraction of sites on either side of a grain boundary, they are termed coincident site lattices (CSL). CSLs are characterized by Σ, where Σ is the ratio of the size of the CSL unit cell to the standard unit cell. An example of special boundaries is shown in Fig. 2.3.17. In this figure grain boundary positions are superimposed on the pattern quality image (a). The boundaries are color coded according to the histogram of misorientation angle (b).

Texture and deformation analysis: Grains are seldom oriented randomly in polycrystalline materials. The preferred crystallographic orientation or texture of

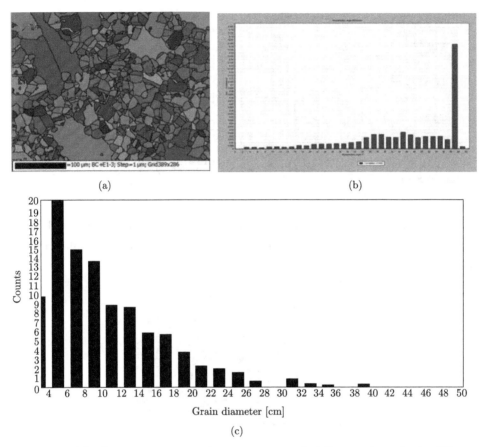

Fig.2.3.16 (a) The orientation map showing grains with different orientation in different color, (b) grain boundary angle distribution and (c) grain size distribution in a Nickel alloy.

polycrystalline materials influences many properties of the bulk material, because physical properties are often anisotropic with respect to crystal direction. Material processing methods are frequently deliberately chosen to produce certain desired textures.

The individual crystal orientation measurements collected by crystal orientation mapping can be used to show the crystallographic textures developed in the sample. Fig. 2.3.18 shows a texture map of a deformed Cu sample, in which various

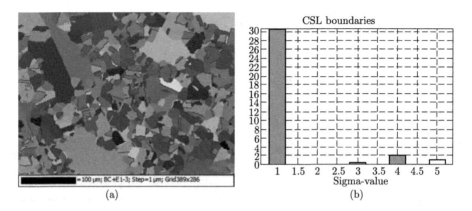

Fig.2.3.17 (a) Coincident site lattice (CSL) boundary positions superimposed on the pattern quality image. (b) The boundaries are color coded by CSL type shown in the histogram of CSL.

Fig.2.3.18 Texture map of a deformed Cu sheet: 52% cubic fiber (yellow), 24% gamma fiber (teal) and 9% ⟨110⟩ fiber (purple) with 20 degree deviation, image courtesy of Oxford Instruments application notes.

textures in the sample can be separated automatically, their volume fractions could be calculated, and the regions of the sample from which they originate is shown. Texture is usually measured by X-ray diffraction. EBSD is an ideal technique for microtexture analysis. In characterizing texture, compared with XRD, EBSD provides not only the types and percentages of textures, but also the microstructure information, such as distribution and grain size for certain texture. However, advances in EBSD processing speed can make the technique competitive with X-ray diffraction for texture measurements for large samples. EBSD and X-ray diffraction are complementary techniques for texture analysis.

Traditionally, texture is usually measured by X-ray diffraction and illustrated by pole figure. The X-ray intensity from a particular diffracting plane is measured, while the sample is stepped through a series of orientations to complete the pole figure. The individual crystal orientations measured with EBSD can also be displayed as pole figures for preferred orientation analysis, as shown in Fig. 2.3.19.

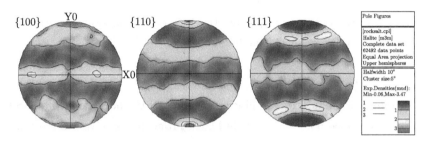

Fig.2.3.19   Pole figures calculated from EBSD data from a nickel sample.

EBSD mapping can also be used to illustrate the deformation within a grain. As shown in Fig. 2.3.20(a), the inverse pole figure (IPF) color coded orientation map shows the lattice rotation within a grain. Since the orientation at each pixel is measured by EBSD in an orientation map, a local misorientation map (misorientation with respect to neighboring pixels) can be used to visualize subtle low angle grain boundaries, which is beyond the capability of pure grain boundary mapping, after removing orientation noise by certain filters. Fig. 2.3.20(b) illustrates these low angle sub-grain boundaries within a grain.

Orientation relationship verification by EBSD: After the orientations of different phases is measured by EBSD, the orientation relationship between these phases can be determined or verified. The interface between FCC and BCC crystals can be found in many important metallic alloys. The orientation relationships (OR) between these two phases often show a deviation from the exact Kurdjumov-Sachs relationship (K-S OR), Nishiyama-Wassermann relationship and so on; the close packed planes in the FCC and BCC phases are usually parallel to or nearly parallel to each other. To verify the OR between two phases, interphase boundaries can

Fig.2.3.20 (a) IPF orientation map of deformed nickel, and (b) local misorientation map showing the sub-grain low angle boundaries.

be plotted overlayed onto the image, as shown in Fig. 2.3.21. In Fig. 2.3.21(b) the K-S OR interface boundaries are plotted in white. If the deviation from K-S OR is over 7° the interface boundaries are in black. From the map we can see that most of particles show a K-S OR matrix.

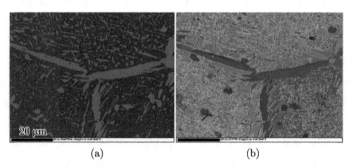

Fig.2.3.21 (a) Phase map and (b) orientation map of the sample.

4. Spatial resolution and measurement accuracy

The electrons contributing to the diffraction pattern originate within nanometres of the sample surface. Hence, the spatial resolution will be related to the electron beam diameter, and this depends on the type of electron source and probe current used. Typical beam diameters at 0.1 nA probe current and 20 kV accelerating voltage are 2 nm for a FEG source, and 30 nm for a tungsten source. The beam profile on the sample surface will also be elongated in the direction perpendicular to the tilt. The spatial resolution achieved in practice will depend on the sample, SEM operating conditions and electron source used, and under optimum conditions, grains as small as 10 nm can be identified.

Errors in crystal orientation measurements from the diffraction pattern will depend principally on the accuracy of the Kikuchi band position measurement and the system calibration, and are generally in the range $\pm 0.3°$. Recently, with the advance of the EBSD pattern indexing algorithm, a method called refined accuracy indexing brings the orientation measurement error down to below $\pm 0.1°$, which will be very useful to reveal the subgrain structures in materials. An example in Fig. 2.3.22 shows data from a single grain in an Al alloy. This material has been deformed 20% and annealed at 200 °C for 0.5 hours. EBSD data has been collected from a

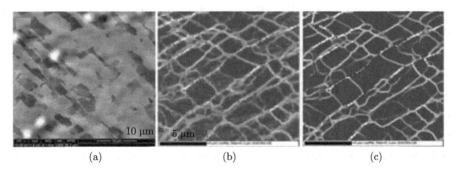

Fig.2.3.22 (a) The BSE image shows that the single crystal has developed deformation bands with very low angle boundaries; (b) Local misorientation map processed using conventional EBSD indexing method, and (c) Local misorientation map processed using refined accuracy showing the clear subgrain structures. Image courtesy of Oxford Instruments application notes.

part of this grain and indexed using refined accuracy. Local average misorientation maps are used to show the detailed microstructure. With refined accuracy the images are less noisy and more detail is visible.

5. Sample preparation for EBSD

For EBSD analysis, the sample surface should be prepared very carefully since it is very sensitive to crystalline perfection, and some treatments may be needed to remove any surface damage. A well prepared sample is a prerequisite for obtaining a good diffraction pattern. Surfaces must be sufficiently smooth to avoid forming shadows on the diffraction pattern from other parts of the sample.

Different materials may require suitable techniques to prepare. For metals and insulators the typical procedure should be mounting in conductive resin, mechanical grinding, diamond polishing and final polishing with colloidal silica. For metals electropolishing is also a useful method for final polishing. Brittle materials such as ceramics and geological materials can often be fractured to reveal surfaces immediately suitable for EBSD. Ion milling can be used for materials which are not amenable to conventional metallography such as zirconium and zircalloy. Dual focused ion beam–electron beam microscopes fitted with EBSD can perform in situ specimen preparation for EBSD. Plasma etching can be used for microelectronic devices.

Charging in non-conductive samples can be eliminated, as for X-ray microanalysis, by the deposition of a conducting layer. The deposited layer must be very thin, for example 2 to 3 nm of carbon, otherwise a diffraction pattern will not be obtained. It may be necessary to increase the electron accelerating voltage to penetrate the conducting layer. Charging can be reduced when the sample is tilted for EBSD experiments and can also be reduced by analyzing the sample in an environmental or low vacuum SEM.

References

Amelinckx, S., van Dyck, D., van Landuyt, J., van Tendeloo, G. (1997) Handbook of Microscopy. VCH Verlagsges., 539–561.

Andrasko, J. and Maehly, A. C. (1977) Detection of gunshot residues on hands by scanning electron microscopy. J. Forensic Sci., 279–287.

Ardenne, M. V. (1938). Das Elektronen-Rastermikroskop (in German). Zeitschrift für Physik **109** (9-10), 553–572.

ASTM (2007) Standard Guide for Gunshot Residue Analysis by Scanning Electron Microscopy/energy dispersive X-ray Spectrometry, Designation: E1588–07.

Everhart, T. E.; Thornley, R. F. M. (1960). Wide-band detector for micro-microampere low-energy electron currents. Adv. Imag. Elect. Phys., **133**, 147–152.

FEI announces new verios extreme high resolution SEM. (2012) Nanotechnology Now.

Goldstein, J., Newbury, D. E., Echlin, P., Joy, D. C., Romig, Jr. A. D., Lyman, C. E., Fiori, C., Lifshin, E. (1992) Scanning Electron Microscopy And X-Ray Microanalysis. Plenum Press, 2nd Edition.

Goldstein, J., Newbury, D.E., Echlin, P., Joy, D. C., Romig, Jr. A. D., Lyman, C. E., Fiori, C., Lifshin, E. (1992) Scanning Electron Microscopy and X-ray Microanalysis, Chapter 2. Springer US.

Hitachi Launches World's Highest Resolution FE-SEM. (2011) Nanotechnology Now.

Hough, P. V. C. (1962) Methods and means for recognizing complex patterns, US patent 3069654.

Joy, D. C. (1995) Monte Carlo Modeling For Electron Microscopy And Microanalysis, Oxford University Press.

Ketek, GmbH (2013) http: www. ketek. net/products/sdd-technology/working-principle/.

Knott, G., Marchman, H., Wall, D., Lich, B. (2008) Serial section scanning electron microscopy of adult brain tissue using focused ion beam milling. J. Neurosci., **28** (12), 2959–2964.

Li, H., Li, Y. M., Lei, Z. C., Wang, K. Y. ang Guo A. K. (2013) Transformation of odor selectivity from projection neurons to single mushroom body neurons mapped with dual-color calcium imaging. P. Natl. Acad. Sci. USA, **110** (29), 12084–12089.

Oxford Instruments (2002), Technique brief, Energy dispersive X-ray microanalysis hardware.

Oxford Instruments (2004), Technique brief, Wavelength dispersive X-ray microanalysis.

Oxford Instruments (2012), Silicon Drift Detectors Explained.

Parallax Research Inc. (2013), http://www.parallaxray. com/wds.html.

Schwartz, A. J., Kumar, M., Adams, B. L., Field, D. P. (2009), Electron Backscatter Diffraction in Materials Science, Springer US, 2nd Edition.

Smith, K. C. A., (1972) Proceedings of the 5th annual SEM Symposium (Ed: Johari O.), IITRI, Chicago.

Zhou, W. L., Wang, Z. L. (2007) Scanning Microscopy for Nanotechnology. Springer New York.

3
Transmission Electron Microscopy

3.1 Introduction

Rongming Wang and Jing Tao

Conventional TEM (CTEM) is referring to the basic TEM techniques, i.e., electron diffraction and imaging techniques such as bright-field (BF), dark-field (DF) and high-resolution TEM (HRTEM). These techniques, in general, provide morphology and structural analysis of the specimen material. CTEM has been used to distinguish the analytical TEM tools that appeared much later in the history of TEM, including scanning transmission electron microscopy (STEM), electron energy-loss spectroscopy (EELS) and energy-dispersive X-ray spectroscopy (EDX) techniques. Analytical TEM techniques are sensitive to the chemical and electronic structures of the specimen, sometimes are capable in providing crystal structures as well. However, TEM techniques have been developing quickly in recent years and often utilize the advantages of both CTEM and analytical TEM. For example, electron diffraction techniques are considered to be CTEM. However, a newly developed technique, scanning electron nanodiffraction (SEND), has combined features of the CTEM and STEM (Tao et al., 2009). Using SEND, the variation of certain type of electronic structures can be mapped in real-space (Zuo and Tao, 2010). Moreover, some techniques that were never used in TEM have emerged and show interesting potential in a number of research areas. One of them is the secondary electron microscopy (SEM) installed in STEM, benefited from the development of the SEM detectors (Zhu et al., 2009). The capability of atomic resolution EELS and EDX has exhibited a promising future that the chemical or electronic structures may be obtained simultaneously with the crystal structure. All the above accomplishments are suggesting that the difference between CTEM and analytical TEM could be diminishing in the fast development of the TEM techniques.

Among the CTEM techniques, electron diffraction is the base, formed directly by the interfered electrons after their scattering from the specimen (Cowley, 1995). Electron diffraction can be formed and recorded in different data patterns, which depends on the details of electron probes or illumination conditions, including selected-area electron diffraction (SAED), nanoarea or nanobeam electron diffraction (NAED) (Zuo et al., 2004) and convergent-beam electron diffraction (CBED) (Spence and Zuo, 1992). In particular, SAED and NAED have sharp reflections in the diffraction patterns, while each disk in CBED patterns contains beams from many scattering directions with rich information. Details of CBED technique can be found in Section 3.5. The interpretation of electron diffraction results can be straightforward by knowing the exact crystal structure. However, in many research projects, there are uncertainties or defects in the known crystal structures which make the analysis of the electron diffraction results difficult and even more complicated by considering the dynamic scattering effects. References (Cowley, 1995; Zuo et al., 2004; Zuo and spence, 1992) provide very knowledgeable details in studying the electron diffraction mechanism and using the technique to solve the problems in materials science.

Unlike the electron diffraction data, imaging results are in real space and sometimes more impressive than electron diffractions to users. However, the interpretation of the TEM images requests careful analysis and simulations, especially for HRTEM. A number of things could affect the image contrast. As introduced in Section 1.3.3, BF images and DF images are determined by the position of objective apertures. The magnetic lenses are far from being perfect, introducing aberrations and other imperfections in the images. As discussed in the Section 3.2, HRTEM images' contrasts are very sensitive to the specimen's thickness, electron beam defocus and a number of other parameters. Many details regarding HRTEM can be found in following sections in this chapter and books of Spence (2003) as well. Recent developments in HRTEM techniques provide better interpretation of the image contrast and new application opportunities in materials science. For instance, HRTEM results are powerful in determining crystal structures at interfaces and defect analysis, especially at the atomic resolution level, as the readers will find in Section 3.3 and 3.4.

In addition, Section 3.6 gives a comprehensive introduction of Lorentz TEM including Fresnel imaging, Foucault imaging and Differential phase contrast imaging and their applications. Most of the recent applications of Lorentz imaging are in magnetic materials, providing direct observations of the magnetic domain configuration with further quantitative analysis. Another TEM technique of electron holography, with its principle and applications, is introduced in Section 3.6. Electron holography, particularly off-axis holography, has been widely used to differentiate the electric and magnetic projected potential distribution in materials such as magnetic and/or ferroelectric specimens. Finally, more applications using a newly emerged TEM technique, electron tomography, are shown in Section 3.7. Moreover, most of these techniques can be combined with in situ capabilities, granting even more richness and versatility of TEM techniques in the scientific research.

3.2 High-resolution transmission electron microscopy imaging

Kui Du

3.2.1 Image formation and simulation of HRTEM

1. Multislice method

By the use of high-resolution transmission electron microscopy (HRTEM) techniques, we can image crystalline specimens along one of their low index zone axes. Under this circumstance, the interference between sets of diffracted beams forms intensity contrast in HRTEM images which corresponds to certain crystallographic planes or sets of crystallographic planes and thus reveal information of crystal lattices, particularly on imperfect structures, such as interfaces, dislocations and precipitates in the specimens (Spence, 2003).

On the theoretic aspect of HRTEM imaging, based on the weak phase object approximation and an assumption of optimal focus setting, the image intensity can be directly related to the projected potential of atom columns in the crystalline specimens along the viewing direction. This provides a simple concept for extracting crystal structures from HRTEM images. Nevertheless, due to the strong scattering effect of matters to electrons, the weak phase object approximation only apply to very thin specimens (i.e. less than a couple of nanometers) made of light elements, for instance, a single gold atom in specimens will defy the weak phase object approximation. In addition, a slight change of objective lens focus can change the feature of HRTEM images. Therefore, a direct interpretation of HRTEM images as the projection of crystal structures is rarely applicable and image simulation is usually necessary to resolve atomic structures of materials from HRTEM images.

The image formation process of high-resolution transmission electron microscopy can be described as following. Firstly, incident electrons enter the specimen and travel through the crystal, where the electrons are scattered by atoms. Since the scattering effect of atoms is very strong to electrons, normally electrons are scattered multiple times before they exit the crystal, hence the dynamical scattering needs to be taken into account. The electrons escaping the exit plane of the crystal will go through lens of the electron microscope, which magnify the electron micrographs. Noteworthy, aberrations of the lens, mainly the objective lens, will influence trajectories of electrons thus change the electron wave function. When the electrons finally reach the image plane, they are registered by record media such as negative film, CCD camera or imaging plates to form images. Therefore, the simulation of HRTEM images comprises two procedures: one for the calculation of dynamic scattering of electrons in the crystal, the other for the image formation through the lens of electron microscopes.

Multislice and Bloch-wave methods are widely used for calculating electron dynamic scattering in specimens. Since HRTEM image simulation, especially the simulation of defect structures, needs the calculation of a large amount of diffracted beams, and Bloch-wave method needs much longer time than fast-Fourier transformation multislice method, hence multislice method is commonly employed in HRTEM image simulation.

The multislice method was first introduced based on physical optics theory (Cowley, 1995). This algorithm was realized in 1970s and the simulated images agreed well with experimental images, since then the multislice method has been widely used for image simulation. Under the high-energy electron approximation, the formulae of the multislice method have been derived from the time-independent Schrödinger equation, which provides a sound physics base for the method. In the practical aspect, the time-consuming convolution has been replaced by multiplication with the introduction of fast Fourier transformations, and it significantly decreases the calculation time and accelerates the wide application of multislice method.

In the multislice calculation, a crystal is treated as a series of two-dimensional slices along the incident beam direction, thus each slice is thin enough and can be assumed as a pure phase grating (Cowley, 1995). That is to say, when electron wave passes through this slice, only the phase of the electron wave will be modified by the phase grating, while the amplitude of the electron wave remains the same. This process can be described by the multiplication of the incident electron wave function with the transmission function of the phase grating in real space, where the transmission function corresponds to the projected potential of all atoms in the slice along the incident beam direction. The slice thickness is usually chosen as about 0.2 nm and corresponding to the distance between crystallographic planes normal to the incident beam direction. When the electron wave propagates from one slice to the next slice, the propagation process can be described by Fresnel equation. This is applied by the convolution of the propagation factor with the electron wave function exiting the previous slice. Since the complex convolution in real space corresponds to a simple multiplication in reciprocal space, the propagation calculation is normally carried out in reciprocal space, while fast Fourier transformations are utilized to transform the electron wave function between the real and reciprocal spaces. Therefore, the whole dynamic scattering of electrons through the crystal can be described as iterative processes of the phase grating and propagation.

During the numerical computing, discrete samplings are used to describe the wave function both in the real and reciprocal spaces. When the specimen includes defect structures, for example a dislocation core, which is non-periodic in $x-y$ plane, the corresponding diffraction intensity will have continuous distribution. Nevertheless, the fast Fourier transformation multislice method will have discrete sampling in the reciprocal space. This introduces an artificial periodic condition in the real space, which means there is one dislocation core in each "supercell" and the crystal is composed by infinite periodic "supercells" along the $x-y$ plane perpendicular to the incident beam. In order to avoid the influence from the artificially introduced "neighbor" dislocation cores on the HRTEM image simulation of the "real" dislocation core, usually a sufficiently large supercell is necessary for the image simulation of defect structures. Similarly, the discrete sampling in the real space will also introduce an artificial periodic spectrum in the reciprocal space, which is called the "aliasing" effect. To avoid this effect, the sampling distance should be sufficiently small, and a sample rate lower than 0.01 nm/pixel in the $x-y$ plane is usually used in the calculation.

The total intensity of electron wave at exit face decreases with the increase of specimen thickness. It is mainly caused by inelastic scattering of electrons in the specimen as well as some of electrons being scattered to very high angles, and this is usually described by the absorptive factor in the dynamic calculation of electron scattering. When historically the absorptive factor was roughly estimated as a constant from experiments, a sounder algorithm has been proposed by Weickenmeier and Kohl (1991) to take into account the absorptive effect. Meanwhile, the influence of temperature on the scattering of electrons in the specimen is normally considered by including Debye-Waller factor (Peng et al., 1996) in the simulation.

2. Image formation with the lens of electron microscopes

In high-resolution transmission electron microscopy, the phase and amplitude of the electron wave, $o(r)$, at exit face of the specimen carries the true structure information about the specimen. In propagation from the exit face to image plane, unfortunately, the phase and amplitude are mixed up by the lens transfer function, which acts as a phase plate and a complex spatial frequency filter in the back focal plane of the objective lens. In the case of the isoplanatic approximation, one finds the complex object wave spectrum $O(g)$, which is Fourier transform of the object wave $o(r)$, multiplied by the lens transfer function $WTF(g)$ and form the spectrum of the image wave $I(g)$. The intensity of the image wave $i(r)$, which is the Fourier transformation of the image wave spectrum $I(g)$, is calculated as the simulated HRTEM image. Here WTF is usually written as

$$WTF(\boldsymbol{g}) = A(\boldsymbol{g})e^{i\chi(\boldsymbol{g})}E_{\rm s}(\boldsymbol{g})E_{\rm c}(\boldsymbol{g}) \tag{3.2.1}$$

$A(\boldsymbol{g})$ can be defined by an aperture stop in the back focal plane. The wave aberration $\chi(\boldsymbol{g})$ acts like a phase plate in that plane. Its effect is to shift the different frequency components of $O(\boldsymbol{g})$ differently in phase, which is, in the rotationally symmetric case, given by

$$\chi(\boldsymbol{g}) = (\pi/2)C_{\rm s}\lambda^3 g^4 + \pi\Delta f \lambda g^2 \tag{3.2.2}$$

where Δf is the defocus, $C_{\rm s}$ the coefficient of spherical aberration and λ the wavelength of incident electrons. Usually, all the aberration coefficients except defocus and spherical aberration are assumed to be corrected approximately during the experiment. Astigmatism can be corrected by using stigmators and coma may be removed by coma-free alignment. Although a three-fold astigmatism remains to be corrected, its effect may be neglected except at very high-resolution.

An envelope approximation has been used to take into account the effect of spatial and temporal coherences in image formation, where envelope functions of spatial coherence $E_{\rm s}(\boldsymbol{g})$ and chromatic aberration $E_{\rm c}(\boldsymbol{g})$ lead to a damping of high resolution information. Here $E_{\rm s}(\boldsymbol{g})$ is written as

$$E_{\rm s}(\boldsymbol{g}) = \exp\{-(\pi\alpha/\lambda)^2[\nabla\chi(\boldsymbol{g})]^2\} \tag{3.2.3}$$

where α is the half-angle of beam convergence, $\nabla\chi(\boldsymbol{g})$ the spatial gradient of $\chi(\boldsymbol{g})$, $E_{\rm c}(\boldsymbol{g})$ given as

$$E_{\rm c}(\boldsymbol{g}) = \exp\{-[\pi^2\Delta^2\lambda^2 g^4]/2\} \tag{3.2.4}$$

where focus spread Δ is defined by

$$\Delta = C_{\rm c}\left[\frac{\sigma^2(V_0)}{V_0^2} + \frac{4\sigma^2(I_0)}{I_0^2} + \frac{\sigma^2(E_0)}{E_0^2}\right]^{\frac{1}{2}} \tag{3.2.5}$$

where $\sigma^2(V_0), \sigma^2(I_0)$ and $\sigma^2(E_0)$ are the variance in the statistically independent fluctuation of accelerating voltage V_0, objective lens current I_0 and the energy distribution of electrons leaving the filament E_0, respectively, and $C_{\rm c}$ is the coefficient of chromatic aberration. The validity of the envelope approximation can be checked by simulations with more accurate models (Du et al., 2007), for example, one can utilize incoherent addition of a tilted beam series to achieve the simulation for the converged beam.

3. Image contrast of simulated electron micrographs

It is noteworthy that simulated images show significantly stronger contrast than the experimental micrographs by a generally believed factor (the so-called Stobbs factor) of about 3 (Boothroyd, 1998, Hijtch and Stobbs 1994). The discrepancy in contrast obviously hampers the reliable interpretation of HRTEM micrographs by the use of image simulation techniques. While discussions have addressed possible reasons for the discrepancy in the contrast, which include the interaction between the incident electrons and solid matter as well as the image formation in the microscope (Boothroyd, 1998), dedicated experiments have also been performed to reveal the origin of this mismatch factor.

Cleaved single crystal of silicon has been chosen as the sample for HRTEM experiments (Du et al., 2007). The cleaved Si makes the specimen thickness accurately determined at each point in the micrographs from the known geometry of the wedge. Additionally, this sample has negligible surface contamination as compared with specimens prepared by conventional ion-milling techniques. After obtaining HRTEM micrographs for a series of focal settings, contrast and pattern are compared systematically between experimental and simulated images for a large range of focus values and specimen thicknesses. The focus value, the specimen thickness and the modulation transfer function of the CCD camera were determined independently or semi-independently. After applying zero-loss energy filtering, as well as taking the damping effect of the CCD camera and the sample vibration and drift into account in the simulation, the mismatch factor is measured as small as 1.1–1.7 for the image contrast (1.0 corresponds to perfect match), where higher mismatch appears around the focus value that mainly transfers the amplitude contrast into the image intensity. Meanwhile, the simulations of the Si crystal show excellent agreement in the image pattern with the experiments for all focal values and specimen thicknesses (Fig. 3.2.1). Here, the image agreement factor is higher than 0.95 for most of the experimental conditions, the image agreement factor of 1 stands for complete match and 0 for unrelated.

Since inelastic scattering of electrons mainly contributes an intensity background to the experimental images, which consequently diminishes the contrast,

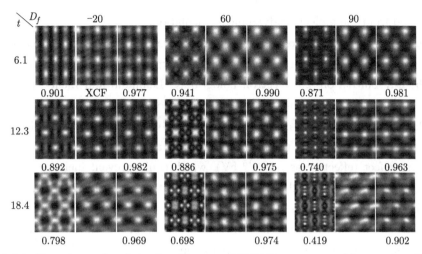

Fig.3.2.1 Experimental and simulated images of Si [110] are presented for a series of focus values (D_f, unit: nm) and specimen thickness (t, unit: nm). For each focus value and specimen thickness, the left image shows the conventional simulation without considering the dampling effects of sample instability and the CCD camera, on the contrary the right image presents the simulation with the sample vibration, the modulation transfer function of the CCD camera as well as specimen tilt and beam tilt taken into account, and the middle image shows the corresponding experimental zero-loss energy filtering image. The image agreement factor is defined by the cross-correlation factor between experimental and simulated images.

zero-loss energy filtering was applied in the experiment. However, thermal diffuse scattering of electrons, which originates from the individual vibration of Si atoms, has an energy-loss less than 1 eV, therefore cannot be removed from the image formation by the energy filtering. This effect needs to be considered in the simulation. In the following work, it was taken into account by the Debye-Waller factor and absorptive factor. Mobus et al., (1996) have proved that images simulated with Debye-Waller factor and absorptive factor have a good agreement on image pattern with the calculation of a more realistic model for thermal diffuse scattering. That model utilizes a summation of images calculated with structure models of snapshots of the thermally vibrating object to simulate the integration effect of recording time, where the structure models of individual atom vibration were obtained from the molecular dynamical calculation.

Quantitative comparisons of the image contrast as well as individual Fourier components have also been performed between experimental and simulated images on high-resolution off-axis electron holography (Lehmann and Lichte, 2003). Although the mismatch factor for each individual Fourier component varies from 0.5 to 2, the overall contrast of the simulation is only slightly lower than the experimental result. Reasons for the match of the contrast were addressed as the energy filtering effect (effective as $\sim 10^{-15}$ eV), which could prevent inelastic scattering electrons including thermal diffuse scattering electrons from image-formation, and the linear image-formation of electron holography, although details have not been provided on modulation transfer function of the recording device or mechanical instability of specimen and electron microscope.

The above results show that the contrast discrepancy between experimental and simulated images is caused by a combination of inelastic scattering of electrons, the modulation transfer function of record media and the sample vibration and drift during the experiment. By taking these effects into account of the simulation, quantitative comparison is reliable between experimental and simulated high-resolution transmission electron micrographs.

3.2.2 Quantitative high-resolution transmission electron microscopy

1. Iterative digital image matching

In recent times, quantitative high-resolution transmission electron microscopy has developed into a routine analysis tool for resolving the structure and chemical composition of materials at an atomic scale. A generally applicable approach is that, by iteratively comparing simulated images with the experimental micrographs, imaging parameters and a model structure of the object are refined until a best match is achieved between the simulated and experimental images (Mobus 1996). This technique has been performed successfully to determine various structures in materials. For instance, Kienzle et al. (1998) observe that the experimental HRTEM image of $\Sigma 3$ (111) grain boundary in $SrTiO_3$ does not match with the simulated image of structure model based on the coincident site lattice theory. By the use of iterative digital image matching (IDIM) technique, they refine the grain boundary structure model and achieved a best match between experimental and simulated HRTEM images. The structure model shows a 61 pm expansion at the grain boundary from the theoretical coincident site lattice structure. Xu et al. (2005) similarly determine that the averaged distance between two adjacent atom columns across a stacking fault in Be-doped GaAs crystal is 17 pm larger than that of the perfect crystal. With the aid of electron energy loss spectroscopy, they reveal that Be atoms segregate at the stacking fault and occupy Ga sites while Ga atoms take As sites thus it causes the expansion of atom distances across the stacking fault.

It is generally known that image features of HRTEM micrographs depend on experimental variables such as specimen thickness, crystal orientation, focus value, beam misalignment and astigmatisms. Nevertheless, these variables normally could not be determined independently by image matching between experimental micrographs and simulations. Therefore, IDIM usually employs numerical optimization strategies to search for the best match between the experimental and simulated images in the space of multiple variables with the aim of escaping from local optima (Mobus, 1996). For the same reason, the initial values of the experimental variables need to be assigned carefully for the calculation. By this means, the experimental parameters are determined from experimental HRTEM images of the perfect crystal region. After that, these imaging parameters are applied to the image simulation of defect structure region. By refining the type and position of individual atoms (in case of grain boundary or interface, also the rigid translation of one half-crystal against the other half-crystal), a best match can be approached between the simulation and the experimental HRTEM image of defect structures. The structure model corresponding to the best match will reveal the structure of defects.

Generally, the final object of quantitative image matching is to resolve the structure of specimens from HRTEM images. Therefore, Nadarzinski and Ernst (1996) have proposed a method, which was recently expanded by Kauffmann et al. (2005) to estimate the accuracy of atom positions determined by quantitative HRTEM analysis. This method defines the accuracy of the atom positions by the extent of the atom column displacements that will not generate a difference from the best-match simulation greater than the residual image discrepancy between the best-match simulation and the experimental image. By the use of this analysis, it is possible to assess the reliability of the structure determination from experimental HRTEM images.

2. Strain analysis directly measured from HRTEM images

Nevertheless, iterative comparison between the experimental and simulated images is generally a time-consuming procedure, particularly when large-area micrographs are processed. A different approach is possible for objects with coherent structures, that is, structures without extended defects. Under favorable conditions, the method is able to derive local information on chemical composition (Schwander and Rau, 1998) or lattice strain (Su et al., 2008) directly from high-resolution micrographs. As this approach does not require the simulation of images at each evaluation procedure, it is fast and particularly suitable for the analysis of large-area micrographs.

Direct strain mapping relies on the assumption that a HRTEM micrograph of a coherent structure taken at proper conditions represents the geometry of the crystal lattice. Although the positions of the image maxima/minima (peaks for short) depend on specimen thickness and orientation, imaging parameters such as defocus and microscope alignment, etc., and, thus, do not necessarily coincide with the projections of the atom columns (Spence, 2003), a constant spatial relationship between the image peaks and the projected atom columns can be assumed on a local scale. Thus, local lattice spacings can be determined directly from the micrograph without the exact knowledge of the projected atom column positions. Special care has to be taken when recording the images: specimen thickness has to be homogeneous and a defocus value has to be chosen such that the image pattern does not change within the analyzed area (Su et al., 2008). The image peak positions can then be determined either by fitting a Gaussian function for the image intensity or by calculating the cross-correlation factor between the experimental micrograph and a template motif. From an area of the image, which shows the undistorted lattice, a reference lattice is created and extrapolated over the entire image. The deviations of the peak positions in the experimental image from this reference lattice represent the local lattice distortions. To eliminate the influence of the noise, which comes from amorphous surface layers on the object, in the micrographs, image processing tools such as Wiener filter are employed occasionally prior to the distortion analysis.

The detection limit (Phillipp et al., 2003) of the analysis method can be estimated by lattice-distortion analysis performed on an experimental micrograph taken from a distortion-free crystalline object. The distances between the peaks along certain direction can be measured within the micrograph. The deviation of the distances measured for different pairs of peaks is not due to the structure of the material, but to the noise in the micrograph and to the limit of the peak position measurement. The standard deviation σ of the distances between image peaks, l, which have been measured from the micrograph of the distortion-free crystal, along the direction gives the detection limit of the distortion analysis as $\varepsilon_{\min} = 2\sigma/l$.

Effects of experimental variables on the accuracy of the direct strain analysis have been estimated with the use of image simulation techniques (Seitz et al., 1998). Here, the variables are focal setting and variation, local thickness and orientation of the sample, as well as misalignments of the sample and the incident beam. When there is a significant variation in local focus values or specimen thicknesses, the determined peak positions may vary at the same scale as the detection limit of the measurement. Also, the image feature or contrast would change notably within the micrographs. For this reason, consistency of image features and contrast within the micrographs is desired for the analysis to eliminate effects of the variations of local focus value and specimen thickness. After proper orientation of a crystalline specimen, even though a local bending may exist in the sample, the misorientation of the object will not have a notable influence on the measurement. However, the incident beam of the microscope needs to be aligned carefully even with the use of a computer-aided approach, since the beam misalignment may introduce strong artefacts around the interface region.

Strain mapping directly from HRTEM micrographs has been applied in numerous studies in various materials systems (Phillipp et al., 2003; Rosenauer and Gerthsen, 1999). For instance, the strain distribution in multilayer InP quantum dot structures has been evaluated. A direct correlation between the red shift of the photoluminescense peak position and the local strain distribution is established. Direct distortion measurements from a HRTEM image of a screw dislocation in Mo revealed three sectors of large deviations of the image peaks from a regular lattice. This yielded, for the first time, direct experimental evidence for the non-planar dissociation of the dislocation core suggested since long time in order to explain the high flow stress of body-centred cubic metals at low temperatures (Phillipp

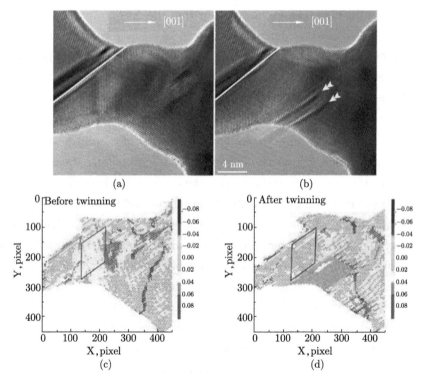

Fig.3.2.2 Deformation twinning with their quantitative strain analysis. (a) and (b) Sequential images captured before and after the twinning partials are emitted; the double arrowheads indicate the location of the deformation twins. (c) and (d) The strain mapping of the HRTEM images of (a) and (b), respectively, showing a strain relaxation immediately after the twin formation. Likewise, the open circles are drawn to represent the twinning partials (Zheng, 2010).

et al., 2003). Recently, lattice strain analysis has been performed on in situ observations of Au nano-crystal deformation (Zheng et al., 2010). Under the tensile loading, twinning was observed nucleating from the free surface and entering the metal nano-crystals (Fig. 3.2.2). The lattice strain distributing along the loading direction in the HRTEM images before and after the twinning are shown in Figs. 3.2.2(c) and 3.2.2(d), respectively. Quantitative strain profiles of the black boxed area (used as lattice strain gage) in the strain maps indicate that before and after twin formation, the mean elastic strain is 0.030 and 0.013, respectively. From the strain analysis, a tensile stress of 1.28 GPa (corresponding to 0.29 GPa of shear stress) is estimated for the nucleation of the twinning partials, while after twinning the tensile stress is significantly relieved.

3.2.3 Outlook

High-resolution images, especially obtained with a field emission gun electron microscope, usually exhibit delocalization of object information, which may displace the image peaks against the true positions of the atom columns they represent. This hampers the direct strain measurement from HRTEM images. Owing to the recent development of aberration-corrected electron microscopes, these artifacts can be effectively reduced by adjusting the spherical aberration coefficient of the objective lens. Therefore, aberration-corrected imaging with the aid of exit wave function reconstruction will be greatly beneficial to quantitative strain analysis of high-resolution electron micrographs.

3.3 A new approach to image analysis in HRTEM

Yumei Wang, Binghui Ge and Fanghua Li

3.3.1 Introduction

The development of HRTEM offers a new approach to crystal structure determination different from the diffraction one. However, images taken at an arbitrary defocus condition may not reflect the projected structure of the examined sample. Only images taken near the so-called Scherzer focus (Scherzer, 1949), can represent the structures intuitively, while most images show the periodicity and symmetry rather than the atomic structure. Traditionally, the trial-and-error method, also named the model method is utilized for crystal structure determination. For this purpose, a through-focus series of images are taken at the same place of the sample, a tentative structure image is selected from among them by referring to some preliminary structure information, and then several possible structure models were proposed according to the selected structure image. For each proposed model several through-focus series of images are simulated with different thicknesses. Finally, the corrected model is picked out by matching the experimental image series with all simulated series one by one. Evidently, such method would be available only when the examined structure is partly known in advance and the sample is strong enough under the electron beam irradiation. In addition, the resolution of such determined structures is limited by the point resolution of the microscope, and in most cases not all atoms can be resolved in the obtained structure maps. Some alternate methods for crystal and defect determination have been developed for removing the weakness of the model method. From the viewpoint of physics, it is essential to solve the inverse problem in HRTEM not lying on any prior knowledge and any guessed model of the examined structure. The methods can be divided into two kinds, one is the exit wave reconstruction that corresponds to the reciprocal process of electron-optical imaging. The exit wave can be reconstructed with the linear and nonlinear imaging theory from two or more images taken under different focus conditions (Schiske, 1968; Kirkland et al., 1984; Kirkland, 1985; Saxton, 1986). It can be applied to periodic and aperiodic material. When the examined crystal is as thin as a phase object, the phase of the exit wave is linear to the projected potential of the crystal, and hence the phase map of the reconstructed exit wave display the projected crystal structure. The readers are recommended to learn more about the exit wave reconstruction from the section 3.4 in this chapter written by Liu and Wang. The other kind of methods, skipping the exit wave, goes directly from the image to the restored structure. These methods pay a great attention to the matter that the examined samples are crystals, and the electron diffraction intensity as well as the diffraction analysis method developed in X-ray crystallography could benefit the image in solving the structure. Therefore, such kind of methods (Li, 2010; Li, 1977; Li and Fan, 1979; Ishizuka, 1982; Fan et al., 1985; Hovmöller, 1992; Dong, 1992; Sinkler, 1999) can be classified in electron crystallography. However, the X-ray diffraction is kinematical, while the electron diffraction is dynamical. Hence, it is necessary to have a common theoretical base that supports both the electron diffraction and imaging, and can also be accepted by X-ray diffraction.

In the present section, a new approach to crystal structure determination based on the combination of HRTEM and diffraction crystallography is briefly introduced. A practical image contrast theory is expounded in Section 3.3.2. The image processing methods including deconvolution and phase extension as well as the related analysis methods are demonstrated in Section 3.3.3 and 3.3.4. A method of atom recognition based on the practical image contrast theory and deconvoluted image is described in Section 3.3.5. Some typical examples of application to crystal structure and defect studies are given in Section 3.3.6. The readers are recommended to refer the review article (Li, 2007) that introduced the present approach in some more detail and with some more examples and figures.

3.3.2 A practical image contrast theory

1. Weak-phase object approximation (WPOA) and multislice theory

When the sample is as thin as a weak-phase object, after interacting with the electron wave the image intensity $I(r)$ is proportional to the convolution of the projected electrostatic potential function (hereafter it is simplified as the potential function) $\varphi(r)$ of the sample with the contrast transfer function (CTF) of the objective lens. At the Scherzer focus condition, the image intensity formula is approximately expressed as

$$I(\boldsymbol{r}) = 1 - 2\sigma\varphi(\boldsymbol{r}) \tag{3.3.1}$$

when C_s is positive. Eq. (3.3.1) interprets the one to one correspondence of the structure image and the projected structure. The corresponding image-contrast theory is named weak-phase object approximation (WPOA). From the viewpoint of the diffraction the WPOA corresponds to the kinematical electron diffraction.

Different from WPOA, the multislice theory (Cowley and Moodie, 1957) widely used in HRTEM is based on the dynamical electron diffraction. It was derived by the physical-optic approach with the sample divided into thin slices perpendicular to the incident beam, and all slices are treated as phase objects, of which the image contrast is due to the phase change of the electron wave. The transmission of electrons through the sample is represented by transmission through each slice and the propagation of electrons from one slice to the next. Though the relation between the image intensity and sample thickness cannot be expressed analytically, it becomes possible to calculate the image intensity numerically with the multislice image simulation. Hence, the WPOA and multislice theory are supplementary to each other, but a room does exist between them for studying the behavior of image contrast more perfectly than before.

2. Derivation of pseudo weak-phase object approximation (PWPOA)

Following the multislice theory the sample is divided into thin slices, but all slices are assumed to be weak-phase objects and identical to one another, and the secondary electron scattering between adjacent slices is assumed to be negligible. The first assumption is reasonable because the incident beam is usually parallel to the shortest crystallographic axis. Though the second assumption corresponds to the kinematical electron diffraction, the thickness effect is taken into consideration in the Fresnel diffraction. The expression of image intensity at the Scherzer-focus condition (Li and Tang, 1985) is obtained as

$$I(r) = 1 - 2\sigma[\varphi(r) + \Delta\varphi(r)] = 1 - 2\sigma\varphi'(r) \qquad (3.3.2)$$

The crystal thickness is a parameter included in the potential increment $\varphi'(r)$. Function $\varphi'(r)$ can be understood as the potential function of an artificial crystal isomorphic to the examined one. For simplify, hereafter $\varphi'(r)$ is named the pseudo potential. Accordingly, the image-contrast theory based on Eq. (3.3.2) is named the pseudo weak-phase object approximation (PWPOA). For the crystal consisting of $n+1$ slices $\Delta\varphi(r)$ is expressed as

$$\Delta\varphi(r) = \varphi(r) \times \sum_{j=1}^{n} S_j(r) - \sigma\varphi(r) \left[\varphi(r) \times \sum_{j=1}^{n} S_j(r)\right] - \frac{\sigma}{2}\left[\varphi(r) \times \sum_{j=1}^{n} S_j(r)\right]^2$$

$$-\frac{\sigma}{2}\left[\varphi(\bm{r}) \times \sum_{j=1}^{n} C_j(\bm{r})\right]^2 \qquad (3.3.3)$$

where

$$S_j(\bm{r}) = (1/jt)\sin(\pi/jt)r^2 \text{ and } C_j(\bm{r}) = (1/jt)\cos(\pi/jt)r^2$$

with j denoting the slice number and t slice thickness. All terms in the right-hand side of Eq. (3.3.3) are significant because they depend upon the crystal thickness and modify the image contrast appreciably with increase in thickness.

To evaluate the overall influence of all terms in $\Delta\varphi(\bm{r})$ the convolution of $\varphi(\bm{r})$ with $S_j(\bm{r})$ and $C_j(\bm{r})$ are analyzed with $\varphi(\bm{r})$ replaced by a Gaussian function. A schematic diagram illustrating the relative change of image contrast with crystal thickness for atoms with different atomic weights is shown in Fig. 3.3.1. The atomic weight decreases progressively from atom A_1 to A_3. When the crystal is a weak-phase object, $\Delta\varphi(\bm{r})$ equals zero and function $\varphi'(\bm{r})$ degenerates into $\varphi(\bm{r})$, so that the PWPOA degenerates into WPOA (see the top picture in Fig. 3.3.1). For crystals of thickness bigger than the weak-phase object the positions of all peak in $\varphi'(\bm{r})$ remain the same as in $\varphi(\bm{r})$, but the peak heights in $\varphi'(\bm{r})$ are different from those of the corresponding peaks in $\varphi(\bm{r})$. For very small thickness, although the heights of all peak in $\varphi'(\bm{r})$ increase with increasing thickness, the increase in speed depends on the atomic weight such that the peak heights increase more rapidly for atoms having a smaller atomic weight (hereafter named light atoms) than for atoms having a bigger atomic weight (hereafter named heavy atoms) (compare the top second and middle pictures in Fig. 3.3.1). In general, with the increase of crystal thickness the peak height of the heaviest atoms A_1, reaches a maximum value at a certain thickness, then decreases, becomes zero and negative (see the bottom picture in Fig. 3.3.1). The critical thickness t_c of a crystal is defined as the thickness when the contrast of the heaviest atoms becomes zero. Evidently, the contrast of heaviest atoms becomes negative above the critical thickness. The prediction about the image contrast given by PWPOA has been proved by experiments and image simulation (Li and Hashimoto, 1984; Tang, 1986). The critical thickness depends on the electron wavelength and the atomic weight of the heaviest atoms in crystal.

It is bigger for a smaller electron wavelength and smaller atomic weight. Usually it is below or about 10 nm for medium energy electrons and not very large atomic weights.

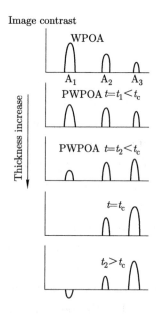

Fig.3.3.1 Schematic diagrams showing the relative change of image contrast with crystal thickness for three differently weighted atoms. Atom A_1 is the heaviest one, A_2 is of medium weight, and A_3 is the lightest one.

3. Validity and practical value of PWPOA

The above argument demonstrates that the PWPOA is available for samples of thickness below the critical value. This is valuable because such thickness scope is appropriate for taking image of good quality.

First, The PWPOA indicates that a possibility of revealing preferentially light or heavy atoms in the images by choosing different sample thickness. Fig. 3.3.2(a) shows the [001] projected structure model of $Li_2Ti_3O_7$ with Li atoms inside the big channels, and Fig. 3.3.2(b) the [001] images taken from a wedge-shaped crystal with a JEM 200CX microscope of point resolution about 0.25 nm. The micrographs enlarged from regions A, B, and C in Fig. 3.3.2(b) are shown in Figs. 3.3.2(c)–

(e) together with the images simulated for thicknesses 0.21, 0.44 and 0.68 nm, respectively, inserted on the right. No contrast of Li atoms can be observed in Fig. 3.3.2(c), but with the increase of thickness the big white dots representing the channels where Li atoms are situated at the center contract gradually and finally split into two separate dots (Figs. 3.3.2(d) and (e)). The contrast of Ti atoms remains almost the same (Tang, 1986). This proves that atoms as light as Li can be observed preferentially by choosing a thicker image region.

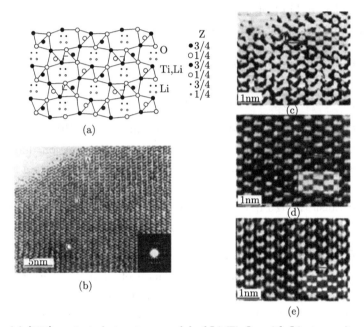

Fig.3.3.2 (a) [001] projected structure model of $Li_2Ti_3O_7$ with Li atoms situated inside the big channels, (b) [001] images of point resolution about 0.25 nm, (c), (d), and (e) micrographs enlarged from regions A, B, and C in (b), respectively. The insets in (c), (d), and (e) are simulated images corresponding to thickness 0.21, 0.44, and 0.68 nm, respectively (Tang, 1986).

Secondly, the crystals with thickness below the critical value can be treated as pseudo weak-phase objects. Thus, Eqs. (3.3.2) and (3.3.3) are valid in practice, and it is reasonable to write

$$I(\boldsymbol{r}) = 1 - 2\sigma\varphi'(\boldsymbol{r}) \times F^{-1}[T(H)] \qquad (3.3.4)$$

where $T(H)$ denotes the CTF of the objective lens. Eq. (3.3.4) degenerates into the formula for the WPOA, when $\varphi'(r)$ is replaces by $\varphi(r)$. Hence, the image analysis methods developed based with the WPOA would be available for practical samples with the thickness below the critical value. This means that the PWPOA has laid a theoretical foundation to the development of image-analysis methods based on the combination of HRTEM and diffraction analysis.

Thirdly, the PWPOA indicates the possibility of setting up a method to recognize different kind of atoms according to the image contrast change with the sample thickness. This will be demonstrated in Section 3.3.5.

3.3.3 Image deconvolution

1. Principle of image deconvolution

The image deconvolution was proposed by Li and Fan in 1979 as a special kind of image processing in HRTEM to restore the image distortion due to the lens aberrations. More accurately, it aims at transforming an image not representing the examined structure intuitively into the structure map.

Fourier transforming Eq. (3.3.4), ignoring the transmitted electron beam yields

$$F'(H) = i(H)/2\sigma T(H) \qquad (3.3.5)$$

where $i(H)$ is the diffractogram of image, and $F'(H)$ denotes the FT of $\varphi'(r)$. $F'(H)$ can be understood as the structure factor of an artificial crystal that is isomorphic to the examined one and hereafter named pseudo structure factor. Hence, starting from a single image one can obtain $F'(H)$ if the $T(H)$ is known. Thus, the pseudo structure map $\varphi'(r)$ that shows the atoms of the examined structure at correct positions would be obtained by inversely Fourier transforming $F'(H)$. $T(H)$ depends on various electron optical parameters, and all parameters can be determined more or less accurately except the defocus value. Therefore, the principle of image deconvolution is simple, but the defocus determination is critical.

2. Methods of deconvolution for ab initio crystal structure determination

Defocus determination: There are different methods to determine the defocus. In the following, the discussion will be concentrated on the methods developed specially as a component part of deconvolution processing (Han, 1986; Tang, 1988; Hu, 1991). First, the defocus value Δf is evaluated roughly based on the experimental record, and an appropriate focus regions, for instance, sized about 100 nm is selected with the evaluated value at the middle. Secondly, a series of trial Δf are assigned in this focus region with a small interval, for instance, 0.5–1 nm. Then a series of trial pseudo structure factors $F'_{\text{trial}}(\boldsymbol{H})$ are calculated from Eq. (3.3.5), and Fourier transforming the series of $F'_{\text{trial}}(\boldsymbol{H})$ yields the series of corresponding trial pseudo potential $\varphi'_{\text{trial}}(\boldsymbol{r})$. Different criteria were proposed, for instance, based on the direct methods (Han, 1986) developed in X-ray crystallography and on the principle of maximum entropy (Hu, 1991) in information theory, to determine the correct $F'(\boldsymbol{H})$ and $\varphi'(\boldsymbol{r})$, respectively, from among the two series of trial functions. The correct $\varphi'(\boldsymbol{r})$ gives a structure map with atoms sitting at correct positions and appearing black. The deconvolution processing methods based on the direct methods and the principle of maximum entropy have also been applied successfully to experimental images for a number of compounds when the thickness is below the critical value (Hu et al., 1992; Fu et al., 1994; Lu et al., 1997; Liu et al., 1998; Jiang et al., 1999)

Peculiarity of maximum-entropy image deconvolution: According to the principle of maximum entropy developed in information theory (Shannon and Weaver, 1949; Jaynes, 1957), a best structure can be obtained from a single high-resolution electron microscope image by entropy maximization (Dong, 1992). For each of the above-mentioned trial defocus, the entropy of the corresponding trial structure map is calculated from the formulas

$$S = -\sum_{i=1}^{N} p_i \ln p_i \qquad (3.3.6)$$

and

$$p_i = \phi_i / \sum_{i}^{N} \phi_i \qquad (3.3.7)$$

where ϕ_i denotes the value of the projected potential of the ith pixel. The structure map that corresponds to the maximum value of entropy would be the best one. Such an image-deconvolution technique is a kind of constrained maximum-entropy method accomplishing the constraint via selecting a best defocus. The method was first tested with a series of simulated images for chlorinated copper phthalocyanine treated as a weak-phase object and also as a pseudo weak-phase object, and successively applied to unknown crystal structures including the incommensurate modulated structures (IMS) (Liu et al., 2001; Wang et al., 2002; Wang et al., 2005; Ge et al., 2008). After a slight modification the method could also be applied to crystals with a big unit cell, for instance, protein crystals (Yang and Li, 2000).

The maximum-entropy image deconvolution method has been studied in detail to show its functions and problems (JAYNES et al., 1957; Wang et al., 2004). It should be mentioned that some electron optical parameters such as the accelerating voltage, spherical aberration coefficient, focus spread due to the chromatic aberration, beam divergence, and so on must be assigned together with the trial defocus values in the calculation of trial pseudo potentials. In principle, the parameters other than the defocus should be known in advance and assigned as accurately as possible. However, when various electron optical parameter errors were given by design, a series of tests in deconvolution processing for simulated images indicated that the errors led to a deviation of the determined defocus value from its true value, but did not influence the quality of the determined best structure map (Wang et al., 2004). Although the determination of the defocus value is an important part of image deconvolution, one cannot expect too much that the determined defocus value would equal the true value even when all electron optical parameters are assigned exactly correctly. It was found that for the same electron optical parameters the determined defocus value would vary with the sample thickness, and again the quality of the determined best structure map is guaranteed. All this means that the errors of the assigned electron optical parameters as well as the thickness effect are compensated automatically to some extent in the process of defocus determination.

Multiple peaks were observed in the entropy–defocus curves for crystals with a small unit cell and simple structure (Huang et al., 1996). At first glance, this phenomenon indicates a problem of multiple solutions in the maximum-entropy image deconvolution method. However, after a careful study it comes to a satisfied interpretation and conclusion that only one of the peaks corresponds to the true structure, while others are caused by the Fourier image effect (Cowley and Moodie, 1957; Cowley and Moodie, 1960) or correspond to some fictitious structures that are isomorphic to the true structure or are reverse structures (Huang et al., 1996). Therefore, it should not be difficult to determine the correct solution via selecting the best structure map for crystallographers when the chemical formula of the sample is known.

The above argument indicates that the deconvolution method based on the principle of maximum entropy is a reliable and powerful tool to transform a single image into the best structure map via removing the combination effect of spherical aberration and defocus. In addition, the dynamical scattering effect can be reduced when the thickness effect is compensated.

3. Methods of deconvolution for crystal defect study at atomic level

Defocus refinement: The defocus determination methods given above are available only for images of perfect crystals. For crystal-defect investigation, first, the defocus value is determined roughly from a perfect crystal image region not far from the examined defect with the methods given above or from a nearby amorphous image region at the edge of the sample by matching the CTF curves calculated for different defocus values with the Thon diffractogram (Thon, 1971), i.e., the FT of the amorphous image. Then, a refinement is needed because the height and thickness of the observed areas in the same sample may be different from one another. For this purpose several trial defocus values are assigned on both under and over focus sides of the roughly determined one with a small focus step, say 0.5 nm. For each trial defocus value the image deconvolution is carried out to deduct the CTF modulation according to Eq. (3.3.5). Like the case of defocus determination for perfect crystals, a set of $F'_{trial}(H)$ and of $\varphi'_{trial}(r)$ are obtained, but here $F'_{trial}(H)$ denotes the structure factor corresponding to an artificially constructed unit cell of large size, in which the examined defect is included at the center. Then, the best structure map is selected from among the trial maps. In this map, atoms in both the perfect and defect regions should be resolved best. The details of the procedure including the construction of a large unit cell, treatment of delocalization effect, subtraction of background, and so on, are described in the literature of He et al. (1997).

Empirical method of reflection amplitude correction: It was noticed that the quality of the structure map obtained by deconvolution processing decrease with increase in sample thickness, especially in the case of defect investigation (Li et al., 2000). To improve the structure map quality, the amplitude and phase change with the increase of crystal thickness was studied with simulated images of Si having a 60° dislocation. It was found that the reflection intensity profiles in image diffractograms for different thicknesses are very similar to one another with the main peaks and shoulders unchanged in position, while the phases of reflections in diffractograms change drastically from place to place inside the same reflection disk, but the relative phase distribution inside the corresponding reflection disks is more or less constant for different crystal thicknesses.

Based on the above analysis results, a method was proposed as follows to correct the determined deconvoluted image in Fourier space, namely, to correct its FT. Hereafter, the FT of deconvoluted image is named the deconvoluted diffractogram, and the FT of the corrected deconvoluted image named, for simplicity, the corrected diffractogram and expressed as $F'^c(H)$. The integral amplitude of each reflection in the deconvoluted diffractogram is calculated by summing up the values in a disk centered at the point of maximum value with a radius of appropriate size. The value of amplitude in the ith pixel for reflection H measured from the deconvoluted diffractogram is expressed, for simplicity, as F'_{H_i}, and the integral amplitude of reflection H as

$$F'^I(H) = \sum_i F'_{H_i} \qquad (3.3.8)$$

Assign a multiplier K_H equal the ratio of amplitude of the structure factor for perfect crystal $F(H)$ to the integral amplitude $F'^I(H)$:

$$K_H = \frac{|F(H)|}{F'^I(H)} \qquad (3.3.9)$$

Then, in the corrected diffractogram the reflection amplitude for the ith pixel $F'^c(H)$ is expressed as

$$F'^c_{H_i} = K_H F'_{H_i} \qquad (3.3.10)$$

The combination of Eqs. (3.3.8)–(3.3.10) means that the integral amplitudes of all the reflections in the corrected diffractogram $F'^c(H)$ are constrained to equal the corresponding structure-factor amplitudes of a perfect crystal. The correction coefficient K_H remains constant inside each diffuse reflection disk but changes from one disk to the other. Finally, Fourier transforming $F'^c(\boldsymbol{H})$ together with the original phase yields the corrected structure map. It was proved that the method is effective in reducing the dynamical effect so that the quality of the determined structure map can be improved, and the sample thickness is permitted to be bigger. This is reasonable because, physically, the present method of reflection amplitude correction plays a role in reducing the discrepancy between the structure-factor amplitude and the square root of diffraction intensity, or say, in compensating the thickness effect. The method has been successfully applied to study the dislocation core structures and other defects at the atomic level (Wang et al., 2002; Wang et al., 2004; Tang et al., 2007; Wang et al., 2008; Wen et al., 2009).

Obviously, the method of deconvolution processing for crystal defect study is different from that for perfect crystal. In practice, the reflection-amplitude correction is performed in the process of image deconvolution, joining with the defocus refinement.

4. Quality of deconvoluted image

The pseudo potential $\varphi'(\boldsymbol{r})$ obtained by deconvolution processing from images of perfect and defect crystals is named the deconvoluted image. It shows a structure map where all atoms appear black for positive C_s, but comparing with the map of weak-phase objects the light atoms in $\varphi'(\boldsymbol{r})$ appear darker, while heavy atoms are brighter when the sample is not very thin. In addition, the quality of the derived structure maps is always better than that of Scherzer-focus images taken experimentally, especially taken with field-emission gun (FEG) electron microscopes. This is because the spherical aberration and defocus effects are removed in the process of deconvolution processing. Fig. 3.3.3 shows the CTF curves for non-FEG and FEG electron microscopes of 200 kV calculated for the Scherzer-focus condition. In the case of non-FEG microscopes, the resolution of experimental structure images is limited by the point resolution of microscopes (0.2 nm for 200 kV). Although the resolution of structure map obtained by image deconvolution does not improve much, the portion of high-resolution, or high spatial frequency structure information are restored so that the quality of derived structure maps is improved.

As in the case of FEG microscopes, the examined structure can never be revealed faithfully even in images taken at the Scherzer focus condition owing to the strong oscillation of CTF in the high spatial frequency region. After deconvolution processing the high-resolution structure information can be restored up to 0.14 nm, thus the quality of the obtained structure map is greatly better than the corresponding experimental image, although the point resolution of FEG microscopes is the same as that of non-FEG ones. Therefore, the deconvolution processing is always beneficial to improving the quality of structure maps even for images taken near the Scherzer-focus condition, either with FEG or non-FEG microscopes.

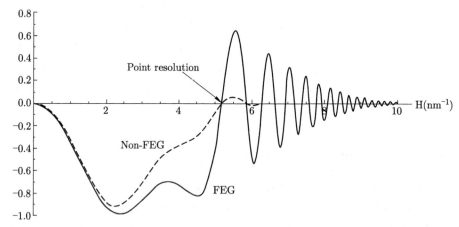

Fig.3.3.3 CTF curves for FEG (solid) and non-FEG (dotted) electron microscopes of 200 kV.

3.3.4 Resolution improvement for images of perfect crystals

1. Principle of phase extension

The structure map can be obtained from a single image by means of deconvolution processing or directly from the image taken at the Scherzer-focus condition. In the structure map atoms or atomic clusters appear as black dots sitting at correct positions, although the blackness does not correspond linearly to the atomic weight. Generally, for medium-voltage microscopes not all atoms can be resolved in the map because of insufficient point resolution and insufficient information limit of the electron microscope. For instance, only metallic atoms in oxides are seen but oxygen atoms are not. Hence, it would be meaningful to develop a method of enhancing the image resolution beyond the point resolution and information limit of the microscope.

One can easily take both the images and electron diffraction patterns (EDPs) with high-resolution transmission electron microscopes from the same area of the same sample. The structural information included in the image wave and diffracted wave are related by FT and supplementary to each other. The amplitudes of both are recorded in experiments, while the phases are not. In the early 1970s, Gerchberg and Saxton (1971) derived the phase of the image wave or electron-diffracted wave from the amplitudes of both. This was the pioneering work of deriving additional structural information by combining the image and corresponding EDP.

In 1977, Li proposed an idea to enhance the image resolution by combining the electron-diffraction data. Fig. 3.3.4 shows two circles in the Fourier space. The structural information carried in an image, for instance, of resolution 0.2 nm, is inside the small circle, and that carried in an EDP is inside the big circle with a resolution of 0.1 nm or better. One can have both the amplitudes and phases inside the small circle from the FT of an experimental structure image or of the deconvoluted image $\varphi'(r)$. The amplitudes inside the big circle can be obtained from the corresponding EDP. Therefore, it becomes possible to enhance the structure resolution, if the phases between the two circles can be derived. There are different methods to derive the unknown phases in X-ray diffraction crystallography. For

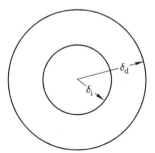

Fig.3.3.4 Two circles in the Fourier space. The structure information included in the image is distributed inside the small one and that in the EDP is inside the big one. δ_i and δ_d denote the reciprocals of image resolution and diffraction resolution limit, respectively.

instance, they can be derived by phase extension using the direct methods (Dorset and Hauptman, 1976; Woolfson and Fan, 1995) developed in X-ray crystallography. For the present purpose, taking the square roots of diffraction intensities as amplitudes inside the big circle and the phases of $F'(\boldsymbol{H})$ as the initial phases (phases inside the small circle) (Fig. 3.3.4), the phases between the two circles can be derived by means of Sayre's equation (Sayre, 1952) or the tangent formula (Karle and Hauptman, 1956).

However, it is a problem that the structure factors are proportional to the square roots of diffraction intensities for X-ray, while for electrons the square relationship between the diffraction intensities and amplitudes of pseudo structure factors $F'(\boldsymbol{H})$ breaks down owing to the dynamical scattering effect. In order to have the high-angle diffraction information as perfect as possible, one has to take EDPs from an area bigger in size and in thickness than for taking images. This leads to an increase of the dynamical scattering effect in diffraction data than in images. In addition, some other factors, such as the Ewald sphere curvature due to the small electron wavelength, crystal bend and tilt, crystal thickness fluctuation, structure distortion caused by the electron irradiation and so on also contribute to the diffraction intensity modulation. To get a set of diffraction intensities qualified for carrying out the phase extension, a method was developed in 1996 (Huang, 1996) to reduce simultaneously all kinds of intensity modulation.

2. Partial structure factor method of diffraction-intensity correction

By referring to the heavy-atom method (Woolfson, 1956) and Wilson statistic (Wilson, 1949) developed in X-ray crystallography, a partial structure factor containing only the contribution of relatively heavy atoms in the examined crystal is calculated with the atomic coordinates read from the structure map obtained by image deconvolution (Huang et al., 1996). The Fourier space is divided into a number of circular zones. The diffraction intensity after the correction is expressed as

$$I_c(\boldsymbol{H})_{|H_i \pm \Delta H_i|} = \frac{\langle |F_p(\boldsymbol{H})|^2 \rangle_{|H_i \pm \Delta H_i}}{\langle |I_o(\boldsymbol{H})| \rangle_{|H_i \pm \Delta H_i}} I_o(\boldsymbol{H}) \qquad (3.3.11)$$

where I_o and I_c denote the observed and corrected diffraction intensities, respectively, H_i is the average value of $\boldsymbol{H}$ in the ith zone, ΔH_i the half-width of the ith

zone, and ⟨ ⟩ denotes averaging. The method was proved to be effective with the compounds $YBa_2Cu_3O_{7-\delta}$ and chlorinated copper phthalocyanine (Dorset, 1997), and applied successively to crystal-structure determination. A careful analysis of the corrected diffraction amplitudes indicates that at very low scattering angle the observed amplitudes are much bigger than the corrected ones in value. The phenomenon of ultra-strong low-angle diffraction intensities should be due to the Ewald-sphere curvature effect. Moreover, very like a reverse process of multiple scattering, some of the biggest observed amplitudes become even bigger after the correction, and some observed small amplitudes become even smaller. This means that after the diffraction-intensity correction the Ewald-sphere curvature effect could be removed more or less and the dynamical effect removed partially. Therefore, in order to improve the resolution of the determined structure, it is necessary to utilize the corrected electron-diffraction data in phase extension.

3. Deriving high-resolution structure maps

The phases between the two circles are derived with the square roots of corrected diffraction intensities as amplitudes and the phases obtained from the deconvoluted image $\varphi'(r)$ as initial phases. After carrying out the first cycle of phase extension, more atoms will appear in the obtained new structure map than in the deconvoluted image. Then, a new set of partial structure factors is calculated and the observed intensity is scaled for the next cycle, until all atoms appear in a stable structure map. Finally, the Fourier synthesis widely used in X-ray crystallography is employed in the structure refinement. It must be emphasized that the finally obtained map is a high-resolution structure map of the examined crystal rather than a pseudo structure map. This is because the diffraction intensity is pressed on toward the kinematical value during the calculation of partial structure factors in every cycle of diffraction-intensity correction. Besides, in the final stage the structure refinement by Fourier synthesis modifies the peak heights toward the true values to some extent. It is obvious that in the process of phase extension the missed structure information due to the zero crosses of CTF is restored, either that between the two circles or inside the small circle. Therefore, the resolution of the determined structure is close to the diffraction-resolution limit.

3.3.5 Atom recognition

According to the discussion given in subsection 3.3.2, when the sample thickness is below the critical value, the blackness of light atoms in the high resolution electron microscope image increases with the increase of thickness more rapidly than that of heavy atoms, and with the further increase of crystal thickness light atoms may appear darker than heavy atoms. Such a feature about the image-contrast change with the sample thickness provides a method to recognize different sorts of atoms in images. For clarity, the method will be illustrated with the [110] image of 3C-SiC as an example to show how Si and C atoms can be distinguished from each other.

Fig. 3.3.5(a) shows the [110] image of 3C-SiC taken with a 200 kV electron microscope of point resolution about 0.2 nm equipped with LaB_6 filament. The crystal thickness increases from the top to the bottom. Three framed regions are labeled R_1, R_2, and R_3, respectively. R_1 is a perfect crystalline region. A 30° partial dislocation and a microtwin are included in R_2 and R_3, respectively. Fig. 3.3.5(b) is the magnified micrograph of region R_1 with the sample thickness increasing from the left to right (the left end corresponds the top end in Fig. 3.3.5(a) and vice versa) as indicated by the arrow. The white dots in the image do not represent atoms, and hence the image does not represent the [110] projected structure of 3C-SiC intuitively. Fig. 3.3.5(c) shows the pseudo potential map obtained from Fig. 3.3.5(b) by deconvolution processing combined with the reflection-amplitude correction (subsection 3.3.3). Every two black dots group in a dumbbell to represent the Si and C atomic pair, indicating that the sample thickness is below the critical value for atom Si and C. Three framed areas in Fig. 3.3.5(c) are magnified and arranged in Fig. 3.3.5(d) to show the image contrast more clearly. In order to discern the contrast difference between the two ends in dumbbells, the gray levels were measured along the three horizontal line segments in regions I, II, and III, and the corresponding contrast profiles are shown in Fig. 3.3.5(e) from the left to right, respectively. The gray value 0–255 is assigned to correspond to the contrast change from white to black. The gray level of the left ends in dumbbells are lower than, almost equal to and slightly higher than that of the corresponding right ends for regions I, II, and III, respectively. This indicates that with the increase of sample

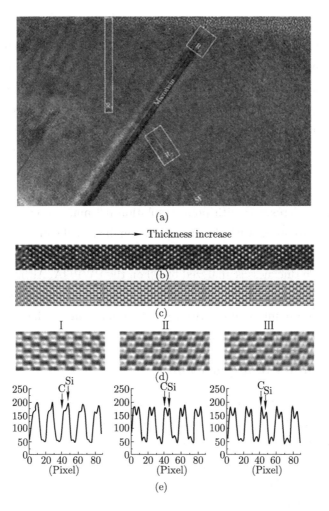

Fig.3.3.5 (a) [110] image of 3C–SiC with thickness increasing from the top to the bottom, (b) magnified image of region R1 in (a), (c) pseudo potential map obtained from (b) by deconvolution processing combined with the reflection-amplitude correction, (d) magnified micrographs of the three framed areas in (c) and (e) curves showing the profiles of gray level obtained by scanning the line segments in the three regions I, II, and III in (d), respectively. The gray value 0–255 is assigned to correspond to the contrast change from white to black.

thickness the gray level of the left ends in dumbbells increases more rapidly than that of the right end. Because region I is near the amorphous area and must be the thinnest one among the three regions, it is reasonable to approximate region I as a weak-phase object, while regions II and III are pseudo weak-phase objects. Thus, according to PWPOA, the above argument leads to the conclusion that the left ends in the dumbbells represent the atomic columns of C while the right ends the Si. Regions R_2 and R_3 shown in Fig. 3.3.5(a) will be discussed in subsection 3.3.6.

The above results indicate that starting from an image taken with a 200 kV LaB$_6$ electron microscope with point resolution 0.2 nm, one can distinguish and recognize the two atoms with the distance as small as 0.109 nm, when the image is processed by methods of electron crystallography, and the image-contrast change with the sample thickness is analyzed based on the PWPOA. Although the present section aims at demonstrating the method of atoms recognition, it shows a way to get the structure information with resolution much higher than point resolution and information limit of the microscope.

3.3.6 Examples of applications

The above-mentioned image-analysis methods can be grouped into two sets of electron crystallographic image processing and analysis methods. One is for the ab initio crystal-structure determination, and the other for crystal-defect study at the atomic level. Some methods, for instance, the method of atom recognition, can join with either set. In the following, some typical examples will be given to show the applications of the two sets.

1. ab initio crystal-structure determination

Commensurate modulated structure of $Bi_2(Sr_{0.9}La_{0.1})_2CoO_y$

The crystal structure of $Bi_2(Sr_{0.9}La_{0.1})_2CoO_y$ is commensurate modulated. The basic structure is orthorhombic with the unit-cell parameters $a_0 = 0.547$ nm, $b_0 = 0.545$ nm, and $c_0 = 2.33$ nm, close to that of $Bi_2Sr_2CuO_y$. The superstructure lattice parameters are $a = 4a_0, b = 4b_0$, and $c = c_0$, and the space group is *Pnnn*. The characteristic of commensurate structural modulation can be identified from the

[100] EDP shown in Fig. 3.3.6.

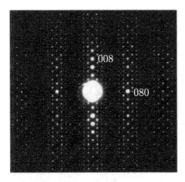

Fig.3.3.6 [100] EDP of $Bi_2(Sr_{0.9}La_{0.1})_2CoO_y$.

Fig. 3.3.7 shows two experimental [100] images taken with different focus conditions. The inset rectangular frames indicate the unit cell of the superstructure. The symmetry of the projected structure in the [100] direction is $C2mm$. Figs. 3.3.8(a) and (b) show the images obtained after Fourier filtering and symmetry averaging from the thin regions in Figs. 3.3.7(a) and (b), respectively. The unit-cell origins

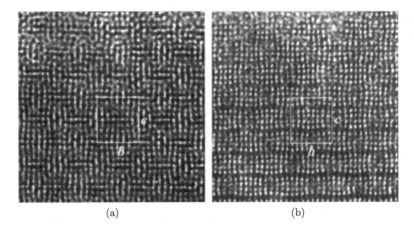

Fig.3.3.7 Two [100] images of $Bi_2(Sr_{0.9}La_{0.1})_2CoO_y$ taken from the same place of the same sample with different focus conditions.

are determined by minimizing the total phase residual. The two images shown in

Figs. 3.3.8(a) and (b) are almost reverse to each other in contrast.

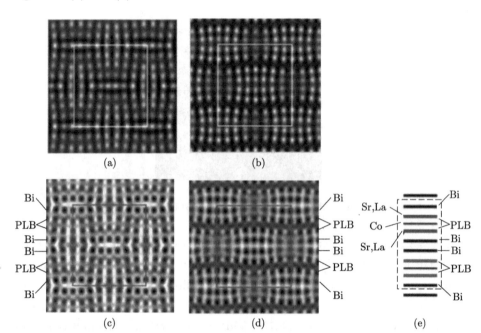

Fig.3.3.8 (a) and (b) Fourier filtered and symmetry-averaged images obtained from Figs. 3.3.7(a) and (b), respectively, (c) and (d) deconvoluted images obtained from (a) and (b), respectively, (e) schematic diagram showing the stacking sequence of different atomic layers in [100] projected structure. PLB represents the perovskite-like block.

The maximum-entropy image deconvolution method (Hu and Li, 1991) was utilized to restore the examined structure. A series of trial defocus values were assigned in the range from −100 to 0 nm with the focus step of 1 nm. Two series of trial deconvoluted images corresponding to Figs. 3.3.8(a) and (b), respectively, were obtained following subsection 3.3.3. The values of entropy were calculated according to eqs. (3.3.6) and (3.3.7) given in subsection 3.3.3 for every trial deconvoluted image. Two entropy-defocus curves corresponding to the two series of trial deconvoluted images were obtained to show the dependence of entropy on the defocus values (Figs. 3.3.9(a) and (b)). In both curves there are two broad peaks and many sharp peaks. According to Huang's reference (1996), only one of broad

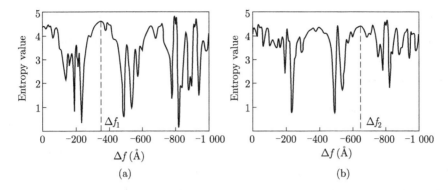

Fig.3.3.9 (a) and (b) Entropy–defocus curves corresponding to images shown in Figs. 3.3.7(a) and (b), respectively.

peaks in each curve corresponds to the true solution, and other sharp peaks are regarded as fictitious. Since the image shown in Fig. 3.3.7(b) was taken at a farther underfocus side than that shown in Fig. 3.3.7(a). Therefore, it is reasonable to pick the defocus value −35 nm (Δf_1) as the correct solution for Figs. 3.3.7(a) and 3.3.8(a) and −64 nm (Δf_2) for Figs. 3.3.7(b) and 3.3.8(b). The former corresponds to the left smooth peak shown in Fig. 3.3.9(a) and the latter corresponds to the right smooth peak shown in Fig. 3.3.9(b). The corresponding deconvoluted images, namely, the obtained structure maps are shown in Figs. 3.3.8(c) and (d), respectively. It can be seen that Figs. 3.3.8(c) and (d) are in agreement with each other in contrast. The stacking sequence of different atomic layers in the projected structure is shown in Fig. 3.3.8(e). By comparing the two structure maps (Figs. 3.3.8(c) and (d)) with Fig. 3.3.8(e), it can be seen that all Bi atoms located in the double Bi-O layers appear as individual black dots in Figs. 3.3.8(c) and (d), but the three atomic layers, a Co-O layer sandwiched in between two Sr(La)-O layers to form the perovskite-like block (PLB), appear in Fig. 3.3.8(c) as two wavy rows of black dots instead of three. Though every PLB appears as three wavy black dot rows in Fig. 3.3.8(d), the arrangement of small black dots is not in agreement with that in the Bi-Sr-Cu-O superconducting oxides. This is due to the insufficient resolution of structure maps. Such a phenomenon as also observed in many experimental images of Bi-based high-temperature superconductors of different composition taken with medium-voltage (up to 400 kV) microscopes. In view of crystallography, the structures revealed in the obtained structure maps are of low resolution comparing with the diffraction resolution limit, although it comes from the high-resolution electron microscope image.

To improve the resolution of structure maps, phase extension was carried out by means of the VEC program (Wan et al., 2003) based on the direct methods. The initial set of structure-factor phases is obtained from the deconvoluted image given in Fig. 3.3.8(c) and the amplitudes are obtained from the electron diffraction data after correction by means of the method introduced in subsection 3.3.4. Phases of 204 reflections are obtained from the FT of the deconvoluted image, among which 55 reflections are independent of symmetry. There are altogether 586 reflections in the observed EDP within the resolution up to 0.1 nm, among which 156 are independent of symmetry. Fig. 3.3.10 shows the final structure map obtained from Fig. 3.3.8(c) after five cycles of phase extension in combination with diffraction-intensity correction and subsequently the Fourier synthesis. It is a high-resolution structure map, because first, as it is discussed in subsection 3.3.4, the diffraction intensity is pressed on toward the kinematical value in the process of diffraction-intensity correction, and secondly, the Fourier synthesis modifies the peak heights

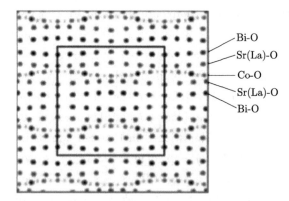

Fig.3.3.10 Final structure map obtained from Fig. 3.3.8(c) after five cycles of phase extension in combination with diffraction-intensity correction and subsequently the Fourier synthesis.

toward the true potential values. All atoms including the oxygen atoms lying in the Co-O layers are resolved individually, and both the displacement and occupation modulation of different atoms are clearly seen.

1D Incommensurate modulated structure of Pb-1223

Fundamental four-dimensional (4D) structural information. The crystal structure of $(Pb_{0.5}Sr_{0.3}Cu_{0.2})Sr_2(Ca_{0.6}Sr_{0.4})_2Cu_3O_y$ (hereafter named as Pb-1223) is one-dimensional (1D) incommensurate modulated with the basic structure isomorphic to that of $HgBa_2Ca_2Cu_3O_y$. Fig. 3.3.11(a) shows the [010] EDP where reflections are indexed with four indexes $hklm$. For main reflections $m=0$, while for satellites $m \neq 0$. The [010] high-resolution electron microscope image (Fig. 3.3.11(b)) was taken with a JEM-4000EX electron microscope operated at 400 kV. The spherical aberration coefficient of the objective lens is 1.0 mm, focus spread due to the chromatic aberration about 10 nm and point resolution 0.17 nm. The basic structure of Pb-1223 is tetragonal with the space group $P4/mmm$ and lattice parameters $a = b = 0.382$ nm, and $c = 1.53$ nm. The modulation vector $\boldsymbol{q}$ determined from the [010] EDP is:

$$\boldsymbol{q} = 0.322\boldsymbol{a}^* + 0.5\boldsymbol{c}^* \tag{3.3.12}$$

The 4D reciprocal lattice vector $\boldsymbol{H}$ is defined by four indices h, k, l, m:

$$\boldsymbol{H} = h\boldsymbol{a}^* + k\boldsymbol{b}^* + l\boldsymbol{c}^* + m\boldsymbol{q} \tag{3.3.13}$$

According to the high-dimensional space description of IMS given by De Wolff (1974) the 1D IMS can be treated as a three-dimensional (3D) hypersection obtained by cutting a 4D periodic structure with the physical space. The projection of the 4D periodic structure onto the 3D physical space yields the average structure of the IMS. The average structure corresponds to the main reflection of IMS and hence has the same unit cell as the basic structure, but it is orthorhombic and the space group is $Pmmm$. The plane group of the corresponding [010] projected structure is $P2mm$. Following the description of reflection conditions for the 24 Bravais classes given by De Wolff et al. (1981) and Janssen et al. (2006), and according to the extinction rule determined from the EDP, the superspace group of 4D periodic structure corresponding to Pb-1223 was determined to be $Pmmm(v01/2)$.

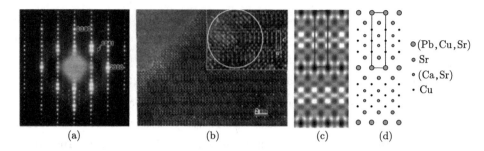

Fig.3.3.11 (a) [010] EDP of Pb-1223 and (b) the corresponding image with the magnified micrograph of a thin area inset on the top right. The circular area was selected for image processing. (c) Deconvoluted image with the defocus value −34 nm and (d) structure model of metallic atoms constructed based on (c). The rectangle indicates the unit cell of the basic structure.

Image deconvolution and low-resolution average structure. First, the FT was performed on the framed circular area in Fig. 3.3.11(b) together with the noise filtering. Then, the image was averaged in the Fourier space according to the plane group $P2mm$. Phases of all reflections were forced to be 0 or π. Generally, the unit cell origin is determined by minimizing the total phase residual. However, in the present case, the correct unit-cell origin corresponds to a small total phase residual, but not the minimum one because of the incomplete image symmetry. After ignoring the very weak $h0l0$ reflections and reflections corresponding to a big phase residual, amplitudes and phases of 17 independent main reflections were picked out for performing the inverse FT to obtain the [010] average image of Pb-1223. The defocus value of the average image was searched by using the program VEC. The trial defocus values were assigned from −100 to 0 nm with interval 1 nm. A set of trail structure factors were obtained for every trial defocus from eq. 3.3.5 with the FT of image $i(\boldsymbol{H})$ replaced by the corresponding electron-diffraction data. Then, the reasonable set of the structure factors was selected by the combined figure of merit (CFOM) used in direct methods (Debaerdemaeker et al., 1985). The defocus value −34 nm corresponding to the maximum value of CFOM, is used to perform the image deconvolution. The deconvoluted image shown in Fig. 3.3.11(c) represents the average structure of Pb-1223, in which all metallic atoms are clearly resolved as black dots, but the image resolution is insufficient to resolve all oxygen atoms. The small black dots between every two (Pb, Sr, Cu) atoms represent oxygen atoms. Fig. 3.3.11(d) is the low-resolution average structure model consisting of metallic atoms only. The FT of the average structure image yields low-resolution average structure factors including amplitudes and phases of 17 independent main reflections with a resolution up to 0.17 nm. These phases would serve as the initial phase information for the phase extension.

Two ways are available to obtain the IMS structure by phase extension starting from the above-mentioned average structure factors. One is to derive the phases of all reflections simultaneously including main reflections and satellites. The other is in two steps: in the first step the phases of main reflections are derived, and in the second step phases of satellites are derived with the phases of main reflections derived in the first step joined in the initial data. In the following, the two steps procedure will be introduced.

Phase extension and high-resolution average structure. The procedure of phase extension based on the multidimensional direct methods (Hao et al., 1987; Fu and Fan, 1994) to derive the phases of main reflections up to 0.1 nm in combination with the diffraction intensity correction is as follows. First, a set of partial structure factors including only the contribution of metallic atoms Pb, Sr, Ca, and Cu was calculated up to 0.1 nm based on the average structure model (Fig. 3.3.11(d)) obtained from the average structure image (Fig. 3.3.11(c)). Although the weak dots between every two (Pb, Sr, Cu) atoms seen in Fig. 3.3.11(c) might represent oxygen atoms, for the reason of safety they were ignored in the partial structure-factor calculation. The scaling factor for correcting the diffraction intensity of each reflection is equal to the ratio of the average square partial structure factor to the observed diffraction intensity. Then, the square roots of corrected diffraction intensities are assigned to be the structure-factor amplitudes. The phase extension was carried out with the program DIMS, starting with the phases of 17 independent main reflections to derive phases of other main reflections up to 0.1 nm. The inverse FT yields a high-resolution average structure map shown in Fig. 3.3.12(a). It can be seen that weak dots located between every two (Pb, Sr, Cu) atoms appear again although they are splitting. Weak peaks also appear in Sr layers. These two kinds of weak dots can be interpreted as oxygen atoms. A new partial structure factor was calculated with all metallic atoms as well as these oxygen atoms. Then, the second cycle of the diffraction-intensity correction and phase extension were carried out. Oxygen atoms in the Cu layer appear in the new map shown in Fig. 3.3.12(b). The third cycle of diffraction intensity correction and phase extension did not lead to an obvious improvement (Fig. 3.3.12(c)). Three cycles of Fourier synthesis starting from the result of phase extension was employed for the structure refinement, and the final map of the average structure with resolution 0.1 nm is shown in Fig. 3.3.12(d).

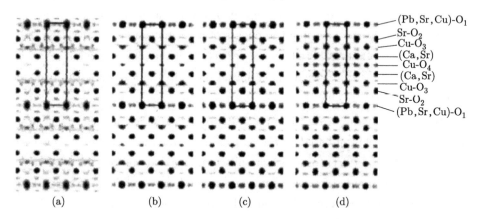

Fig.3.3.12 [010] high-resolution average structure maps of Pb-1223 obtained after phase extension in combination with diffraction intensity correction (a) once, (b) twice, and (c) three times, respectively, and (d) after three cycles of Fourier synthesis. Rectangles indicate unit cells of average structure.

Incommensurate modulated structure (IMS). The FT of the map given in Fig. 3.3.12(d) generates 54 independent $h0l$ reflections of the average structure that are identical to the $h0l0$ main reflections of the IMS with 0.1 nm resolution. The phases of the 54 main reflections now serve as the initial data together with the amplitudes obtained from the corrected diffraction intensities to perform the second step of phase extension based on the multidimensional direct methods. Phases of 88 $h0lm(m \neq 0)$ satellite reflections were derived. The FT of structure factors with the amplitudes from the corrected diffraction intensities and phases obtained from the phase extension yields the projection of 4D potential distribution function $\varphi(x_1, x_2, x_3, x_4)$ along the x_2-axis, $\varphi(x_1, 0, x_3, x_4)$, where x_1, x_2, x_3 and x_4 are fractional coordinates in the 4D unit cell. The hypersection of function $\varphi(x_1, 0, x_3, x_4)$ with the 3D real space yields the projected potential of the determined IMS along the b-axis, namely, the [010] projected structure in Pb-1223. Finally, three cycles of Fourier synthesis were calculated to refine the function $\varphi(x_1, 0, x_3, x_4)$, and the obtained [010] IMS map of resolution 0.1 nm is shown in Fig. 3.3.13. In this IMS map each black dot represents an atomic column. The blackness variation of dots representing (Pb, Sr, Cu) atoms indicates a strong occupation or substitution modulation. Because the [010] projected IMS in Pb-1223 is aperiodic, it cannot be shown overall but only partly in Fig. 3.3.13. The modulation character can be seen more clearly and perfectly in two hypersections $\varphi(x_1, 0, 0, x_4)$ and $\varphi(x_1, 0, 0.5, x_4)$ of the 4D structure shown in Figs. 3.3.14(a) and (b), respectively. The potential at the Pb site varies with x_4 indicates a strong compositional modulation in the Pb sites due to the replacement of Pb atoms by Sr or Cu atoms. The compositional modulation of Sr and Cu atoms is weaker than that of Pb, but the displacive modulation is seen. The modulation of oxygen atoms is also weak.

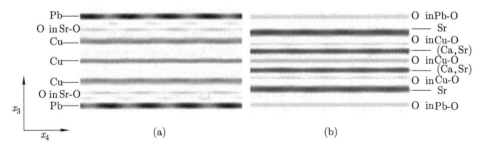

Fig. 3.3.13 Hypersections (a) $\varphi(x_1, 0, 0, x_4)$ and (b) $\varphi(x_1, 0, 0.5, x_4)$ of 4D structure maps corresponding to [010] IMS in Pb-1223.

2. Defect-structure determination

Dislocation core structure in $Si_{0.76}Ge_{0.24}/Si$

The $Si_{0.76}Ge_{0.24}$ epilayer was grown on a Si (001) substrate. The crystal structure of SiGe is isomorphic to that of Si. In the [110] projected structure of

$Si_{0.76}Ge_{0.24}$, the distance between two atoms in the pair, named a dumbbell, is about 0.14 nm. It was expected to distinguish atoms individually in the pair from images taken with a 200 kV field-emission high-resolution electron microscope by deconvolution processing. The [110] cross-sectional specimens were observed with a JEM–2010F electron microscope having the point resolution about 0.2 nm and information limit toward 0.14 nm. To restore SiGe atomic columns individually, the image should contain the structure information with the spatial frequency up to $(0.14 \text{ nm})^{-1}$. In this spatial-frequency region the reflections independent of symmetry are indexed 111, 222, 311, and 400. To avoid a considerable information loss, all these reflections need to appear in the image diffractogram. Two qualified images shown in Figs. 3.3.14(a) and 3.3.15(a), respectively were selected to restore the defect structures at the atomic level. All reflections are clearly seen in the corresponding diffractograms (see insets in Figs. 3.3.15(a) and 3.3.16(a)). This implies that the information loss in the two images due to the zero crosses of CTF is negligible.

Lomer dislocation. The frame area in Fig. 3.3.14(a) with a dislocation inside is magnified and is shown in Fig. 3.3.14(b). Two extra terminated {111} planes running from the top left to the bottom right and from the top right to the bottom left are indicated by two arrows, respectively. Since the image resolution is 0.2 nm and white dots do not represent atoms, the atomic configuration is not revealed intuitively in the image, although the dislocation can be recognized as the Lomer type. A composite diffractogram was obtained by Fourier transforming a thin image region near the defect and containing both the amorphous and crystalline images. It consists of diffuse rings that come from the Thon diffractogram (Thon, 1971) and sharp reflections from the SiGe crystal. The defocus value is roughly determined to be −39 nm by matching the intensity profile of the Thon diffractogram with several CTF curves calculated for different focus values at a small interval. Other parameters for CTF calculation are accelerating voltage 200 kV, spherical aberration coefficient 0.5 mm, and defocus spread due to the chromatic aberration 3.8 nm. The composite diffractogram together with the matched CTF curve is inset on the bottom left of Fig. 3.3.14(a).

The defocus value was refined and the reflection amplitude was corrected in

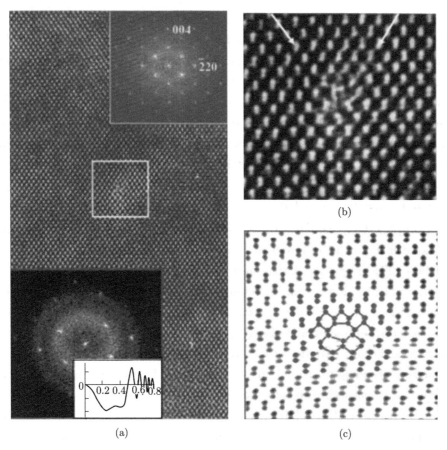

Fig.3.3.14 (a) [110] image of $Si_{0.76}Ge_{0.24}/Si$ with the diffractogram inserted on the top right and the composite diffractogram together with the matched CTF curve on the bottom left, (b) magnified image from the framed area in (a) with two arrows pointing out the terminated {111} planes and (c) deconvoluted image corresponding to (a). The structure model of a Lomer dislocation core is superimposed on (c).

the process of the deconvolution as described in Section 3.3.3. Fig. 3.3.14(c) shows the restored structure map where each black dot represents an atomic column projected in the [110] direction. The dislocation core structure model was obtained by linking the adjacent atoms near the center. A five-membered ring and a seven-membered ring form the symmetric undissociated core without dangling bonds.

This is in agreement with the Hornstra model of Lomer dislocations. The reliability of the obtained atomic configuration for the Lomer dislocation was verified by image simulation using the multislice method, and the matched defocus and sample thickness are −40 and 6.14 nm, respectively. Several models have been proposed for the Lomer dislocation. Bourret et al. (1982) proposed two symmetric and two asymmetric core models in Si and Ge. McGibbon et al. (1995) claimed to have observed the Hornstra-like and an unexpected core structure in polar compounds like CdTe/GaAs. The present result indicates that in the case of SiGe/Si, the Lomer dislocation with Hornstra structure can be formed in the region of epilayers close to the interface boundary.

Complex 60° dislocation. Another qualified best image taken with SiGe is shown in Fig. 3.3.15(a), and the framed area is magnified and shown in Fig. 3.3.15(b). The arrows and dotted line segments indicate the two terminated {111} planes ending inside the framed rectangle. The same methods as in the case of a Lomer dislocation were utilized to extract the structure map from Fig. 3.3.15(b). The composite diffractogram consisting of diffuse rings comes from the Thon diffractogram and sharp reflections from SiGe crystal together with the matched CTF curve corresponding to the defocus value −31.5 nm is given in Fig. 3.3.15(c). The deconvoluted image corresponding to Fig. 3.3.15(b) is shown in Fig. 3.3.15(d), and the framed area in Fig. 3.3.15(d) is magnified and shown in Fig. 3.3.15(e). The detailed geometry of the dislocation complex can be seen more clearly by linking the adjacent atoms to show the bonding situation in the magnified picture. It is seen that the dislocation complex is composed of a perfect 60° dislocation (in the left) and an extended 60° dislocation (in the right). The two dislocations lie at different glide planes. The extended 60° dislocation dissociates into a 90° partial dislocation and a 30° partial dislocation with a short stacking fault (SF) (only three lattice spacings) between them.

The above results indicate that the deconvolution processing in combination with the reflection-amplitude correction is effective to transform a single image that does not represent the defect structure intuitively into the structure map with the resolution improved from 0.2 up to 0.14 nm so that the (Si, Ge) atoms are resolved individually in the obtained structure map, when images are taken with a

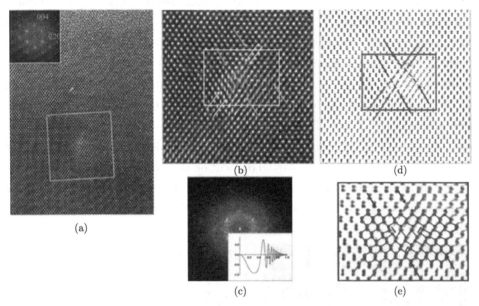

Fig.3.3.15 (a) [110] image of $Si_{0.76}Ge_{0.24}/Si$ with the diffractogram inserted on the top left, (b) magnified image of the framed area in (a), (c) the composite diffractogram together with the matched CTF curve, (d) deconvoluted image corresponding to (b) and (e) magnified picture corresponding to the framed rectangular region in (d). A perfect 60° dislocation is located in the left and an extended 60° dislocation dissociated into two partials, 30° and 90°, in the right in (e). The big arrows in (b), (d), and (e) point out the terminated {111} planes, and small arrows in (e) denote the Burgers vectors.

Structure of dislocation core, microtwin and interfacial boundary in 3C-SiC/Si

30° partial dislocation core structure. The micrograph shown in Fig. 3.3.16(a) is the magnified picture of region R2 in Fig. 3.3.5(a) (Tang et al., 2007). A 30° partial dislocation (the terminated plane is pointed by an arrow and its end is marked with a circle) associated with a SF can be seen. Horizontal streaks caused by the SF are seen clearly on both sides of all reflections included in the diffractogram (see the inset on the top left of Fig. 3.3.16(a)). The deconvoluted image for Fig.

3.3.16(a) with $\Delta f = -47$ nm is shown in Fig. 3.3.16(b). The strong streaks seen in the diffractogram inserted in Fig. 3.3.16(b) indicates that most of the structural information contributed by the SF is retained. This is because the elliptical windows designed especially for retaining the planar defect information (Tang and Li, 2005) were utilized in the Fourier filtering. In Fig. 3.3.16(b) all atoms appear as black and the dumbbells are recognized, though the two atoms of distance 0.109 nm in the dumbbells are not resolved clearly because of the insufficient resolution of the microscope (0.2 nm). The region framed by the dotted rectangle is magnified and shown in Fig. 3.3.16(c) where the structure model is superimposed on the deconvoluted image. For a 30° partial dislocation the inserted plane may terminate either in atom Si or atom C (Blumenau, 2003). To identify the terminal atom pointed by arrows in Figs. 3.3.16(b) and (c), the argument of image-contrast

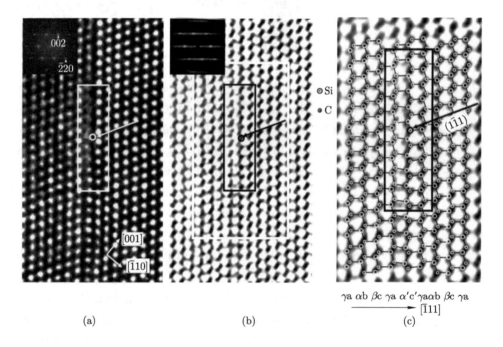

Fig.3.3.16 (a) [110] image of 3C-SiC, (b) deconvoluted image corresponding to (a) and (c) picture magnified from area framed by the dotted rectangle in (b) with the structure model of 30° partial dislocation superimposed.

variation with crystal thickness given in subsection 3.3.5 was referred. Since region R_2 is close to the lower part of region R_1 in Fig. 3.3.5(a), its thickness should not be smaller than that of the right part in Fig. 3.3.5(b). Hence, the contrast profile of segment III shown in Fig. 3.3.5(d) can be applied to Figs. 3.3.16(b) and (c). Thus, the right contrast profile shown in Fig. 3.3.5(e) can be used for Fig. 3.3.16(c) to make out that the gray level for Si is lower than that for C. Thus, the smaller ends in the dumbbells correspond to atomic columns of Si, while the bigger ends correspond to the atomic columns of C. Therefore, all Si and C atoms can be identified in the deconvoluted image shown in Fig. 3.3.16(c) so that it is clear that the partial dislocation terminates in the C atomic column. The atomic configuration of the 30° partial dislocation core and its associated SF was then derived, and the corresponding model is superimposed on the deconvoluted image (Fig. 3.3.16(c)). The stacking sequence is γaαbβcγaα'c'-γaαbβcγa, where αβγ (α') represent atomic layers of Si and abc(c') represent atomic layers of C.

Structures of microtwin and twin boundaries. In 3C-SiC, the {111} twins could be either a 180° rotation twin or a mirror reflection twin (Holt 1964) with the corresponding stacking sequence αaβbγcβbαa... or,... αaβbγγbβaα..., respectively. The micrograph given in Fig. 3.3.17(a) is the magnified picture of region R3 in Fig. 3.3.5(a). It contains a segment of {111} microtwin (marked by two arrows). The corresponding diffractogram is inserted on the top right. Fig. 3.3.17(b) shows the deconvoluted image corresponding to Fig. 3.3.17(a) with $\Delta f = -46.5$ nm. Again, atoms appear black and the two atoms in the dumbbells are not resolved clearly. Since this region is near the thin amorphous area, the thickness should be comparable with the left part in Fig. 3.3.5(b). Hence, the contrast profile of the segment I shown in Fig. 3.3.5(d) can be applied to Fig. 3.3.17(b). Thus, it is reasonable to treat the bigger ends of the dumbbells as atomic columns of Si. The structure model of the microtwin segment together with its boundaries constructed on the basis of the above argument is superimposed on the deconvoluted image (Fig. 3.3.17(b)). It is seen that the stacking sequence is ... αaβbγcααγcβbααγ cβbγcααβb..., and the microtwin segments is sandwiched between two 180° rotation twins. Fig. 3.3.17(c) shows the image simulated according to the model given in Fig. 3.3.17(b) with $\Delta f = -46.5$ nm and crystal thickness of 3.08 nm. The simulated image agrees well with the experimental image.

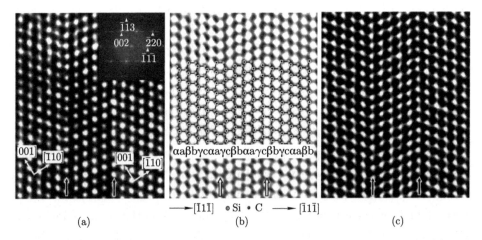

Fig.3.3.17 (a) [110] image of 3C-SiC, (b) deconvoluted image corresponding to (a) with the microtwin structure model superimposed and (c) image simulated based on the model given in (b).

It is seen that even in the case of 200 kV LaB$_6$ microscopes, the Si and C atoms of distance as small as 0.109 nm in 3C-SiC can be recognized in the restored structure map, though they cannot be resolved clearly. Hence, the atomic configurations of dislocation core, microtwin, and twin boundary in 3C-SiC were determined quite uniquely.

3. Discussion

The examples given above show that the structure maps, or the structure and defect models of examined crystals, were derived from the experimental images via deconvolution processing rather than proposed or guessed based on the experimental image contrast. In the case of structure determination, the structure map was improved in both the resolution and contrast by carrying out the phase extension in combination with the diffraction data correction and subsequently the Fourier synthesis. In this stage the accuracy of both atomic positions and contrast in the structure map were constrained to improve toward to the true structure. For defect investigation the derived defect model was improved originally in the process of deconvolution with the reflection-amplitude correction. To confirm the final structure map one may match the experimental images with the simulated ones, but principally, there is no need to do so.

The images shown above were recorded with the electron-sensitive films, for which the linear relationship between the blackness and number of incident electrons is far from perfect. In addition, the image contrast varies with the developing time and temperature. After digitizing the films or micrographs with a scanner, the measured image contrast is modulated further. Principally, all the modulations of image contrast are involved in the deconvolution process and have influence on the derived structure map, mainly in the contrast. However, the influence can mostly be removed with the methods subsidiary to the deconvolution processing, and to the stage of phase extension and structure refinement mentioned above. The electron-diffraction intensities were recorded with multiple exposures and subsequently transformed into a set of data with an approximately linear relationship. Though the data sets are available for the present task, it is recommended to record the diffraction intensity with image plates to simplify the experiments and data treatments. To improve the resolution of the structure map and the accuracy of atomic positions and recognizing atomic sorts, the most important factors are the microscope information limit and then the methods of image analysis rather than the recording materials.

More than a decade ago it was found that the experimental image contrast was generally less than the simulated one by a factor three to five, namely, the so-called Stobbs factor (Hÿtch and Stobbs, 1994). This phenomenon was also found in energy-filtered images of thin and thicker crystalline and amorphous samples over a wide range of spatial frequency, and a number of studies have been focused on thermal diffuse scattering (Boothroyd, 1995; Boothroyd, 1998; Boothroyd, 2000; Meyer et al., 2000; Howie, 2004). The Stobbs factor is no doubt also involved in the deconvolution process. Again, its influence should mostly be removed with the image-analysis methods subsidiary to the deconvolution processing and others.

The developed analysis method should also be beneficial to images taken with spherical-aberration-corrected microscopes that provide a point resolution and an information limit higher than the conventional microscope. Since the portion of phase contrast at the Scherzer-focus condition goes down for a very small spherical aberration coefficient, it is beneficial to take the image deviated from the Scherzer focus and then restore it into a structure map by deconvolution processing. All in all, it is expected to derive the structure map of resolution 0.1 nm or better by means of deconvolution processing only, either for perfect or defected crystals. This would make the task much simpler and easier than before because the phase extension might be ignored. For instance, to determine a structure or an IMS one can derive the high-resolution structure directly by means of deconvolution processing and then refine the structure with Fourier synthesis.

In principle, the electron crystallographic image processing methods can also be applied to the defect study of a crystalline grain fully embedded in the heterocrystalline matrix, i.e., when the strain field is inhomogeneous along the electron-beam path. It would be possible to separate the image diffractogram of the grain from that of the matrix, though maybe not perfectly. However, the reflections in the diffractogram for a grain with very small lateral size are diffuse and blurred, leading to a low resolution of the corresponding deconvoluted image. For the same grain size a smaller lattice parameter corresponds to a bigger numbers of crystalline periods and hence a higher resolution, and vice versa. Besides, in order to get an experimental image of good quality, the sample, including the embedded grain and matrix, must be very thin along the incident beam direction, usually a few nanometers. Obviously, the portion of the grain must be big enough compared with the matrix. Therefore, the success of the study also depends on the sample preparation.

3.3.7 Conclusion

High-resolution electron microscope images may not represent the examined structures intuitively. Fortunately, in most cases such images can be restored to show the projected structures of examined crystals. In this section, a group of research studies including the image-contrast theory, image-analysis methods and their applications is reviewed to show a peculiar new development in HRTEM.

The derived image-intensity formula shows how the image contrast changes with the sample thickness for atoms with different atomic weights. This links up the WPOA and multislice theory, and results in the image-contrast theory, PW-POA. The theory plays an important role in developing image-analysis methods in HRTEM. Apart from pointing out the feasibility of preferentially observing atoms of different atomic weight and providing a method to recognize them, it has laid a theoretical base for developing new image-analysis methods. First, it indicates that the image-intensity formula for WPOA remains valid if the potential were replaced by pseudo potential when the crystal thickness for all constituent atoms is below a critical value. This affirms the use of deconvolution processing as a tool to restore the structure maps of examined crystals, either perfect or defect. Accordingly, it becomes reasonable to introduce some structure analysis methods developed in X-ray crystallography into HRTEM, to utilize them directly or with some modification.

Thus, a group of special image-analysis methods has been developed to form two sets of electron crystallographic image processing and analysis methods. One of them is for the ab initio crystal-structure determination including the determination of IMS, and the other for crystal-defect studies at the atomic level.

In the former case the deconvolution method based on the principle of maximum entropy in information theory is a kind of constrained maximum-entropy method accomplishing the constraint via selecting a best defocus. It is a reliable and powerful tool to transform a single image into the best structure map via removing the combination effect of spherical aberration and defocus, and reducing the dynamical scattering effect. In addition, the resolution of structure map can be enhanced up to 0.1 nm or better if the phase extension is performed subsequently in combination with the electron-diffraction-data correction.

In the case of crystal-defect studies the method of reflection-amplitude correction is effective to reduce the dynamical effect and improve the quality of the derived structure map. In principle, the resolution of the structure map can be enhanced up to the microscope information limit (to 0.14 nm for 200 kV FEG microscopes). Practically, one can get some structure details of resolution much better than the information limit. For instance, the Si and C atoms of distance as small as 0.109 nm in 3C-SiC can be recognized in the restored structure map derived from the image taken with a 200 kV LaB_6 microscope. This enables the atomic configurations of dislocation core, microtwin, and twin boundary in 3C-SiC to be determined uniquely.

All the image-analysis methods are available for images taken with different kinds of high-resolution transmission electron microscopes, including spherical-aberration-corrected ones.

3.4 Focal series reconstruction

Wei Liu and Rongming Wang

3.4.1 Basic principles of HRTEM imaging

From the introduction of former session concerning on the high resolution imaging theory of weak phase object, we learned that the defects in the electron optic system, including spherical aberration (C_s) of the objective lens, chromatic aberration (C_c) as well as the imaging conditions, such as defocus value (Δf), beam convergence angle (α), give out the origin of the phase contrast in single image through a form of contrast transfer function. Among these parameters, the spherical aberration and chromatic aberration are determined inherently by the microscope (either the objective lens or the electron gun type). Then there exist two major parameters that can be adjusted and performed in different ways in order to improve the spatial resolution of the recorded images further. As an estimation of the microscope ability on synchronously focusing electrons of different energy, chromatic aberration originates from the monochromaticity of the electron emission gun, and thus determines the information limit, by compressing high frequencies through introducing a damping envelope (temporal envelope) into the contrast transfer function (CTF). Noting that this information limit corresponds to the smallest object detail that can be resolved by a particular microscope. While, for the spherical aberration, it is caused by the objective lens together with defocusing effects, a single high resolution image represents a highly encoded mixture of the properties of the object with those belonging to the observation instrument.

To push the point resolution approaching to the information limit, the most direct way is to implement aberration correction to the objective lens of the microscope (Haider et al., 1998, Haider et al., 1998). However, this is an expensive way for most users. To achieve the highest potential resolution of a FEG HRTEM instrument, another effective way is to employ the focal-series exit-wave function reconstruction or shorted for focal series reconstruction (FSR) technique, which will be discussed after the introduction of HRTEM imaging theory.

According to the high resolution TEM imaging theory, the weak phase object approximation (WPOA) works for a very thin object ($i\sigma V_p(r) \ll 1$)) with a weak scattering ability of electrons. Then gives an incident electron plane wave form, and the exit wave function $\Psi_e(K)$ at the lower surface of a thin sample is described as Eq. (3.4.1), where σ is the interaction coefficient between atom and electrons, $V_p(r)$ is the projection potential of the sample, t is the thickness.

$$\Psi_e(r) \approx 1 + i\sigma V_p(r)t \tag{3.4.1}$$

Then on the image plane, the wave function can be described as the convolution of the exit wave function Ψ and the transfer function T of the microscope as shown in Eq. (3.4.2):

$$\Psi_i(r) = \Psi_e(r) \otimes T(r) \tag{3.4.2}$$

Then, we can obtain the visible intensity description of the HRTEM image by calculating the square of modulus (Eqs. (3.4.3) and (3.4.4)).

$$I(r) = |\Psi_i(r)|^2 = \Psi_i(r)\Psi_i^*(r) \tag{3.4.3}$$

$$I(r) \approx 1 + 2\text{Re}\{i\sigma t V_p(r) \otimes T(r)\} + |i\sigma t V_p(r) \otimes T(r)|^2 \tag{3.4.4}$$

In linear imaging theory, the object is assumed as a weak scatter, thus the quadratic term in Eq. (3.3.4) is small and can be neglected, so that we obtain a simplified form as Eq. (3.4.5).

$$I(r) \approx 1 + 2\text{Re}\{i\sigma t V_p(r) \otimes T(r)\} \tag{3.4.5}$$

Since V_p is a real quantity, $i\sigma t V_p$ is imaginary and no contrast would be visible without the transfer function of the microscope (Eq. (3.4.5)). The lens aberrations result in a transfer of some imaginary part information at the exit plane into the real part at the image plane, which can be imaged by HRTEM. Besides, to discuss the influence of aberrations in the HRTEM image formation process, it is more convenient to work in Fourier space, where the real-space quantities $I(r)$, $\Psi(r)$, $V_p(r)$, and $T(r)$ are related to their counterparts $I(q)$, $\Psi(q)$, $V_p(q)$, and $T(q)$ by a Fourier transformation. Distances r in direct space correspond to spatial frequencies q in Fourier space. Then, we can transform the imaging descriptions given out above into the following ones. (Eqs. (3.4.6)–(3.4.10)).

$$\Psi_e(q) \approx \delta + i\sigma V_p(q) \tag{3.4.6}$$

$$\Psi_i(q) = \Psi_e(q) T(q) \tag{3.4.7}$$

As introduced in the former session, the transfer function of a microscope consist of four parts, the aperture function $A(q)$, the objective lens aberration term $\chi(q)$, the spatial incoherence envelope $E_s(q)$, and the temporal incoherence envelope $E_t(q)$. Focusing on basic principles of linear imaging theory and within the frequency range of retrievable information details through reconstruction, we first neglect the impact of the spatial and temporal envelop terms for simplicity since their compression on the high frequencies of this range is not so remarkable, and meanwhile, set the aperture function to 1 thus making sure all frequencies contribute to image forming. Therefore, the transfer function $T(q)$ is in form of Eq. (3.4.8):

$$T(q) = \exp[-i\chi(q)] \tag{3.4.8}$$

In the WPOA and only considering the linear contributions in coherent imaging, the observed intensity I can be simplified by combining Eq. (3.4.4) and (3.4.6)–(3.4.8) as:

$$I(q) \approx 2\sigma t V_p(q) \sin[\chi(q)] \tag{3.4.9}$$

here, the phase shift based on a weak phase object comes from the aberration of the objective lens (C_s) and the defocus condition (Δf), meanwhile depends on the spatial frequency (q). Without the phase shift χ due to the lens aberrations, a weak phase object would not be visible in HRTEM. Therefore, the transfer function of objective lens $\sin[\chi(q)]$ is also known as the contrast transfer function (CTF)

$$\chi(q) = 2\pi \left[\frac{C_s \lambda^3 q^4}{4} + \frac{\Delta f \lambda q^2}{2} \right] \tag{3.4.10}$$

According to principles of imaging theory of a HRTEM image, under a Scherzer defocus $-1.2\sqrt{C_s\lambda}$ (also called as extended Scherzer focus), the "normally" transferred frequency range, where atomic positions might appear dark (CTF=−1), is maximized, the highest frequency at the edge corresponds to the smallest interatomic distance that can be resolved in a single HRTEM image. In this condition, for a sample satisfying WPOA with potential function $\phi(\boldsymbol{r}) = tV_p(\boldsymbol{r})$, the image intensity formula comes to Eq. (3.3.1). For low spatial frequencies, no noticeable differences are transferred into the image contrast; for high frequencies, where the CTF varies rapidly with defocus, the variations will start to cancel each other out, thereby limiting the information that is transferred into the image. Only spatial frequencies from a weak-phase object locate in the proper range (CTF=−1) is intuitively interpretable. However, frequencies beyond this point up to the information limit are also available through utilizing special microscopy processing on the basis of known imaging parameters.

3.4.2 Focal series reconstruction

Although the amplitude and phase information of the exit wave function cannot be retrieved from a single HRTEM image, they can be obtained by combining several HRTEM images recorded at different defocus settings (Rose, 1990; Rose and Preikszas, 1992; Haider et al., 1995). On this motivation, the focal series reconstruction (FSR) technique was developed to retrieve the aberration-free exit wave function involving the amplitude/phase information. Upon several years' development (Coene et al., 1992; Thust et al., 1994; Thust et al., 1996), FSR has become one of the most important techniques to extend the microscope resolution from its point resolution up to the information limit. In this architecture, the information contained in a complete focal series of HRTEM images is combined in order to retrieve the exit-plane wave function. This procedure makes it possible to eliminate the well-known imaging artifacts which complicate the interpretation of single images. Due to spherical aberration and defocusing effects caused by the objective lens, a single high-resolution image represents a highly encoded mixture of the properties of the object with those belonging to the observation instrument. In contrast to a single image, the retrieved exit-plane wave function is ideally free from any imaging artifacts and is determined exclusively by the object itself and the wavelength of the incident electron beam.

In last twenty years, field-emission electron gun provides a highly coherent electron beam with high brightness, which makes it possible to extend the range of transferred spatial frequencies beyond that obtained with a conventional thermionic source (Humphreys and Spence, 1981; Otten et al., 1992). Moreover, as quantitative image acquisition devices, slow-scan CCD cameras are superior to photographic film plates (Daberkow et al., 1991), completing the experimental "environment" of the focal-series reconstruction.

1. The algorithms

The idea for focal series reconstruction was originally developed by Schiske (2002) and Kirkland (1984) who realized that although amplitude and phase cannot be retrieved from a single HRTEM image, they can be obtained by combining several HRTEM images recorded at different defocus settings. This idea has become the basis for various focal series reconstruction schemes, one of which is the PAM-MAL method developed by Coene, Thust and coworkers (1996). It is implemented in TrueImage, a software package widely adapted for focal series reconstruction in HRTEM. The method has been further developed by Thust (1996) and allows other improvements of the determination and correction of residual aberrations. These improvements are also implemented in the TrueImage software package.

Another representative method concerned on exit wave reconstruction and resolution extension in both real and reciprocal space are developed in recent years (Hsieh et al., 2004). In their mind the electron direct method in real space involves non-interferometric phase (retrieval by a transport of intensity equation/maximum entropy method (TIE/MEM)) to retrieve phase information from each HRTEM image of the image series, and then a self-consistent propagation method is adapted for exit wave reconstruction. In the reciprocal space, the solution method is to using a complex MEM and the Gerchberg–Saxton algorithm to extend the structural information in the exit wave beyond the information limit determined by the microscope and by wave propagation inside the crystal. Both methods they proposed are dealing with a complex signal related to the projected crystal structure. A resolution better than 1.1 Å (the information limit of the microscope as simulated) can be achieved using these electron direct methods. In such a MEM/TIE approach the focal step between the experimental image series is not necessarily equal across the entire image series, but it strongly relies on precise pre-determination of focus levels and lens aberrations inputed in order to achieve a better convergence.

Taking PAM-MAL method as an example, we will give an overall demonstration on the primary algorithms as well as some practical application issues. This algorithm mainly utilized in TrueImage was developed by Coene et al. (2001) in the project of BRITE-EURAM. Within this framework, two different types of focal series reconstruction algorithms have been optimized and have achieved the application level: the so-called "paraboloid method" (PAM) (Vandyck et al., 1993; Saxton 1994) and the "maximum likelihood" method (MAL) (Kirkland et al., 1985). It is well known that the linear contribution of an object frequency q to the image contrast repeats periodically with defocus Δf varying. The focal repetition period ΔZ, often called "Fourier period", amounts to $2/\lambda q^2$, where λ is the electron wave length. Let the intensity of an image taken at a defocus Δf be represented by $I(r, \Delta f)$. The application of a three-dimensional Fourier transform (with respect to $x, y, \Delta f$) to a focal image series reveals that the linear contributions to the transformed intensity $I(q, \xi)$ are located on two paraboloids satisfying $\xi = \pm \lambda q^2/2$. Here, ξ is the conjugate coordinate to the defocus Δf. This result is a reformulation of the above-stated focal repetition behavior.

In practice, the linear contributions is achieved by a back-propagation of the linear contrast contributions lying on one of the two dimensional paraboloids onto a common reference plane (as the exit plane where $\Delta f = 0$). The back-propagation is performed by means of complex filter functions F_n (de Beeck et al., 1996). A focal series of N HRTEM images are represented by intensity distributions $I_n(q)$. Behind the sample the exit plane wave function $\Psi(q)$ is splited into its "dc" component c_o and its diffracted part $\Phi(q)$ with $\Psi(q) = c_o \delta(q) + \Phi(q)$, it can be retrieved when $q \neq 0$ as

$$c_0^* \phi(\boldsymbol{q}) = \sum_{n=1}^{N} F_n(\boldsymbol{q}) I_n(\boldsymbol{q}) \qquad (3.4.11)$$

After application of Eq. (3.4.11) one can use the obtained wave function $\Phi(q)$

as a first guess $\Phi^1(q)$ and calculate its estimated non-linear intensity contributions $I_{n,j}^{N.L.}(q)$ to every single image. By subtraction of the intensity contributions from the initial intensity I_n, Eq. (3.4.11) can be reformulated in a recursive way as Eq. (3.4.12),

$$c_{j+1}^* \phi^{j+1}(q) = \sum_{n=1}^{N} F_n(q)[I_n(q) - \gamma_j I_{n,j}^{N.L.}(q)] \qquad (3.4.12)$$

where γ_j is a feedback parameter which avoids resonances due to overestimated nonlinear terms and it should be optimized by trying test of values that yield smallest misfit between the input images with intensity I_n and the resulting non-linearly estimated images with intensity $I_{n,j+1}$, and then used for the recursive scheme.

Compared to the PAM formalism, the MAL method follows an almost complementary approach with respect to the treatment of non-linear contrast contributions. Whereas the PAM formalism aims to eliminate non-linear contrast contributions by means of an intrinsic averaging process, the MAL formalism makes explicit use of these nonlinear terms. Hence, not only information on the two-dimensional paraboloid surfaces but also information contained in the full three-dimensional space $(q_x, q_y, \Delta f)$ is exploited in order to retrieve the exit-plane wave function. On principle, algorithms following the maximum likelihood concept are designed to minimize an error functional S^2 with

$$S^2 = \frac{1}{N} \sum_{n=1}^{N} \int dq |\delta I_n(q)|^2 \qquad (3.4.13)$$

where N is the total number of images within a focal series and n refers to the defocus Δf_n of a particular image. The function δI_n denotes the difference between the experimentally observed intensity $I_{n,E}$ and the image intensity $I_{n,\psi}$ calculated from an estimated wave function ψ with

$$\delta I_n(q) = I_{n,E}(q) - I_{n,\psi}(q) \qquad (3.4.14)$$

The functional S^2 is the averaged squared difference between the experimental image intensity and the intensity calculated from an estimated wave. The averaging extends over complete Fourier space and over all images of a focal series.

2. Implementation of FSR in practices

A typical FSR project mainly divides into two procedures, the images series acquisition procedure where 10~20 high resolution TEM images of proper focus difference are obtained and the exit-wave reconstruction procedure which removes most aberrations introduced by the optical system of the microscope and pushes the resolution up to the information limit resulting in a more intuitively interpretable image. Issues of these two procedures will be introduced in the following text.

Images series acquisition

To perform a successful exit wave reconstruction and retrieve an aberration-free amplitude/phase image with the resolution extended to the information limit, 10–20 high-resolution TEM images with a fixed defocus difference are automatically collected by either the operation system of the microscope (for most types of FEI microscopes) or the third-party commercial software package, such as the well-known TrueImage. Based on the focal series, the reconstruction is performed off-line and yields the exit wave function of the specimen.

To achieve a successful reconstruction, the quality of HRTEM images series, which performed as data source of the whole reconstruction procedure, should be guaranteed first under the optimized acquisition condition. Since a certain FSR project is on the basis of a series of HRTEM images acquired on a particular microscope, the microscope parameters as well as the imaging parameters of the image series, including the start defocus, focus step, convergence angle, CCD sampling (pixel resolution of CCD) et al., will originally determine the final quality of FSR result. In the following session, the optimization of these parameters will be discussed in a practical point of view.

Start defocus

Although a Scherzer defocus ensures the resolution of the highest frequency (smallest distance) in a single HRTEM image, however, the higher frequencies beyond the direct interpretable ones (CTF=−1) are seriously compressed and result in little linear component can be transferred into the final individual HRTEM image. This is undoubtedly harmful for the following reconstruction procedure if images series were acquired in this range. To estimate an optimized defocus range suitable for focal series reconstruction, one should try to find out a bandwidth range, where spatial frequencies nearly oscillate in a manner of equal amplitude so as to make a equal contribution to the final image contrast. According to scientists' research (Lichte, 1991), in such an optimized defocus, spatial frequencies interested are modulated through CTF and generate a smallest modulus of the wave aberration gradient $d\chi/dq$. Since this gradient is a monotone function of spatial frequency q, then to minimize the modulus, the optimized defocus value is given by

$$(d\chi/dq)_{q_{min}} = -(d\chi/dq)_{q_{max}} \qquad (3.4.15)$$

where q_{min} and q_{max} are the edges of the interested spatial frequency range. By combining Eqs. (3.4.10) and (3.4.15), the optimized defocus value, namely Lichte defocus, is obtained as

$$\Delta f_{\text{Lichte}} = -0.75 C_s (q_{max}\lambda)^2 \qquad (3.4.16)$$

Focus step optimization

In the practice, determined by either the image parameters of the microscope or the required sufficient resolution in a peculiar application case, the spatial incoherence provides an upper bound on the defocus step size, while the requirement that phase information manifests itself in sufficient variation in the defocused images provides a lower bound on the defocus step size, which confines the optimized value of focus step used in each case. These two limits exist for the retrievable bandwidth of spatial frequencies q. These limitations are closely related to the focal step size δ between successive images and to the total focal length of the series $L = (N-1)\delta$. That is, to avoid artificial resonances, the phase shift for the largest q_{max} should be no more than half of the focal repetition period $(1/\lambda q^2)$ under the smallest defocus interval (δ), thus yielding the limitation for the highest frequency as Eq. (3.4.17); while, the lowest reconstructable frequency q_{min} can be roughly determined by the criterion that the total range L of the series should be wide enough in order to detect the linear contrast change induced by propagation over the distance L, Assuming a propagation-induced phase change of $\pi/4$ as a detection limit, then the lowest accessible frequency can be estimated as Eq. (3.4.18):

$$q_{max} \approx \sqrt{1/\lambda\delta} \qquad (3.4.17)$$

$$q_{min} \approx \frac{1}{2}\sqrt{1/\lambda L} \qquad (3.4.18)$$

Taken a FEI Titan 80–300 microscope (300 kV) for instance, for 20 series images and a 5 nm defocus step, the retrievable bandwidth of spatial frequency will located among 0.99~1.56, corresponding a creditable resolution of 0.10 nm ~0.865 nm. A smaller defocus interval will bring a higher q_{max}. However, it should be noted that the defocus step should be larger than the focus spread of the microscope to make sure a detective contrast change of HRTEM image series. Besides, although it is implied in Eq. (3.4.14) that a larger number of image series can ensure a broader retrievable bandwidth of spatial frequency, however it propose a strict requirement on the stability of microscope condition (high voltage, sample drift, beam irradiation tolerance) as well as increase the calculation workload. Therefore, the image series number and the defocus interval value should be carefully determined on the basis of both the reconstruction requirement and the microscope condition.

Convergence angle and CCD sampling

Besides the defocus selection, the illumination condition also makes a significant impact on the reconstruction quality mainly by the contrast delocalization due to the beam convergence effect. This impact is integrated into a multiple term of the transfer function $T(\boldsymbol{q})$, amounting to Bessel function of χ_α in Eq. (3.4.19), where χ_α is the function of convergence angle α and defocus Δf, and D which is the gaussian standard deviation of the chromatic aberration.

$$T_\alpha(\boldsymbol{q}) = \frac{2J_1(\chi_\alpha)}{\chi_\alpha} \qquad (3.4.19)$$

$$\chi_\alpha(\boldsymbol{q}) = 2\pi\alpha[q\Delta f + (\lambda C_s - i\pi D^2)q^3] \qquad (3.4.20)$$

In practice, the convergence angle is mainly determined by the magnification of the microscope. In an assumption of point resolution of 0.2 nm, when the magnification is below 500 k the convergence angle will not be larger than 0.10 mrad. Then the high frequency range is mainly modulated by the temporal envelope and there will be no necessary to worry about the convergence effect. Therefore, it should be reasonable to maintain in a lower magnification when acquiring the HRTEM series. However, another factor confining the lower limit of magnification is CCD sampling (the pixel resolution), which should ensure the record of smallest information needed to be resolved through reconstruction. In practice, the maximum sampling should be smaller than the one out of fifth of the expected resolution in reconstruction. For example, for a normal CCD camera (1024 pixel ×1024 pixel and the real size for one pixel is ∼2 μm) then to obtain a 0.10 nm resolution through FSR, the minimum magnification can be estimated by $M_{min} = 2000/(0.2 d_{recons}) \approx 5 \times 10^5$. That is a minimum magnification of 500 k required for a 0.1 nm reconstruction resolution. In practice, since a spectrum collection CCD camera (such as the GIF camera) is always below the normal one, it will present an additional magnification factor of ∼15, which provides an effective way to acquire focal series images in higher coherent illumination (a smaller convergence angle), meanwhile maintain a enough CCD sampling in the HRTEM images.

Fig. 3.4.1 represents the intuitionistic impact of defocus value and convergence angle upon the CTF. The results is achieved by simulating a traditional 200 kV FEG Super Twin microscope, where $C_s = 1.2$ mm, $C_c = 1.2$ mm, $\Delta f_{scherzer} \approx -60$ nm, $\Delta f_{Lichte} \approx -815$ nm, $\alpha \approx 0.1$ mrad.

Images series processing

The reconstruction of the exit wave can be divided into four steps: it starts with an alignment step in which the images of the focal-series are aligned with respect to each other. In the second step, an analytical inversion of the linear imaging problem is achieved by using the paraboloid method (PAM) to generate a first approximation to the exit wave function. This approximated exit wave function is then refined in the third step by a MAL approach that accounts for the non-linear image contributions. Finally, the exit wave is corrected for residual aberrations of the microscope. With the microscope aberrations eliminated, the exit wave function can be directly understood only in terms of the electron beam–specimen interaction.

Image series alignment

The primary algorithms for PAM and MAL approach have been discussed in the previous session. It should be emphasized that the image alignment procedure is much necessary prior to reconstruction because no microscope can record a series of

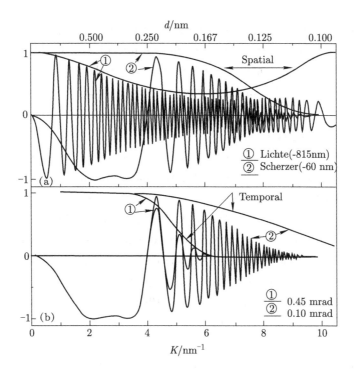

Fig.3.4.1 The contrast transfer function (CTF) curves of 200 kV TEM with different defocus and radiation angle. (a) CTF curves under Scherzer and Lichte focus; (b) CTF curves with different radiation angle (Wang, 2006).

10–20 HRTEM images without mechanical drift. The initial alignment of the images is normally performed by cross-correlation between successive images, which is only a rough alignment because the contrast of the images in the focal series is changing due to the defocus change. A poor image alignment always results in a reduction of the FSR quality, such as the resolution reduction and dislocations around the retrieved image.

The wrap-around problem is commonly encountered in the exit wave reconstruction. In general the range of the CTF is much shorter than that of the signal (the information included in a single HRTEM image), the resulting convolution of these two functions will be falsely "polluted" at the far edges with some wrapped-around data from the far end of the signal. A new padding scheme is introduced —

pading the input HRTEM images with zero or averaged image intensities and providing a good estimate of the image intensities immediately outside the boundaries of the HRTEM images, which can reduce the estimated image intensity discontinuity across the boundaries and therefore providing a larger undistorted view of the exit wave.

Exit Wave Reconstruction

Up to now, either proprietary or commercial reconstruction programs have been developed to perform an efficient FSR in the materials studies. One of the free program named REW is developed for the exit-wave reconstruction from a through focal series of HRTEM images (Lin et al., 2007). It is a stand-alone program with a user friendly interface running at Windows system and users can easily perform the exit-wave reconstruction by using MAL method or Iterative Wave Function Reconstruction (IWFR) (Allen, 2004) Also, REW provides Cross Correlation Function, Mutual Correlation Function (MCF) and Phase Correlation Function (PCF) to align the focal series of images because the specimen drift is unavoidable during the image recording (Meyer et al., 2002). Especially, by adding a numerical aperture to filter the high-frequencies information, the alignment precision is improved. Based on the accurate alignment, REW resolves the Si dumbbell structure from only six HRTEM images in [110] orientation.

Additionally, REW provides the HRTEM image simulation using the multi-slice method. Users can build a crystal according to its space group and lattice parameters, or import a complex crystalline structure from a PDF file, e.g. nanotube and defect structures. In REW, three-dimensional structures can be displayed and rotated for view. Using the HRTEM image simulation, users can compare their experimental images with the ideal HRTEM images to confirm the crystal structure. On the other hand, it is a convenient function to assist users to perform the exit-wave reconstruction if the crystal structure and imaging orientation are known for some regions in the focal series of HRTEM images. According to the TEM parameters and a larger range of focuses, the focal series of ideal HRTEM images are simulated. By comparing the image characteristics in the simulated images and the experimental images, users can determine whether the absolute focus of the experimental HRTEM image departs from its true value, the non-coherent parameters about TEM are reasonable, and there is alignment deviation between the first image and the last image in the focal series of images, et al. Based on these estimations, the exit-wave reconstructions can be performed with ease. Fig. 3.4.2 presents an atom-resolved phase image of a perfect hexagonal lattice of BN monolayer reconstructed with REW. It clearly shows the hexagonal network of the bright spots, which indeed correspond to the individual atoms in the h-BN network. The monolayer region is in the middle and the bilayer can be also seen at the left and right sides of the image.

Fig.3.4.2 The reconstructed phase image of h-BN monolayer discriminates the B (darker) and N (brighter) elements distribution.

Except for some proprietary reconstruction program like REW, there are still commercial ones specially designed for FSR, among which TrueImage is a highly professional software package for ultra-high-resolution electron microscopy applications. It is a numerical method to reconstruct the electron wave function at the exit-plane of an object, based on a series of high-resolution electron microscopy images. It breaks the barrier of the point-resolution of the electron microscope and pushes the resolution up to the information limit. Meanwhile, it removes most aberrations introduced by the optical system of the microscope, resulting in a more intuitively interpretable image.

In order to achieve true high-end results the microscope parameters have to be measured accurately for each microscope and the microscope needs to be aligned very well. Besides an ultra-thin (< 10 nm) sample will be necessary (satisfy the WPOA).

A typical FSR action is divided into three main steps, load and display image series; edit/input microscope, imaging, and reconstruction parameters; reconstruction and aberration correction. In the follow content, we will give a primary introduction of these three steps.

Load and Display Image Series

In this step all consecutively numbered image series should be loaded into the program firstly. For TrueImage all available acquisition parameters embedded in HRTEM images of particular format (MRC, Tietz-EM, DM2, or DM3) will also be automatically loaded. The first image of the focal series is automatically displayed in the Display Area after loading a new series. A Histogram is generated and can be used to adjust the display contrast. The size of display area, display mode (discrete images or movie), and the display type (HRTEM or FFT pattern) are all optional in the program interface.

Edit Parameters

Microscope Parameters: Microscope parameters of most microscope types are stored in the program and are editable in the Set Microscope Parameters. HT is defined as high tension of the microscope in kV, C_s is defined as spherical aberration of the objective lens in mm. The Focus Spread is defined as one half of the $(1/e)$-width of a Gaussian distribution of foci in nm and it is equivalent to the Information Limit of the microscope in nm. MTF is defined as modulation transfer function of the CCD provided by the CCD producer.

Imaging Parameters: Imaging parameters below are normally stored in the heading information of image series and are editable in the Set Imaging Parameters. Sampling is defined as pixel size in nm. Semiconvergence Angle is defined as convergence semi-angle of the illumination system.

Reconstruction Parameters: Reconstruction parameters below are editable and should be optimized in the Set Reconstruction Parameters (Fig. 3.4.3).

First Image: Number of the first image being used for the reconstruction.

Number of Images: Total number of images used for the reconstruction, needs to be between 10 and 20.

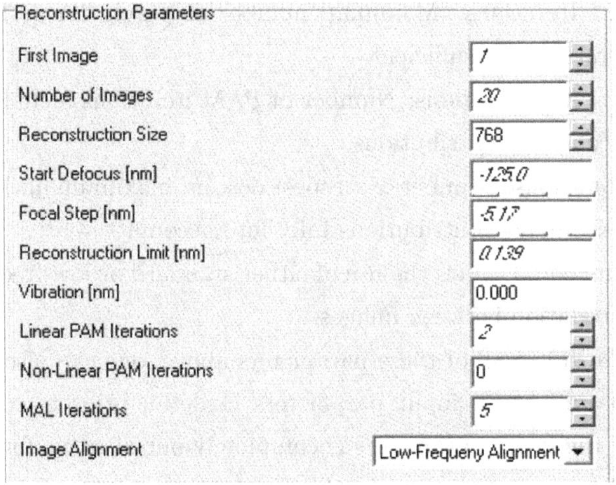

Fig.3.4.3 The panel interface for setting the reconstruction parameters.

Reconstruction Dimension: Size of the reconstruction area in pixels.

Starting Focus: Estimated defocus value of the first image of the focal-series (not the starting image for the reconstruction) in nm. This value should be optimized during the image series recording on the microscope so as to ensure the focus variation range pass through the minimized delocalization point known as the Lichte defocus.

Focal Increment: Focal distance, or focal step, between successive images (nm). This is critically important to achieve a high quality FSR result in TrueImage while it is not forced to be consistent in other specially programed platform, e. g. REW. Since there is always slight turbulence on either lense condition or specimen shift, the real defocus value of each HRTEM image might deviate a little from the expected, which affect the real focal step. Therefore, one should firstly determine the exact value of each image and then the mean focal step could be obtained by linearly fitting of these defocus values.

Reconstruction Limit: Resolution limit of the reconstructed wave function in nm (should be no higher than the information limit of the microscope).

Vibration Amplitude: Amplitude (one half of the $(1/e)$-width) of mechanically or electrically induced image vibrations in nm.

Linear PAM Iterations: Maximum number of purely linear PAM iterations. Typically 2 iterations are sufficient.

Non-Linear PAM Iterations: Number of PAM iterations including a correction for non-linear contrast contributions.

SD-MAL Iterations: Number of steepest descent maximum likelihood iterations taking non-linear image contributions fully into account.

Image Alignment: Defines the use of either standard or low-frequency mode for initial cross-correlation between images.

After restore/input all of these parameters above, one can also perform Check Parameters to analyze all input parameters in detail prior to reconstruction to verify if one of the input parameters is completely out of range for this FSR.

Reconstruction and Aberration Correction

During the reconstruction procedure the program calculating status can be monitored in real time, such as FSR convergence, image alignments, and focal steps analysis et al. The output wave function can be displayed in various forms, including the Amplitude, the Phase, the Real Part, the Imaginary Part, and the Fourier Transform.

The minimum contrast focus visually selected on the microscope typically differs slightly from the real minimum contrast focus setting. The reconstructed wave function therefore needs to be propagated to the exit-plane of the specimen. Furthermore, residual aberrations due to the limited microscope alignment affect the exit-wave function and need to be corrected before attempting any quantitative interpretation of the result. The aberration correction module of TrueImage offers the possibility to correct for the most significant aberrations (Fig. 3.4.4). The wave function can be well propagated to exit-plane of the specimen through adjusting the defocus value (C_{20}) and the modulus and azimuth of the 2-fold astigmatism (C_{22}), coma (C_{31}), and 3-fold astigmatism (C_{33}). The output display will be continuously updated if Auto Update is checked. An automatic estimate of the microscope aberrations is optional, one can use Minimize Real Part to estimate the defocus and 2-fold astigmatism and Estimate Coma to estimate coma. After the entire FSR procedure, one can output the reconstruction results in various form, such as the project (it stores the whole reconstruction with all parameters and optimizations), the exit wave function or aligned image series et al. For more details on the practical application, please refer to the Step-by-Step Guide of TrueImage provide by FEI Company.

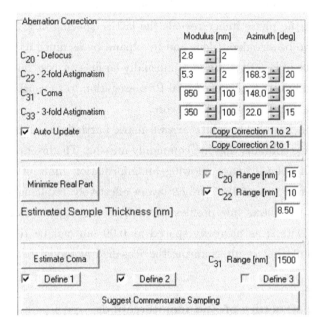

Fig.3.4.4 The panel interface of the residual aberration correction.

3.4.3 Applications

In recent years, microscopy with extended spatial resolution benefiting for FSR technique has been serving as a significant factor in the research area of nanomaterials, such as investigate the elements distribution and surface lattice relaxation of a single nanoparticle, atoms arrangement determination of complex 3D nanostructures, the atomic structure and defects at the grain boundaries, and even the crystallography relationship between nanoprecipitates and the matrix et al. As the representative, some recent works will be introduced to present a delicate illustration of FSR application.

1. FePt icosahedral nanoparticles

The periodic shell structure and surface reconstruction of metallic FePt nanoparticles with icosahedral structure has been quantitatively studied by HRTEM with focal series reconstruction (Fig. 3.4.5) (Wang et al., 2008). The quantitative information from the phase image reveals the lattice spacing of (111) planes in the surface region to be size dependent and to expand by as much as 9% with respect to the bulk value of $Fe_{52}Pt_{48}$. Besides unusually large layer wise outward relaxation is also observed in terms of preferential Pt segregation to the surface forming a Pt enriched shell around a Fe-rich Fe-Pt core.

As shown in Fig. 3.4.5, white arrows mark partially occupied shells. Dark arrows mark edge columns that are commonly missing. The inset of (a) is a Fourier transform of the phase image showing an information limit of 0.090 nm. Note the successive removal of the surface layers effectively reducing the size of the particle. Panel (d) shows line profiles taken from the indicated areas of the real space images. Details as narrowly spaced as 0.09 nm can be resolved. Further microscopy simulation result confirms the icosahedral model revealed from FSR image.

2. 3D atomic structure of ZnS nanotetrapods

The knowledge on the complex 3D structures, such branched tetrapods are mainly remains on phase distinguish and models interpretations. Direct investigations on the atoms arrangement especially at the core/arm interface are rarely reported suffering from the difficulties encountered in microscopy analysis, such as the stereo morphology. By using FSR from a proper angle of the sample, the atom-resolved phase image (Fig. 3.4.6) of ZnS nanotetrapods is obtained (Liu et al., 2011). The tetrapod grows from the octahedron core: the four arms are selectively grown on the Zn-terminated surfaces by changing the atomic stacking order at core/arm interface. As the arms grow longer, the cross section turns into hexagon due to degeneration of atoms on the lateral edges.

3. Twin boundary in a BaTiO$_3$ thin film

Single crystal and bulk $BaTiO_3$ exhibits a sharp paraelectric-to-ferroelectric

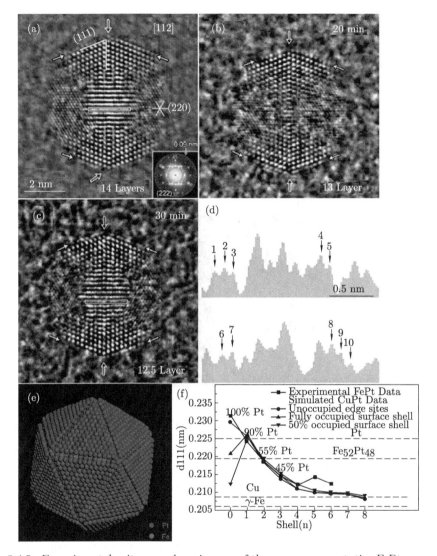

Fig.3.4.5 Experimental exit wave phase images of the same representative FePt nanoparticle exposed to a current density of 20 A/cm^2 at 300 kV for 0, 20, and 30 min (panels (a), (b), and (c), respectively). Panel (d) shows line profiles taken from the indicated areas of the real space images. Panel (e) is the proposed icosahedron 10179 atom sites and Pt atoms that segregated to the surface. Panel (f) is the comparison of experimentally measured (FePt) and simulated (CuPt) shell spacings.

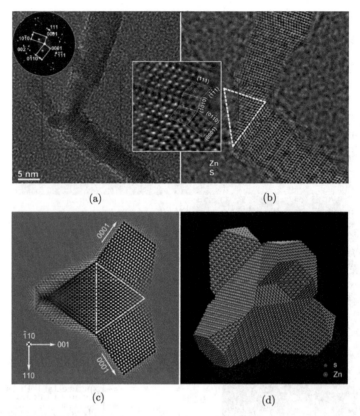

Fig.3.4.6 (a) HRTEM image of a single tetrapod and its corresponding FFT pattern (inset); (b) phase image reconstructed from marked area of image (a); (c) simulated HRTEM image of a tetrapod with 41993 atoms indexed by $[\bar{1}10]_{ZB}$ of the octahedral core; (d) atoms stacking model of an entire tetrapod with four wurtzite arms.

transition at 393 K, which is absent in the presence of submicron grains. Since the twin boundaries along the four crystallographically equivalent {111} planes constitute the main lattice defects, the local atomic arrangement of the core of twin intersections was critical for the interpretation of the electric-property transition. Focal-series reconstruction is utilized to give a deep insight of this problem (Jia and Thust, 1999). Fig. 3.4.7 shows the comparison of high-resolution images representative of a focal series and the corresponding reconstructed phase images after numerical residual aberration correction from the same junction area between twin boundaries in BaTiO$_3$. In the focal series, the structure of the central junction can only be recognized coarsely due to contrast delocalization in HRTEM imaging. After focal-series reconstruction, the resulting exit wave function reveals a sharp phase image of the central twin junction with two distinct types of phase maxima (Fig. 3.4.7) corresponding to the positions of the atomic columns. The exit wave function can be interpreted almost intuitively, yet simulations of the exit wave function were used to confirm the structure. Based on simulations of the single-crystalline areas, the sample thickness was estimated to be 9 nm and this thickness was used to simulate the central twin junction. In their following works, quantitative observation of the oxygen atoms distribution in oxide materials with an aberration corrector is reported (Jia et al., 2003; Jia and Urban, 2004).

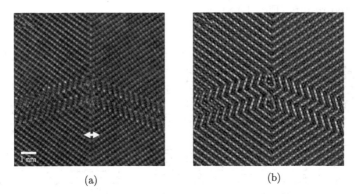

Fig.3.4.7 (a) HRTEM images ($\Delta f = -123$ nm) of $\Sigma 3$ twin boundaries in BaTiO3. The contrast delocalization at the interface is indicated by arrows; (b) shows the phase after correction for residual 2-, 3-fold astigmatism, and coma (Kübel and Thust, 2003). The difference clearly illustrates that correction of aberrations is crucial for a quantitative analysis.

4. The Silicon nanoprecipitates in AlMgSi alloys

The introduction of precipitated phase in a pure metal through alloying with small amounts of other elements can provide the strength of steel at only half of the weight. The nanometer-sized precipitates, which act as obstacles to dislocation movement in the crystal (atomic matrix), strengthen significantly the matrix. A detailed investigation of the crystallography relationship between nanoprecipitates and the matrix under different thermal treatment is of great practical importance for the understanding of the precipitation hardening phenomenon. Chen et al. perform a remarkable FSR reconstruction in the atomic-resolution electron microscopy investigation on the nanoprecipitates strengthen mechanism of AlMgSi alloys (Chen et al., 2006). In their work, 20 HRTEM focal series images were used to retrieve the exit-wave function of the atomic structure of early-stage needles. The obtained phase image clearly resolves all of the atom columns in the precipitate (Fig. 3.4.8). To deduce the precipitate structure, initial structure model was firstly established according to some empirical assumptions of AlMgSi alloys. These models were then confirmed by the good agreement of phase images between the calculated and reconstructed results (Fig. 3.4.8).

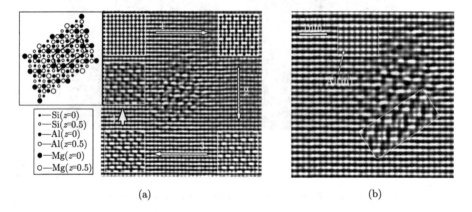

Fig.3.4.8 Deduction of the precipitate structure. (a) The phase-image of the reconstructed exit wave of a precipitate in the matrix. The bright dots represent atom columns. The insets are the calculated images from a structure model (left) with varying specimen parameters: 1) changing the Si-Si distances in the model; 2) adjusting the crystal tilt and thickness; 3) adjusting the positions and compositions of atomic columns and 4) including the Debye-Waller factors of atoms. (b) Refinement results: The calculated images (insets) match with the reconstructed images for both the precipitate and the matrix [Al(m)].

3.5 Convergent beam electron diffraction

Lijun Wu

3.5.1 Introduction

In transmission electron microscope (TEM), when a parallel beam illuminates the specimen, a diffraction pattern with sharp spots forms at the back focal plane of the objective lens. With a convergent beam focuses on the specimen, the diffraction beams expand to the disks in the back focal plane, forms convergent beam electron diffraction (called as CBED). The concept of convergent beam comes from the discovery of Kossel patterns in X-ray diffraction in 1937 and of Kikuchi patterns in 1928. The first CBED pattern was obtained by Kossel and Mollenstedt in 1939. In 1940, MacGillavry attempted to measure the structure factors by fitting experimental CBED patterns using dynamical electron diffraction theory in two beam condition. In 1957, Kambe, in his study of three beam theory, showed that the intensity depends on a certain sum of structure factors and could be measured (Kambe, 1957). The crystallographic studies by CBED originated with Goodman and Lehmpfuhl in 1965. By the seventies, systematic procedures for point-group and space-group determination by CBED had been established (Goodman, 1975; Buxton et al., 1976; Tanaka et al., 1983). Quantitative work on structure determination, e.g. atomic position and structure factors, started in eighties. Vincent et al. applied CBED method to determine atomic positions of AuGeAs using a quasi-kinematical theory (Vincent 1984). The recent development of quantitative CBED (Zuo, 1993; Zuo et al., 1993; Zuo et al., 1999; Zuo et al., 1997, Zuo et al., 1997) and parallel recording of dark field images has opened the door to mapping valence electron distributions by accurately determining the structure factors of the innermost reflections. In this Section, we will first introduce the experimental setup and Bloch wave calculation for the CBED, and then its applications in the determination of symmetry and valence electron distribution.

3.5.2 Experimental setup

Most modern analytical TEM provide a CBED mode. The convergent angle α can be easily changed by the condenser aperture. More advanced TEM may be equipped with field emission gun and energy filter which significantly improve the quality of the CBED. Fig. 3.5.1 shows a simplified ray diagram for a CBED pattern. The electron source is focused on the surface of a thin crystalline sample. Each point within the condenser aperture defines an incident beam, giving rise to a set of scattered points in the back focal plane of the objective lens. For example, the point A in the condenser aperture defines beam AA_0, which forms diffracted points A_0, A_{-g}, A_g, etc. Considering all beams within the condenser aperture, a disk is formed in the back focal plane for each reflection, and each point in the disk corresponds to a beam direction. The beam direction for each point in the disk can be calculated based on the central beam direction OO_0. For example, beam direction of AA_0 is: $\boldsymbol{K} = \boldsymbol{K_O} + \boldsymbol{V}$, where $\boldsymbol{K_O}$ is the wavevector of the central

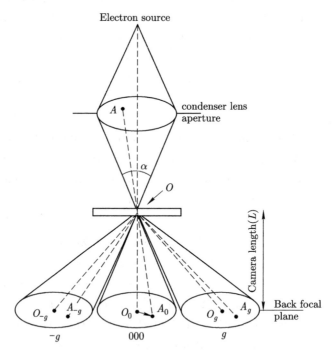

Fig.3.5.1 Simplified ray diagram for CBED with convergent angle of α. Source point A defines an incident beam AA_0, giving rise to conjugate points A_0, A_{-g} and A_g in the 000, $-g$, and g disks of the CBED pattern. Each source point defines a beam direction which can be determined by the central beam OO_0.

beam. The beam direction can be further quantified by the component K_t of the incident beam K in the zero-order Laue zone (ZOLZ), and the excitation errors S_g which is defined as a vector from the reciprocal point g to the Ewald's sphere along the beam direction (Fig. 3.5.2). If a small deviation from the Bragg condition

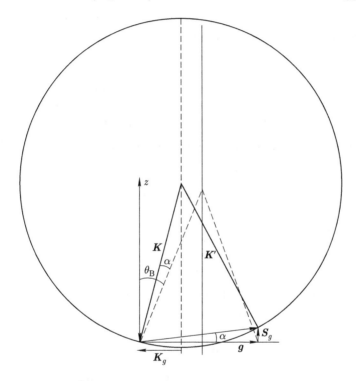

Fig.3.5.2   The diagram of Ewald's sphere.

is allowed, so that the scattered wavevector $K' = K + g + S_g$ (g is the reciprocal vector). For elastic Bragg scattering, energy conservation requires $|K| = |K'|$, for small g and S,

$$2KS_g = K'^2 - (K+g)^2 = -2K \cdot g - g^2 \qquad (3.5.1)$$

This equation applies for both ZOLZ and high-order Laue zone (HOLZ) reflections. For ZOLZ reflections, Eq. (3.5.1) can be simplified to

$$2KS_g = -2K_t \cdot g - g^2 \qquad (3.5.2)$$

An excitation error can be assigned to every point in the disks of the CBED pattern. It varies in the disk, resulting in the variation of diffraction intensity in the disk.

3.5.3 Bloch wave calculation for CBED

For a collimated incident electron beam of the form $\exp(2\pi i \mathbf{K}_0 \cdot \mathbf{r})$, the "relativistically corrected" Schrödinger equation describing high energy electron diffraction is

$$\frac{-h^2}{8\pi^2 m}\nabla^2 \Psi(\mathbf{r}) - |e|V(\mathbf{r})\Psi(\mathbf{r}) = \frac{h^2 K_0^2}{2m}\Psi(\mathbf{r}) \qquad (3.5.3)$$

Where h is Planck's constant, m the relativistic electron mass, e the charge on electron, $\Psi(\mathbf{r})$ the wave function inside the crystal, $V(\mathbf{r})$ the crystal potential in volts. To solve the equation, we expand the wave function as a sum of Bloch waves and reciprocal lattice vectors g

$$\Psi(\mathbf{r}) = \sum_j c_j \exp(2\pi i \mathbf{k}^{(j)} \cdot \mathbf{r}) \sum_g C_g^j \exp(2\pi i \mathbf{g} \cdot \mathbf{r}) \qquad (3.5.4)$$

and expand the crystal potential as a Fourier series.

$$V(\mathbf{r}) = \sum_g V_g \exp(2\pi i \mathbf{g} \cdot \mathbf{r}) \qquad (3.5.5)$$

The result of substituting these two equations into Eq. (3.5.3) yields the standard dispersion equation of high-energy electron diffraction

$$[K^2 - (\mathbf{k}^{(j)} + \mathbf{g})^2]C_g^{(j)} + \sum_h U_{g-h} C_h^{(j)} = 0 \qquad (3.5.6)$$

where U_g is the complex electron "structure factor" in angstroms^{-2}.

$$U_g = \frac{2m|e|V_g}{h^2} = \frac{\gamma F_g}{\pi \Omega} \qquad (3.5.7)$$

here, γ is the relativistic constant, Ω the volume of the unit cell. F_g is the electron structure factor in angstrom

$$F_g = \sum_j f_j^e \exp(-2\pi i \mathbf{g} \cdot \mathbf{r}_j) \qquad (3.5.8)$$

where, f_j^e is the atomic scattering amplitude for electrons, can be obtained from International Tables for Crystallography C (Pages 259–262). An additional absorption component U_g' may be added to the structure factor for inelastic scattering, thus the total structure factor is given by

$$U_g = \frac{\gamma F_g}{\pi \Omega} + iU'_g \qquad (3.5.9)$$

The wave mechanical boundary condition which ensures conservation of the electron density is that both the wave function and its derivative normal to the boundary should be continuous. This requires that the Bloch wavevectors, $\boldsymbol{k}^{(j)}$, inside the crystal have the same tangential components as the incident beam at the crystal entrance surface. Let the Bloch wavevectors

$$\boldsymbol{k}^{(j)} = \boldsymbol{K} + \gamma^{(j)} \boldsymbol{n} \qquad (3.5.10)$$

where, $\boldsymbol{n}$ is a unit vector of the entrance surface normal of the crystal, then

$$\begin{aligned} K^2 - (\boldsymbol{k}^{(j)} + \boldsymbol{g})^2 &= K^2 - (\boldsymbol{K} + \gamma^{(j)}\boldsymbol{n} + \boldsymbol{g})^2 \\ &= K^2 - (\boldsymbol{K} + \boldsymbol{g})^2 - 2(\boldsymbol{K} + \boldsymbol{g}) \cdot \boldsymbol{n}\gamma^{(j)} - \gamma^{(j)2} \\ &= 2KS_g - 2(\boldsymbol{K} + \boldsymbol{g}) \cdot \boldsymbol{n}\gamma^{(j)} - \gamma^{(j)2} \end{aligned} \qquad (3.5.11)$$

Defining $g_n = \boldsymbol{g} \cdot \boldsymbol{n}$, $K_n = \boldsymbol{K} \cdot \boldsymbol{n}$ and neglecting small γ^2 term, we get

$$2KS_g C_g^{(j)} + \sum_h U_{g-h} C_h^{(j)} = 2K_n(1 + g_n/K_n)\gamma^{(j)} C_g^{(j)} \qquad (3.5.12)$$

If the surface normal is approximately antiparallel to the beam, so that $K_n \gg g_n$, then g_n/K_n is negligible, Eq. (3.5.12) becomes

$$2KS_g C_g^{(j)} + \sum_h U_{g-h} C_h^{(j)} = 2K_n \gamma^{(j)} C_g^{(j)} \qquad (3.5.13)$$

This is an eigenvalue equation to be solved. The equation can be further written in matrix form

$$A \begin{pmatrix} C_0 \\ C_g \\ C_h \\ \vdots \end{pmatrix} = \begin{pmatrix} 0 & U_{-g} & U_{-h} & \cdots \\ U_g & 2KS_g & U_{gh} & \cdots \\ U_h & U_{hg} & 2KS_h & \cdots \\ \vdots & \vdots & \vdots & \ddots \end{pmatrix} \begin{pmatrix} C_0 \\ C_g \\ C_h \\ \vdots \end{pmatrix} = 2K_n\gamma \begin{pmatrix} C_0 \\ C_g \\ C_h \\ \vdots \end{pmatrix} \qquad (3.5.14)$$

Where the "structure matrix" A is $n \times n$ matrix if n-beams are included, and in general complex. For crystals without absorption, A is Hermitian. For centrosymmetric crystals without absorption, A is real, symmetric and Hermitian. Eq. (3.5.14) gives n eigenvalues and n eigenvectors. The wave amplitude at crystal thickness t is

$$\phi_g(t) = \sum_{j=1}^{n} c_j C_g^j \exp(2\pi i \gamma^j t) \qquad (3.5.15)$$

and

$$I_g = \phi_g(t) \times \phi_g^*(t) \qquad (3.5.16)$$

The excitation coefficients c_j can be determined based on the boundary condition at the entrance surface, e.g. $t = 0$, where the wave function $\phi_g(0)$ are known. From Eq. (3.5.15)

$$\begin{pmatrix} \phi_0(0) \\ \phi_g(0) \\ \vdots \end{pmatrix} = \begin{pmatrix} C_0^1 & \cdots & C_0^n \\ C_g^1 & \cdots & C_g^n \\ \vdots & \ddots & \vdots \end{pmatrix} \begin{pmatrix} c_1 \\ \vdots \\ c_n \end{pmatrix} = C \begin{pmatrix} c_1 \\ \vdots \\ c_n \end{pmatrix} \qquad (3.5.17)$$

so

$$\begin{pmatrix} c_1 \\ \vdots \\ c_n \end{pmatrix} = C^{-1} \begin{pmatrix} \phi_0(0) \\ \phi_g(0) \\ \vdots \end{pmatrix} \qquad (3.5.18)$$

Where C^{-1} is the inverse matrix of C. For each point in the CBED disk, we need to diagonalize the structure matrix A to obtain the eigenvalues and eigenvectors, then calculate the intensities of all reflections. Fig. 3.5.3 shows an energy filtered CBED pattern of Si with incident beam along $[1\bar{1}\bar{4}]$ direction and calculated CBED patterns with different thickness. It is seen that there are coarse fringes and fine lines in the CBED pattern. The former fringes are called as rocking curves which are due to the dynamic coupling among zero-order Laue zone (ZOLZ). The number of the fringes increases with the crystal thickness. The fine line in the incident-beam disk is the locus of the Bragg condition for high-order Laue zone (HOLZ) reflection, called as HOLZ line. The HOLZ lines occur in pairs, a maximum of intensity in the outer HOLZ reflection disk (excess line, as indicated by the arrow in Fig. 3.5.3(d)) and a corresponding minimum of intensity in the incident-beam disk (deficiency line). In the kinematic approximation, the line positions are given by the simple Bragg law. The geometry of HOLZ lines is therefore the geometry of the Bragg condition projected onto the plane of observation, thus sensitive to the lattice parameter and crystal orientation. In later, we will show that the HOLZ line positions are affected by accelerating voltage as well.

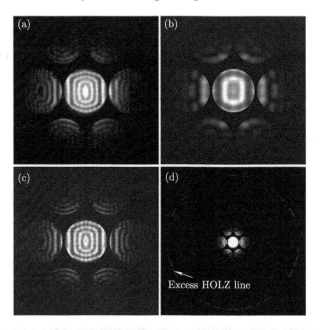

Fig.3.5.3 (a) shows an experimental CBED pattern of Si with beam along [1$\bar{1}$4] direction. (b)–(d) are calculated CBED patterns with thickness of 100 nm and 225 nm, respectively. (d) The same pattern as (c) but with large field view, showing outer excess HOLZ line ring.

3.5.4 Applications of CBED

1. Thickness measurement

The CBED have many applications. Its sensitivity to the crystal thickness has been widely used to accurately measure the sample thickness which is important for quantitative calculations, e.g. measurement of the mean potential and quantitative high resolution annular dark field scanning electron microscopy. By comparing the experimental CBED pattern with a series of calculated patterns with different thickness, we find that the calculated CBED pattern with thickness=225 nm (Fig. 3.5.3(c)) is the best fit to the experimental observation (Fig. 3.5.3(a)), thus we determine the sample thickness to be 225 nm.

2. Symmetry determination

The CBED pattern symmetries can be described by the 31 diffraction group, which are related with the point groups of the crystals. Determination of the point group of a crystal usually involves the following steps: (a) identification of diffraction group of the crystal, along one close-packed direction by study of the CBED pattern for whole pattern (WP) symmetry, bright field (BF) symmetry and dark field (DF) symmetry along $+g$ and $-g$ directions; (b) repetition of (a), if required, along different zone axes; and (c) derivation of the point group symmetry of the crystal based on the diffraction symmetries. It should be noted that the ZOLZ reflections in WP symmetry exhibits the symmetry elements of the crystal projected along the zone axis. For three dimensional symmetries, HOLZ reflections are necessary to get correct diffraction group, as shown in Fig. 3.5.4(a). It is clearly seen that the ZOLZ reflections show the projected $6mm$ symmetry along [111] direction for Si, while the HOLZ lines show three dimensional $3m$ symmetry along [111] direction.

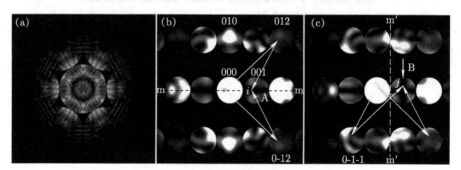

Fig.3.5.4 (a) Si [111], 200 kV, t=250 nm. (b) [100] BeO, 297 kV, t=100 nm. (c) [100], the center of Laue circle at (0,0,0.5), BeO, 297 kV, t=100 nm.

The glide planes and screw axes in the crystal can be determined based on the extinction rule in the approximation of kinematical diffraction. The kinematical forbidden reflections by these symmetry elements can gain certain intensity due to

Umweganregung of dynamical diffraction. However, the extinction of intensity still occurs in these reflection disks for specific directions of the incident beam. Such an effect appears as dark lines in the CBED disks which are called as dynamical extinction lines or Gjonnes-Moodie (GM) lines. The dynamical extinction effect is similar to the interference phenomenon in the Michelson interferometer. The incident beam is split into two beams by Bragg reflection in a crystal. These beams follow different paths, in which they suffer a relative phase shift when reflected by crystal planes, and are then superposed on a kinematical forbidden reflection to cancel out each other. Fig. 3.5.4(b) shows the calculated CBED pattern for BeO in [100] zone (for experimental pattern, see Spence and Zuo, 1992). GM lines are marked by the arrows. Detailed methods to determine space group from diffraction group and GM lines are discussed in many references (Steeds, 1983b; Tanaka, 1994; 1983a).

3. Accelerating voltage and lattice parameters

The lattice parameters of the crystal can be measured using the Bragg law in diffraction experiments. The accuracy in the lattice parameter measurement from an ordinary electron diffraction pattern is, however, rather poor in comparison with X-ray and neutron diffraction due to the small angle scattering of electrons and the geometry distortion induced by the lens. With the use of the HOLZ lines in the CBED pattern, it has been demonstrated that the lattice parameters can be measured with an accuracy comparable to the X-ray and neutron diffraction. Although many lattice parameter measurements by HOLZ lines are based on a comparison of HOLZ line patterns simulated by Bragg law in kinematical approximation, high accurate measurements require dynamical calculations since the HOLZ line positions often deviate from their kinematical positions due to multiple scattering. Fig. 3.5.5(a) shows an energy-filtered CBED pattern of Si taken in the JEM-3000F microscope equipped with Gatan energy filter and accelerating voltage reading at 300 kV. The beam is slightly off the [133] zone with the center of the Laue circle at (− 0.14, 0.023, 0.023). Sharp HOLZ lines are shown in the incident beam disk. HOLZ lines ($7\bar{7}5$), ($\bar{7}7\bar{5}$) and ($11\bar{1}3$) intersect at a point, as shown in the inset. Calculated CBED pattern (Fig. 3.5.5(b)) with accelerating voltage being 300 kV shows the above three HOLZ lines does not intersect at one point. We noted that the Gatan energy filter reduces the accelerating voltage. After adjust the accelerating voltage to 297 kV, the calculated HOLZ lines (Fig. 3.5.5(c)) fit the experimental observation very well. In Fig. 3.5.5(d), we use the same diffraction conditions as Fig. 3.5.5(c) but increase the lattice parameter by 1%. It is seen that the HOLZ lines shift significantly.

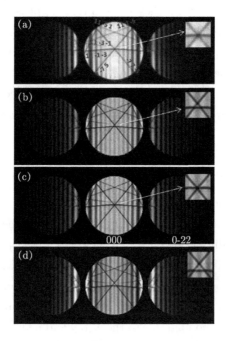

Fig.3.5.5 (a) Experimental, (b) 300 kV at t=300 nm, (c) 297 kV, (d) 297 kV, a=0.5485 nm

4. Charge density by quantitative CBED

The ability to accurately measure charge distribution in crystal is of vital importance since they often control the physical properties of a material. The experimental charge density in materials was usually measured by single crystal X-ray diffraction. However, it requires large single crystal, and suffers from large extinction error in measuring low-order structure factors which are crucial to the valence electron distribution of the crystal. In contrast, in a transmission electron microscope (TEM), a very small electron probe can be used to study a defect-free nanometer region in the sample. Moreover, electron diffraction is more sensitive to the valence electron distribution in crystal in the small scattering angle than X-ray diffraction. Measuring accurately just a few of the low-order electron structure factors also provides a useful test of electronic structure calculations by density functional theory. The recent development of energy-filtered quantitative CBED and parallel recording of dark-field images (PARODI) has demonstrated that electron diffraction can accurately measure low-order structure factors and map valence electron distributions. Here, we will show the valence electron distribution of perovskite $CaCu_3Ti_4O_{12}$ (CCTO) by accurate measurement of the low-order structure factors using quantitative CBED and PARODI.

CCTO has been extensively studied due to its unusually high dielectric constant ($\varepsilon \sim 10^5$) in kilohertz radio frequency region across a wide range of temperatures (ε remains almost constant between 100 K and 600 K). Density functional theory (DFT) calculations based on the ideal stoichiometric single crystal suggest that the material can only have an intrinsic dielectric constant four orders of magnitude less than that observed experimentally. After an exhaustive search for its intrinsic mechanism, attention was then shifted to defects that form the "internal barrier layer capacitance" (IBLC) responsible for this anomalous behavior. Various morphological models were proposed. The puzzling aspect is that the long-expected twin boundaries in the system acting as the IBLC were never truly observed. Furthermore, the type and density of the defects are very different in single crystals, bulk ceramics, and thin films of CCTO (Goodman, 1975), yet all exhibit similar dielectric behavior (Zuo, 1999). Clearly, there is a general interest as well as an urgent need to understand the origin of the unusual dielectric property in CCTO. Experimental bonding-electron distribution for this material would provide important information to understand the origin of the unusual dielectric property in CCTO.

CCTO has a cubic cell that is $2 \times 2 \times 2$ times the size of perovskite subcell ABO_3, with Ca/Cu at A sites and Ti at B sites. Because of the fairly large tilt of the TiO_6 octahedral, 3/4 of the A atoms (i.e., the Cu atoms, hereafter denoted as an A''

site with an *mmm* symmetry) are coordinated with four oxygen atoms forming a square with a Cu atom at the center. The remaining A atoms, Ca (A′ site with a $m-3$ symmetry), have a bcc arrangement, and each is surrounded by a 12-oxygen icosahedral environment. To measure the low-order structure factors, people usually tilt the crystal to the systematic row or "one-dimensional" condition. That is only a row of reflections are strongly excited. Fig. 3.5.6(a) shows an energy-filtered CBED pattern of hkk systematic row with (000), ±(211) and ±(422) reflections of CCTO. The orientation is determined to be $[1\bar{1}\bar{1}]$ zone with the center of the Laue circle at (0.745, −54.065, 54.810) based on the Kikuchi lines. In order to extract structure factors of (211) and (422) (F_{211} and F_{422}) from this pattern, dynamical Bloch wave calculations of the diffraction intensities is performed and compared with the experimental intensities using χ^2 as a goodness of fit. Fig. 3.5.7 shows the refinement procedure. The initial calculations usually start from electron structure factors of the procrystals (i.e., a hypothetical crystal with atoms

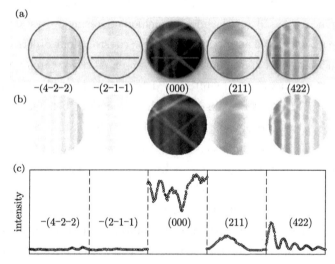

Fig.3.5.6 (a) Energy-filtered CBED of $CaCu_3Ti_4O_{12}$ at hkk systematic row $[1\bar{1}\bar{1}]$ zone with Laue circle center at (0.745, −54.065, 54.810). (b) Calculation based on Bloch-wave method. (c) Line scan of the intensity profile from the experiment pattern in (a). The open circles are experiment while the red line curve is calculation after refining the structure factors of (211) and (422).

having the electron distribution of free atoms; also so-called independent atom model (IAM)). Then iterative calculations were carried out by adjusting values of structure factors until the χ^2 meet the preset value or reaches minimum. To ensure that this minimum is not a saddle local minimum, we test it by selecting several different starting points and re-calculating structure factors iteratively until we reach the same lowest minimum. Fig. 3.5.6(b) shows the calculated CBED pattern based on the refined structure factors. It agrees the experimental pattern (Fig. 3.5.6(a)) very well. Quantitative comparison from the line profiles are shown in Fig. 3.5.6(c).

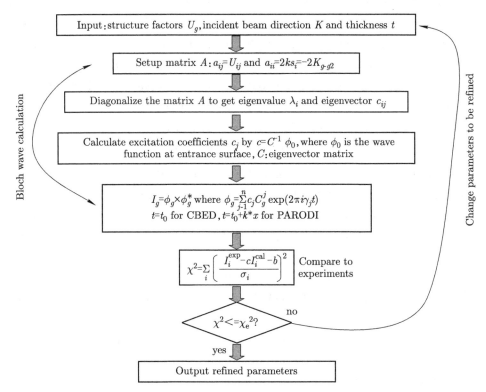

Fig.3.5.7. Procedure of the structure factors refinement.

The convergent angle of the CBED may be limited for low-order reflections to avoid the overlapping of the reflections when the unit cell is large. In this case, the information in the conventional CBED (cCBED) may not be enough to accurately retrieve the low-order structure factors. Therefore, the Parallel recording of dark field images (PARODI) method is introduced by adding an additional dimension, i.e. thickness. This can be done by focusing the electron probe above a wedge sample. In PARODI, we form shadow images (mainly dark-field images) of a large illuminated area within the diffraction disks that contain not only information on orientation (as with cCBED), but also thickness profiles, or Pendellösung plots, for the many simultaneously recorded reflections. The discrete data points within each reflection (corresponding to different thicknesses of the sample) are independent for defect-free sample areas or incoherent illumination; this results in a high ratio of experimental observations versus the number of fitting parameters in model calculations, giving correspondingly higher levels of confidence. The calculation and refinement of the PARODI pattern is similar to the cCBED. For a wedge sample, the thickness changes linearly, thus only one thickness parameter is required for refinement. For a general area, the thickness distribution can be routinely mapped using off-axis electron holography. Fig. 3.5.7 demonstrates the refinement of the low-order structure factors (200 and 400) of the CCTO using PARODI pattern.

Quantitative electron diffraction (cCBED and PARODI) measures the low-order structure factors (110), (200), (211), (220), (400), and (422) that are most sensitive to the charge and orbital electrons, but negligibly influenced by the precision of atomic positions and thermal parameters. For reflections further out in reciprocal space, synchrotron X-ray diffraction (XRD) was used. The charge distribution can then be obtained based on multiple refinement. Fig. 3.5.8(b) is the resulting experimental density map of CCTO showing bonding valence states in the (001) plane, along with that calculated by DFT (Fig. 3.5.8(c)) within the LDA+U approximation. The magnetically active and orbital-ordered Cu-d states exhibit a well-defined t_{2g} symmetry (pointing toward the [110] directions), and form antibonds with neighboring O-p states, in good agreement with calculations. Surprisingly, the states at A' sites that are supposed to be occupied by Ca also show orbital anisotropy with weak e_g symmetry in Fig. 3.5.8(b), in clear contrast to the spherical density profile of Ca^{2+} obtained from the theoretical density map (Fig. 3.5.8(c)). Since Cu atoms are the only ones that have open d shell in this system, this immediately points to a considerable amount of Cu substitution of Ca atoms. This finding points out the existence of nanoscale disorder of Cu/Ca and suggests its remarkable effects on the dielectric properties of CCTO. First-principles-based theoretical analysis indicates that the Ca-site Cu atoms possess partially filled degenerate e_g states, suggesting significant boost of dielectric response from additional low-energy *electronic* contributions.

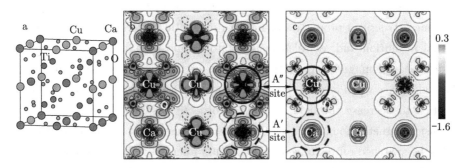

Fig.3.5.8 The bonding-electron distribution of CCTO in (001) basal plane containing Cu, Ca, and O atoms. (a) Structure model. (b) Experimental observation extracted from structure factor measurements using combined electron and X-ray diffraction data (i.e. the static deformation electron density map using neutral atoms as a reference). (c) DFT calculation based on the ideal crystal structure. The color legend indicates the magnitude of the charge, and the contour plot has an interval of 0.1 e/Å^3. The orbital-ordered Cu 3d states of xy symmetry are clearly visible, as well as its antibonding with O 2p states. Note the significant difference near the A′ sites (circled by solid line) between (b) and (c), where an anisotropic density pattern of e_g electrons is observed only experimentally, suggesting some degree of disorder with Cu replacing Ca in real material.

3.6 Lorentz electron microscopy

Haihua Liu and Ying Zhang

3.6.1 Introduction

Lorentz microscopy has been extensively used to study domain structures in magnetic related materials (Hale et al., 1959; Chapman and Scheinfein, 1999; Chapman,

1989; Hirsh et al., 1965; Saito et al., 2006). As we know, the local microstructure of the material will have big influence on the applicable magnetic properties, which makes magnetic properties extrinsic rather than intrinsic. Hence a detailed knowledge of both the physical and magnetic microstructure is crucial to understand the structure-property relationship and thus materials with optimized properties will be produced. The high resolution of the modern transmission electron microscope (TEM) makes it an attractive tool for the study of micromagnetic and microstructural properties of modern magnetic materials. It offers very high spatial resolution and, detailed insight into compositional, electronic, as well as structural and magnetic properties because of the large number of interactions that take place when a beam of fast electrons hits a thin solid specimen. The modification to decrease the field strength of objective lens is required to study the intrinsic magnetic domains in the specimen. Therefore, this objective lens must either be switched off in conventional Lorentz microscopy or be designed specially as in JEOL dedicated Lorentz TEM with sample immersed in the upper pole piece instead of the traditional gap area between the upper and lower pieces. Therefore, the strong field from the lens is routed around the region where the sample sits by mu-metal shielding. This unique design for the dedicated Lorentz TEM allow us to study the magnetic domains in low residual magnetic field of about 4 Oe compared to ~200 Oe of conventional TEM with the objective lens (OL) off (Hale et al., 1959). It should be mentioned that the OL should be switched off prior to sample insertion even for the dedicated Lorentz TEM. Although the measured field is 4Oe with the OL on and 0 Oe with OL off after the sample is fully inserted into the column, the sample is subjected to about 350 Oe field during insertion, which will change the magnetic domain structures of the soft magnetic materials (Schofield et al., 2007). The resolution is then limited by the Lorentz microscopy requirements. The resolution that is achievable also depends largely on the specimen itself. Typical resolutions achievable for structural imaging are 0.1–1.0 nm, for extraction of compositional information 1–3 nm and for magnetic imaging 2–20 nm (Chapman and Scheinfein, 1999).

When imaging magnetic structures in TEM, the information sought can include 1) the domain geometry, 2) the magnetization direction within individual domains,

3) the structure of domain walls, 4) the distribution of stray fields beyond the specimen, and 5) the response of the specimen to applied magnetic fields, heating, etc. In addition, to understand what is being observed, simultaneous information on the physical microstructure is required, and this can encompass the local crystal structure, defects, local elemental composition and surface topography (Chapman, 1989).

After more than 50 years extensive development of Lorentz microscopy, especially with fast development of modern high resolution microscopy equipment and the high performance computer, a number of different imaging and diffraction modes have been developed including Fresnel, Foucault and coherent Foucault, low angle electron diffraction as well as differential phase contrast imaging. An introduction to relevant aspects of the electron beam-specimen interaction in magnetic materials will be given in Section 3.6.2. A brief description of each method, along with its principal advantages and drawbacks is given in Section 3.6.3. The in situ magnetic field generation methods will be shown in Section 3.6.4. And then some in situ examples of Lorentz TEM study on different magnetic related functional materials will be given in Section 3.6.5.

3.6.2 Electron beam-magnetic specimen interaction in TEM

Magnetic structures are most commonly revealed in the TEM using one of the modes of Lorentz microscopy. This generic name is used to describe all imaging modes in which contrast is generated as a result of the electron deflection experience caused by Lorentz Force as they pass through a region of magnetic induction (Ploessl et al., 1993). The Lorentz deflection angle β_L is given by

$$\beta_L = e\lambda t(B \Lambda n)/h \tag{3.6.1}$$

where B is the induction averaged along the electron trajectory, n is a unit vector parallel to the incident beam, t is the specimen thickness and λ is the electron wavelength (Chapman and Scheinfein, 1999). Substituting typical values into eq. 3.6.1 suggests that β_L rarely exceeds 100 μrad and, indeed, can be < 1 μrad in the case of some magnetic multilayers or films with low values of saturation magnetization, that is when the magnetization vector lies perpendicular to the plane of the film (Aharanov and Bohm, 1959). Clearly, under these conditions β_L is zero unless the specimen is tilted with respect to the electron beam to introduce an induction component perpendicular to the direction of electron travel. Given the small magnitude of β_L there is no danger of confusing magnetic scattering with the more familiar Bragg scattering where angles are typically in the range 1–10 mrad.

The description given so far is essentially classical in nature and much of Lorentz imaging can be understood in these terms. However, for certain imaging modes and, more generally if a full quantitative description of the spatial variation of induction is required, a quantum mechanical description of the beam-specimen interaction must be sought. Using this approach, the magnetic film should be considered as a phase modulator of the incident electron wave, the phase gradient $\nabla \phi$ of the specimen transmittance being given by

$$\nabla \phi = 2\pi e t (B \wedge n)/h \qquad (3.6.2)$$

e and h are the electronic charge and Planck's constant, respectively. Substituting typical numberical values shows that magnetic films should normally be regarded as strong, albeit slowly varying, phase objects (Chapman, 1984). For example, the phase change involved in crossing a domain wall usually exceeds π rad.

So far it has been assumed that the only interaction suffered by the beam is due to the magnetic induction. However, other than in the free space region beyond the edges of a magnetic specimen, allowance must also be made for the effect of the electrostatic potential. Indeed, the electrostatic interaction is generally much the stronger of the two. Hence Eq. (3.6.2) is incomplete and a fuller form is

$$\nabla \phi = 2\pi e t (B \phi n)/h + \pi (V t)/\lambda E_o \qquad (3.6.3)$$

Here E_o is the electron energy and V denotes the inner potential. The latter has a rapidly varying component corresponding to the periodicity of the lattice, but there are also variations at boundaries and defects. Further there is a variation, which might be either slow or rapid if there are compositional changes throughout the specimen. Whilst for many thin film samples t does not vary appreciably, and hence does not contribute to the second term in Eq. (3.6.3) due to the gradient operator, contributions can also be expected here when there are variations in local specimen topography. A good example of where this is the case is at the edges of small magnetic elements where the thickness changes abruptly and the large "electrostatic" signal so generated can conceal important "magnetic" structure in the nearby region.

3.6.3 Imaging modes in Lorentz microscopy

The principal difficulty encountered when using a TEM to study magnetic materials is that the specimen is usually immersed in the high magnetic field (typically 0.50–2.0 T) of the objective lens (Reimer, 1984). This is sufficient to completely eradicated or severely distort most domain structures of interest. A number of strategies have been devised to overcome the problem of the high field in the specimen region (McFadyen and Chapman 1992). These include 1) simply switching off the standard objective lens, 2) changing the position of the specimen so that it is no longer immersed in the objective lens filed (Hitachi Technical Data EM Sheet No. 47), 3) retraining the specimen in its standard position but changing the pole-pieces (Tsuno and Taoka, 1983; Tsuno and Inoue, 1984), once again to provide a non-immersion environment, or 4) adding super mini-lensed in addition to the standard objective lens which is once again switched off (Chapman et al., 1993; Chapman et al., 1994).

1. Fresnel imaging

The most commonly used techniques for revealing magnetic domain structures are the Fresnel (or defocus) and Foucault imaging modes (Chapman et al., 1984). Both are normally practiced in a fixed-beam (or conventional) TEM. Fresnel-mode Lorentz microscopy has the similarity of the mechanism with that of the optical Fresnel interference experiment. Out-of-focus will delineate the boundaries contrast of uniformly magnetized domains because of the Lorentz force. Schematics of how magnetic contrast is generated are shown in Fig. 3.6.1. For the purpose of illustration, a simple specimen comprising three domains separated by two 180° domain walls is assumed. In Fresnel microscopy the imaging lens is simply defocused so that the object plane is no longer coincident with the specimen. Narrow dark and bright bands, delineating the positions of the domain walls, can then

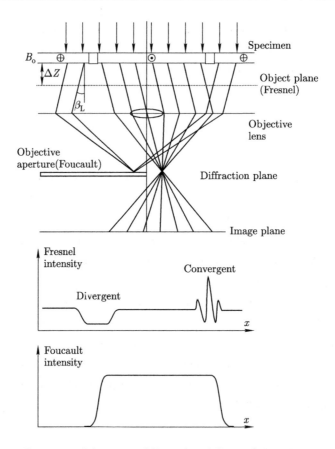

Fig.3.6.1 Schematic of Fresnel and Foucault imaging.

be seen in an otherwise contrast-free image. The structure shown in the bright (known as a convergent domain wall image) reflects the fact that a simple ray treatment is incomplete and that detail, in this case interference fringes, can arise due to the wave nature of electrons. Some software such as QPt package for Digital Micrograph based on the "transport-of-intensity" (TIE) equation has been widely used to quantitatively compute the phase changes of propagating wave and magnetic components of the local magnetization. Three images of in-focus, under-focus and over-focus with almost the same defocus are recorded using Lorentz microscope Fresnel method and then by inputting them into QPt software, the map of projected magnetic induction, presented both by color and vector code, will be calculated out from the electron-wave phase shift.

2. Foucault imaging

For Foucault microscopy, a contrast forming aperture must be present in the plane of the diffraction pattern and this is used to obstruct one of the two components into which the central diffraction spots is split due to the deflections suffered as the electrons pass through the specimen. Note that in general, the splitting of the central spot is more complex than for the simple case considered here. As a result of partial obstruction of the diffraction spot, domain contrast can be seen in the image. Bright areas correspond to domains where the magnetization orientation is such that electrons are deflected through the aperture and dark areas to those where the orientation of magnetization is oppositely directed.

The principal advantages of Fresnel and Foucault microscopy together are that they are generally fairly simple to implement and they provide a clear picture of the overall domain geometry and a useful indication of the directions of magnetization in (at least) the larger domains. Such attributes make them the preferred techniques for in situ experimentation. This theme is developed further in the next section. However, a significant drawback of the Fresnel mode is that no information is directly available about the direction of magnetization within any single domain whilst re-producible positioning of the contrast-forming aperture in the Foucault mode is difficult. Moreover, both imaging modes suffer from the disadvantage that the relation between image contrast and the spatial variation of magnetic induction is usually non-linear (Chapman et al., 1984). Thus extraction of reliable quantitative data especially from regions where the induction varies rapidly is problematic.

A part solution to the problem is provided by the novel coherent Foucault imaging mode which requires a TEM equipped with a field emission gun (FEG) for its successful implementation. Here the opaque aperture is replaced by a thin phase-shifting aperture, an edge of which must be located on the optic axis. An amorphous film containing a small hole works well and the thickness is chosen so that the phase shift experienced is π rad. The technique can only be applied close to a specimen edge as a reference beam passing through free space is required. Provided this is the case, the image takes the form of a magnetic interferogram of the kind obtained using electron holography.

3. Low angle diffraction

Although not an imaging technique, another way of determining quantitative data from the TEM is by directly observing the form that the split central diffraction spot assumes. The main requirement for low angle diffraction (LAD) is that the magnification of the intermediate and projector lenses of the microscope is sufficient to render visible the small Lorentz deflections. Camera constants (defined as the ratio of the displacement of the beam in the observation plane to the deflection angle itself) in the range 50–1000 m are typically required. Furthermore, high spatial coherence in the illumination system is essential if the detailed form of the low angle diffraction pattern is not to be obscured. In practice this necessitates that the angle subtended by the illuminating radiation at the specimen is considerably smaller than the Lorentz angle of interest. The latter condition is particularly easy to fulfill in a TEM equipped with a FEG. Thus whilst LAD moves some way to supplementing the deficiencies of Fresnel and Foucault imaging, the fact that it provides global information from the whole of the illuminated specimen area rather than local information means that alternative imaging techniques must also be utilized.

4. Differential phase contrast imaging

Differential phase contrast (DPC) microscopy overcomes many of the deficiencies of the imaging modes discussed above. Its normal implementation requires a scanning TEM (Dekkers and de Lang, 1974; Rose, 1977; Chapman and Morrison, 1983). It can usefully be thought of as a local-area LAD technique using a focused probe. Fig. 3.6.2 shows a schematic of how contrast is generated. In DPC imaging the local Lorentz deflection at the position about which the probe is centered is determined using a segmented detector situated in the far-field. Of specific interest are the difference signals from opposite segments of a quadrant detector as these provide a direct measure of the two components of β_L. As the probe is scanned in a regular raster across the specimen, directly quantifiable images with a resolution approximately equal to the electron probe diameter are obtained by monitoring the

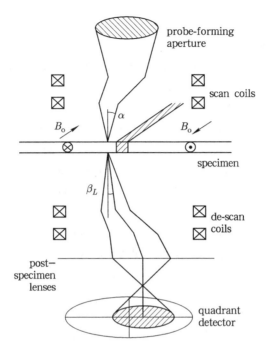

Fig.3.6.2 Schematic of DPC imaging.

difference signals. Providing a FEG microscope is used, probe sizes < 10 nm can be achieved with the specimen located in field-free space (Morrison et al., 1998). A full wave-optical analysis of the image formation process confirms the validity of the simple geometric optics argument outlined above for experimentally realizable conditions (Morrison and Chapman, 1983). The two difference-signal images are collected simultaneously and therefore in perfect registration. From them a map of the component of magnetic induction perpendicular to the electron beam can be constructed. In addition a third image formed by the total signal falling on the detector can be formed. Such an image contains no magnetic information (the latter being dependent only on variations in the position of the bright-field diffraction disc in the detector plane) but is a standard incoherent bright-field image as would be obtained using an undivided spot detector. Thus a perfectly registered structural image can be built up at the same time as the magnetic images, a further distinct advantage of DPC imaging (Chapman et al., 1990).

The specific advantages offered by the technique include: 1) the direction and relative magnitude of the in-plane component of induction in each domain is determined; 2) stray field distributions can be determined quantitatively; 3) to an excellent approximation it is a linear imaging mode (Morrison and Chapman, 1983) so that domain wall profiles can be determined with little or no computer processing (Chapman et al., 1985; Morrison et al., 1988); 4) a spatial resolution better than 3 nm has been demonstrated; 5) microstructural information can be extracted using different combinations of the signals which provide the micromagnetic information (Chapman et al., 1988); and 6) the technique is compatible with such techniques as microdiffraction, X-ray microanalysis and electron energy loss spectroscopy (Chapman et al., 1986). Recently, this technique is commercialized in newly-designed Talos TEM by FEI Company, which will speed up its full application in magnetic domain analysis. The main disadvantages against which this must be set are the undoubted increase in instrumental complexity, the operational difficulty compared with the fixed-beam imaging modes and the longer image recording times inherent to all scanning techniques. In addition, whilst it does not share any of the specific disadvantages listed in connection with the other techniques, there are a number of disadvantages inherent to all TEM investigations of magnetic materials.

3.6.4 In situ magnetic field for Lorentz TEM

The ability to observe domains during the magnetization reversal process is crucial if hysteresis is to be understood at a microscopic level. There are two different approaches to subjecting the specimen to varying magnetic fields.

1. Using a magnetizing stage to induce in-plane magnetic field to specimen

In the first, a horizontal field in the plane of the specimen is generated by magnetizing coils (Tonomura et al., 2001; Uhlig et al., 2003) or new design of "magnetizing stage" (Yi et al., 2004; Inoue et al., 2005), i.e. a specimen holder equipped with a small electromagnet or coils. Two variants are possible. A magnetizing stage can be built into the specimen mounting rod or, if spare ports are available at appropriate positions in the microscope column, one can be introduced as a separate entity.

As the maximum available field relates strongly to the volume of the exciting coils and the nature of the magnetic circuit, greater fields can usually be generated using an independent stage. However, irrespective of which variation is adopted, the presence of a horizontal field over a vertical distance many times greater than the specimen thickness results in the electron beam being deflected through an angle orders of magnitude greater than typical Lorentz angles. Under these conditions the illumination normally disappears from the field of view and can only be restored using compensating coils. This introduces an additional element of complexity and the inevitable result is that in such experiments there is limited opportunity to observe changes as they occur. Rather the changes that took place due to a change of field are normally observed sometime after the event. But this problem is solved in JEOL dedicated Lorentz TEM by installing deflector coil inside TEM automatically coupling with its magnetic field holder and makes the in-plane field as large as ± 600 Oe (Inoue et al., 2005). Fig. 3.6.3 shows a magnetizing stage and its schematic model.

2. Using objective lens to induce perpendicular and in-plane magnetic field to specimen

In order to in situ study a process of technical magnetization by TEM, the simplest way is to apply a magnetic field produced by the objective lens as a source of magnetic field given that the specimen remains located somewhere close to its center (Chapman et al., 1995; McVitie et al., 1995; Harrison and Leaver, 1973; Wang et al., 2007; Liu et al., 2008). The objective lens now acts as a source of vertical field whose excitation is under the control of the experimenter. Variation in the excitation of the mini-lens compensates for the additional focusing effect introduced. As the field of the objective lens is parallel to the optics axis, it is clearly suitable for studying magnetization processes in perpendicular magnetic materials. However, by tilting the specimen, a component of field in the plane of the specimen can also be introduced. Furthermore, given that demagnetizing effects perpendicular to the plane of a thin film are very large, the presence of even moderately large perpendicular fields can often be ignored and it is the much smaller fields in the plane of the specimen that are of interest. When this is the

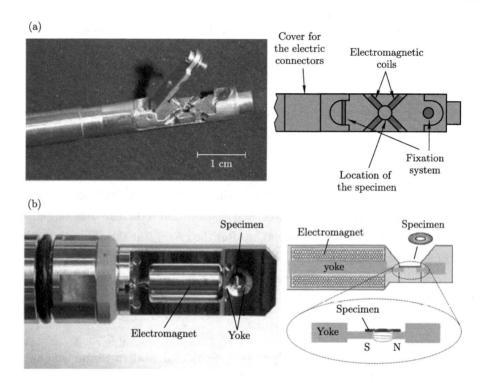

Fig.3.6.3 A magnetizing stage and its schematic model (a) by Uhlig et al. (2003), (b) by JEOL Inoue et al. 2005.

case the objective is set to an appropriate fixed excitation giving a constant vertical field and the sample is simply tilted from a positive angle to a negative one and back again to take it through a magnetization cycle. The principal attraction of the second approach is that the electron optical conditions do not vary during the experiment. Hence, the specimen can be observed throughout and the experimenter can devote full attention to the changing magnetization distribution.

Fig. 3.6.4(a) shows the schematic diagram of the double tilting stage (changing α angle for tilt along y direction and β angle for tilt along x direction). A component magnetic field ($\boldsymbol{H}\sin\alpha$) in the x direction is induced by tilting of the stage along the y axis as seen in Fig. 3.6.4(b), and a component magnetic field ($\boldsymbol{H}\sin\beta$) in the y direction by tilting of the stage along x axis in Fig. 3.6.4(c), where the vector

H denotes the perpendicular magnetic field produced by the objective lens (Liu et al., 2008).

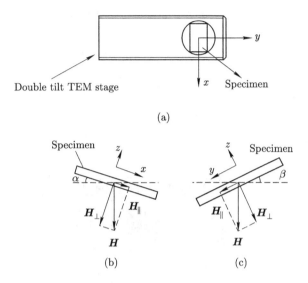

Fig.3.6.4 (a) A schematic model of the TEM double tilt stage. (b) Tilting of the stage along the y axis induces a component of the magnetic field in the x direction. (c) Tilting of the stage along x axis, a component of the magnetic field in the y direction.

3. Measurement of the remnant magnetic field in Lorentz mode

Magnetic microstructure observations should be carried out using an objective lens that can be switched off or a magnetically shielded objective lens (a so-called Lorentz lens), in order to avoid the loss of the original domain structure due to interaction with the strong magnetic field produced by a conventional objective lens (Inoue et al., 2005). The magnetic field in the specimen position can be decreased to zero by using a Lorentz lens installed in a TEM. However, in most cases, a small remnant magnetic field still remains, even though the objective lens is already switched off or shielded, and this remnant field may be too strong to allow observation of magnetic domains in soft magnetic materials. Measurement of the remnant magnetic field is essential therefore for Lorentz TEM investigations. Various methods have been used to characterize the remnant magnetic field in the specimen position in a TEM (Lau et al., 2007; Volkov et al., 2002). A simple yet cost-effective method to characterize the magnetic field at sample position is put forward. A standard TEM holder was modified with Hall probe sensor mounted in place of sample position for this purpose as seen in Fig. 3.6.5(a). The schematic of sensor circuit is displayed in Fig. 3.6.5(b) with output voltage read with a Keithley multi-meter. The detailed magnetic field was measured by simply inserting this modified TEM holder, and the remnant magnetic field is 4 Oe with the OL on and 0 with the OL off after the sample is fully inserted into the column in 200 kV JEOL 2100 F dedicated Lorentz TEM and the field is around 300 Oe with OL off in 300 kV JEOL 3000 F TEM.

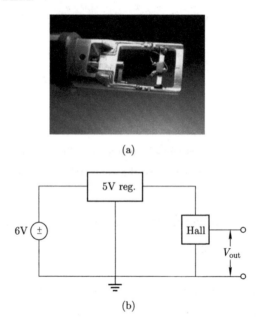

Fig.3.6.5 Hall probe installed in the modified TEM holder (a) and schematic of the sensor circuit (b).

Here, a novel method to measure the remnant magnetic field at the exact specimen position through investigating the movement of a circle Bloch line along the main wall of the cross-tie wall by tilting the specimen of a soft magnetic material (Liu et al., 2009). Fig. 3.6.6(a) shows an under-focused Lorentz microscopy image of a cross-tie wall in Permalloy obtained with the objective lens completely switched off. The main wall is close to the y-axis of the holder and the direction of the magnetization vector M, is denoted by long black arrows in Fig. 3.6.6(a). The remnant magnetic field is measured using a two steps procedure as follows. In the first step, we tilt the holder along y-axis (increasing α angle), so that an in-plane component of the remnant magnetic field is applied to the specimen in the x direction (Fig. 3.6.6). Fig. 3.6.6(b)–(d) show an example where doing such thing that the circle Bloch line marked B moves along the main wall. In the second step, we keep α amount constant, and increase the objective lens current by a very small amount. Fig. 3.6.6(e)–(h) show the resulting movement of the circle Bloch line B along the main wall, but now in the opposite direction compared with that shown in Fig. 3.6.6(a)–(d), which indicates that the remnant magnetic field is opposite to that of the objective lens field.

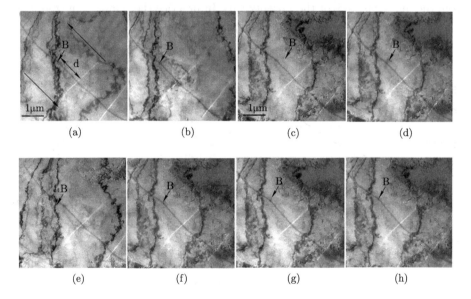

Fig.3.6.6   In situ Lorentz microscopy images of cross-tie wall movements. By tilting specimen holder along y axis: (a) $\alpha = 0°$, (b) $\alpha = 0.5°$, (c) $\alpha = 1.0°$, (d) $\alpha = 1.32°$. By increasing the current of the objective lens: (e) 0.05%, (f) 0.10%, (g) 0.128%, (h) 0.139%.

The plot in Fig. 3.6.7(a) shows the displacement, Δd, of the circle Bloch line

B along the main wall as a function of the tilting angle α. The displacement Δd is defined to be positive for a downward movement of circle Bloch line B. A linear relationship is seen. However, the best fit straight line does not pass through the origin, which implies that the specimen plane is not exactly horizontal at $\alpha = 0$. For α-tile angle 1.32°, the displacement Δd of the circle Bloch line B is about 200 nm. Fig. 3.6.7(b) shows the plot showing the displacement Δd of the circle Bloch line B as a function of objective lens current (expressed as a percentage of the fully saturated value). From this figure we can see that an increasing of the objective lens current of about 0.087% can move the the circle Bloch line B back to its origin position, i.e. $\Delta d = 0$ nm. Assuming the objective lens magnetic field when 100% excited is about 2 T (20000 Oe), the remnant magnetic field is therefore estimated to be about 17 Oe. Since the remnant magnetic field lies opposite to that of the objective lens magnetic field, it is expected to arise not from the objective lens, but from the mini-lens and/or the Lorentz lens. In practice the remnant magnetic field can be compensated by increasing the objective lens current by a known amount.

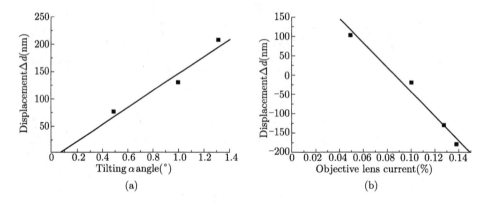

Fig.3.6.7 Plots of the displacement of circle Bloch line along the main wall as a function of (a) tilt angle α, and (b) objective lens current (%).

From an investigation of a test specimen of Permalloy, we have confirmed the existence of a remnant magnetic field in the Lorentz mode of a FEI Tecnai F20 TEM. By using a novel method of only adjustment of the sample tilt and the objective lens current, the remnant magnetic field was estimated to be about 17

Oe, in a direction opposite to that of the objective lens, such that the remnant field can be compensated by increasing the objective lens current in a controlled manner.

3.6.5 Examples of in situ Lorentz microscopy

Lorentz TEM together with different TEM holders such as magnetizing holder, cooling holder, heating holder, electric current holder and tomography holder are good at all kinds of magnetic domain observation and in situ magnetic phase evolution study under different external field. The relationship between magnetic microstructure and physical properties will be established by this kind of study, which will further help to understand the physical mechanism.

1. In situ observation of permalloy under applied magnetic field produced by objective lens

Permalloy, one kind of soft magnetic alloys, has quite good magnetic properties under weak magnetic field, such as very high initial magnetic permeability μ_i and maximal magnetic permeability μ_m, higher magnetic saturation induction B_s and low coercive field H_c (Cui et al., 2005; Li, 2000). It can be easily processed as sheet strip, and is widely used as magnetic head working under weak ac magnetic field conditions, electronmagnet yoke, magnetic conductor, magnetic screen, and magnetic sensor etc. The permalloy TEM specimen was prepared by mechanically thinning, and then ion-milling in an argon atmosphere to electron transparency using a Gantan ion milling system (model 600 B) with a liquid nitrogen cold-stage. Three examples of in situ Lorentz microscopy observation of permalloy will be given in this sections according to the above two types of in situ inducing external magnetic field methods.

The cross-tie wall in Permalloy film was first observed by E. E. Huber using bitter pattern methods in 1958. The magnetic flux closure within the plane of the film has been studied by Huber, as shown in Fig. 3.6.8(a). The main wall is cut at regular intervals by short, right-angle "cross ties" which terminate in free, single ends. The domain wall creeping of the cross-tie walls is connected with circle Bloch line movements along the main wall between two cross-tie walls under the influence of an hard-axis field normal to the wall in Fig. 3.6.8(b) and a variation of the main wall curvature by buckling of located parts of the wall under the influence of an easy-axis field parallel to the wall in Fig. 3.6.8(c), observed using the in-plane magnetic field produced by the Helmholtz coils (Burger, 1971; 1972).

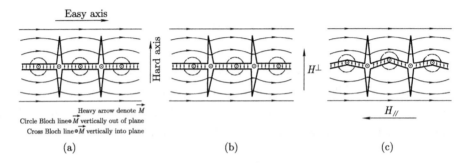

Fig.3.6.8 (a) Diagram of 180° cross-tie walls structure in the Permalloy thin film showing flux closures within the planes of the film. (b) Under the influence of the transverse field in the hard axis direction. (c) Under an easy axis direction field (Liu et al., 2008).

A Philips CM200 field emission Gun (FEG) TEM with Free Lens Control system has been used to carry out in situ Lorentz microscopy within the low magnification (LM) mode. In order to avoid the magnetic saturation of the magnetic specimen, the objective lens must be switched off using the Free Lens Control system before the TEM specimen holder was inserted into the TEM. By increasing the objective lens current and tilting the TEM stage along y axis or x axis in Fig. 3.6.8, a magnetic field parallel to the specimen foil plane is applied.

Fig. 3.6.9 shows Lorentz microscopy images of the cross-tie walls in the Permalloy thin film by tilting the TEM holder. The main wall is close to y direction of the holder and the direction of the magnetization vector M is denoted by the long dash white arrows in Fig. 3.6.9(a). When tilting the TEM stage along y axis, a hard direction field $H_{//} = H \sin \alpha$ is applied in the foil plane along the x direction, where the magnetic field H is about 4 Oe. With the increase of the tilting α angle of the holder, the hard direction field is increased. Fig. 3.6.9 shows the motion of circle Bloch line A indicated by a small white arrow. Under the hard direction field, sections of the Neel wall with magnetization direction against the hard direc-

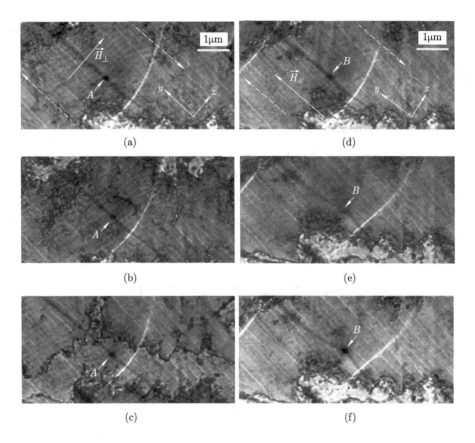

Fig.3.6.9 Lorentz microscopy images by tilting the specimen along y axis, (a) $\alpha = 0.5°$, (b) $\alpha = 0.73°$, (c) $\alpha = 1.72°$; by tilting specimen along x axis, (d) $\beta = 1.43°$, (e) $\beta = 1.49°$, (f) $\beta = 1.60°$.

tion field will reverse into the applied field direction, which causes the circle Bloch line to move along the main wall (Fig. 3.6.8(b)). In Fig. 3.6.9(d) tilting the stage along x axis induces an easy direction field $H_{//} = H \sin \beta$ in the y direction, where the magnetic field H is about 8 Oe. With the increase of β angle of the holder, the buckling of the main wall can be seen at the position of circle Bloch line B (Fig. 3.6.9(e) and (f)), which is in consistent with experiment results by W. Burger (1971).

In summary, the fields in the hard and easy directions were induced by tilting the specimen along y axis and x axis of the TEM holder, respectively. And then in situ observations of the motion of the cross-tie walls in the Permalloy film with applying field was carried out.

2. In situ investigation of permalloy under applied magnetic field produced by magnetizing stage

The in situ observation of technical magnetizing process in permalloy film was carried out with applied in-plane magnetic field produced by a homemade magnetizing stage. Other than cross-tie walls (Huber, 1958) in permalloy, there are two other types of microstructures including chain walls in Fig. 3.6.10(a) (Goodenough, 1959), and displaced types of cross-tie walls and chain walls in Fig. 3.6.10(b) (Yuzi and Yuzo, 1960).

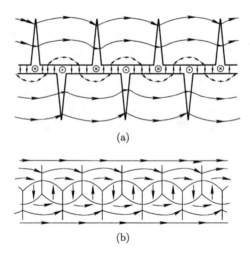

Fig.3.6.10 (a) Magnetic structure of the displaced type of cross-tie walls. (b) The magnetic structure of displaced chain walls. The small black arrows indicate the magnetization vector direction.

Fig. 3.6.11 shows Fresnel Lorentz microscopy images of the modified type of displaced chain wall structure consist of two reversed comb-like cross-tie walls denoted with capital A and B, respectively, in the permalloy thin film in the absence of applied field (Liu et al., 2009). Magnetic domain walls are clearly imaged as white lines and black lines due to deflection of the incident electron beam by the magnetic domains. Figs. 3.6.11(a) and (b) show the same area at over-focused and under-focused conditions, and the contrast of the domain walls is reversed. The magnetization vector direction of the magnetic domains can be revealed by the contrast of the domain walls. The extinction contour breaking at the chain domains in Fig. 3.6.11(c) indicates that the orientations of the domains on the two sides are almost the same, and the chain domains are mis-oriented. Fig. 3.6.12 shows a schematic magnetic flux model of the displaced chain walls observed here. The big magnetic domains with the same magnetization vector direction are separated by the chain domains whose magnetization directions are zigzag and opposite to that of the big domains on two sides of the scratch. The buckling of the chain walls occurs at the position of the Bloch lines due to the long-range stray field interaction between Neel walls. The contrast of the two comb-like cross-tie walls in the Lorentz microscopy images in Fig. 3.6.11 is reversed, which is consistent with this magnetic flux model.

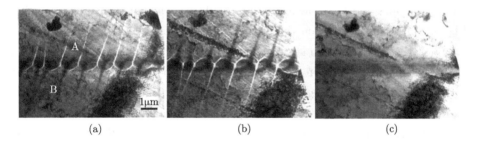

Fig.3.6.11 Lorentz images of displaced chain wall in the permalloy film at different defocused conditions. (a) over-focused, (b) under-focused, (c) in-focused.

Fig. 3.6.13 shows in situ Lorentz microscopy observation results with applied in-plane magnetic field produced by a homemade magnetizing stage. In Fig. 3.6.13(a), there are 7 pairs of white wall A and black wall B numbered by 1 to 7 in the origin magnetic state of the permalloy thin film before applying magnetic field. Under the influence of gradually increasing applied in-plane magnetic field H with direction labeled by the black arrow in Fig. 3.6.13(b), the 2 and 6 pairs of walls disappear in Fig. 3.6.13(b), the white and black walls move towards each other in the 3,

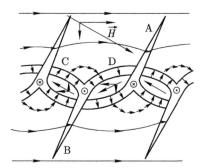

Fig.3.6.12 A schematic illustration of the magnetic flux closure of the modified type of chain walls in Fig. 3.6.11.

4, 5 pairs of walls in Fig. 3.6.13(c). And then 1 and 3 pairs of walls disappear in Fig. 3.6.13(d), and 4 and 5 vanish in Fig. 3.6.13(e). Finally, all the domain walls vanish in the view field as the field up to 11 Oe in Fig. 3.6.13(f). With the decreasing of the applied in-plane magnetic field, the effect of anisotropic inner stress plays an important role again and the displaced chain walls appear again in Fig. 3.6.13(g) and (h). When the applied magnetic field decreases to zero in Fig. 3.6.13(i), the magnetization state of the permalloy thin film do not come back to the initial magnetic state in Fig. 3.6.13(a) due to the magnetic hysteresis effect. In Fig. 3.6.13(j)–(l), the negative direction magnetic field increases gradually, and more pairs of cross-ties walls A and B appear. The number of the cross-tie walls is almost the same as initial magnetic domain microstructure in Fig. 3.6.13(a) as the magnetic field up to –16 Oe in Fig. 3.6.13(l) (Liu et al., 2009).

This magnetization reversal process behavior of the displaced chain walls under applied in-plane magnetic field can be explained by the illustrate diagram in Fig. 3.6.12. As seen from Fig. 3.6.12, the chain domains are separated into a series of subdomains C and D by cross-ties walls A and B. The magnetization vectors of the subdomains C and D both have two components, perpendicular to and antiparallel to that of the two adjacent side magnetic domains, respectively. Because the magnetization vector of the subdomain C is more opposite to the applied magnetic field than that of the subdomain D, the subdomain C disappears first, and the subdomain D gradually reduces which causes the cross-tie walls A and B to move

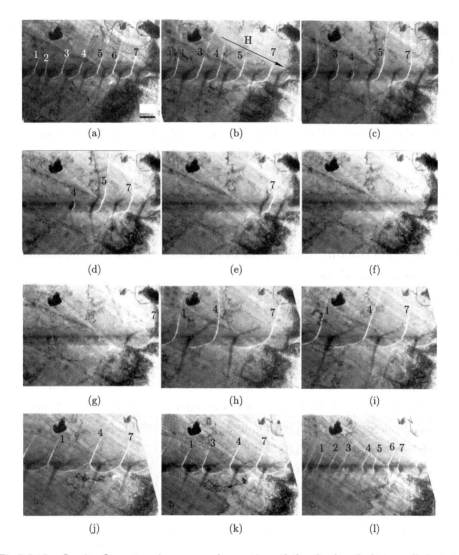

Fig.3.6.13 In situ Lorentz microscopy observation of the displaced chain wall domain structures in permalloy with applied in-plane magnetic field denoted by the black arrow with capital H. The in-plane magnetic field at the position of the specimen: (a) 0 Oe, (b) 7 Oe, (c) 8 Oe, (d) 10 Oe, (e) 10.5 Oe, (f) 11 Oe, (g) 9 Oe, (h) 5 Oe, (i) 0 Oe, (j) –11 Oe, (k) –13 Oe, (l) –16 Oe.

towards each other, and disappears when the circle Bloch line on cross-tie walls B and cross Bloch line on A meet. Then there is only one big magnetic domain in the Permally film. When the applied magnetic field removes, the subdomain D grows up first, and then the subdomain C comes forth.

In summary, in situ Lorentz microscopy investigation of one modified type of chain walls structure in the permalloy film and explained its magnetization reversal process with a schematic illustration of its magnetic flux closures within the plane of the film has been reported. In situ Lorentz microscopy observation was carried out by applying in-plane magnetic field to the permalloy film using the homemade magnetizing stage.

3. Lorenz microscopy application in spintronics, multiferroic and transition-metal oxides

Besides the above magnetic domain study in traditional magnetic materials, Lorentz microscopy has also been widely used in spintronics, multiferroics, transition-metal oxides and superconductivity materials in recent years. The skyrmion named after physicist Tony Skrme for his first theoretical skyrmion identification was originally used to describe a localized, particle-like, configuration in field theory (Skyrme, 1961; 1962). But in magnetic materials, the skyrmions demonstrate a topological spin structure with spin-direction distribution wrapping around a sphere as shown in its schematic configuration in Fig. 3.6.14(a) (Romming, 2013), which are relevant to important physical phenomena such as the topological Hall effects (Neubauer et al., 2009; Lee et al., 2009; Huang and Chien, 2012) and special spin dynamics (Mochizuki 2012, Onose et al. 2012) in spintronics. The real-space imaging of two-dimensional nanometer-scale skyrmions has been well analyzed by Lorentz microscopy under certain magnetic field and low temperature in some B20 type metallic alloys such as MnSi (Tonomura et al., 2012; Yu et al., 2013), Fe-CoSi (Yu et al., 2010) and FeGe (Yu et al., 2011), et al. The similar skyrmions evolution process in an insulating chiral-lattice multiferroic Cu_2OSeO_3 (Seki et al., 2012) was then fully studied by high-resolution Lorentz TEM imaging together with transport-of-intensity (TIE) equation calculation technique enabling mapping the clear lateral magnetization distribution in real space. Figs. 3.6.14 (d)–(g) show the lateral magnetization distribution map for a thin-film sample of Cu_2OSeO_3 obtained through the TIE analysis of Lorentz TEM data taken at 5 K with the direction and relative magnitude distinguished by the color. The magnetic ground state with stripe patterns of lateral magnetization began to change into skyrmions when the magnetic field of about 500 Oe is applied normal to the sample plane and fully occupied skyrmions are observed with $H \sim 800$ Oe and then skyrmions disappear for $H > 1800$ Oe, implying transition into the collinear (ferrimagnetic) spin state. Under a certain magnetic field of about 400 Oe, the magnetic structure evolves from stripe at 5 K to skyrmions occupied state at 40 K and then the skyrmions disappear with the temperature beyond 60 K as shown in Figs. 3.6.14 (h)–(k). Based on these real space observations, the H-T phase diagram for thin-film Cu_2OSeO_3 is summarized (Fig. 3.6.14(c)) by the measurement of skyrmion density. Compared with the narrow A-phase (Skyrmion) region in the bulk sample, the skyrmion phase over a wider T and H rang in the thin film diagram indicates the good stability due to the dimensional restriction. The observed magnetoelectric coupling may potentially enable the manipulation of the skyrmion by an external electric field. The new discovery of writing and deleting single magnetic skyrmions (refers to1 and 0 logic state) by local spin-polarized currents from scanning tunneling microscope (Fig. 3.6.14(b)) makes the application goes further. Lorentz TEM together with in situ electric current TEM holder was used to demonstrate near-room-temperature motion of skyrmions driven by electrical currents in helimagnet FeGe microdevice(Yu et al., 2012). The rotational and translational motions of skyrmion crystal was observed by Lorentz TEM under critical current densities far below 100 A · cm^{-2}, which is 5 orders of magnitude smaller than the driven domain wall current in ferromagnets. Recently, biskyrmion as defined by a molecular form of two bound skyrmions with the total topological charge of 2 has been observed in $LaSrMn_2O_7$ by Lorentz TEM under external fields and the biskyrmions electrically driven behavior is also studied with lower current density ($< 10^8$ A · m^{-2}) (Yu et al. 2014). The electric controllability that low current density can drive skyrmion motion, as well as its nanometer size particle-like nature, points to potential application of skyrmions in high density magnetic storage devices. And in situ Lorentz TEM is definitely the best method so far to see the skyrmions and study its evolution under different external field.

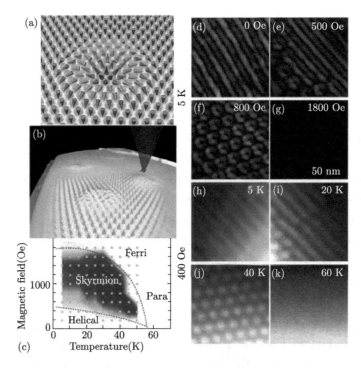

Fig.3.6.14 Magnetic skyrmions studied by Lorentz TEM (a) schematic configuration, (b) manipulation of the skyrmion by scanning tunneling microscopy (Romming et al. 2013), (c) phase diagram for thin-film Cu_2OSeO_3, (d)–(g) lateral magnetization distribution maps for magnetic domains evolution at 5 K under different magnetic field, and (h)–(k) magnetic domains evolution at 400 Oe under different temperatures (Seki et al., 2012).

Spin-transfer torques offer great promise for the development of spin-based devices and has become a hot topic recently. Lorentz microscopy combined with gigahertz excitations clearly maps the orbit of a magnetic vortex core with < 5 nm resolution in permalloy squares (Pollard et al., 2012). The home-made TEM specimen holder was designed to deliver high-frequency electrical signals (0.1 MHz– 1 GHz) and pulsed excitations into a sample region with minimal signal loss and waveform distortion. High-frequency magnetic fields and/or spin-polarized currents near the vortex gyrotropic resonance frequency excite elliptical motion of the vortex around its central equilibrium position. Fig. 3.6.15 shows Lorentz images of the vortex core precession orbit at different driving frequencies and the AC current was applied along the horizontal (X) direction. The precession orbits of the vortex core are directly observed along an elliptical path and reaches the largest at the resonance frequency around 180 MHz. Based on the clear Lorentz images and further analysis, the subtle changes in the ellipticity, amplitude and tilt of the orbit as the vortex is driven through resonance provide a robust method to determine the non-adiabatic spin torque parameter of Landau–Lifshitz–Gilbert (LLG) equation to be $\beta = 0.15° \pm 0.02°$ with unprecedented precision, independent of external effects.

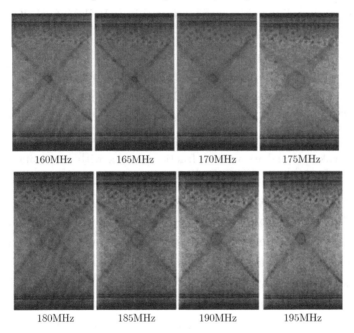

Fig.3.6.15 Lorentz micrographs of the vortex core precession orbit at various driving frequencies. An AC current was applied along the horizontal direction, with the current density maintained at 7.7×10^{10} A · m^{-2}.

4. Lorenz microscopy application in transition-metal oxides

Transition-metal oxides have been attracting a lot of attention due to their peculiar electron-transport, magnetic properties and their unclear mechanism. Manganese oxides, for example, have colossal magnetoresistance (CMR) effect and metal-to-insulator transition (MIT) accompanying the charge, orbital ordering and phase separation et al (Dagotto, 2005). Although, various techniques such as magnetic force microscopy (Lu et al., 1997), scanning Hall probe microscopy (Fukumura et al., 1999), and Kerr microscopy (Gupta et al., 1996) have been used to observe the magnetic domains in perovskite-type manganese oxides, Lorentz TEM is becoming a powerful tool because of its high resolution on the nanometer scale, high sensitivity to small variations in magnetization and its in situ dynamic behavior observation of domain walls under different external field. Lorentz TEM with in situ heating and cooling holder was conducted in LaPrCaMnO (LPCMO) films to examine both the microstructure and magnetic behavior around MIT(He et al., 2010). The stability and conversion of the charge order (CO) and ferromagnetic (FM) states at different temperatures plays an important role for the insulator-to-metal phase transition. Using fixed conditions for Lorentz Fresnel imaging, He et al. found CO and FM two phases coexisted below MIT temperature at ∼164 K, exhibiting mesoscale phase separation as in Figs. 3.6.16 (a)–(c). The elliptic like FM state weakened and its volume fraction shrank with increasing temperature and it finally vanished above the Curie point ∼170 K, which can be clearly seen by Lorentz TEM (Figs. 3.6.16(b)–(d). Mixed CO-FM state area was observed to transform to a pure FM phase when the sample is cooled to 150 K. These Lorentz TEM data provide clear evidence and are interpreted as an interface wetting phenomenon and the CO phase is partially magnetic melted at its interface with FM phase rather than creating new stable CO-FM phases. For LPCMO single crystal, Lorentz TEM with cooling TEM holder also provides reliable method to study the evolution dynamics of complex nanoscale domain structure under magnetic field. The nucleation and growth process of FM phase is demonstrated in Figs. 3.6.17(a)–(f) for conditions of zero-field cooling (ZFC) and 120 Oe field cooling (FC), separately. The Fresnel imaging contrast for ZFC shows that the FM phase initially occurs in a single domain then changes to a double domain (Fig. 3.6.17(b)) with the volume increase induced by cooling to reduce the stray field energy. For the Lorentz micrographs of the FM phase produced in FC (120 Oe), FM domain nucleates and grows on cooling but remains a single domain Fig. 3.6.17(d)–(f) even

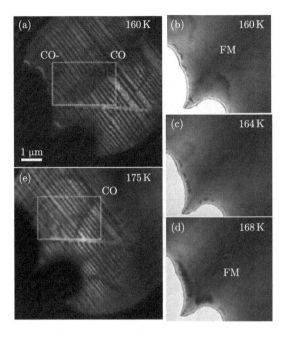

Fig.3.6.16 Heating experiment in TEM for LPCMO illustrating the melting effect for the FM phase close to magnetic Curie point. Dark-field images at (a) 160 K and (e) 175 K, Fresnel images (b)–(d) recorded from the same LPCMO area show the gradual shrinkage and vanishing of the FM cluster for magnetic domain walls, outlined by the dashed contour, when the magnetic Curie point is reached.

after growing to the order of 100 nm. The applied magnetic field appears to assist nucleation and growth of an ordered ferromagnetic phase and the boundary motion in addition to the thermal driving force (Murakami et al., 2009). For multiferroic $BiMnO_3$ polycrystalline sample, ferromagnetic domains was successfully observed by using dedicated Lorentz TEM and liquid-helium cooling holder and the magnetic components of the local magnetization at 19 K was quantitatively computed by the commercial software QPt for Digital Micrograph (Asaka et al., 2012). Lorentz TEM with cooling system works as an effective way to figure out the relationship between the microstructure and the physical properties in transition-metal oxides. Similar studies can also be found in $LaSrMnO_3$ (Koyama et al., 2012), $NdSrMnO_3$ (Asaka et al., 2002), $SrRuO_3$ (Marshall et al., 1999), $LaSrMn_2O_7$ (Asaka et al., 2008) et al.

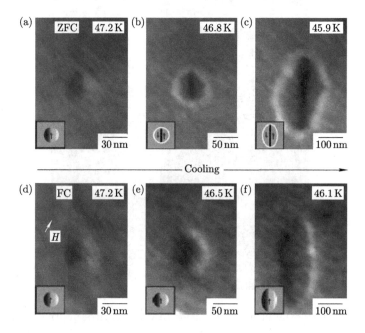

Fig.3.6.17 Nucleation and growth of the FM phase on cooling. (a)–(c) Lorentz microscope images of the FM phase during ZFC. (d)–(f) Lorentz microscope images of the FM phase during FC (120 Oe).

5. Vector field electron tomography by Lorentz TEM

Although the vector potential is central to condensed matter physics, such as superconductivity and magnetism, a precise measurement and visualization of the vector potential in three dimensions has been first experimentally achieved in and around Permalloy thin film by phase-reconstructed Lorentz TEM and tomographic reconstruction recently (Charudatta et al., 2010). The method can probe the vector potential of patterned structures with a resolution of about 13 nm. The dedicated Lorentz TEM with the magnetic field less than 10 Oe was used to characterize the soft magnetic material. Two tilt series about orthogonal axes with tilt range ±70° were recorded and under-, in-, and over-focus Fresnel images were acquired for each angle and then the vector potential was reconstructed based on the images. The 3D visualization of the magnetic induction, clearly showing the vortex configuration and the 90° domains, and vector potential displaying the rotational character for the square island were demonstrated in Fig. 3.6.18. There are deviations between the theoretical and reconstructed vector potentials especially the edges region caused by defocus blurring and limited angle, but this technique definitely expands the arsenal of TEM-based characterization tools to include 3D magnetic characterization.

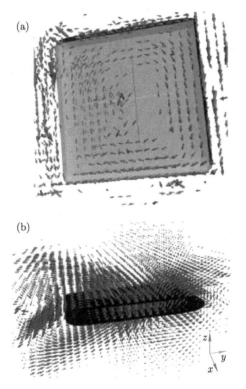

Fig.3.6.18   3D visualization of (a) magnetic induction and (b) vector potential for the square island.

3.6.6 Discussion

Although Lorentz microscopy has been used to study magnetic materials for five decades, it remains a very active area and a number of new imaging modes have been introduced quite recently. The evolution of the technique has been necessary as the complexity of technologically important magnetic materials has grown. In this section we have discussed the principal techniques currently in use and commented on their advantages and disadvantages. Among the most important recent advances are 1) the development of imaging modes capable of yielding more quantitative information on the variation of magnetic induction at high spatial resolution, 2) some success in separating contributions to the image contrast arising from the physical and magnetic microstructures and 3) the development of experimental methods to allow in situ studies to be made of the evolution of the magnetization distribution as a function of field, temperature and/or time. The wide application of Lorentz TEM on magnetic domain study in not only traditional magnetic materials but also other advanced materials such as spintronics, multiferroics, transition-metal oxides, et al. were introduced in the last part. Whilst these developments have turned TEM into one of the most powerful means of studying magnetic structures at high spatial resolution, a number of problems still remain and further development can be anticipated in the future. We end with a brief discussion of some of the major remaining problems. Inherent to the techniques discussed here is the fact that the magnetic signal measured relates directly to components of magnetic induction perpendicular to the electron trajectory and integrated along its length. The problem is particularly severe in multilayer systems where the properties of layers at different physical locations in the stack can differ considerably. It is further exacerbated in cases where structural (as opposed to magnetic) contrast is dominant and this is frequently the case near to defects and edges. Whilst we foresee no straightforward solution to these problems we anticipate that continuing progress will be made due to the increasing power of computers. Thus instead of recording single images, sequences at various tilt angles will be recorded and analyzed. At the same time that the more powerful computers allow us to handle large image sequences, they also facilitate improved micromagnetic modelling by allowing larger array sizes to be used. It therefore seems likely that the inclusion of a powerful workstation as a standard adjunct to the microscope, together with a realisation of the complementary advantages of imaging and modelling, will allow yet more information to be accurately extracted from complex magnetic materials

3.7 Electron holography

Yuan Yao and Xiaofeng Duan

3.7.1 Introduction

Electron holography was first introduced to the electron microscopy community by Gabor (1949) to describe the method which can record all information required to reconstruct the object. In fact, the basic idea of electron holography can retrospect to the concept of the electron interference, which emerged in the 1930s. Based on the Born's interpretation, the interference can occur even for one electron, so it can be used to detect the electromagnetic filed. When an electron crosses the electromagnetic filed, the amplitude and the phase of the electron wave can be altered by the electromagnetic filed. Especially, the electron phase change reflects the distribution of the electromagnetic filed. But Gabor's motivation was to overcome the *notorious* sphere aberration of the transmission electron microscope which was thought unavailable at that time. When two coherent electron beams, one penetrates the specimen (object wave) and another is reference wave with the known form, overlap in the observe plane, the inference patterns record the object wave completely. The amplitude and phase of the objective wave then can be reconstructed with optical light and appreciate phase-plate, or digital processing now, to compensate the sphere aberration and then the resolution could be improved.

Although Haine and Mulvey (1951) showed the 1 nm resolution in the first electron holograms, the limited coherent electron beam restricted the development of the electron holography because of the poor spatial coherent of the electron beam. The intrinsic value of the electron holography was not been realized until the invention of the field-emission electron gun, which emits the high coherent electron beam shining the dawn on the electron holography. (In contrast, the optical holograph, or laser hologram, has been widespread since 1960s.)

Nowadays, the application of electron holography is not confined in the improvement of resolution, but is used widely to research the electric and magnetic properties of the condensed materials. Electron holography can relate the electromagnetic field space distribution with the structure morphology directly, which is important to the application of function materials. Especially combined with in situ technology, electron holography may reveal the physical process of the materials in the external forces, such as electric and magnetic potential, heating and cooling processing, stress and strain field, with or without the structure variation. These researches are far-reaching but are just at the initial stage.

There are some good textbooks and review papers for electron holography, therefore this chapter will summarize the recently development in this research field. For the details of the history and principle of electron holography, the readers are recommended to read the literatures of Völkl et al. (1999), Lichte et al. (2008), Tonomura (1999) and Midgley (2001).

3.7.2 Principle and scheme

Inference phenomenon is the key feature of the wave, from optical light to electron beam. It relates many famous experiments in the history of physics. Generally, assuming that ψ_r is the reference wave with the unit amplitude $|\psi_r|^2 = 1$ and ψ_o is the object wave $\psi_o = ae^{i\varphi_o}$ with small amplitude $a \ll 1$, then the inference wave of these two waves can be written as

$$\psi = \psi_r + \psi_o \tag{3.7.1}$$

the intensity of the combined wave is

$$\begin{aligned} I &= |\psi_r + \psi_o|^2 = (\psi_r + \psi_o)^*(\psi_r + \psi_o) \\ &= |\psi_r|^2 + |\psi_o|^2 + \psi_r^*\psi_o + \psi_r\psi_o^* \\ &= 1 + a^2 + \psi_r^*\psi_o + \psi_r\psi_o^* \\ &\approx 1 + \psi_r^*\psi_o + \psi_r\psi_o^* \end{aligned} \tag{3.7.2}$$

where the second order of a is neglected. So the signals registered on record media reflect not only the intensity, but also the phase information of the object wave. If using the same reference wave ψ_r illuminating the hologram, the object wave could be reconstructed in the free space

$$\psi_r I = \psi_r + \psi_o + \psi_r^2 \psi_o^* \tag{3.7.3}$$

The configurations of the reference and object beams form different holography schemes: in-line and off-axis holography (Cowley, 1992).

1. In-line holography

When penetrating into the sample in the conventional TEM, the electron beam can be classified as transmitted and scattered beams. If the objective lens is defocused to form the image of the wave front just beneath the sample plane, the relative phase shift between the transmitted and the scattered waves is modified by the defocus and the aberration of objective lens, and then interference patterns can be portrayed in the image plane of the objective lens, which was the idea of Gabor in 1951.

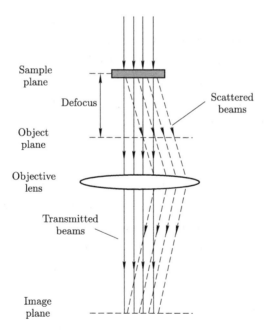

Fig.3.7.1   The ray diagram of in-line electron holography.

If the object wave is put as $\psi_o(r) = 1 - p(r)$ where $p(r)$ is assumed to be very small, the intensity distribution in image plane of objective lens becomes

$$I(r) = |\psi_o(r) \otimes t(r)|^2 = |[1 - p(r)] \otimes t(r)|^2 \qquad (3.7.4)$$

where $t(\mathbf{r})$ is the inverse Fourier transform of the contrast transfer function (CTF) of objective lens $T(\mathbf{q})$. Neglecting the second order of $p(\mathbf{r})$, the intensity can be expanded as

$$I(\mathbf{r}) = 1 - p(\mathbf{r}) \otimes t(\mathbf{r}) - p^*(\mathbf{r}) \otimes t^*(\mathbf{r}) \qquad (3.7.5)$$

Its Fourier transform is

$$\mathfrak{I}\{I(\mathbf{r})\} = \delta(\mathbf{q}) - P(\mathbf{q}) \cdot T(\mathbf{q}) - P^*(-\mathbf{q}) \cdot T^*(-\mathbf{q}) \qquad (3.7.6)$$

where $P(\mathbf{q})$ is the Fourier transform of $p(\mathbf{r})$. If this is multiplied by $T^*(\mathbf{q})$ and inverse Fourier transformed, one can obtain

$$I'(\mathbf{r}) = 1 - p(\mathbf{r}) - \mathfrak{I}^{-1}\{T^{*2}(\mathbf{q})\} \otimes p^*(\mathbf{r}) \qquad (3.7.7)$$

It presents the object wave and its conjugate image. The conjugate image locates two times defocus distance away from the reconstructed image along the beam direction and perturbs the reconstructed image. When the defocus is small, referred to as Fresnel holography, the conjugate image will mess the fine structure of the sample. To overcome this obstacle, a large defocus is suggested to obtain the electron holography, referred to as Fraunhofer holography. In the Fraunhofer mode, the conjugate image is separated to a long distance and indeed forms the merged background but with the expense of the resolution of the reconstructed image. Another way to improve the resolution of the reconstructed image is the "through-focus series" method which changes the defocus during recording the hologram in the Fresnel mode. This method is equal to moving the conjugate image along the beam direction and altering the background of the reconstructed image, therefore the reconstructed image could be enhanced if summing all images. However, this method is not valid for the thicker samples because weak phase object approximation (WPOA) is not satisfied there. More sophisticated methods have been developed to resolve the complex situation (Van Dyck and Op de Beeck 1999).

2. Off-axis holography

Principle

Off-axis holography was first proposed by Leith and Upatnieks with laser beam in 1962, and realized by Mollenstedt and Wahl (1968) with electron beam later. In the ray diagram of off-axis holography (Fig. 3.7.2), the object wave is the electrons passing through the sample, which bring the information of the sample, while the reference wave is the electrons propagate in the vacuum. A biprism filament with positive potential forces two electron beams overlap and form interference fringes in the observe plane. If the relative tilt angle between the reference wave and the object wave is 2α corresponding to a carrier frequency

$$|\boldsymbol{q}_c| = 2|\boldsymbol{k}_0|\sin\alpha \approx 2|\boldsymbol{k}_0|\alpha$$

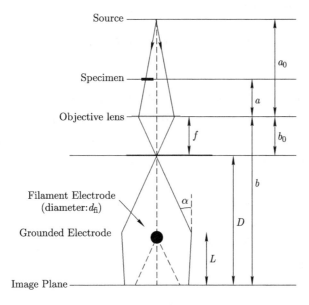

Fig.3.7.2 The schematic diagram of the optical system for conventional off-axis electron holography with the parameters.

where $\boldsymbol{k}_0$ is the wave vector of the incident electron beams, the reference wave can be considered as a plane wave $\psi_r(\boldsymbol{r}) = e^{2\pi i \boldsymbol{q}_c \cdot \boldsymbol{r}}$ and the hologram intensity should be

$$\begin{aligned} I(\boldsymbol{r}) &= |\psi_r(\boldsymbol{r}) + \psi_o(\boldsymbol{r}) \otimes t(\boldsymbol{r})|^2 \\ &= |e^{2\pi i \boldsymbol{q}_c \cdot \boldsymbol{r}} + \psi_o(\boldsymbol{r}) \otimes t(\boldsymbol{r})|^2 \\ &= 1 + e^{-2\pi i \boldsymbol{q}_c \cdot \boldsymbol{r}}[\psi_o(\boldsymbol{r}) \otimes t(\boldsymbol{r})] + e^{2\pi i \boldsymbol{q}_c \cdot \boldsymbol{r}}[\psi_o(\boldsymbol{r}) \otimes t(\boldsymbol{r})]^* + |\psi_o(\boldsymbol{r}) \otimes t(\boldsymbol{r})|^2 \end{aligned}$$

$$= 1 + |\psi_o(r) \otimes t(r)|^2 + 2|\psi_o(r) \otimes t(r)| \cos[2\pi i q_c r + \phi(r)] \quad (3.7.8)$$

where $\phi(r)$ is the phase of convolution image $\psi_o \otimes t(r)$. Obviously, the amplitude and the phase shift $\phi(r)$ of the cosine fringes present the amplitude and the phase of the sample image, respectively, though $\phi(r)$ is not the accurate phase of object wave.

When a plane wave illuminates the hologram in the reconstruction process, the intensity distribution of the Fraunhofer diffraction, or the Fourier transform of the off-axis hologram, is given as

$$\Im\{I(r)\} = \delta(q) + [\psi_o(q) \cdot T(q) \otimes \psi_o^*(q) \cdot T^*(q)] \quad (3.7.9)$$
$$+\delta(q + q_c) \otimes \psi_o(q) \cdot T(q) + \delta(q - q_c) \otimes \psi_o^*(q) \cdot T^*(q)$$

The first and second terms correspond with the auto-correlation function which is the normal diffraction patterns, third and fourth terms present the Fourier transform of the convolution image $\psi_o \otimes t(r)$ and its conjugation, respectively. Here, the Fourier transforms of the convolution image and the conjugate centering on different position $q = -q_c$ and $q = q_c$ are called sidebands which can be separated easily with a proper aperture. Thus the confusion from the conjugate image can be erased off. So off-axis holography is now widely used for the materials research due to this advantage. The following sections will focus on the reconstruction and the application of off-axis electron holography.

Generally, the reconstructed image from the off-axis electron hologram is the convolution image $\psi_o \otimes t(r)$. Assuming the transfer function $t(r)$ or $t(q)$ is known, for convenience, the "reconstructed image" and the "object wave" are equivalent to each other and the transfer function is neglected in the following statement, especially for the moderate and low spatial resolution cases where Gauss defocus is used to acquire the electron hologram and the restoration of the object wave is not necessary (Gajdardziska-Josifovska, 1995).

Elliptical illumination

An important experimental factor in electron holography is the coherence of the electron beam. The coherence can be improved as far as possible if minimizing the beam convergence angle. In practice, the electron beam is demagnified in one direction which is perpendicular to the biprism filament to provide the maximum lateral coherence, namely elliptical illumination, while another direction parallel to the biprism is shrunk to keep enough brightness since the electrons in this direction are not contributed to the holographic recording. The highly elliptical beam is produced by adjusting the condenser lens stigmators and focus setting, so the major axis of the ellipse is normal to the interference fringes.

Fringe contrast, spacing and width of view

The interference fringes (Fraunhofer fringes) spacing s as well as the width of the hologram w are determined by the bias U_{fi} on the biprism when the diameter and the setting of the filament are fixed

$$s = \frac{1}{q_c} = \frac{D\lambda}{2\alpha M(D-L)} = \frac{D\lambda}{2kMU_{fi}(D-L)} \propto \frac{1}{U_{fi}} \quad (3.7.10)$$

$$w = \frac{1}{M}\left(2\alpha L - \frac{Dd_{fi}}{D-L}\right) = \frac{1}{M}\left(2kU_{fi}L - \frac{Dd_{fi}}{D-L}\right) \propto U_{fi} \quad (3.7.11)$$

where D is the distance between the crossover and the image plane, M is the magnification of the objective lens, L is the distance from the biprism and the image plane (Fig. 3.7.2), α is the deflection angle, d_{fi} is the diameter of the biprism, and λ is the wavelength of the incident electron beams. Obviously, s and w cannot be controlled independently. Higher U_{fi} means narrower fringe spacing and broader width, improving the spatial resolution of the phase and the field of view. However, due to the resolution limit of the recording medium, the fringe spacing has to be larger than the four times of the critical pixel size of the recording medium, which actually restricts the minimum fringe spacing. In the case of CCD camera, each fringe should be sampled at least by 4 CCD-pixels, so the biprism bias as well as the field of view is restricted in a confined range. At the same time, the fringes contrast C will be worse with the increasing U_{fi} because of the limited lateral coherence of the electron waves. Considering that the fringe contrast is very important to determine the precision of the phase measurement, the fringe contrast also set an up limit to the biprism bias. Thus, a proper bias is important to balance the view field, the spatial resolution and the phase measurement precision.

The trade-off among the view field, fringe spacing and the contrast will bring the inconvenience for some samples, especially for the case which needs large field of view and high spatial resolution at the same time. One method overcoming this obstacle is to change the electron ray diagram in the "free control" mode of the TEM (Sickmann et al., 2008). Recently, Lichte et al. (2011) have reported a set-up ray scheme with a field of view of 200 nm and a lateral resolution of 3.3 nm. Another way is to use Lorentz lens to acquire the hologram with a large field of view but the reliable spatial resolution is no more than 2 nm due to the strong aberration of Lorentz lens.

Fresnel fringes

The Fresnel diffraction caused by the edge of the biprism can result in the Fresnel fringes at the edge of the interference region (Fig. 3.7.3). Obviously, the Fresnel fringes add a phase shift on the reconstructed object wave near the edge of the inter-

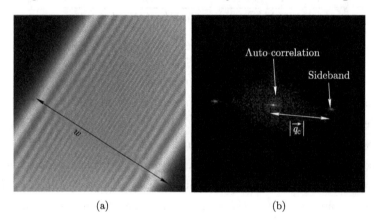

(a) (b)

Fig.3.7.3 (a) Electron interferogram (hologram) for biased biprism is showed where the width of the view field w is indicated. (b) Fast Fourier transform (FFT) of the hologram shows the autocorrelation and two sidebands, where the distance between autocorrelation and one sideband is shown.

ference region. Such effect is difficult to remove even using the reference hologram. A practical way is to adapt a broaden interference region whose edge is outside the CCD camera because the middle of the interference region is affected less by the Fresnel diffraction than the edge (Fig. 3.7.4). Yamamoto et al have proposed a correction method which uses the top envelope curve and the bottom envelope curve of the interference fringes to normalize original hologram and achieved the phase measurement precision of $2\pi/420$ rad from 100 holograms and 100 reference holograms, nearly four times of the common $2\pi/100$ (Yamamoto et al., 2003). They have also put a fine filament on the object plane as a beam stopper whose shadow covers the biprism and suppress the Fresnel diffraction from the biprism edges. The experimental results indicated that about 70% of the phase error was eliminated (Yamamoto et al., 2004).

Fig.3.7.4 The influence of Fresnel fringes could be suppressed by increasing the prism bias and just using the middle region of the interferogram, with the risk of degeneration of fringe contrast.

Multiple-biprism

Tonomura et al. introduced the double-biprism off-axis electron holography to control the s and w independently. Two biprisms are set on the beam axis serially, as shown in Fig. 3.7.5. In the double-biprism mode, the upper biprism is installed just on the image plane of the objective lens and the filament of the lower biprism is set between the crossover point and image plane of the magnifying lens, the fringe spacing and the field of view are

$$s = \frac{a_2 D_l \lambda}{2 M_l M_u [\alpha_l a_2 (D_l - L_l) + \alpha_u b_2 D_u]} \quad (3.7.12)$$

$$w = \frac{2\alpha_l L_l}{M_l M_u} - \frac{d_u}{M_u} \quad (3.7.13)$$

where the parameters are depicted in Fig. 3.7.5. As w is independent of upper deflection angle α_u (or U_{fu}), this leads to s and w being independently controlled.

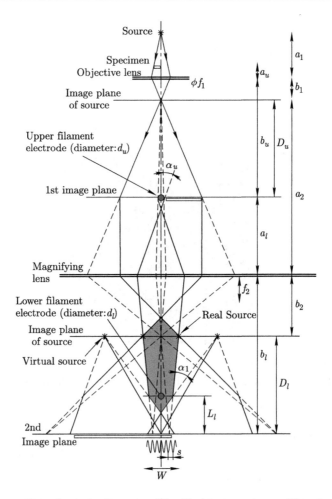

Fig.3.7.5 The schematic diagram of double-biprism system (Harada, 2004).

Furthermore, when the lower biprism is located on the crossover point of the magnifying lens, $(D_l - L_l = 0)$, s also becomes independent of lower deflection angle α_l. Consequently, individual control of both s and w can be completed (Fig. 3.7.6). Additionally, if the lower biprism stays inside the shadow of the upper one, the Fresnel fringes are not generated. It is another advantage of the double-biprism off-axis electron holography. But an adventure in the double-biprism interferometry is that the azimuth angle between the two biprism may complicate the interference fringes in the hologram and make it difficult to interpret. Tonomura et al. also developed a triple-biprism electron interferometry to overcome this disadvantage (Harada, 2006, Ikeda et al., 2011).

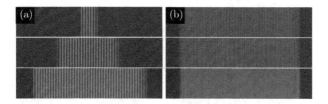

Fig.3.7.6 The width of view field w (a) and fringes spacing s (b) can been controlled independently with flexibility, respectively, in double biprism system, where the Fresnel fringes disappear (Harada, 2005).

3.7.3 Reconstruction of the electron holography

The reconstruction of the electron hologram was realized with optical scheme in the early works. With the development of computer technology, the digital processing is the mainstream technique to retrieve the object wave recently. Comparing with the optical reconstruction road, the digital method displays many advantages, such as the preservation of the full amplitude and phase information, a great diversity of digital filters, etc. The digital reconstruction processing is based on the Fourier transform method. Mathematically, the Fourier transform of the hologram contains three parts: the auto-correlation part and two sidebands, as mentioned in previous section. The distance between the centers of auto-correlation and one sideband equals $|q_c|$ (Fig. 3.7.3b). Each sideband brings the information of object but conjugate each other. Inverse Fourier transform of one sideband can recover the object wave. Such processing is more convenient than the optical frame.

The digital reconstruction contains following steps:

1) Make FFT of the hologram with an appropriate Hann or Haming window function.

2) Determine the center of the auto-correlation and the sidebands.

3) Select one sideband and centering it.

4) Use digital aperture to define the inverse transform range.

5) Make inverse FFT to obtain the amplitude and phase image separately.

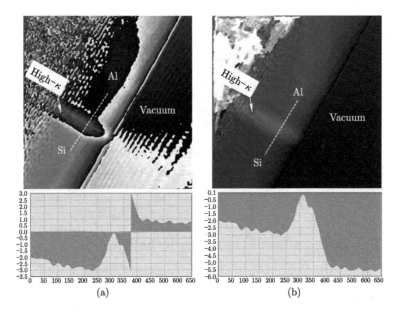

Fig.3.7.7 The phase image of a high-κ dielectric layer between Al metal gate and Si substrate, (a) before and (b) after phase unwrapping with corresponding line profile cross the dielectric film.

1. Reference hologram

Reference hologram is favorable to the reconstruction processing. In the digital FFT, the center of the sideband is not easy to confirm because sometimes it may not fall on a pixel exactly. The error of determination of the sideband center will cause an additional phase shift in the reconstructed image and disturb the phase measurement. The reference hologram acquired without any sample in the field of view could be introduced to verify the center of sideband. In addition, there are always some image distortions from many sources during the hologram recording. These distortions have a significant impact on the image phase as stored in the hologram. The reference hologram is very important to compensate the image distortions in the phase image because the reference hologram without the specimen records the most distortion information. However, since the Fresnel fringes in the object image is influenced by the phase change of the sample, the reference hologram can only corrects the phase distortion of the vacuum region accurately.

2. Phase jump and unwrapping

Since the restored phase is calculated from arctan $\{\text{Im}(\varphi_o)/\text{Re}(\varphi_o)\}$, thus the retrieved phase will be limited in $[-\pi, \pi]$ in the case of standard C-code arctan() function. It will introduce a phase jump in the retrieved phase image if the real phase range spans $-\pi$ or π. For example, a continuous change between $[0, 2\pi]$ in real phase will be separated to $[0, \pi]$ and $[-\pi, 0]$, where a jump from π to $-\pi$ is shown in the obtained phase image. This appearance is known as "phase wrapping" and the technique to remove the phase jump is called "unwrapping". Although some methods have been employed by electron microscopists, the authors here recommend those methods which are used thoroughly in synthetic aperture radar (SAR) research because their algorithms are more general and robust for two-dimensional unwrapping processing (Ghiglla and Pritt, 1998).

3. Spatial and phase resolution

Because of the sampling law the digital aperture radius confining the inverse FFT range of the sideband should not exceed $|q_c|/3$, while for the strict weak phase object, this constraint could be relaxed to $|q_c|/2$. The aperture ensures that the reconstructed objective wave avoids the disturbance from the auto-correlation but the cost is the spatial resolution which is limited to $2/|q_c|$ or $3/|q_c|$. In other words, the spatial resolution of the retrieved image is restricted by the smallest distinguishable fringe spacing of the hologram.

The phase resolution is important because it determines the measurement accuracy of the restored potential value or its distribution. The analysis of the phase resolution is a complicate work. Generally, phase detection limit is the basic conception for the resolution of single hologram. The smallest phase difference between two adjacent pixels reconstructed from hologram fringes is the phase detection limit. It is given by

$$\delta\varphi_{\lim} = \frac{\sqrt{2}snr}{C\sqrt{N}} \qquad (3.7.14)$$

where C is the fringe contrast, N is the electron number collected per constructed pixels and snr is the wanted signal/noise ratio; $snr = 3$ is a reasonable choice. This equation is simple but the measurement of $\delta\varphi_{\lim}$ is not easy. The abbreviated form

$$\delta\varphi_{\lim} = (\text{Noise figure})snr\frac{n_{\text{rec}}}{C_{\text{inel}}} \qquad (3.7.15)$$

is more convenient where n_{rec} is the number of reconstructed pixels, C_{inel} is the contrast loss due to inelastic scattering and Noise figure is defined as

$$\text{Noise figure} = \frac{\sqrt{2\pi}}{|\mu^{sc}|C_{\text{inst}}C_{\text{MTF}}\sqrt{-\ln(|\mu^{sc}|)\frac{B}{ek^2}\varepsilon\tau DQE}} \qquad (3.7.16)$$

with the degree of spatial coherence μ^{sc} of the electron wave, the contrast reduction C_{inst} stemming from the system instability, the contrast damping C_{MTF} by the modulation transfer function (MTF) of CCD camera, the brightness B, the wave number k, the ellipticity ε of the illumination, the exposure time τ and the detection quantum efficiency DQE of the CCD camera. Noise figure delineates the influence of the holographic TEM system which changes little for a stable system and can be regulated periodically. A practical way is using the reference hologram acquired in the vacuum to obtain the value of Noise figure in same condition and then to estimate the phase detection limit of the reconstructed phase of the specimen. Higher spatial coherence, much stabler system, brighter illumination, longer exposure time can improve the phase detection limit. Cooper has reported a phase resolution 0.080 ± 0.008 rad in Titan microscope with a long exposure time 128 seconds to reveal the potential steps of less than 0.030 ± 0.003 V in the B-dope Si (Cooper et al., 2007). An equal method is to take series holograms in same conditions and sum them to enhance the snr. A recent literature shows the accuracy can be increase to $2\pi/1000$ with 100 holograms (Voelkl and Tang, 2010).

Another method to improve the measurement accuracy of the phase shift is to take series holograms in different conditions. Duan (1998) has proposed a method by using two biprism biases to increase the measurement of phase shift 10 times.

Yamamoto and his colleagues have shown that the Fresnel fringes can be removed from the holograms using Lagrange's third-order interpolation and so the phase distortion from Fresnel diffraction can be corrected. Thus the phase change can be detected up to $2\pi/300$ with this method in the phase-shifting electron holography (Yamamoto et al., 2000; Fujita et al., 2006; Yamamoto et al., 2000) and the crystal polarity in a HgCdTe single crystal has been constructed (Yamamoto et al., 2010).

3.7.4 Application of electron holography

Considering only the elastic scattering events and neglecting the absorption, the phase shift of an electron passing through a static electric and magnetic fields is described by the resolution of the Schrödinger equation in Wentzel-Kramers-Brillouin (WKB) approximation

$$\begin{aligned}\varphi_o &= \frac{\pi}{\lambda E}\int_S V(x,y,z)ds - \frac{2\pi e}{h}\int_S \boldsymbol{A}(x,y,z)\cdot d\boldsymbol{S} \\ &= \frac{\pi}{\lambda E}\int_L V(x,y,z)dz - \frac{2\pi e}{h}\int_L A_z(x,y,z)\cdot dz\end{aligned} \quad (3.7.17)$$

where V and $\boldsymbol{A}$ are the static electric potential and magnetic vector potential, respectively, A_z is the z component of $\boldsymbol{A}$, λ and e are the electron wave length and charge, respectively, h is the Plank constant, L is the integration path which parallel to the optical axis and S is the corresponding area with the electric and magnetic field. The phase shift can be written as $\varphi_o = \varphi_{el} + \varphi_{mag}$ with the electric phase shift $\varphi_{el} = C_E \int_L V(x,y,z)dz$ where C_E is the energy-dependent interaction constant and with the magnetic phase shift $\varphi_{mag} = -\frac{2\pi e}{h}\int_l A_z(x,y,z)dz$. Thus the phase shift of the transmitting electrons carries the information of the static electric and magnetic potentials in its trajectory.

1. Electric field

Mean inner potential
When electrons penetrate the samples, the intrinsic electric potential of the materials could influence the behavior of the electrons. The volume average of this intrinsic electron potential is known as its mean inner potential V_0, corresponding

to the zero-order Fourier coefficient of the crystal potential (Reimer and Kohl, 2008) V_0 has several important physical meaning, such as proportional to the moment of the atom charge density, sensitive to the redistribution of the out valance electrons due to bonding. Mean inner potential could cause a phase shift of the transmitting electrons relative to the reference waves which travel the vacuum only, while the refraction effect of the mean inner potential is usually negligible due to the large difference between the value of the energy of the incident electron beam E and the mean inner potential V_0. The phase shift $\Delta\varphi$ depends on the project thickness of the sample thickness t and its mean inner potential

$$\Delta\varphi = C_E V_0 t \qquad (3.7.18)$$

So using electron holography to acquire the phase change of the electron beam may obtain the value of mean inner potential directly for the uniform thickness materials. On the other hand, the phase shift from the mean inner potential should be removed if trying to observe the magnetic potentials variation induced by spontaneous or external potential in the materials. Thus the measurement of the mean inner potential is an important part of the electron holography study.

Table 3.7.1 Experimental values of mean inner potential. Values without notations are from Gajdardziska-Josifovska et al. (1999).

Materials	V_0 (V)	Materials	V_0 (V)
Be	7.8 ± 0.4	C	7.8 ± 0.6
Al	13.0±0.4 12.4±1 11.9 ±0.7	Si	9.26±0.05 12.1 ±1.3
Be	7.8 ± 0.4	C	7.8 ± 0.6
Cu	23.5 ±0.6 20.1 ±1.0	a–Si	11.9 ± 0.9
Ge	15.6 ± 0.8	Ag	20.7 ± 2 17.0–21.8

Continued

Materials	V_0 (V)	Materials	V_0 (V)
MgO	13.01 ± 0.08 13.2 13.5 13.9	a-SiO$_2$	10.1 ± 0.6
ZnS	10.2 ± 1	GaAs	14.53 ± 0.17 14.18 ± 0.07 (Kruse et al. 2003)
PbS	17.19 ± 0.12	InAs	14.50 ± 0.07 (Kruse et al. 2003)
GaP	14.36 ± 0.12 (Kruse et al. 2003)	InP	14.49 ± 0.07 (Kruse et al. 2003)
SiC nanowire	2–2.5 (Yoshida et al. 2007)	Cu Silicide	17–19 (Kohno et al. 2004)
a-C	9.09 (Wanner et al. 2006)	GaN	11 ± 3 (Cherns et al. 1999)

However, the reported experimental values of mean inner potential are limited for a few of materials, including aluminum, silicon, copper, gold, etc, and the published data for the same material from different research teams often showed higher scatter than the error bars (Gajdardziska-Josifovska et al., 1999). Recently, mean inner potential of more materials, including III–V semiconductors (Kruse et al., 2003, Cherns et al., 1999), SiC nanowires (Yoshida et al., 2007), silicon/silica nanoparticles (Kohno et al., 2004), amorphous carbon films (Wanner et al., 2006), clathrate-II phases K$_x$Ge$_{136}$ and Na$_x$Si$_{136}$ (Simon et al., 2011), Hf-based high-κ dielectric film (Yao et al., 2011), etc, have been measured by off-axis electron holography. Although the measurement of mean inner potential is straightforward, there is still a big handicap to accurately determine the thickness of the sample when calculating V_0 from the phase shift. Both the reconstructed amplitude from the same electron hologram and the low-loss spectrum of the EELS can be used to measure the sample thickness (McCartney and Gajdardziska-Josifovska, 1994; Egerton, 2011). But the accuracy of these two methods is often limited by the accuracy of the reported mean free path which is also different for same material. For the thick sample, CBED is a good method to estimate the thickness (Spence and Zuo, 1992).

Electrostatic potential and charge distribution in p-n junction and interface

p-n junction is the core structure in the semiconductor device. The charge distribution of and nearby the p-n junction decides the properties of transistors. The hole doped district and the electron doped district with the different potential can be mapped by electron holography, especially for the nanostructure devices due to the high lateral resolution of the electron holography. Rau and his colleagues first reported the two-dimensional mapping of the potential difference and the depletion region in the transistor with 100 nm gate (Rau et al., 1999). The source and drain can be distinguished in the potential image restored from the phase image with spatial resolution of 10 nm and potential sensitivity of 0.1 V. Ikarashi et al have characterized the potential distribution in 30 nm gate mosfets with the ultra-shallow junctions and revealed that the potential change at the p-n junction is more abrupt in the source-drain extensions formed by a co-implantation technique than those formed by a conventional BF_2 implantation technique (Ikarashi et al., 2010; Gribelyuk, 2008). Gribelyuk et al have studied the As diffusion in the silicon substrate and the position of p-n junction in bulk Si, Silicon-on-insulator structures. The retrieved potential profile from the electron hologram correlated with SIMS result very well (Gribelyuk et al., 2008). Kittler et al have found the evidence of an internal Schottky barrier of 90 mV at the interface between $NiSi_2$ precipitates and silicon matrix, which is about five times smaller than the dark barrier due to the screen of the excess charge carries generated by the incident electron beam or other artifacts, like variation of mean inner potential, dead layers and dynamical diffraction (Formanek and Kittler, 2005).

Electron holography has also been used to quantify two-dimensional electrostatic fields and dopant distribution within the compound semiconductor devices with nanometer resolution. The spontaneous polarity of the ZnO film has been investigated (Xu et al., 2009). The averaged piezoelectric field strength ~ 2.2 MV/cm at InGaN/GaN quantum-well structure has been investigated, which is strongest at the central part of the quantum-well structure (Deguchi et al., 2010). p-n junction has been observed clearly in GaAs, as well as the dopant concentration regions of 1.3×10^{16} cm^{-3} and 3.0×10^{18} cm^{-3} (Sasaki et al. 2006, Sasaki et al. 2008). Strong internal electric field of -2.2 ± 0.6 MV/cm with a sheet charge density of 0.027 C/m^2 has been surveyed in the GaN/In$_{0.18}$Ga$_{0.82}$N/GaN quantum well, and those free electron and hole densities of 10^{20}/cm^3 are confined separately in the quantum well (Cai and Ponce, 2002). The electrostatic potential and dopant concentration have been profiled in AlGaAs/GaAs diode, in agreement with the simulation result (Chung et al., 2009). Polarization fields of 6.9 MV/cm and a two-dimensional electron gas (2DEG) of ~2.1×10^{13}/cm^3 have been measured for the AlInN/AlN/GaN heterostructure (Smith et al., 2010; Zhou et al., 2009), as well as for the AlGaN/InGaN/AlGaN heterostructure (McCartney, 2000). The effect of polarization charge on the conduction band profile of a AlGaN/GaN high electron mobility transistor have been investigated through a detailed comparison of numerical computer simulation to experimental electron holography observation, but a complex surface compensation effect should be considered to understand the obtained potential profile of the heterointerface (Marino et al., 2010). A carrier-trap concentration of 6×10^{16} cm^{-3} has been measured in the graded AlGaAs barriers, which is located away from the InGaAs regions in InGaAs/AlGaAs light-emitting diode (Chung et al., 2010). Moreover, 2DEG and 2DHG (two-dimensional hole gas) have been observed in n-type and p-type AlGaN/AlN/GaN, respectively (Wu et al., 2007; Wei et al., 2009; Wei et al., 2010). The charged dislocations have been surveyed in n-GaN and n-ZnO and the charge density between 5×10^{19} cm^{-3} and 5×10^{20} cm^{-3} were found (Müller et al., 2006). A negatively charged dislocation core has also been observed in threading screw dislocations in n-doped 4H-SiC epitaxial layers and comparison between experimental and simulated potential profiles indicated that the density of trapped charges increases for a higher doped epilayer with the ionization energy 0.89 ± 0.22 eV of the trap (Chung et al., 2011). Electron holography has determined that the cluster formed by silicon atoms and the interstitial carbon in the Si$_{1-x}$C$_x$ films are negative though the measured charge density is inconclusive due to the dynamical electron diffraction effect (Gribelyuk et al., 2011).

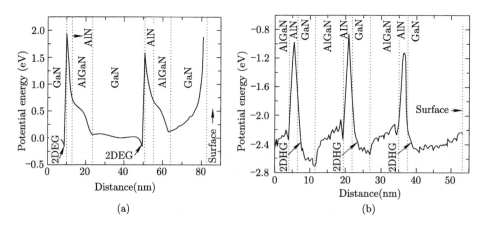

Fig.3.7.8 Electrostatic potential energy profile for the top two periods of n-type (a) and p-type (b) AlGaN/AlN/GaN heterostructure, extracted from a high-magnification electron hologram (Wei et al., 2009, with permission).

Electron holography is also a powerful tool to characterize the potential variation at the crystal boundaries. Small positive phase changes have been assessed at both Σ5(310) and Σ13(510) grain boundaries of SrTiO$_3$, with the core width 1.6±0.4 nm and 2.0±0.5 nm, respectively (Ravikumar et al., 1995; 1997), while the negative phase changes have been demonstrated in the Mg doped grain boundary, corresponding to double Schottky barrier (Ravikumar et al., 1997). The potential barrier of ∼ 3 V at Σ13(510) tilt boundary of SrTiO$_3$ bicrystals has been detected by electron holography and the measurement coincided with a relaxed Σ13(510) tilt boundary supercell model with half-occupied atom columns at the interface (Liu et al., 2004). A barrier height of about 0.7 V with an electrically inactive 24°, [001] tilt grain boundary in SrTiO$_3$ has been found and attributed to a grain-boundary charge superimposed on any other residual effects such as local lattice strain and density reduction in the grain-boundary core associated with the intrinsic nature of grain boundaries in electrical ceramics (Wang and Dravid, 2002). In situ electron holography has displayed that a high current can induce the boundary barrier breakdown in Nb doped Σ5(310) boundary of SrTiO$_3$ (Johnson and Dravid, 1999). Using electron holography, the reduction of negative charge at grain boundary dislocations has been shown in Ca-doped YBa$_2$Cu$_3$O$_{7-x}$, which can lead to a depletion of electron holes available for superconductivity and the improvement of interfacial superconductivity (Schofield, 2004). The experimental measurements have demonstrated that electron diffusion from the n-type SrNb$_x$Ti$_{1-x}$O$_3$ to both La$_{0.9}$Sr$_{0.1}$MnO$_3$ (in p-n junction) and La$_{0.7}$Ce$_{0.3}$MnO$_3$ (in n-n junction). In situ cooling experimental measurement on n-n junctions reveals an apparent change of potential along with the ferromagnetic phase transition (Tian et al., 2005).

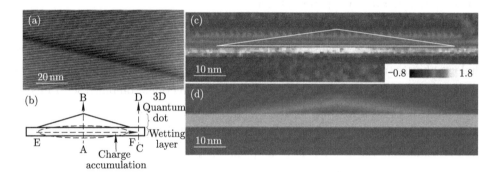

Fig.3.7.9 (a) Electron hologram of individual Ge quantum dot with [110] projection embedded in Si [001] substrate. (b) Sketch of the pyramid shaped dot and wetting layer. (c) Reconstructed phase image of the Ge quantum dot. Phase bar calibrated in radians is shown at bottom right. (d) Simulated phase image of Ge quantum dot based on the pyramid model sketched in (b), the same phase bar as in (c) is applied. (Li et al., 2009 with permission)

Because of the high spatial resolution, electron holography is favorable to characterize the electrostatic potential or charge distribution in nanomaterials. Li et al. have inferred that charge density of 0.03 holes/nm^3 which corresponds to about 30 holes localizing on individual Ge quantum dot embedded in boron-doped silicon. Zhou et al. (2011) have observed the large phase shift of the individual GaN quantum dot due to the spontaneous polarization and piezoelectric field of 7.8±2 MV/cm in AlN/GaN quantum dot supperlattice structure. den Hertog et al.

have mapped active dopant concentration of 10^{19} and 10^{20} cm^{-3} in single silicon nanowires with a detection limit as low as 10^{18} cm^{-3} and estimated the charge density of -1×10^{12} cm^{-2} at the wire-oxide interface (den Hertog et al., 2009; 2010). Li et al. (2011) have measured the phase shift across Ge/Si core/shell nanowires, which indicates the hole density of 0.4 ± 0.2 cm^{-3} inside the Ge cores.

Even there are abundant literatures in charge density measurement by electron holography, it always should be emphasized that the source of potential change across the interface is very complex because many factors, such as work function difference, beam charge aggregation, stray field, surface depletion layer etc are involved to generate such potential change.

In situ electrostatic field

Besides mapping the intrinsic electrostatic potential in the specimen, electron holography can also characterize the potential distribution responding to the external force, such as the external bias on the p-n junction, combined with in situ technique in TEM. Twitchett et al. have measured electrostatic potential profiles across reverse-biased Si p-n junctions in situ in TEM (Twitchett et al., 2002), and at the same time Zettl et al. have mapped the potential field near the tip of an emitting carbon nanotube (Cumings et al., 2002). Beleggia et al. have measured the charge distribution along a biased carbon nanotube bundle and shown the charge density increases linearly with distance from its base, reaching a value of ~0.8 electrons/nm near its tip (Beleggia et al., 2011). Chou et al. have found that the long range electric field created by the biased WO_3 nanowire can perturb the reference wave when field emission occurs. It will degrade the quantifying interpretation of the reconstructed phase image (Chou et al., 2006). They have also proposed that employing a tiny tungsten anode can shield the reference wave against the stray field (Kim et al., 2007). Yamamoto and his colleagues have directly observed the changes of electric potential in an all-solid-state lithium ion batteries during charge-discharge cycles and quantified the 2D potential distribution resulting from movement of lithium ions near the positive-electrode/electrolyte interface, which confirms that lithium extraction from the positive electrode during charging results in oxidation of cobalt from Co^{3+} to Co^{4+}, combined with EELS data (Yamamoto et al., 2010). Recently, Yao et al. have mapped the negative charge distribution under positive bias in the charge trapping memory and discovered that the charges accumulates near the interface between the Al_2O_3 block layer and the HfO_2 trapping layer clarifying a confusion in the memory derive study (Yao et al., 2013)

Polarization in ferroelectrics

Ferroelectric domain wall plays an important role in some functional materials and electron devices. Although TEM observation of ferroelectric structures has been achieved for many years, electron holography is effective to visualize the domain wall in a high-resolution picture. 90° domain has been demonstrated in $BaTiO_3$ film (Zhang et al., 1993, 1992; Spence et al., 1993), the optimal condition has been discussed (Lichte et al., 2002) and some influences, such as diffraction effect or beam-induced polarization, should be minimized to obtain a consistent value for the spontaneous polarization (Matsumoto et al., 2008). At low temperatures, the ferroelectric polarization results in a remarkable positive potential on the interface and a negative potential on the surface of epitaxial $Ba_{0.5}Sr_{0.5}TiO_3/LaAlO_3$ film, and at room temperature, certain charges could only accumulate at the interfacial dislocations and other defective areas (Tian et al., 2005). This research is still in development.

Electron holographic tomography

As shown in Fig 3.7.10, substantial modification of the potential is apparent near the top and bottom surfaces of the reconstruction arising from a combination of sample preparation effects and properties of semiconductors (such as surface depletion) near surfaces. The amorphous and crystalline surface layers have been represented schematically and were not reconstructed tomographically. Equipotential contours spaced every 0.2 V have been superimposed onto the reconstructed tomogram, highlighting the electrostatic potential distribution in the $y - z$ cross-section of the specimen. (Twitchett-Harrison et al., 2007)

Directed by the concept of electron tomography, electron holography has been extended to draw the three-dimensional distribution of the electrostatic field in nanostructrures. Twitchett-Harrison et al have composed a tomographic potential of a biased Si p-n junction from series of tilted 2D maps (Twitchett-Harrison et al., 2006; 2007; 2008). Chen et al. have demonstrated quantitative electron graphic tomography using a latex sphere and shown the promising future of this technique

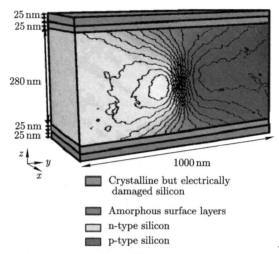

Fig.3.7.10 Tomographic reconstruction of the electrostatic potential arising from a 3-D section of a thin FIB-prepared membrane containing an electrically biased p-n junction.

(Fujita and Chen, 2009). Wolf et al. have developed a software package to retrieve the 2D information and reconstruct the tilt series images automatically (Wolf et al., 2010). They have also employed this method to map the 3D potential distribution of pure GaAs and GaAs/AlGaAs core/shell nanowires, respectively, and the measured mean inner potential of GaAs nanowires agrees well with the bulk material whereas about 1 V lower for GaAs/AlGaAs nanowires (Wolf et al., 2011). This interdisciplinary technique will bring the prospective advantages in future for analyzing the materials intensively.

2. Magnetic field

Magnetic flux

Eq. (3.7.17) indicates the magnetic vector potential can lead to a phase shift of the electrons passing through the magnetic field and then can be visualized by electron hologram although the interpretation is more complicated than the electric case. If the specimen is of uniform composition and of a constant thickness within the observed region, or the phase shift from electric potential is known, the phase shift caused by magnetic field in a phase image can be written in the form of a closed loop integral

$$\begin{aligned}\varphi_m &= -\frac{2\pi e}{h}\int_{L_1} A_z(x,y,z)\cdot dz + \frac{2\pi e}{h}\int_{L_2} A_z(x,y,z)\cdot dz \\ &= -\frac{2\pi e}{h}\oint \boldsymbol{A}\cdot d\boldsymbol{L} \\ &= \frac{2\pi e}{h}\iint \boldsymbol{B}\cdot \hat{n} dS \\ &= \frac{\pi}{\phi_0}\Phi(S) \end{aligned} \qquad (3.7.19)$$

for a loop formed by two electron trajectories, L_1 and L_2, where Stokes' theorem is utilized and $\phi_0 = h/2e = 2.07\times 10^{-15}$ T·m^2 is the quantized flux. The phase difference between any two points is therefore a measure of the magnetic flux through the whole region of space bounded by two electron trajectories crossing the sample at these two points. Thus a graphical representation of the magnetic flux distribution throughout the sample can be obtained by measuring the phase difference in the reconstructed phase image. A phase difference of 2π corresponds to an enclosed magnetic flux of 4.14×10^{-15} T·m^2. The relationship between the phase shift and the magnetic induction $\boldsymbol{B}$ can be established by considering the gradient of phase difference

$$\nabla \varphi_m = \frac{\pi}{\phi_0}[B_y^p(x,y) - B_x^p(x,y)] \qquad (3.7.20)$$

where $B_i^p(x,y) = \int B_i(x,y,z)dz (i=x \text{ or } y)$ are the components of the magnetic induction perpendicular to the incident electron beam direction. In the case where both the thickness and magnetic indution do not vary in z direction, a simplified form

$$\nabla \varphi_m = \frac{\pi t}{\phi_0}[B_y(x,y) - B_x(x,y)] \qquad (3.7.21)$$

can interpret the distribution of lateral component of magnetic induction directly in the phase image.

A mandatory precondition of validity of Eq. (3.7.21) is the phase shift contributed from mean inner potential must be deducted from the retrieved phase image before analyzing the magnetic contribution according to Eq. (3.7.17). Several methods have been proposed in order to accomplish this purpose. "Time-reversal operation" reverses the sample against the incident electron beam to flip the sign of phase from magnetic field but the mean inner potential unchanged (Tonomura et al., 1986). External magnetic field has also been applied to change the sign of magnetic phase shift during in situ observation of magnetic interactions within patterned cobalt nanostructures (Dunin-Borkowski et al., 1998). The hologram of the nonmagnetic phase obtained above the transition temperature T_c, which possessed the mean inner potential only, has been used to acquire the pure magnetic information of the magnetic phase below T_c too (Yoo et al., 2002; Loudon et al., 2002).

An important work of magnetic flux measurement with electron holography was the disclosure of Aharonov-Bohm (AB) effect accomplished by Möllenstedt and Bayh (1962) first and then confirmed by Tonomura and his colleagues who have used a magnetic ring which was embedded in superconducting material to shield the magnetic flux perfectly from leaking outside and measured the Ehrenberg–Siday–Aharonov–Bohm effect (ESAB-effect) in excellent agreement with theory (Tonomura et al., 1986). Later, the quantized fluxons (Matsuda et al., 1989) and their dynamic evolvement have also been revealed with the electron holography (Matsuda et al., 1991; Bonevich et al., 1993; Yoshida et al., 1992).

Magnetic domains and magnetic nanostructures

Magnetic domain has been characterized by both Lorentz electron microscope (or Lorentz mode in conventional TEM) and electron holography. Compared with Lorentz image mode, electron holography can map longer-range magnetic fields and the fine details in the reconstructed image can be related to the specimen structure directly because it is acquired in focused condition. That is especially important for the interpretation of magnetic nanostructures which need avoid the contrast from the edges of the specimen. Another advantage is that electron holography can measure the flux density value absolutely if the specimen thickness is known.

The cross-tie and Neel wall have been investigated by electron holography in the permalloy film first by Tonomura et al (Tonomura et al., 1982). Anisotropic magnetic films of Fe-Pt film with the anisotropic $L1_0$ order deposited on MgO substrate have been characterized and the induction map confirmed that adjacent domains within the FePt film have opposite polarity (McCartney et al. 1997). The

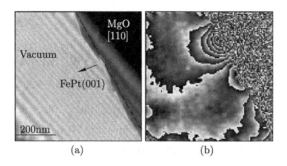

Fig.3.7.11 (a) Off-axis electron hologram of MgO/FePt with [001] easy axis parallel to film normal. (b) Reconstructed phase image showing phase shifts in vacuum consistent with flux leakage associated with domain wall (McCartney et al., 1997 with permission).

direction of the recorded magnetization for in-plane and perpendicular magnetic recording film has been pictured respectively (Osakabe et al., 1983; Matsuda et al., 1982). The representative magnetic induction maps recorded from pseudo-spin-valve elements when applying external field, which were prepared from a polycrystalline $Ni_{79}Fe_{21}/Cu/Co/Cu$ film, have shown the directions of the two magnetic layers were inferred to be parallel where the phase contours in the elements were closely-spaced, while the magnetic layers were inferred to be antiparallel where the contours were widely-spaced. The magnetic configurations did not have end domains or vortices, and magnetic interactions between the neighboring elements are not significant (Kasama et al., 2005). The magnetic induction maps of magnetic tunnel junction Au/Co/Fe/MgO/FeV/MgO indicated no magnetic coupling between FeCo and FeV when specimen was tilted in the magnetic field created by slightly excited objective lens, a simulation of the switch on and off, and the local hysteresis loop drawn from the phase image was consistent with the magnetometry measurement (Javon et al., 2010). Using electron holography and Fresnel imaging, micrometre-sized ferromagnetic regions spanning several grains coexisting with similar-sized regions with no local magnetization have been founded and the ferromagnetic regions have a local magnetization of 3.4 ± 0.2 Bohr magnetons per Mn atom (the spin-aligned value is 3.5 μ_B per Mn). A diversity of unexpected interaction phenomena between crystallographic twins and magnetic domain walls in monoclinic magnetite has been revealed by dynamic observations of magnetite during heating and cooling across the Verwey transition (Kasama et al., 2010). In situ electron holography has also uncovered the magnetic domain and induction orientation movement in metal (Junginger et al., 2007; Bromwich et al., 2006) and complex composites (Shindo et al., 2002; 2004; Kasahara et al., 2007).

The magnetic structures of the nanoparticles allure many attention due to their widely utilization in the biosensor (Wang and Li, 2008; Haun et al., 2010), drug delivery (Arruebo et al., 2007), and medical diagnosis or therapy (Mornet et al., 2004). Three-dimensional magnetic vortices in chains of FeNi nanoparticles have been characterized using electron holography and micromagnetic simulation, which revealed that the diameters of the vortex cores depend sensitively on the orientation with respect to the chain axis and that vortex formation can be controlled by the presence of smaller particles in the chains (Hÿtch et al., 2003). The magnetic rings have been pictured in the isolated polyhedral Co nanoparticles whose magnetic dipoles were aligned into a closed circuit and produced a net moment of zero to minimize both the magnetic energy and the field outside of the ring (Kasama et al., 2008; Tripp et al., 2003). The hologram of an isolated 50-nm-diameter single crystal of magnetite (Fe_3O_4) has been examined at room temperature and 90 K (below the Verwey transition), respectively, and uniformly magnetized single-domain magnetic states, including the characteristic return flux of an isolated magnetic dipole, have been found consistent with shape anisotropy dominating the magnetic state of the crystal (Kasama et al., 2011). The nanorod composed of crystalline FePt nanoparticles showed residual magnetic flux density about 1.53 T which parallel to the axis of the nanorod, estimated via electron holography (Che et al., 2005). Phase distributions of α-Fe nanorod encapsulated in carbon nanotubes have depicted the saturated magnetization of the α-Fe nanorod with a CNT outer shell was very close to the bulk materials, which suggested that the core-shell nanostructure effectively prevented the surface oxidation which may significantly degrade the magnetic properties of nanostructured α-Fe (Fujita et al., 2007). Similar phenomenon has also been found for the Fe-Co-P nanoparticles encapsulated in carbon nanotubes (Jourdain et al., 2006). The magnetic induction map of chains of nano-crystals of magnetite (Fe_3O_4) or greigite (Fe_3S_4) in magnetotactic bacteria has illustrated the highly optimized linear nature of the magnetic field lines associated with a linear chain of magnetite magnetosomes, with the help of electron holography (Dunin-Borkowski et al., 1998; 2001; Simpson et al., 2005). The dipolar ferromagnetic 180° domain structures imaged in Fe_3O_4 nanoparticle ordered arrays and the domain wall evolvement at different temperature have revealed a true phase transition at 575 °C, implying a spontaneous development of long-range magnetic order (Yamamoto et al., 2011).

3. Strain measurement: dark holography

The core idea of electron holography is the interference of the electron waves: the object wave and the reference wave. If both of these waves are diffracted beams and form a dark field image superposed by the interfered fringes, it is called "dark holography". The phase difference between the two diffracted beams depends on their dynamic elastic scattering and the geometric phase. If the thickness is uniform, the former will be a constant phase term, while the later encodes the strain information through phase gradients (Hÿtch et al., 2008; 2011). To achieve dark holography, the specimen is tilted to double-beams condition and then TEM is changed to the "center dark image" mode to ensure the enough contrast and avoid the phase distort from optical system. The dark holography is extremely suited for the 2D strain field mapping of the nano-electronic device, such as SiGe/Si strain transistors (Hüe et al., 2009; Cooper et al., 2009, 2010; Wang et al., 2012), because of its nanometer-scale resolution, a large field of view and excellent sensitivity. Béché and his colleagues also have investigated some experimental parameters, influence to optimize the dark field holography technique (Béché et al., 2011). Although in the infancy, dark holography opens up a new application field of the electron interference and shows the profound significance for the materials research.

4. High resolution electron holography

The initial concept of electron holography was to obtain the total information of the objective wave to improve the resolution of Cs-uncorrected TEM. As the development of high coherent field-emission electron gun and stable optical systems, the high resolution off-axis electron holography has been achieved with the help of optical or digital correction of aberrations. Lichte and his colleagues have demonstrated Si [110]-dumbbell in the retrieved phase image containing information up to 0.104 nm with digital aberration correction (Orchowski et al., 1995). They have also revealed the mis-tilt in Σ13 grain boundary in Au which cannot be observed in common diffractogram due to Friedel's law (Orchowski and Lichte, 1996). The Ga ($Z = 31$) and As ($Z = 33$) can be distinguished using high-resolution electron holography and the structure polarity can be determined due to the phase sensitivity of the retrieved image (Lehmann and Lichte, 2005).

As the emergence of Cs-corrected TEM, electron holography has been enhanced by the current density increasing, signal-to-noise ratio improvement and collective efficiency elevation, enough to measure the phase shift in atom dimension more accurately (Lichte et al., 2009). Liu and his colleagues (2010) have detected the phase fluctuation across the stacking faults in a non-polar ($11\bar{2}0$) GaN using off-axis electron holography on Cs-corrected TEM, suggesting the presence of a built-in electric field.

3.7.5 Outlook

Benefited from the development of electron microscope, electron holography is feasible nowadays to interpret the electric and magnetic field in solid state physics and materials research, spanning micrometers to atomic scale, without any changes in conventional TEM. Digital image processing is flexible for data analysis and software are available on market. Combined with HRTEM, diffraction, EELS and in situ techniques, electron holography can be used to characterize the relationship among crystal structure, chemistry and physical functions, especially in nanomaterials research. Therefore the application of electron holography is growing rapidly worldwide.

Since electron holography may reveal the atomic information directly in the Cs-corrected TEM with high brightness and signal/noise ratio, it is possible to picture the potential distorted by chemical bond or electrons fluctuation patterns in atomic dimension. Moreover, in situ, three-dimensional and real-time electron holography are emerging and should be developing fast in the future.

References

Aharanov, Y., Bohm, D. (1959) Significance of electromagnetic potentials in the quantum theory. Phys. Rev. **115**, 485.

Allen, L.J., McBride, W., O'Lear,y N. L., Oxley, M. P. (2004) Exit wave reconstruction at atomic resolution. Ultramicroscopy, **100**, 91.

Arruebo, M., Fernández-Pacheco, R., Ibarra, M. R., Santamaría, J. (2007) Magnetic nanoparticles for drug delivery. Nanotoday, **2**, 22.

Asaka, T., Anan, Y., Nagai, T., Tsutsumi, S., Kuwahara, H., Kimoto, K., Tokura, Y., and Matsui, Y. (2002) Ferromagnetic domain structures and nanoclusters in Nd1/2Sr1/2MnO$_3$. Phys. Rev. Lett. **89**, 207203.

Asaka, T., Kimura, T., Nagai, T., Yu, X. Z., Kimoto, K., Tokura, Y., and Matsui, Y. (2005) Observation of Magnetic Ripple and Nanowidth Domains in a Layered Ferromagnet. Phys. Rev. Lett. **95**, 227204.

Asaka, T., Nagao, M., Yokosawa, T., Kokui, K., Takayama-Muromachi, E., Kimoto, K., Fukuda, K. and Matsui, Y. (2012) Magnetocrystalline anisotropy behavior in the multiferroic BiMnO$_3$ examined by Lorentz transmission electron microscopy. Appl. Phys. Lett, **101**, 052407.

Béché, A., Rouvière, J. L., Barnes, J. P. and Cooper, D. (2011) Dark field electron holography for strain measurement, Ultramicroscopy, **111**, 227.

Beleggia, M., Kasama, T., Dunin-Borkowski R.E., Hofmann S. and Pozzi G. (2011) Direct measurement of the charge distribution along a biased carbon nanotube bundle using electron holography. Appl. Phys. Lett., **98**, 243101.

Blumenau, A. T., Fall, C. J., Jones, R., Oberg S., Fravenheim T. and Briddon P R. (2003) Structure and motion of basal dislocations in silicon carbide Phys. Rev. B. **68** , 174108.

Bonevich, J. E., Harada, K., Matsuda, T., Kasai, H., Yoshida, T., Pozzi, G., and Tonomura, A. (1993) Electron holography observation of vortex lattice in a superconductor. Phys. Rev. Lett. **70**, 2952.

Boothroyd, C. B. (2000) Quantification of high-resolution electron microscope images of amorphous carbon. Ultramicroscopy, **83**, 159.

Boothroyd C. B. (1998) Why don't high-resolution simulations and images match?. J. Microsc, **190**, 99.

Boothroyd, C. B., Dunin-Borkowski, R. E., Stobbs, W. M. and Humphreys. C. J. (1995) Quantifying the effects of amorphous layers on image contrast using energy filtered transmission electron microscopy, Beam-Solid Interact. Mater.

Res. Soc. Symp. **354**, 495.

Bourret, A., Desseaux, J., Renault, A. (1982) Core structure of the lomer dislocation in germanium and silicon. Philos. Mag. A. **45**, 1.

Bromwich, T. J., Kasama, T., Chong, R. K. K., Dunin-Borkowski, R. E., Petford-Long, A. K., Heinonen, O. G. and Ross C. A. (2006) Remanent magnetic states and interactions in nano-pillars. Nanotechnology, **17**, 4367.

Burger, W. Phys. Status Solidi., 1971, 4: 713.

Burger, W. Phys. Status Solidi., 1971, 4: 723.

Burger, W. Phys. Status Solidi., 1972, 13: 429.

Buxton, B. F., Eades, J. A., Steeds, J. W. and Rackham, G. M. (1976) The Symmetry of Electron Diffraction Zone Axis Patterns. Philosophical Transactions of the Royal Society of London Series A-Mathematical Physical and Engineering Sciences, **281**, 171.

Cai, J. and Ponce, F. A. (2002) Study of charge distribution across interfaces in GaN/InGaN/GaN single quantum wells using electron holography. J. Appl. Phys. **91**, 9856.

Chapman, J. N. and Morrison, G. R. J. Magn. Magn. Mater., 1983, 35: 254.

Chapman, J. N., McFadyen, I. R. and Bernards, J. P. C. (1986) Investigation of Cr segregation within rf-sputtered CoCr films. J. Magn. Magn. Mater., **62**, 359.

Chapman, J. N., McVitie, S. and McFadyen, I. R., in Kirschner, J., Murata, K. and Venables, J. A. (eds.) (1987) Scanning Microscopy Supplement 1. Chicago: AMF O'Hare, 221.

Chapman, J. N., Morrison, G. R., Jakubovics J. P. and Taylor, R. A. (1985) Determination of domain wall structures in thin foils of a soft magnetic alloy. J. Magn. Magn. Mater., **49**, 277.

Chapman J. N. et al. (1984) Electron Microscopy and Analysis 1983, Proceeding of the Iop conference **68**, 197-200.

Chapman, J. N., Rogers, D. J. and Bernards, J. E C. (1965) High resolution imaging of magnetic structures in the transmission electron microscope. J.Phys. Paris Colloq. C8, Suppl. **12**, 49.

Chapman, J. N. (1989) High resolution imaging of magnetic structures in the transmission electron microscope. Materials Science and Engineering, **B3**, 355.

Chapman, J. N. (1984) The investigation of magnetic domain structures in thin foils by electron microscopy. J. Phys. D, **17**, 623.

Chapman, J. N., Ferrier, R. P., Heyderman, L. J., McVitie, S., Nicholson, W. A. P. and Bormans, B. (1993). Micromagnetics, microstructure and microscopy In: Electron microscopy and analysis. 1993. Proceedings of the Institute of Physics Electron and Analysis Group Conference pp. 1-8.

Chapman, J. N., Heyderman, L. J., McVitie, S., Nicholson, W. A. P., in: Bando, Y., Kamo, M., Haneda, H., Aizaw, T. (Eds.). (1995) Proceedings of the 2nd International Symposium on Advanced Materials (ISAM'95), 123.

Chapman, J. N., Johnston, A.B., Heyderman, L.J., McVitie, S., Nicholson, W.A.P. and Bormans, B. (1994) Coherent magnetic imaging by TEM. IEEE Trans. Magn. **30**, 4479.

Chapman, J. N., McFadyen, I. R., McVitie, S. (1990) Modified differential phase contrast Lorentz microscopy for improved imaging of magnetic structures. IEEE Trans. Magn., **26**, 1506.

Chapman, J. N., Morrison, G. R., Jakubovics, J. P., Taylor, R. A., in: Doig, P. (Ed.) (1983) Investigation of micromagnetic structures by STEM Electron Microscopy and Analysis, IOP Conf. Ser. No. 68, **1984**, 197.

Chapman, J. N., Scheinfein, M. R. (1999) Transmission Electron Microscopies of Magnetic Microstructures. Journal of Magnetism and Magnetic Materials, **200**, 729.

Charudatta et al. (2010) phys Rev lett **104**, 253901.

Che, R. C., Takeguchi, M., Shimojo, M., Zhang, W. and Furuya, K. (2005) Fabrication and electron holography characterization of FePt alloy nanorods. Appl. Phys. Lett., **87**, 223109.

Chen, J. H., Costan, E., van Huis, M. A., Xu, Q. and Zandbergen, H. W. (2006) Atomic pillar-based nanoprecipitates strengthen AlMgSi alloys. Science, **312**, 416.

Cherns, D., Barnard, J., Ponce, F. A. (1999) Measurement of the piezoelectric field across strained InGaN/GaN layers by electron holography. Solid State Commun,

111, 281.

Chou, L. J., Chang, M. T., Chueh, Y. L., Kim, J. J., Park, H. S. and Shindo, D. (2006) Electron holography for improved measurement of microfields in nano-electrode assemblies. Appl. Phys. Lett., **89**, 023112.

Chung, S., Berechman, R. A., McCartney, M. R. and Skowronski, M. (2011) Electronic structure analysis of threading screw dislocations in 4H–SiC using electron holography. J. Appl. Phys. **109**, 034906.

Chung, S., Johnson, S. R., Ding, D., Zhang, Y. H., Smith, D. J. and McCartney, M. R. (2009) Quantitative analysis of dopant distribution and activation across p-n junctions in AlGaAs/GaAs light-emitting diodes using off-axis electron holography. IEEE Trans. Electron Dev., **56**, 1919.

Chung, S., Johnson, S. R., Ding, D., Zhang, Y. H., Smith, D. J. and McCartney, M. R. (2010) Quantitative dopant profiling of p-n junction in InGaAs/AlGaAs light-emitting diode using off-axis electron holography. J. Vac. Sci. Technol. B, **28**, C1D11.

Coene, W., Janssen, G., Debeeck, M. O., Vandyck, D. (1992) Phase retrieval through focus variation for ultra-resolution in field-emission transmission electron-microscopy. Physical Review Letters, **69**, 3743.

Coene, W. M. J., Thust, A., de Beeck, M., VanDyck, D. (1996) Maximum-likelihood method for focus-variation image reconstruction in high resolution transmission electron microscopy. Ultramicroscopy, **64**, 109.

Cooper, D., Barnes, J. P., Hartmann, J. M., Béché, A. and Rouvière, J. L. (2009) Dark field electron holography for quantitative strain measurements with nanometer-scale spatial resolution. Appl. Phys. Lett., **95**, 053501.

Cooper, D., Béché, A., Hartmann, J. M., Carron, V. and Rouvière, J. L. (2010) Strain evolution during the silicidation of nanometer-scale SiGe semiconductor devices studied by dark field electron holography. Appl. Phys. Lett., **96**, 113508.

Cooper, D., Béché, A., Hartmann, J. M., Carron, V. and Rouvière, J. L. (2010) Strain mapping for the semiconductor industry by dark-field electron holography and nanobeam electron diffraction with nm resolution. Semicond. Sci. Technol., **25**, 095012.

Cooper, D., Truche, R., Rivallin, P., Hartmann, J. M., Laugier, F., Bertin, F., and

Chabli, A. Rouviere J. L. (2007) Medium resolution off-axis electron holography with millivolt sensitivity. Appl. Phys. Lett., **91**, 143501.

Cowley, J. M. and Moodie, A. F. (1957) The scattering of electrons by atoms and crystals. I. A new theoretical approach. Acta Cryst., **10**, 609.

Cowley, J. M. (1992) Twenty forms of electron holography. Ultramicroscopy, **41**, 335.

Cowley, J. M. (1995) Diffraction Physics, Third Edition. North-Holland Personal Library.

Cowley, J. M., Moodie, A. (1957) Fourier images: I–The point source Proc. Phys. Soc. B, 70, 486.

Cowley, J. M., Moodie, A. (1960) Fourier images IV: The phase grating Proc. Phys Soc., **76**, 378.

Cowley, J. M. (1995) Diffraction Physics, Third Edition. North-Holland, Amsterdam.

Cui, Y. R., Liu, H., Shi, Y., Li, X. M., Ma, H. Z. (2005) Improvement of annealed method to enhance magnetic properties of 1J85 alloys. Foundry Technology, **26**, 1145.

Cumings, J., Zettl, A., McCartney, M. R. and Spence, J. C. H. (2002) Electron holography of field-emitting carbon nanotubes. Phys. Rev. Lett., **88**, 056804.

Daberkow, I., Herrmann, K. H., Liu, L. B., Rau, W. D. (1991) Performance of electron image converters with yag single-crystal screen and ccd sensor. Ultramicroscopy, **38**, 215.

Dagotto, E. (2005) Complexity in Strongly Correlated Electronic Systems. Science, **309**, 257.

De Wolff, P. M., Janssen, T., Janner, A. (1981) The superspace groups for incommensurate crystal structures with a one-dimensional modulation. Acta Cryst. A, **A37**, 625.

De Wolff, P. M., (1974) Pseudo-Symmetry of Modulated Crystal-Structures. Acta Cryst. A, **A30**, 777.

Debaerdemaeker, T., Tate, C. and Woolfson, M. M. (1985) On the application of phase relationships to complex structures. XXIV. The Sayre tangent formula.

Acta Crystallogr. A, **A41**, 286.

de Beeck, M. O., Van Dyck, D., Coene, W. (1996) Wave function reconstruction in HRTEM: The parabola method. Ultramicroscopy, **64**, 167.

Deguchi, M., Tanaka, S. and Tanji, T. (2010) Determination of piezoelectric fields across InGaN/GaN quantum wells by means of electron holography. J. Electron. Mater., **39**, 815.

Dekkers, N. H and de Lang, H. (1974) Differential phase contrast in a STEM. Optik, **41**, 452.

den Hertog, M. I., Rouviere, J. L., Schmid, H., Cooper, D., Björk, M. T., Riel, H., Dhalluin, F., Gentile, P., Ferret, P., Oehler, F., Baron, T., Rivallin, P., Karg, S. and Riess, W. (2010) Off axis holography of doped and intrinsic silico nanowires: interpretation and influence of fields in the vacuum. J. Phys. Conf. Ser. **209**, 012027.

den Hertog, M. I., Schmid, H., Cooper, D., Rouviere, J. L., Björk, M. T., Riel, H., Rivallin, P., Karg, S. and Riess, W. (2009) Mapping active dopants in single silicon nanowires using off-axis electron holography. Nano Lett., **9**, 3837.

Dong, W., Baird, T., Fryer, J. R., Gilmore, C. J., Macnicol, D. D., Bricogne, G., SMITH, D. J., O'KEEFE, M. A. and HÖVMOLLER, S. (1992) Electron microscopy at 1-Å resolution by entropy maximization and likelihood ranking. Nature, **355**, 605.

Dorset, D. L., Hauptman, H. A. (1976) Direct phase determination for quasi-kinematical electron-diffraction intensity data from organic microcrystals. Ultramicroscopy, **1**, 195-201.

Dorset, D. L. (1997) The accurate electron crystallographic refinement of organic structures containing heavy atoms. Acta Cryst. A **53**, 356.

Du, K., Phillipp, F. (2006) On the accuracy of lattice-distortion analysis directly from high-resolution transmission electron micrographs. J. Microsc., **221**, 63.

Du, K., von Hochmeister, K., Phillipp, F. (2007) Quantitative comparison of image contrast and pattern between experimental and simulated high-resolution transmission electron micrographs. Ultramicroscopy, **107**, 281.

Duan, X. F., Gao, M. and Poeng, L. M. (1998) Accurate measurement of phase shift in electron holography. Appl. Phys. Lett., **72**, 771.

Dunin-Borkowski, R. E., McCartney, M. R., Frankel, R. B., Bazylinski, D. A., Posfai, M. and Buseck, P. R. (1998) Magnetic Microstructure of magnetotactic bacteria by electron holography. Science, **282**, 1868.

Dunin-Borkowski, R. E., McCartney, M. R., Kardynal, B. and Smith D. J. (1998) Magnetic interactions within patterned cobalt nanostructures using off-axis electron holography. J. Appl. Phys., **84**, 374.

Dunin-Borkowski, R. E., McCartney, M. R., Posfai, M., Frankel, R. B., Bazylinski, D. A. and Buseck, P. R. (2001) Off-axis electron holography of magnetotactic bacteria: magnetic microstructure of strains MV-1 and MS-1. Eur. J. Mineral., **13**, 671.

Egerton, R. F. (2011) Electron Energy-Loss Spectroscopy in the Electron Microscope, 3rd, New York: Springer Science+Business Media.

Fan H. F., et al., Acta Crystallogr. A, 1985, 41: 163.

Formanek, P. and Kittler, M. (2005) Application of electron holography to extended defects: Schottky barriers at $NiSi_2$ precipitates in silicon. Phys. Stat. Sol. C, **2**, 1878.

Frank, J. (1974) Optik, **41**, 90.

Fu, Z. Q., Fan, H. F. (1994) DIMS-a direct-method program for incommensurate modulated structures. J. Appl. Crystallogr. **27**, 124.

Fu, Z. Q., Huang, D. X., Li, F. H., Li, J. Q., Zhao, Z. X., Cheng, T. Z., Fan, H. F. (1994) Incommensurate modulation in minute crystals revealed by combining high-resolution electron microscopy and electron diffraction. Ultramicroscopy, **54**, 229.

Fujita, T. and Chen, M. (2009) Quantitative electron holographic tomography for a spherical object. J. Electron Microsc., **58**, 301.

Fujita, T., Chen, M., Wang, X., Xu, B., Inoke, K. and Yamamoto, K. (2007) J. Appl. Phys. **101**, 014323.

Fujita, T., Yamamoto, K., McCartney, M. R. and Smith, D. J. (2006) Reconstruction technique for off-axis electron holography using coarse fringes. Ultramicroscopy, **106**, 486.

Fukumura, T., Sugawara, H., Hasegawa, T., Tanaka, K., Sakaki, H., Kimura, T. and Tokura, Y. (1999) Spontaneous Bubble Domain Formation in a Layered

Ferromagnetic Crystal. Science, **284**, 1969.

Gajdardziska-Josifovska, M. and Carim, A. H., ed. by Völkl, E., Allard, L. F. and Joy, D. C. (1999) Applications of electron holography, in Introduction to electron holography. New York: Kluwer Academic/Plenum Publishers, 267.

Gajdardziska-Josifovska, M. (1995) Off-axis electron holography of hetero-interfaces. Interf. Sci., **2**, 425.

Ge, B. H., Wang, Y. M., Li, X. M., Li, F. H., Song, H. L., Zhang, H. J. (2008) A study of an incommensurately modulated structure in $Ca_{0.28}Ba_{0.72}Nb_2O_6$ by electron microscopy. Philos. Mag. Lett. **88**, 213.

Gerchberg, R. W., Saxton, W. O. (1971) Phase determination from image and diffraction plane pictures in the electron-microscope, Optik **34**, 277.

Ghiglia, D. C. and Pritt, M. D. (1998) Two-dimensional phase unwrapping: theory, algorithms, and software. Wiley-Interscience Publication.

Goodenough, J. B. (1959) On the influence of 3d4 ions on the magnetic and crystallographic properties of magnetic oxides. J. Phys. Radium, **20**, 155.

Goodman, P. (1975) A practical method of three-dimensional space-group analysis using convergent-beam electron diffraction. Acta Crystallographica Section A, **A31**, 804.

Goodman, P. and Lehmpfuh, (1965) G. Zeitschrift Fur Naturforschung Part a-Astrophysik Physik Und Physikalische Chemie A, 20.

Gribelyuk, M. A., Domenicucci, A. Ronsheim, G. P. A., McMurray, J. S. and Gluschenkov, O. (2008) Application of electron holography to analysis of sub-micron structure. J. Vac. Sci. Technol. B, **26**, 408.

Gribelyuk, M. A., Adam, T. N., Ontalus, V., Ronsheim, P. R., Kimball, L. and Schonenberg, K. T. (2011) Electron holography of Si:C films and Si:C-based devices. J. Appl. Phys., **110**, 063522.

Gupta, A., Gong, G. Q., Xiao, G., Duncombe, P. R., Lecoeur, P., Trouilloud, P., Wang, Y. Y., Dravid, V. P., and Sun, J. Z. (1996) Grain-boundary effects on the magnetoresistance properties of perovskite manganite films. Phys. Rev. B. **54**, R15629.

Haider, M., Braunshausen, G., Schwan, E. (1995) Correction of the spherical-aberration of a 200-kv tem by means of a hexapole-corrector. Optik, 99, 167.

Haider, M., Rose, H., Uhlemann, S., Schwan, E., Kabius, B., Urban, K. (1998) A spherical-aberration-corrected 200 kV transmission electron microscope. Ultramicroscopy, **75**, 53.

Haider, M., Uhlemann, S., Schwan, E., Rose, H., Kabius, B., Urban, K. (1998) Electron microscopy image enhanced. Nature, **392**, 768.

Haine, M. E. and Mulvey, T. (1951) The formation of diffraction image with electrons in the Gabor diffraction microscope. J. Opt. Soc. Am., **42**, 763.

Hale, M. E., Fuller, H. W., Rubinstein, H. (1959) Magnetic domain observations by electron microscopy. J. Appl. Phys., **30**, 789.

Han, F. S., Fan, H. F., Li, F. H. (1986) Image Processing in High-resolution Electron Microscopy Using the Direct Method. II. Image Deconvolution. Acta Cryst. A, **42**, 353.

Hao, Q., Liu, Y. W., Fan, H. F. (1987) Direct Methods in Superspace. I. Preliminary Theory and Test on the Determination of Incommensurate Modulated Structures. Acta Cryst. A, **43**, 820.

Harada, K., Akashi, T., Togawa, Y., Matsuda, T. and Tonomura, A. (2005) Optical system for double-biprism electron holography. J. Electron Microsc., **54**, 19.

Harada, K., Tonomura, A., Togawa, Y. Akashi, T. and Matsuda, T. (2004) Double-biprism electron interferometry. Appl. Phys. Lett., **84**, 3229.

Harada, K., Matsuda, T., Tonomura, A., Akashi, T. and Togawa, Y. (2006) Triple-biprism electron interferometry, J. Appl. Phys. **99**, 113502.

Harrison, C. G., Leaver, K. D. (1973) The analysis of two-dimensional domain wall structures by Lorentz microscopy. Phys. Status Solidi. A, **15**, 415.

Haun, J. B., Yoon, T.J., Lee, H. and Weissleder, R. (2010) Magnetic nanoparticle biosensors. Wiley Interdiscip. Rev. Nanomed. Nanobiotechnol., **2**, 291.

He, J. Q., Volkov, V. V., Asaka, T., Chaudhuri, S., Budhani, R. C. and Zhu, Y. (2010) Competing two-phase coexistence in doped manganites: Direct observations by in situ Lorentz electron microscopy. Phys. Rev. B. **82**, 224404.

He, W. Z., Li, F. H., Chen, H., Kawasaki, K., Oikawa, T. (1997) Image deconvolution for defected crystals in field-emission high-resolution electron microscopy. Ultramicroscopy, **70**, 1.

Hefferman, S. J., Chapman, J. N., McVitie, S. (1990) In-situ magnetising experiments on small regular particles fabricated by electron beam lithography. J. Magn. Magn. Mater., **83**, 223-224.

Hirsh, P. B., Howie, A., Nicholson, R. B., Pashley, D. W., Whelan, M. J. (1965) Electron Microscopy of Thin Crystals. London: Butterworths, 292.

Hitachi Technical Data EM Sheet No. 47.

Holt, D. B. (1964) Grain boundaries in sphalerite structure J. Phys. Chem. Solids, **25**,1385.

Horiuch, S., Cantoni, M., Tsuruta, T., Matsui, Y. (1998) Direct observation of the interaction between a vortex lattice and dislocations in a superconducting Nb crystal. Appl. Phys. Lett. **31**, 1293.

Hovmöller, S. (1992) CRISP: crystallographic image processing on a personal computer. Ultramicroscopy, **41**, 121.

Hovmöller, S., Sjogren, A., Farrants, G., Sundberg, M., Marinder, B. (1984) Accurate atomic positions from electron-microscopy. Nature. Syst, **311**, 238.

Howie, A. (2004) Hunting the Stobbs factor. Ultramicroscopy, **98**, 73.

Hsieh, W. K., Chen, F. R., Kai, J. J., Kirkland, A. I. (2004) Resolution extension and exit wave reconstruction in complex HREM. Ultramicroscopy, **98**, 99.

Hu, J. J., Li, F. H. (1991) Maximum entropy image deconvolution in high resolution electron microscopy. Ultramicroscopy, **35**, 339.

Hu, J. J., Li, F. H., Fan, H. F. (1992) Crystal-structure Determination of $K_2O \cdot 7 \cdot Nb_2O_5$ by Combining High-resolution Electron-microscopy and Electron-diffraction. Ultramicroscopy, **41**, 387.

Huang, S. X. and Chien, C. L. (2012) Extended Skyrmion Phase in Epitaxial FeGe (111) Thin Films. Phys. Rev. Lett. **108**, 267201.

Huang, D. X., He, W. Z., Li, F. H. (1996) Multiple solution in maximum entropy deconvolution of high resolution electron microscope images. Ultramicroscopy, **62**, 141.

Huang, D. X., Liu, W., Gu, Y. X., Xiong, J. W., Fan, H. F., Li, F. H. (1996) A Method of Electron Diffraction Intensity Correction in Combination with High-resolution Electron Microscopy. Acta Cryst. A, **52**, 152.

Huber Jr, E. E., Smith, D. O. and Goodenough, J. B. (1958) Domain-Wall Structure in Permalloy Films, J. Appl. Phys. **29**, 294.

Hüe, F., Hÿtch, M., Houdellier, F., Bender, H. and Claverie, A. (2009) Strain mapping of tensiley strained silicon transistors with embedded Si1-yCy source and drain by dark-field holography. Appl. Phys. Lett. **95**, 073103.

Humphreys, C. J., Spence, J. C. H. (1981) Resolution and illumination coherence in electron-microscopy. Optik, **58**, 125.

Hÿtch, M. J., Dunin-Borkowski, R. E., Scheinfein, M. R., Moulin, J., Duhamel, C., Mazaleyrat, F. and Champion, Y. (2003) Vortex flux channeling in magnetic nanoparticle chains. Phys. Rev. Lett. **91**, 257207.

Hÿtch M. J., Stobbs W. M. (1994) Quantitative comparison of high resolution TEM images with image simulations. Ultramicroscopy, **53**, 191.

Hÿtch, M., Houdellier, F., Hüe, F. and Snoeck, E. (2011) Dark-field electron holography for the measurement of geometric phase. Ultramicroscopy, **111**, 1328.

Hÿtch, M., Houdellier, F., Hüe, F. and Snoeck, E. (2008) Nanoscale holographic interferometry for strain measurements in electronic devices. Nature, **453**, 1086.

Ikarashi, N., Oshida, M., Miyamura, M., Saitoh, M., Mineji, A. and Shishiguchi, S. (2008) Electron holography characterization of ultra-shallow junctions in 30-nm-gate-length metal-oxide-semiconductor field-effect transistors. Jpn. J. Appl. Phys. **47**, 2365.

Ikarashi, N. Toda, A. Uejima, K. Yako, K Yamamoto, T and Hane, M. (2010) Electron holography for analysis of deep Submicro devicesipresent status and challenges I. Vac Sci Technol B, **28**, CID5.

Ikeda, M., Sugawara, A. and Harada, K. (2011) Twin-electron biprism. J. Electron Microsc. **60**, 353.

Inoue, M., Tomita, T., Naruse, M., Akase, Z., Murakami, Y. and Shindo, D. (2005) Development of a magnetizing stage for in situ observations with electron holography and Lorentz microscopy. J. Electr. Micros. **54**, 509.

Ishizuka, K., Miyazaki, M. and Uyeda, N. (1982) Improvement of electron microscope images by the direct phasing method. Acta Crystallogr., **A38**, 408.

Janssen, T., Janner, A., Looijenga-vos, A., Wolff, P. M. D. (2006) International Tables for Crystallography. Kluwer Academic Publishers: Dordrecht, Vol. C.

Javon, E., Gatel, C., Masseboeuf, A. and Snoeck, E. (2010) Electron holography study of the local magnetic switching process in magnetic tunnel junctions. J.

Appl. Phys. **107**, 093D310.

Jaynes, E. T. (1957) Information theory and statistical mechanics, Phys. Rev. **106**, 620.

Jia, C. L., Lentzen, M. (2003) Urban K. Atomic-resolution imaging of oxygen in perovskite ceramics. Science, **299**, 870.

Jia, C. L., Thust, A. (1999) Investigation of atomic displacements at a Sigma 3 {111} twin boundary in BaTiO(3) by means of phase-retrieval electron microscopy. Phys. Rev. Lett. **82**, 5052.

Jia, C.L., Urban, K. (2004) Atomic-resolution measurement of oxygen concentration in oxide materials. Science, **303**, 2001.

Jiang, H., Li, F. H., Mao, Z. Q. (1999) Electron Crystallographic Study of Bi2(Sr 0.9 La 0.1) 2CoOy. Micron, **30**, 417.

Johnson, K. D. and Dravid, V. P. (1999) Direct evidence for grain boundary potential barrier breakdown via in situ electron holography. Microsc. Microanal. **5**, 428.

Johnson, K. D. and Dravid, V. P. (1999) Grain boundary barrier breakdown in niobium donor doped strontium titanate using in situ electron holography. Appl. Phys. Lett. **74**, 621.

Johnston, A. B., Chapman, J. N. (1995) The development of coherent Foucault imaging to investigate magnetic microstructure. J. Microsc., **179**, 119.

Jourdain, V., Simpson, E. T., Paillet, M., Kasama, T., Dunin-Borkowski, R. E., Poncharal, P., Zahab, A., Loiseau, A., Robertson, J. and Bernier, P. (2006) Periodic inclusion of room-temperature-ferromagnetic metal phosphide nanoparticles in carbon nanotubes. J. Phys. Chem. B, **110**, 9759.

Junginger, F., Kläui, M., Backes, D., Rüdiger, U., Kasama, T., Dunin-Borkowski, R. E., Heyderman, L. J., Vaz, C. A. F. and Bland, J. A. C. (2007) Spin torque and heating effects in current-induced domain wall motion probed by transmission electron microscopy. Appl. Phys. Lett., **90**, 132506.

Kambe, K. (1957) Study of Simultaneous Reflexion in Electron Diffraction by Crystals I. Theoretical Treatment. Journal of the Physical Society of Japan, **12**, 13.

Karle, J., Hauptman, H. (1956) A theory of phase determination for the 4 types of non-centrosymmetric space groups 1P222, 2P22, 3P12, 3P22 Acta Crystallogr.

9, 635.

Kasahara, T., Shindo, D., Yoshikawa, H., Sato, T., Kondo, K. (2007) In situ observations of domain structures and magnetic flux distributions in Mn-Zn and Ni-Zn ferrites by Lorentz microscopy and electron holography. J. Electron. Microsc. **56**, 7.

Kasama, T., Barpanda, P., Dunin-Borkowski, R. E., Newcomb, S. B., McCartney, M. R., Castaño, F. J. and Ross, C. A. (2005) Off-axis electron holography of individual pseudo-spin-valve thin film magnetic elements. J. Appl. Phys., **98**, 013903.

Kasama, T., Church, N. S., Feinberg, J. M., Dunin-Borkowski, R. E. and Harrison, R. J. (2010) Direct observation of ferrimagnetic/ferroelastic domain interactions in magnetite below the Verwey transition. Earth Plane. Sci. Lett. **297**, 10.

Kasama, T., Dunin-Borkowski, R. E. and Beleggia, M. (2011) Electron Holography of Magnetic Materials, in Holography-Different Fields of Application. ed. F. A. M. Ramírez, InTech, 53.

Kasama, T., Dunin-Borkowski, R. E., Scheinfein, M. R., Tripp, S. L., Liu, J. and Wei, A. (2008) Reversal of flux closure states in cobalt nanoparticle rings with coaxial magnetic pulses. Adv. Mater., **20**, 4248.

Kauffmann, Y., Recnik, A., Kaplan, W. D. (2005) The accuracy of quantitative image matching for HRTEM applications. Mater. Charact., **54**, 194.

Kienzle, O., Ernst, F., Mobus, G. (1998) Reliability of atom column positions in a ternary system determined by quantitative high-resolution transmission electron microscopy. J. Microsc. **190**, 144.

Kim, J. J., Shindo, D., Murakami, Y., Xia, W., Chou, L.J. and Chueh, Y. L. (2007) Direct observation of field emission in a single $TaSi_2$ nanowire. Nano Lett. **7**, 2243.

Kirkland, E. J. (1984) Improved high-resolution image-processing of bright field electron-micrographs. I. theory. Ultramicroscopy, **15**, 151.

Kirkland, E. J., Loane, R. F. and Silcox, J. (1987) Simulation of annular dark field stem images using a modified multislice method. Ultramicroscopy, **23**, 77.

Kirkland, E. J., Siegel, B. M., Uyeda, N., Fujiyoshi, Y. (1985) Improved high-resolution image-processing of bright field electron-micrographs. II. experiment.

Ultramicroscopy, **17**, 87.

Kirkland, E. J. (1984) Improved high resolution image processing of bright field electron micrographs: I. Theory. Ultramicroscopy, **15**, 151.

Kohno, H., Yoshida, H., Ohno, Y., Ichikawa, S., Akita, T., Tanaka, K. and Takeda, S. (2004) Formation of silicon/silicide/oxide nanochains and their properties studied by electron holography. Thin Solid Films, **204**, 464.

Kossel, W. and Mollenstedt, G. (1939) Elektroneninterferenzen im konvergenten Bündel. Annalen der Physik **428**, 113.

Koyama, T., Togawa, Y., Takenaka, K. and Mori, S. (2012) Ferromagnetic microstructures in the ferromagnetic metallic phase of $La_{0.825}Sr_{0.175}MnO_3$. J. Appl. Phys. **111**, 07B104.

Krivanek, O. L. (1976) Method for determining coefficient of spherical aberration from a single electron micrograph. Optik, **45**, 97.

Kruse, P., Rosenauer, A., Gerthsen, D. (2003) Determination of the mean inner potential in III–V semiconductors by electron holography. Ultramicroscopy, **96**, 11.

Kübel, C., Thust, A. (2003) True Image——A Software Package for Focal-Series Reconstruction in HRTEM, FEI Company.

Lau, J. W., Schofield, M. A., Zhu, Y. (2007) A straightforward specimen holder modification for remnant magnetic-field measurement in TEM. Ultramicroscopy, **107**, 396.

Lee, M., Kang, W., Onose, Y., Tokura, Y. and Ong, N. P. (2009) Unusual Hall Effect Anomaly in MnSi under Pressure. Phys. Rev. Lett. 102, 186601.

Lehmann, M. and Lichte, H. (2005) Electron holographic material and analysis at atomic dimensions. Cryst. Res. Technol. **40**, 149.

Lehmann M., Lichte H. (2003) Is there a Stobbs-Factor in Off-axis Electron Holography? Microsc. microanal. **9**, 46.

Leith, E. N., Upatnieks, J. (1962) Reconstructed wavefronts and communication theory. J. Opt. Soc. Am. **52**, 1123.

Li, F. H. and Fan, H. F. (1979) Acta Phys. Sin. **28**, 276 (in Chinese).

Li, F. H. and Hashimoto, H. (1984) Use of dynamical scattering in the structure determination of a minute fluorocarbonate mineral cebaite $Ba_3Ce_2(CO_3)_5F_2$

by high-resolution electron microscopy. Acta Crystallogr. B. **40**, 454.

Li, F. H. and Tang, D. (1985) Pseudo-weak-phase-object approximation in high-resolution electron microscopy. I. Theory. Acta Crystallogr. A. **41**, 376.

Li, F. H. (2010) Developing image-contrast theory and analysis methods in high-resolution electron microscopy. Phys. Status Solidi A. **207**, 2639.

Li, J. C. (2000) Southern Iron and Steel, **116**, 7.

Li, L., Ketharanathan, S., Drucker, J. and McCartney, M. (2009) Study of hole accumulation in individual germanium quantum dots in p-type silicon by off-axis electron holography. Appl. Phys. Lett. **94**, 232108.

Li, L., Smith, D. J., Dailey, E., Madras, P., Drucker, J. and McCartney, M. R. (2011) Observation of hole accumulation in Ge/Si core/shell nanowires using off-axis electron holography. Nano Lett., **11**, 493.

Li, F. H. (1977) Determination of crystal structures by high resolution electron microscopy. Acta Physica Sinica (In Chinese) 26, 193.

Li, F. H., Wang, D., He, W. Z., Jiang, H. (2000) Amplitude correction in image deconvolution for determining crystal defects at atomic level. J. Electron Microsc. **49**, 17.

Lichte, H. and Lehmann, M. (2008) Electron holography-basics and applications. Rep. Prog. Phys. **71**, 016102.

Lichte, H. (1991) Optimum focus for taking electron holograms. Ultramicroscopy, **38**, 13.

Lichte, H., Geiger, D. and Linck, M. (2009) Off-axis electron holography in an aberration-corrected transmission electron microscopy. Phil. Trans. R. Soc. A, **367**, 3773.

Lichte, H., Reibold, M., Brand, K. and Lehmanm, M. Ferroelectric electron holography. Ultramicroscopy, **93**, 199.

Lin, F., Chen, Q., Peng, L. M. (2007) REW-exit-wave reconstruction and alignments for focus-variation high-resolution transmission electron microscopy images. Journal of Applied Crystallography, **40**, 614.

Liu, H. H., Duan, X. K., Che, R. C., Wang, Z. F., Duan, X. F. (2008) In situ-investigation of the magnetic domain wall in Permalloy thin film by Lorentz electron microscopy. Materials Lett. **62**, 2654.

Liu, H. R., Wang, Y. G., Liu, Q. X. and Yang, Q. B. (2004) A study of the potential barrier at the Σ13(510) tilt boundary of strontium titanate using electron holography and dynamic simulation. J. Phys. D: Appl. Phys. **37**, 1478.

Liu, H. H., Duan, X. K., Che, R. C., Wang Z. F., Duan, X. F. (2009) Measurement of the remnant magnetic-field in Lorentz mode using permalloy. Acta Metall. Sin. **22**, 435.

Liu, H. H., Duan, X. K., Che, R. C., Wang, Z. F., Duan, X. F. (2009) In situ lorentz microscopy observation of displaced chain walls in permalloy. Mater. Trans. **50**, 1660.

Liu, J., Li, F. H., Wan, Z. H., Fan, H. F., Wu, X. J., Tamura, T., Tanabe, K. (2001) Electron Crystallographic Image-processing Investigation and Superstructure Determination for $(Pb_{0.5}Sr_{0.3}Cu_{0.2})Sr_2(Ca_{0.6}Sr_{0.4})Cu_2Oy$. Acta Cryst. A. **57**, 540.

Liu, L. Z. Y., Rao, D. V. S., Kappers, M. J., Humphreys, C. J. and Geiger, D. (2010) Basal-plane stacking faults in non-polar GaN studied by off-axis electron holography. J. Phys. Conf. Ser. **209**, 012012.

Liu, W., Wang, N., Wang, R.M., Kumar, S., Duesberg, G. S., Zhang, H. Z., Sun, K. (2011) Atom-Resolved Evidence of Anisotropic Growth in ZnS Nanotetrapods. Nano Lett. **11**, 2983.

Liu, J., Li, F. H., Wan, Z. H., Fan, H. F., Wu, X. J., Tamura, T., Tanabe, K. (1998) Incommensurate modulated structure of "Pb"-1223 determined by combining high resolution electron microscopy and electron diffraction. Mater. Trans., JIM **39**, 920.

Loudon, J. C., Mathur, N. D. and Midgley, P. A. Charge-ordered ferromagnetic phase in $La_{0.5}Ca_{0.5}MnO_3$. Nature, **420**, 797.

Lu, B., Li, F. H., Wan, Z. H., Fan, H. F., Mao, Z. Q. (1997) Electron Crystallographic Study of Bi4(Sr0.75La0.25)8Cu5Oy Structure. Ultramicroscopy, **70**, 13.

Lu, Q., Chen, C. C. and de Loznne, A. (1997) Observation of Magnetic Domain Behavior in Colossal Magnetoresistive Materials With a Magnetic Force Microscope. Science, **276**, 2006.

Marino, F. A., Cullen, D. A., Smith, D. J., McCartney, M. R. and Saraniti, M.

(2010) Simulation of polarization charge on AlGaN/GaN high electron mobility transistors: Comparison to electron holography. J. Appl. Phys. **107**, 054516.

Marshall, A. F., Klein, L., Dodge, J. S., Ahn, C. H., Reiner, J. W., Mieville L., Antagonazza L., Kapitulnik, A., Geballe, T. H. and Beasley, M. R. (1999) Lorentz transmission electron microscope study of ferromagnetic domain walls in $SrRuO_3$: Statics, dynamics, and crystal structure correlation. J. Appl. Phys. **85**, 4131.

Matsuda, T., Fukuhara, A., Yoshida, T., Hasegawa, S., Tonomura, A. and Ru, Q. (1991) Computer reconstruction from electron holograms and observation of fluxon dynamics. Phys. Rev. Lett. **66**, 457.

Matsuda, T., Hasegawa, S., Igarashi, M., Kobayashi, T., Naito, M., Kajiyama, H., Endo, J., Osakabe, N., Tonomura, A., and Aoki, R. (1989) Magnetic field observation of a single flux quantum by electron-holographic interferometry. Phys. Rev. Lett. **62**, 2519.

Matsuda, T., Tonomura, A., Suzuki, R., Endo, J., Osakabe, N., Umezaki, H., Tanabe, H., Sugita, Y., and Fujiwara, H. (1982) Observation of microscopic distribution of magnetic field by electron holography. J. Appl. Phys. **53**, 544.

Matsumoto, T., Koguchi, M., Suzuki, K., Nishimura, H., Motoyoshi, Y., and Wada, N. (2008) Ferroelectric 90° domain structure in a thin film of $BaTiO_3$ fine ceramics observed by 300 kV electron holography. Appl. Phys. Lett. **92**, 072902.

McCartney, M. R., Ponce, F. A., Cai, J. and Bour, D. P. (2000) Mapping electrostatic potential across an AlGaN/InGaN/AlGaN diode by electron holography. Appl. Phys. Lett. **76**, 3055.

McCartney, M. R., Gajdardziska-Josifovska, M. (1994) Absolute measurement of normalized thickness, $t/\lambda i$, from off-axis electron holography. Ultramicroscopy, **53**, 283.

McCartney, M. R., Smith, D. J., Farrow, R. F. C. and Marks, R. F. (1997) Off-axis electron holography of epitaxial FePt films. J. Appl. Phys. **82**, 2461.

McFadyen, I. R., Chapman, J. N. (1992) EMSA Bull. **22**, 64.

Mcgibbon, A. J., Pennycook, S. J., Angelo, J. E. (1995) Direct observation of dislocation core structures in CDTE/GAAS(001). Science, **269**, 519.

McVitie, S., Chapman, J. N., Zhou, L., Heyderman, L. J., Nicholson, W. A. P.

(1995) In-situ magnetising experiments using coherent magnetic imaging in TEM. J. Magn. Magn. Mater. **148**, 232.

Meyer, R. R., Sloan, J., Dunin-Borkowski, R. E., Kirkland, A. I., Novotny, M. C., Bailey, S. R., Hutchison, J. L., Green, M, L. H. (2000) Discrete atom imaging of one-dimensional crystals formed within single-walled carbon nanotubes. Science, **289**, 1324.

Meyer, R. R., Kirkland, A. I., Saxton, W. O. (2002) A new method for the determination of the wave aberration function for high resolution TEM 1. Measurement of the symmetric aberrations. Ultramicroscopy, **92**, 89.

Midgley, P. A. (2001) An introduction to off-axis electron holography. Micron, **32**, 167.

Mobus, G. (1996) Retrieval of crystal defect structures from HREM images by simulated evolution I. Basic technique. Ultramicroscopy, **65**, 205.

Mobus, G., Gemming, T., Gumbsch, P. (1998) The Influence of Phonon Scattering on HREM Images. Acta Crystallogr. Sect. A. **54**, 83.

Mochizuki, M. (2012) Spin-Wave Modes and Their Intense Excitation Effects in Skyrmion Crystals. Phys. Rev. Lett. **108**, 017601.

Möllenstedt G. and Bayh W. Kontinuierliche phasenschiebung von elektronenwellen im kraftfeldfreien raum durch das magnetische vektorpotential eines solenoids. Physikalische Bl., 1962, 18: 299.

Mollenstedt, G., Wahl, H. (1968) Elektronenholographie und Rekonstruktion mit Laserlicht. Naturwissenschaften, **55**, 340.

Mornet, S., Vasseur, S., Grasset, F. and Duguet, E. (2004) Magnetic nanoparticle design for medical diagnosis and therapy. J. Mater. Chem. **14**, 2161.

Morrison, G. R., Chapman, J. N. (1983) Optik, **64**, 1.

Morrison, G. R., Gong, H., Chapman, J. N., Hrnciar, V. (1988) The measurement of narrow domain-wall widths in SmCo5 using differential phase contrast electron microscopy. J. Appl. Phys. **64**, 1338.

Müller, E., Gerthsen, D., Brückner, P., Sholz, F., Gruber, T. and Wang, A. (2006) Probing the electrostatic potential of charged dislocations in n-GaN and n-ZnO epilayers by transmission electron holography. Phys. Rev. B.**73**, 245316.

Murakami, Y., Kasai, H., Kim, J. J., Mamishin, S., Shindo, D., Mori, S. and

Tonomura, A. (2009) Ferromagnetic domain nucleation and growth in colossal magnetoresistive manganite. Nature nanotechnology, **5**, 37.

Nadarzinski, K., Ernst, F. (1996) The atomistic structure of a Sigma = 3, (111) grain boundary in NiAl, studied by quantitative high-resolution transmission electron microscopy. Philosophical Magazine A. **74**, 641.

Neubauer, A., Pfleiderer, C., Binz, B., Rosch, A., Ritz, R., Niklowitz, P. G. and Bo"ni, P. (2009) Topological Hall Effect in the A Phase of MnSi. Phys. Rev. Lett. **102**, 186602.

O'Keefe, M. A., Hetherington, C. J. D., Wang, Y. C., Nelson, E. C., Turner, J. H., Kisielowski, C., Malm, J. O., Mueller, R., Ringnalda, J., Pan, M., Thust, A. (2001) Sub-Angstrom high-resolution transmission electron microscopy at 300 keV. Ultramicroscopy, **89**, 215.

Onose, Y., Okamura, Y., Seki, S., Ishiwata, S. and Tokura, Y. (2012) Observation of Magnetic Excitations of Skyrmion Crystal in a Helimagnetic Insulator Cu_2OSeO_3. Phys. Rev. Lett. **109**, 037603.

Orchowski, A. Lichte, H. (1996) High-resolution electron holography of non-periodic structures at the example of a $\Sigma = 13$ grain boundary in gold. Ultramicroscopy, **64**, 199.

Orchowski, A., Rau, W. D. and Lichte, H. (1995) Electron holography surmounts resolution limit of electron microscopy. Phys. Rev. Lett., **74**, 399.

Osakabe, N., Yoshida, K., Horiuchi, Y., Matsuda, T., Tanabe, H., Okuwaki, T., Endo, J., Fujiwar, H. and Tonomura, A. (1983) Observation of recorded magnetization pattern by electron holography. Appl. Phys. Lett., **42**, 746.

Otten, M.T., Mul, P.M., Dejong, M. J. C. (1992) Design and performance of the CM20 FEG field-emission TEM. Microscopy Microanalysis Microstructures, **3**, 83.

Peng, L. M., Ren, G., Dudarev, S. L., Whelan, M. J. (1996) Debye-Waller Factors and Absorptive Scattering Factors of Elemental Crystals. Acta Crystallogr. Sect. A. **52,** 456.

Phatak, C. Petfordtong, A K and Graef M. D (2010) Three Dimensional Study of the vector potential of magnetic structares phys. Rev. Lett., **104**, 253901.

Phillipp, F., Du, K., Jin-Phillipp, N. Y. (2003) Quantitative HRTEM studies on local lattice distortions in strained materials systems. Materials Chemistry and Physics, **81**,205.

Ploessl, R., Chapman, J. N., Scheinfein, M. R., Blue, J. L., Mansuripur, M., Hoffmann, H. (1993) Micromagnetic structure of domains in Co/Pt multilayers. I. Investigations of wall structure, J. Appl. Phys. **74**, 7431.

Pollard, S. D., Huang, L., Buchanan, K. S., Arena, D. A. and Zhu, Y. (2012) Direct dynamic imaging of non-adiabatic spin torque effects. Nat. Commun. **3**, 1028.

Qian, W. Spence, J. C. H. and Zuo, J. M. (1993) Transmission low-energy electron diffraction (TLEED) and its application to the low-voltage point-projection micnscok Acta Crystallogra phica Section **A49**, 436.

Rau, W. D., Schwander, P., Baumann, F. H., Höppner, W. and Ourmazd, A. (1999) Two-dimensional mapping of the electrostatic potential in transistors by electron holography. Phys. Rev. Lett. **82**, 2614.

Ravikumar, V., Rodrigues, R. P. and Dravid, V. P. (1997) Space-charge distribution across internal interfaces in electroceramics using electron holography: II, Doped Grain Boundaries. J. Am. Ceram. Soc. **80**, 1131.

Ravikumar, V., Rodrigues, R. P. and Dravid, V. P. (1995) Direct Imaging of Spatially Varying Potential and Charge across Internal Interfaces in Solids. Phys. Rev. Lett. 75, 4063.

Ravikumar, V., Rodrigues, R. P. and Dravid, V. P. (1997) Space-charge distribution across internal interfaces in electroceramics using electron holography: I, Pristine Grain Boundaries. J. Am. Ceram. Soc. **80**, 1117.

Reimer, L. and Kohl, H. (2008) Transmission Electron Microscopy: Physics of Image Formation, 5th, Springer Science+Business Media, LLC, 233 Spring Street, New York, USA.

Reimer, L. (1984) Springer Series in Optical Sciences. Berlin: Springer, 36.

Romming, N., Hanneken, C., Menzel, M., Bickel, E. J., Wolter, B., Bergmann, V. K., Kubetzka, A., Wiesendanger, R., (2013) Writing and Deleting Single Magnetic Skyrmions. Science **341**, 636.

Rose, H. (1997) Nonstandard imaging methods in electron microscopy. Ultramicroscopy, **2**, 251.

Rose, H., Preikszas, D., (1992) Outline of a versatile corrected LEEM. Optik, **92**,

31.

Rose, H. (1990) Inhomogeneous wien filter as a corrector compensating for the chromatic and spherical-aberration of low-voltage electron-microscopes. Optik, **84**, 91.

Rosenauer, A., Gerthsen, D. (1999) Atomic scale strain and composition evaluation from high-resolution transmission electron microscopy images, in: Hawkes, P. W. (Ed.), Advances in imaging and electron physics. San Diego: Academic Press, 121.

Saito, K., Park, H. S., Shindo, D., Yoshizawa, Y. (2006) Journal of Magnetism and Magnetic Materials. **305**, 304.

Sasaki, H., Ootomo, S., Matsuda, T. and Ishii, H. (2008) Observation of carrier distribution in compound semiconductors using off-axis electron holography. Furukawa Rev. **34**, 24.

Sasaki, H., Yamamoto, K., Hirayama, T., Ootomo, S., Matsuda, T., Iwase, F., Nakasaki, R. and Ishii, H. (2006) Mapping of dopant concentration in a GaAs semiconductor by off-axis phase-shifting electron holography. Appl. Phys. Lett. **89**, 244101.

Saxton, W. O. (1994) What is the focus variation method-is it new-is it direct. Ultramicroscopy, **55**, 171.

Saxton, W. O. (1986) Proceedings 11th International Congress on Electron Microscopy, Kyoto, 1.

Sayre, D. (1952) The Squaring Method: a New Method for Phase Determination. Acta Cryst. **B5**, 60-65.

Scherzer, O. (1949) The theoretical resolution limit of the electron microscope. J. Appl. Phys., 20.

Schiske P. (1968) Proceedings 4th Europegn Conference on Electron Microscopy Rome, 145.

Schiske, P. (2002) Image reconstruction by means of focus series. Journal of Microscopy-Oxford, **207**, 154.

Schofield, M. A., Beleggia, M., Zhu, Y., Guth, K. and Jooss, C. (2004) Direct evidence for negative grain boundary potential in Ca-doped and undoped YBa$_2$Cu$_3$O$_7$-x. Phys. Rev. Lett. **92**, 195502.

Schofield, M. A., Beleggia, M., Lau, J. W. and Zhu, Y. (2007) Characterization of the JEM-2100F-LM TEM for electron holography and magnetic imaging. JEOL News, **42**, 1.

Schwander, P., Rau, W. D., Ourmazd, A. (1998) Composition mapping at high resolution. J. Microsc. **190**, 171.

Seitz, H., Ahlborn, K., Seibt, M., Schröter, W. (1998) Sensitivity limits of strain mapping procedures using high-resolution electron microscopy. J. Microsc. **190**, 184.

Seki, S, Yu, X. Z, Ishiwata, S, Tokura, Y. (2012) Observation of Skyrmions in a Multiferroic Material. Science, **336**, 198

Shannon, C. E., Weaver, (1949) W. The Mathematical Theory of Communication. Univ. of Illinois Press.

Shindo, D., Park, Y. G., Yoshizawa, Y. (2002) Magnetic domain structures of $Fe_{73.5}Cu_1Nb_3Si_{13.5}B_9$ films studied by electron holography. J. Magn. Magn. Mater. **238**, 101.

Shindo, D., Park, Y. G., Gao, Y. H. and Park, H. S. (2004) Electron holography of Fe-based nanocrystalline magnetic materials. J. Appl. Phys. **95**, 6521.

Sickmann, J., Formánek, P., Linck, M. and Lichte, H. (2008) Extended field of view for medium resolution electron holography at Philips CM200 microscope. EMC 2008, **1**, 277.

Sickmann, J., Formánek, P., Linck, M., Muehle, U. and Lichte, H. (2011) Imaging modes for potential mapping in semiconductor devices by electron holography with improved lateral resolution. Ultramicroscopy, **111**, 290.

Simon, P., Tang, Z., Carrillo-Cabrera, W., Chiong, K., Böhme, B., Baitinger, M., Lichte, H., Grin, Y. and Guloy, A. M. Synthesis and electron holography studies of single crystalline nanostructures of clathrate-ii phases K_xGe_{136} and Na_xSi_{136}. J. Am. Chem. Soc. **133**, 7596.

Simpson, E. T., Kasama, T., Pósfai, M., Buseck, P. R., Harrison, R. J. and Dunin-Borkowski, R. E. (2005) Magnetic induction mapping of magnetite chains in magnetotactic bacteria at room temperature and close to the Verwey transition using electron holography. J. Phys.: Conf. Ser., **17**, 108.

Sinkler, W. and Marks, L. D. (1999) Dynamical direct methods for everyone. Ultramicroscopy, **75**, 251.

Skyrme, T. H. R. (1962) A unified field theory of mesons and baryons. Nucl. Phys. **31**, 556.

Skyrme, T. H. R. (1961) Particle states of a quantized meson field. Proc. R. Soc. A. **262**, 237.

Smith, D. J., Cullen, D. A., Zhou, L. and McCartney, M. R. (2010) Application of TEM imaging, analysis and electron holography to III-nitride HEMT devices. Microelectron. Reliab. **50**, 1514.

Spence, J. C. H. and Zuo, J. M. (1992) Electron microdiffraction. New York: Plenum Press.

Spence, J. C. H., Cowley, J. M. and Zuo, J. M. (1993) Comment on "Electron holographic study of ferroelectric domain walls". Appl. Phys. Lett. **62**, 2446.

Spence, J. C. H. (2003) High-resolution electron microscopy, Third Edition ed. Oxford: Oxford University Press.

Su, D. S., Jacob, T., Hansen, T. W., Wang, D., Schlogl, R., Freitag, B., Kujawa, S. (2008) Surface chemistry of Ag particles: Identification of oxide species by aberration-corrected TEM and by DFT calculations. Angew. Chem. -Int. Edit. **47**, 5005.

Tanaka, M., Saito, R. and Sekii, H. (1983) Point-group determination by convergent-beam electron diffraction. Acta Crystallographica Section A. 39.

Tang, D., Li, F. H. (1988) A Method of Image-restoration for Pseudo-weak-phase Objects. Ultramicroscopy, **25**, 61.

Tang, D., Teng, C. M., Zou, J., Li, F. H. (1986) Pseudo-Weak-Phase-Object Approximation in High-Resolution Electron-Microscopy . II. Feasibility of Directly Observing Li+. Acta Cryst. B. **42**, 340.

Tang, C. Y., Li, F. H. (2005) Restoring Atomic Configuration at Interfaces by Image Deconvolution. J. Electron Microsc. **54**, 445.

Tang, C. Y., Li, F. H., Wang, R., Zou, J., Zheng, X. H., Liang, J. W. (2007) Atomic Configurations of Dislocation Core and Twin Boundaries in 3C-SiC Studied by High-resolution Electron Microscopy. Phys. Rev. B. **75**, 184103.

Tao, J., Niebieskikwiat, D., Varela, M., Luo, W., Schofield, M. A., Zhu, Y., Salamon, M. B., Zuo, J. M., Pantelides, S. T., Pennycook, S. J. (2009) Direct imaging of nanoscale phase separation in $La_{0.55}Ca_{0.45}MnO_3$: relationship to colossal magnetoresistance. Phys. Rev. Lett. **103**, 097202.

Telesnin, R. V., Ilycheva, E. N. and Kanavina, N. G., Osukhovskii, V. E. and Shishkov, A. G. (1969) Phys. Stat. Sol. **34**, 443.

Thon, T. (1971) Phase Contrast Electron Microscopy. In Electron Microscopy in Material Science, Valdrè, U., Ed. Academic Press: New York and London, 570.

Thust, A., Coene, W. M. J., deBeeck, M. O., VanDyck, D. (1996) Focal-series reconstruction in HRTEM: Simulation studies on non-periodic objects. Ultramicroscopy, **64**, 211.

Thust, A., Lentzen, M., Urban, K. (1994) Nonlinear reconstruction of the exit plane-wave function from periodic high-resolution electron-microscopy images. Ultramicroscopy, **53**, 101.

Thust, A., Overwijk, M. H. F., Coene, W. M. J., Lentzen, M. (1996) Numerical correction of lens aberrations in phase-retrieval HRTEM. Ultramicroscopy, **64**, 249.

Tian, H. F., Sun, J. R., Lü, H. B., Jin, K. J., Yang, H. X., Yu, H. C. and Li, J. Q. (2005) Electrostatic potential in manganite-based heterojunctions by electron holography. Appl. Phys. Lett. **87**, 164102.

Tian, H. F., Yu, H. C., Zhu, X. H., Wang, Y. G., Zheng, D. N., Yang, H. X. and Li, J. Q. (2005) Off-axis electron holography and microstructure of $Ba_{0.5}Sr_{0.5}TiO_3$ thin films on $LaAlO_3$. Phys. Rev. B. **71**, 115419.

Tonomura, A., Yu, X. Z., Yanagisawa, K., Matsuda, T., Onose, Y., Kanazawa, N., Park, S. H., Tokura, Y. (2012) Real-Space Observation of Skyrmion Lattice in Helimagnet MnSi Thin Samples. Nano lett. **12**, 1673.

Tonomura, A. (1999) Electron holography. Berlin Heidelberg: Springer-Verlag, 2nd Springer Series Opt. Sci., 70.

Tonomura, A., Kasai, H., Kaminuma, O., Matsuda, T., Harada, K., Nakayama, Y., Shomoyama, J., Kishio, K., Hanaguri, T., Kitazawa, K., Sasae, M. and Okayasu, S. (2001) Observation of individual vortices trapped along columnar defects in high-temperature superconductors. Nature, **412**, 620.

Tonomura, A., Matsuda, T., Endo, J., Arii, T. and Mihama, K. (1986) Holographic interference electron microscopy for determining specimen magnetic structure and thickness distribution. Phys. Rev. B. **34**, 3397.

Tonomura, A., Matsuda, T., Tanabe, H., Osakabe, N., Endo, J., Fukuhara, A., Shinagawa, K. and Fujiwara, H. (1982) Electron holography technique for investigating think ferromagnetic films. Phys. Rev. Lett. **25**, 6799.

Tonomura, A., Osakabe, N., Matsuda, M., Kawasaki, T., Endo, J., Yano, S. and Yamada, H. (1986) Evidence for the Aharonov-Bohm effect with magnetic field completely shielded from electron wave. Phys. Rev. Lett. **56**, 792.

Tonomura, A (2005) Direct observatian of thithertoun observable quantum phenomena by using electrons PNAS 102 14952

Tripp, S. L., Dunin-Borkowski, R. E. and Wei, A. (2003) Flux Closure in self-assembled cobalt nanoparticle rings. Angew. Chem. Int. Ed. **42**, 5591.

Tsuno, K., Inoue, M. (1984) Double gap objective lens for observing magnetic domains by means of differential phase contrast electron microscopy. Optik, **67**, 363.

Tsuno, K., Taoka, T. (1983) Magnetic-field-free objective lens around a specimen for observing fine structure of ferromagnetic materials in a transmission electron microscope, J. Appl. Phys. **22**, 1041.

Twitchett, C., Dunin-Borkowski, R. E. and Midgley, P. A. (2002) Quantitative Electron holography of biased semiconductor devices. Phys. Rev. Lett. **88**, 238302.

Twitchett-Harrison, C., Yates, T. J. V., Dunin-Borkowski, R. E. and Midgley, P. A. (2008) Quantitative electron holographic tomography for the 3D characterisation of semiconductor device structures. Ultramicroscopy, **108**, 1401.

Twitchett-Harrison, C., Yates, T. J. V., Dunin-Borkowski, R. E., Newcomb, S. B. and Midgley, P. A. (2006) Three-dimensional electrostatic potential of a Si p-n junction revealed using tomographic electron holography. J. Phys. Conf. Ser. **26**, 29.

Twitchett-Harrison, A C., Yates, T. J. V., Newcomb, S. B., Dunin-Borkowski, R. E. and Midgley, P. A. (2007) High-resolution three-dimensional mapping of semiconductor dopant potentials. Nano Lett. **7**, 2020.

Uhlig, T., Heumann, M., Zweck, J. (2003) Development of a specimen holder for in situ generation of pure in-plane magnetic fields in a transmission electron microscope. Ultramicroscopy, **94**, 193.

Unwin, P. N. T., Henderson, R. (1975) Molecular-structure determination by electron-microscopy of unstained crystalline specimens. J. Mol. Biol. **94**, 425.

Van Dyck, D. and Op de Beeck, M. (1999) Focus variation electron holography, in Introduction to electron holography. ed. by E. Völkl, L. F. Allard and D. C. Joy. New York: Kluwer Academic/Plenum Publishers.

Van Dyck, D., Debeeck, M.O., Coene, W. (1993) A new approach to object wavefunction reconstruction in electron-microscopy. Optik, **93**, 103.

Vincent, R., Bird, D. M. and Steeds, J. W. (1984) Structure of AuGeAs determined by convergent-beam electron diffraction. I: Derivation of basic structure. Philosophical Magazine a-Physics of Condensed Matter Structure Defects and Mechanical Properties, **50**, 745.

Voelkl, E., Tang, D. (2010) Approaching routine $2\pi/1000$ phase resolution for off-axis type holography. Ultramicroscopy, **110**, 447.

Völkl, E., Allard, L. F. and Joy, D. C. (1999) Late Pleistocene megafaunal extinctions: a European perspective. New York: Kluwer Academic/Plenum Publishers.

Volkov, V. V., Crew, D. C., Zhu, Y., Lewis, L. H. (2002) Magnetic field calibration of a transmission electron microscope using a permanent magnet material. Rev. Scient. Instr. **73**, 2298.

Wan, Z. H., Liu, Y. D., Fu, Z. Q., Li, Y., Cheng, T. Z., Li, F. H., Fan, H. F. (2003) Visual computing in electron crystallography. Z. Kristallogr. **218**, 308.

Wang, H. B., Jiang, H., Li, F. H., Che, G. C., Tang, D. (2002) A study on the position of boron atoms in $(Y_{0.6}Ca_{0.4})(SrBa)(Cu_{2.5}B_{0.5})O_7$- delta. Acta Cryst. A. **A58**, 494.

Wang, H. Y., Dai, X. F., Wang, Y. G., Duan, X. F. and Wu, G. H. (2007) In Situ Observation of Magnetic Domain Structure in $Co_{50}Ni_{50}FeGa_{59}$ Alloy under the Applied Magnetic Field, Mater. Trans. **48**, 2255.

Wang, R. M. (2006) Reconstruction of electron exit waves: basic theory and applications. Journal of Chinese Electron Microscopy Society, 1.

Wang, R.M., Dmitrieva, O., Farle, M., Dumpich, G., Ye, H.Q., Poppa, H., Kilaas, R., Kisielowski, C. (2008) Layer resolved structural relaxation at the surface of magnetic FePt icosahedral nanoparticles. Phys. Rev. Lett. **100**, 017205

Wang, S. X. and Li, G. X. (2008) Advances in giant magnetoresistance biosensors with magnetic nanoparticle tags: review and outlook. IEEE Trans. Magn. **44**, 1687.

Wang, Y. G. and Dravid, V. P. (2002) Determination of electrostatic characteristics at a 24 degree, [001] tilt grain boundary in a $SrTiO_3$ bicrystal by electron holography. Phil. Mag. Lett. **82**, 425.

Wang Y. M., Wang H. B., Li F. H., Jia L. S., Chen X. L. (2005) Maximum Entropy Image Deconvolution Applied to Structure Determination for Crystal $Nd_{1.85}Ce_{0.15}CuO_{4-\delta}$. Micron, **36**, 393.

Wang, D., Chen, H., Li, F. H., Kawasaki, K., Oikawa, T. (2002) Atomic Configuration in Core Structure of Lomer Dislocation in $Si_{0.76}Ge_{0.24}/Si$. Ultramicroscopy, **93** , 139.

Wang, D., Zou, J., He, W. Z., Chen, H., Li, F. H., Kawasaki, K., Oikawa, T. (2004) Determination of a misfit dislocation complex in SiGe/Si heterostructures by image deconvolution technique in HREM. Ultramicroscopy, **98**, 259.

Wang, H. B., Wang, Y. M., Li, F. H. (2004) A further discussion on the peculiarity of maximum entropy image deconvolution in HREM. Ultramicroscopy, **99**, 165.

Wang, Y. M., Wan, W., Wang, R., Li, F. H., Che, G. C. (2008) Atomic configurations of twin boundaries and twinning dislocation in superconductor $Y_{0.6}Na_{0.4}Ba_2Cu_{2.7}Zn_{0.3}O_{7--delta}$. Philos. Mag. Lett. **88**, 481.

Wang, Z. Yao, Y., He, X, Yang, Y., Gu, L., Wang, Y., Duan, X. (2012) Invetigation of strain and thin film relaxation in Gexsirx/si Strained-layer superlattice by dark-fied electron holography Mater. Trans. **53** 2019.

Wanner, M., Bach, D., Gerthsena, D., Wernerb, R., and Tesche, B. (2006) Electron holography of thin amorphous carbon films: Measurement of the mean inner potential and a thickness-independent phase shift. Ultramicroscopy, **106**, 34.

Wei, Q. Y., Wu, Z. H., Ponce, F. A., Hertkorn, J., and Scholz, F. (2010) Polarization effects in 2-DEG and 2-DHG AlGaN/AlN/GaN multi-heterostructures measured by electron holography. Phys. Status solidi B, **247**, 1722.

Wei, Q., Wu, Z., Sun, K., Ponce, F. A., Hertkorn, J. and Scholz, F. (2009) Evidence of two-dimensional hole gas in p-type AlGaN/AlN/GaN heterostructures. Appl.

Phys. Express, **2**, 121001.

Weickenmeier A., Kohl, H. (1991) Computation of absorptive form factors for high-energy electron diffraction. Acta Crystallogr. Sect. A, **47**, 590.

Wen, C., Wang, Y. M., Wan, W., Li, F. H., Liang, J. W., Zou, J. (2009) Nature of interfacial defects and their roles in strain relaxation at highly lattice mismatched 3C-SiC/Si (001) interface. J. Appl. Phys. **106**, 073522.

Wilson, A. J. C. (1949) THE PROBABILITY DISTRIBUTION OF X-RAY INTENSITIES Acta Crystallogr. **2**, 318.

Wolf, D., Lichte, H., Pozzi, G., Prete, P. and Lovergine, N. (2011) Electron holographic tomography for mapping the three-dimensional distribution of electrostatic potential in III-V semiconductor nanowires. Appl. Phys. Lett., **98**, 264103.

Wolf, D., Lubk, A., Lichte, H. and Friedrich, H. (2010) Towards automated electron holographic tomography for 3D mapping of electrostatic potentials. Ultramicroscopy, **110**, 390.

Woolfson, M. M. and Fan, H. F. (1995) Physical and Non-Physical Methods of Solving Crystal Structure. Cambridge: Cambridge University Press, 221.

Woolfson, M. M. (1956) AN IMPROVEMENT OF THE HEAVY-ATOM METHOD OF SOLVING CRYSTAL STRUCTURES Acta Cryst. A, **9**, 804.

Wu, Z. H., Stevens, M., Ponce, F. A., Lee, W., Ryou, J. H., Yoo, D. and Dupuis, R. D. (2007) Mapping the electrostatic potential across AlGaN/AlN/GaN heterostructures using electron holography. Appl. Phys. Lett. **90**, 032101.

Xia, W. X., Tohara, K., Murakami, Y., Shindo, D., Ito, T., Iwasaki, Y. and Tachibana, J. (2006) Observation of magnetization of obliquely evaporated Co-CoO magnetic recording tape. IEEE Trans. Magn., **42**, 3252.

Xu, Q. Y., Wang, Y., Wang, Y. G., Du, X. L., Xue, Q. K. and Zhang, Z. (2004) Polarity determination of ZnO thin films by electron holography. Appl. Phys. Lett. **84**, 2067.

Xu, X., Beckman, S.P., Specht, P., Weber, E.R., Chrzan, D.C., Erni, R.P., Arslan, I., Browning, N., Bleloch, A., Kisielowski, C. (2005) Distortion and segregation in a dislocation core region at atomic resolution. Physical Review Letters, **95**, 145501.

Yamamoto, K., Sugwara, Y., McCartney, M. R. and Smith, D. J. (2010) Phase-shifting electron holography for atomic image reconstruction. Jpn. Soc. Microscopy, **59**, S81.

Yamamoto, K., Hirayama, T., and Tanji, T. (2004) Off-axis electron holography without Fresnel fringes, Ultramicroscopy., **101**, 265.

Yamamoto, K., Hirayama, T., Tanji, T. and Hibino, M. (2003) Evaluation of high-precision phase-shift electron holography by using hologram simulation. Surf. Interface Anal. **35**, 60.

Yamamoto, K., Hogg, C. R., Yamamuro, S., Hirayama, T. and Majetich, S. A. (2011) Dipolar ferromagnetic phase transition in Fe_3O_4 nanoparticle arrays observed by Lorentz microscopy and electron holography. Appl. Phys. Lett., **98**, 072509.

Yamamoto, K., Iriyama, Y., Asaka, T., Hirayama, T., Fujita, H., Fisher, C. A. J., Nonaka, K., Sugita, Y. and Ogumi, Z. (2010) Dynamic Visualization of the Electric Potential in an All-Solid-State Rechargeable Lithium Battery. Angew. Chem. Int. Ed., **49**, 4414.

Yamamoto, K., Kawajiri, I., Tanji, T., Hibino, M. and Hirayama, T. (2000) High precision phase-shifting electron holography. J. Electron Micros., **49**, 31.

Yang, S. X., Li, F. H. (2000) Image deconvolution for protein crystals. Ultramicroscopy, **85**, 51.

Yang, Y., Wang, W.X., Yan, Y., Liu, H. F. Naganuma, H. Sakul, T. S., Han, X. F. and Yu, R C (2012) Chemical diffusion: another factor offecting the magnetoresistance ratio in Ta/CoFeB/MgO/CoFeB/Ta magnetic tunnel junction. Appl. Phys Lett., **101**, 012406.

Yang, Y, Niu, N, Li, C., Yao, Y, Piao, G., and Yu, R. C., (2012) Electron holography characterization as a method for measurements of diameter and mean inner potential of hollow wamomaterials Nanoscale, **4**, 7460.

Yao, Y., Yang, Y., Duan, X. F., Wang, Y. G., Yu, R. C. and Xu, Q. X. (2011) Electron holography characterization of the electrostatic potential of thin high-κ dielectric film embedded in gate stack. Appl. Phys. Lett., **99**, 163506.

Yao, Y., Li, C., Huo, Z. L., Liu, M., Zhu, C. X., Gu. C.Z., Duan, X. F., Wang. Y. G., Yu, R. C. (2013) In situ electran holography study of charge distribution in high-kappa charge-trapping memory Nat. Comm. **4**, 2764.

Yi, G., Nicholson, W. A. P., Lim, C. K., Chapman, J. N., Mcvitie, S., Wilkinson, C. D. W. (2004) A new design of specimen stage for in situ magnetising experiments in the transmission electron microscope. Ultramicroscopy, **99**, 65.

Yoo, J. H., Murakami, Y., Shindo, D., Atou, T. and Kikuchi, M. (2002) Behavior of magnetic domains in $La_{0.46}Sr_{0.54}MnO_3$ during the ferromagnetic phase transformation studied by electron holography. Phys. Rev. B. **66**, 212406.

Yoshida, H., Kohno, H., Ichikawa, S., Akita, T., Takeda, S. (2007) Inner potential fluctuation in SiC nanowires with modulated interior structure. Mater. Lett. **61**, 3134.

Yoshida, T., Matsuda, T., Fukuhara, A., Tonomura, A. (1992) Electron holography observation of flux-line dynamics. Proc. Boston: 50th meeting of Electron Microscopy Society of America.

Yu, X. Z., DeGrave, J. P., Hara, Y., Hara, T., Jin, S., Tokura, Y. (2013) Observation of the Magnetic Skyrmion Lattice in a MnSi Nanowire by Lorentz TEM. Nano. lett. **13**, 3755.

Yu, X. Z., Kanazawa, N., Onose, Y., Kimoto, K., Zhang, W. Z., Ishiwata, S., Matsui, Y., and Tokura, Y. (2011) Near room-temperature formation of a skyrmion crystal in thin-films of the helimagnet FeGe. Nature Mater., **10**, 106.

Yu, X. Z., Kanazawa, N., Zhang, W. Z., Nagai, T., Hara, T. T., Kimoto, K., Matsui, Y., Onose, Y. and Tokura, Y. (2012) Skyrmion flow near room temperature in an ultralow current density. Nat. Commun, **3**, 988.

Yu, X. Z., Onose, Y., Kanazawa, N., Park, J. H., Han, J. H., Matsui, Y., Nagaosa, N., Tokura, Y. (2010) Real-space observation of a two-dimensional skyrmion crystal. Nature, **465**, 901.

Yu, X. Z., Tokunaga, Y., Kaneko, Y., Zhang, W. Z., Kimoto, K., Matsui, Y., Taguchi, Y. and Tokura, Y. (2014) Biskyrmion states and their current-driven motion in a layered manganite. Nat. Commun. **5**, 3198.

Yuzi, G. and Yuzo, O. (1960) In situ lorentz microscopy observation of displaced chain walls in permalloy. J. Phys. Soc. Japan., **15**, 535.

Zhang, X., Hashimoto, T., Joy, D. C. (1992) Electron holographic study of ferroelectric domain walls. Appl. Phys. Lett., **60**, 784.

Zhang, X., Joy, D. C., Zhang, Y., Hashimoto, T., Allard, L., Nolan, T. A. (1993)

Electron holography techniques for study of ferroelectric domain walls. Ultramicroscopy, **51**, 21.

Zheng, H., Cao, A., Weinberger, C. R., Huang, J. Y., Du, K., Wang, J., Ma, Y., Xia, Y., Mao, S. X. (2010) Discrete plasticity in sub-10-nm-sized gold crystals. Nat. Commun. **1**, 144.

Zhou, L., Cullen, D. A., Smith, D. J., McCartney, M. R., Mouti, A., Gonschorek, M., Feltin, E., Carlin, J. F. and Grandjean, N. (2009) Polarization field mapping of $Al_{0.85}In_{0.15}N/AlN/GaN$ heterostructure. Appl. Phys. Lett. **94**, 121909.

Zhou, L., Smith, D. J., McCartney, M., Xu, T. and Moustakas, T. D. (2011) Measurement of electric field across individual wurtzite GaN quantum dots using electron holography. Appl. Phys. Lett., **99**, 101905.

Zhu, Y., Inada, H., Nakamura, K., and Wall, J. (2009) Imaging single atoms using secondary electrons with an aberration-corrected electron microscope. Nature Materials, **8**, 808.

Zou, X., Sundberg, M., Larine, M., Hovmoller, S. (1996) Structure projection retrieval by image processing of HREM images taken under non-optimum defocus conditions. Ultramicroscopy, **62**, 103.

Zuo, J. M. (1993) Automated structure-factor refinement from convergent-beam electron diffraction patterns. Acta Crystallographica Section **A49**, 429.

Zuo, J. M., O'Keeffe, M. Rez, P. and Spence, J. C. H. (1997) Charge density of MgO: implications of precise new measurements for theory. Physical Review Letters, **78**, 4777.

Zuo, J. M., O'Keeffe, M. Rez, P. and Spence, J. C. H. (1999) Direct observation of d-orbital holes and Cu–Cu bonding in Cu_2O. Nature, **401**, 49.

Zuo, J. M., Blaha, P. and Schwarz, K. (1997) The theoretical charge density of silicon: experimental testing of exchange and correlation potentials. Journal of Physics-Condensed Matter, **9**, 7541.

Zuo, J. M. and Tao, J. (2010) Scanning Transmission Electron Microscopy: Imaging and Analysis. London, Springer.

Zuo J. M., Gao, M., Tao, J., Li, B. Q., Twesten, R., and Petrov, I. (2004) Coherent Nano-Area Electron Diffraction. Microscopy Research and Technique, **64**, 347.

4
Scanning Transmission Electron Microscopy (STEM)

Binghui Ge

4.1 Introduction

4.1.1 Brief history of STEM

Although Baron Manfred von Ardenne (1938) developed the first scanning transmission electron microscopy (STEM) in 1938, placing the image lens before the specimen instead of after the specimen as in the Ruska TEM design, it is just a sound idea in principle: he did not use a field emission source such that the 10 nm resolution images he achieved were too noisy. Crewe realized the necessity of using a high brightness cold field emission gun (FEG) (Crewe, 1966) to achieve sufficient beam current in a small probe and a new microscope developed could produce a resolution about 0.5 nm (Crewe and Wall, 1970). Using molecule stained with uranium and thorium atoms samples individual atoms were first imaged by an electron microscope. Because of these contributions, Crewe is regarded to be the father of STEM.

The first so-called "Z-contrast" image (Crewe, 1971) was formed from the ratio from the elastic signal collected by annular detector to the inelastic signal collected by the spectrometer, for their cross section ratio is approximately proportional to atomic number Z. But for crystalline materials, unlike the biology systems interested in Crewe group, the Crewe ratio method for Z-contrast imaging is not satisfiying because of the dominant influence of diffraction contrast. Success was achieved during imaging of catalyst particles with a high-angle annular dark field (HAADF) detector, by which thermal diffuse scattering (TDS) electrons are collected. Comparing with the Crewe ratio method, the signals give enhanced Z-contrast due to the minimum diffraction effects included. Since then, HAADF imaging becomes an essential part in the application of STEM.

Crewe's first microscope had no annular detector; it had a spectrometer to allow imaging at zero loss or at a chosen value of energy loss. He felt that the electron energy loss spectroscopy (EELS) signal held the greatest promise for high contrast (Crewe et al., 1968) since it contained much information, and it should be possible to use all the electrons that lose energy. Indeed, the common DNA bases could be easily distinguished in the low loss region of the spectrum (Crewe et al., 1971). The simultaneous detection of elastic scattering with an ADF detector and inelastic scattering with the spectrometer was a feature incorporated into their next microscope. This development leads to another important part of application for STEM ascribed to its flexibility, simultaneous collection of various analytical signals in addition to imaging, such as electron energy loss spectrum (EELS), nanobeam diffraction (NBD), X-ray energy dispersive spectrum (XEDS or EDS) and secondary electron spectroscopy (SES), and so on, as shown in Fig. 4.1.1. Especially when the condenser aberrations are corrected such that electron probe is focused to be below 0.1 nm and much larger current is delivered into nanoscale probe, not only atomic resolution EELS mapping (Kimoto et al., 2007) becomes true, but also EDS mapping (Chu et al., 2010; D'Alfonso et al., 2010) and even secondary electron imaging in STEM (Inada et al., 2010). STEM, now, has become a powerful and comprehensive analytical tool for the study of material and other related sciences.

4.1.2 Instrumentation

There are two types of STEM instruments commonly used by electron microscopists: dedicated scanning transmission electron microscope (DSTEM) and STEM attachment in TEM. With the advance of field-emission technology, there is no much difference nowadays between DSTEM and STEM mode in TEM using STEM attachments: high-brightness electron probe with the small size in STEM mode of TEM can be obtained as that in DSTEM. For the popularity of the STEM

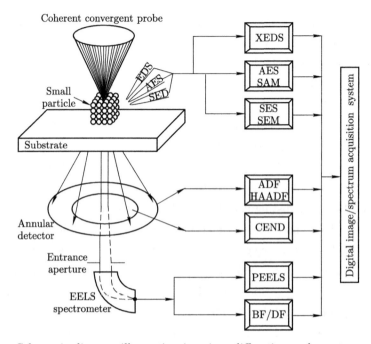

Fig.4.1.1 Schematic diagram illustrating imaging, diffraction and spectroscopy modes commonly used in STEM: X-ray energy dispersive spectroscopy (XEDS); auger electron spectroscopy (AES) and scanning auger spectroscopy (SAM); secondary electron spectroscopy (SES) and secondary electron microscopy (SEM); annular dark-field (ADF) and high-angle annular dark-field (HAADF) microscopy; coherent electron nano-diffraction (CEND); parallel electron energy-loss spectroscopy (PEELS); and bright-field (BF) and dark-field (DF) microscopy. Liu, 2000.

attachment in China, STEM mode in TEM is taken as example in this chapter. Unless stated particularly, the term STEM is used to present the STEM mode in TEM in the rest of this chapter.

A STEM microscope uses an FEG to generate high-brightness electron probes. At least two condenser lenses and an objective lens are usually used to form a small electron probe on the specimen. A condenser aperture is placed between the condenser lens and the objective lens to control the convergent angle of the incident electron probe according to the Ronchigram, which will be introduced in Section 4.3, to keep outside the electron waves with large phase variation. The size and the intensity of the high-energy electron probe can be manipulated by selecting the proper size of the condenser aperture.

In some STEM instruments, the electrons, passing through the specimen, directly reach the detector plane without the use of any post-specimen lenses. It is, however, desirable to have post-specimen lenses to offer great flexibility for effective utilizing various detector configurations, conveniently observing and recording nanodiffraction patterns.

Interchangeable annular detectors can be installed to provide flexibility for ADF imaging or for special imaging modes using configured detectors. The attachment of a series EELS or a parallel EELS (PEELS) detectors at the bottom of the microscope column makes it possible to analyze the composition or electron structure of the sample at an atomic resolution. It also allows bright-field (BF) or dark-field (DF) imaging with only elastically scattered electrons or with other selected energy-loss electrons. A charge couple device (CCD) can be used to quantitatively record nanodiffraction patterns, shadow images or electron holograms.

For effectively collecting characteristic X-rays, a retractable, windowless XEDS spectrometer is usually attached to the column of a STEM instrument. Because of the small volume probed by the electron nanoprobe, one or more XEDS detectors can be placed close to the sample region to increase the strength of the collected X-ray signal.

A stable operation of FEG requires the vacuum in the gun chamber to be almost 10^{-6} Pa for thermal FEG and 10^{-8} Pa for cold FEG to prevent the contamination and oxidation. While even in ultra-high vacuum, surface contamination build up on the tip of gun. Eventually, it becomes necessary to remove the contamination by "flashing" the tip for cold FEG. The column vacuum is generally better than 10^{-6} Pa to prevent significant back streaming of gas molecule into the gun chamber and to reduce the effects of contamination on the specimen surface. Most of the STEM instruments can be baked at moderate temperature for extended periods to obtain a high vacuum.

4.2 The Principle of reciprocity

Before embarking on a discussion of the origins of contrast in STEM imaging, it is firstly important to consider the implications of the principle of reciprocity (Liu,

2000). Consider elastic scattering so that all the electron waves in the microscope have the same energy, the propagation of the electrons is time reversible (Nellist, 2011). That is, the principle of reciprocity developed in the light optics can be equally applied to electron optical systems (Cowley, 1969; Zeitler and Thomson, 1970). The wave amplitude at a point P due to a point source at Q is identical to the wave amplitude at Q due to a point source at P as shown in Fig. 4.2.

The essential components of a STEM imaging system are similar to those of a TEM microscope: the ray diagram of STEM is the reciprocal of that of TEM. This is demonstrated with the aid of the schematic ray diagram of Fig. 4.2.1 The STEM detector replaces the TEM electron source; the STEM gun is placed in the detector plane of the TEM; the plane of the condenser lens in TEM is used to place projector lens in STEM while the plane of projector lens in TEM to place condenser lens. Therefore, for a particular detector configuration, the contrast of STEM images can often be obtained by finding the equivalent TEM geometry. In the following, two kinds of STEM imaging, large angle BF (LABF) and annular BF (ABF) imaging, will be explained briefly by means of the principle of reciprocity.

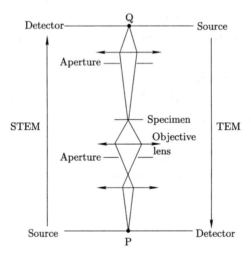

Fig.4.2.1 Schematic ray diagram illustrating the principle of reciprocity in electron optics: the ray diagram of STEM is the reciprocal of that of TEM.

In TEM, although the increase of illumination convergent angle destroys the phase contrast, according to the literature of Mitome et al.(1990), the point resolution of high-resolution TEM images is improved under the convergent-beam illumination (CBI) condition comparing with the parallel-beam illumination condition, because the position of first crossover of phase contrast transfer function (PCTF) shifts to the higher frequency, which means more diffraction beams transmit through the objective lens without phase reversal. According to the principle of reciprocity, the increase of the illumination convergent angle in TEM is equivalent to the increase of collection angle in STEM BF imaging. Thus, the resolution of LABF imaging will be improved comparing with BF imaging according to the results in the literature of Mitome et al.(1990). Fig. 4.2.2(a) is a high-resolution BF STEM image of a GaAs crystal oriented in the [110] zone axis. Comparing with in BF image, a better image resolution can be seen clearly from the LABF image (semi-collection angle of about 30 mrad) with the same area as shown in Fig. 4.2.2(b).

(a)

(b)

Fig.4.2.2 Atomic resolution BF (a) and large-angle BF (b) STEM images of the same area of a GaAs crystal oriented along the [110] zone axis. The large-angle BF STEM image clearly shows a better resolution and a higher contrast. Liu, 2000.

Analogically, due to the principle of reciprocity, a recently developed imaging technique in STEM, ABF imaging (Findlay et al., 2009; 2010; Okunishi et al., 2009) by locating an annular detector within the bright-field region (namely, the direct-beam disc), is equivalent to the hollow-cone illumination (HCI) imaging in TEM, which employs a series of off-axial illuminations over certain angle ranges of incident beams as shown in Fig. 4.2.3. In the early studies of optics (Mathews, 1953; Hanssen, 1971; Rose, 1977), it was already shown as to the HCI imaging that not only the resolution is significantly improved just like that in CBI imaging, but also signal-to-noise ratio of a phase contrast due to minimizing the effect of wavelength fluctuations of the incident beam (Komoda, 1966) so that ABF imaging should also have the enhanced resolution and the phase contrast, comparing with the BF imaging.

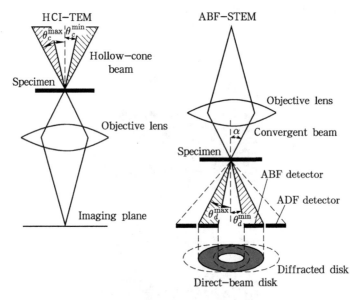

Fig.4.2.3 Schematic ray diagrams of HCI–TEM/ABF–STEM. HCI–TEM with the cone angle ranging from minimum θ_c^{min} to θ_c^{max} is equivalent to ABF–STEM with the detector angle ranging from minimun θ_d^{min} to θ_d^{max}. Ishikawa et al., 2011.

ABF imaging was first reported by Okunishi in 2009, and light atoms such as oxygen and even hydrogen (Ishikawa et al., 2011) atoms in compounds were observed. Due to this advantage combined with its insensitiveness of image contrast as to sample thickness, ABF imaging has been widely used in studying light atoms in functional materials (Gu et al., 2011). In Section 4.5, its imaging contrast theory and applications will be introduced in detail.

4.3 Principle of STEM imaging

4.3.1 Theoretical background

As mentioned above, the electrons was emitted from FEG and then converged by condenser lenses to form the tiny probe. After reaching the materials, the focused electrons, the probe, interact with atomic nucleus of samples and electrons outside of nucleus and they are scattered to different angles with some of them losing their energy or only changing their trajectories, depending on different scattering mechanism. Then the scattered electrons can be collected separately by different detectors through changing the camera length or detector configuration. After integration of collected electrons the intensity at one position in a STEM image is determined. When scan coils are arranged to scan the probe over the sample in a raster, a STEM image is formed.

Therefore, throughout the STEM imaging three steps can be categorized: electron probe formation, electron scattering (interaction between probe and sample), and electron collection, which will be introduced in the following.

1. Electron probe

The full-width-half maximum (FWHM) of the probe beam is one criterion for the resolution of STEM imaging so that electron beams should be focused by condenser lenses. Meanwhile, with the reduction of the size of beam, the brightness of the beam, usually defined as the current density, will be increased, which is necessary in STEM imaging (Fultz and Howe, 2008).

As to the size of the probe several factors are attributed: spherical aberration, chromatic aberration, and diffraction effect and source size.

Spherical Aberration: Spherical aberration changes the focus of off-axis rays. The further the ray deviates from the optic axis, the greater its error in focal length is. Thus, spherical aberration causes an enlargement of the image of a point. The minimum enlargement of the point is termed the "disk of least confusion." The diameter, d_s, of the disk of least confusion caused by spherical aberration is

$$d_s = 0.5 C_s \alpha_p^3 \qquad (4.3.1)$$

where C_s is the spherical aberration coefficient (approximately 1 mm) for a conventional STEM, and α_p is semi angle of convergence.

Chromatic Aberration: Electrons with different energies mainly originated from the gun and specimen come to different focal points, when entering a lens along the same path. The spread in focal lengths is proportional to the spread in energy of the electrons, and it makes the image of the point a disk, too. The disk of least confusion for chromatic aberration corresponds to a diameter at the specimen, d_c

$$d_c = \alpha_p C_c \frac{\Delta E}{E} \qquad (4.3.2)$$

where $\frac{\Delta E}{E}$ is the fractional variation in electron beam voltage, C_c is the chromatic aberration coefficient (approximately 1 mm).

Diffraction Effect: For an aperture that selects a range δk, the smallest spatial features in the image have the size $2\pi/\delta k$. In optics, this effect is explained as "diffraction" from the edge of an aperture. It contributes a disk of confusion of diameter corresponding to a distance at the specimen, d_d

$$d_d = \frac{0.61 \lambda}{\alpha_p} \qquad (4.3.3)$$

where λ is the electron wavelength. This equation is the classic Rayleigh criterion for resolution in light optics. In essence, it states that when the intensity between two point (Gaussian) sources of light reaches 0.81 of the maximum intensity of the sources, they can no longer be resolved.

Source Size: The focused spot on the specimen is, in fact, an image of the source itself, so it should be easy to form a small spot when the source itself has a small size. Assuming perfect lenses, the beam diameter, d_0, can be related with the brightness of the electron gun, β, and the convergence angle of the lens. That is,

$$d_0 = \frac{\sqrt{\frac{4I_p}{\beta}}}{\pi\alpha_p} = \frac{C_0}{\alpha_p} \qquad (4.3.4)$$

where I_p is the beam current. For a given current, small values of the beam diameter are obtained by increasing the brightness (choosing a gun with higher brightness, such as FEG) or by increasing the semi angle of convergence (choosing a larger aperture, which requires a lens with smaller spherical aberration). In practice, however, α_p has a maximum value due to the lens aberrations and β is limited by the design of the electron gun such that to obtain a beam diameter with small size, only one thing can be done, sacrificing the beam current.

Taken into consideration all the contributing factors mentioned above, a general expression for the beam size, d_p, can be obtained by summing in quadrature (this is strictly valid only when all broadenings are of Gaussian shape, so that convolutions of these different beam broadenings have a Gaussian form) all diameters of the disks of least confusion from the spherical aberration, chromatic aberration, diffraction effect and source size, d_s, d_c, d_d, and d_0:

$$d_p^2 = d_s^2 + d_c^2 + d_d^2 + d_0^2 = \frac{C_0^2 + (0.61\lambda)^2}{\alpha_p^2} + 0.25 C_s^2 \alpha_p^6 + \left(\alpha_p C_c \frac{\Delta E}{E}\right)^2 \qquad (4.3.5)$$

For a FEG, $C_0 \ll \lambda$, and the contributions of d_c and d_0 can be neglected. Superposition of the remaining terms, d_s, and d_d, yields a minimum beam size, and the parametric plot of the minimum size versus semi angle of convergence is shown in Fig. 4.3.1 ($C_s = 1.2$ mm, $\lambda = 0.00251$ nm). Furthermore, a defocus Δf needs to be chosen, similar to the Scherzer focus in HRTEM. Crewe and Salzman (1982) solved this problem, and derived that for an optimum defocus Δf_{opt} and for an optimum illumination semi-angle α_{opt}

$$\Delta f_{\text{opt}} = -(C_s \lambda)^{1/2} \qquad (4.3.6)$$

$$\alpha_{\text{opt}} = (4\lambda/C_s)^{1/4} \qquad (4.3.7)$$

under which the resolution limit is defined as

$$d_{\min} = \frac{0.61\lambda}{\alpha_{\text{opt}}} = 0.43 C_s^{1/4} \lambda^{3/4} \qquad (4.3.8)$$

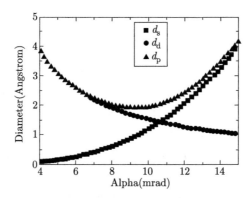

Fig.4.3.1 Parametric plot of the minimum probe size versus semi convergence angle. $C_s=1.2$ mm, $\lambda = 0.00251$ nm. The optimum semi convergence angle is about 9.3 mrad.

which is more than 30% smaller than the Scherzer resolution for coherent imaging in high-resolution transmission electron microscopy (HRTEM).

Actually, the above considerations about the lateral size of an electron probe have clear limits (Erni, 2010). The effects of diffraction, the spherical and chromatic aberrations and the infinite size of the gun cannot be treated independently. They are highly interrelated, which, in particular, becomes apparent if the electron probe is not solely considered as a two-dimensional focused electron spot but as a three dimensional entity, which, apart from the lateral extension, also has a longitudinal or vertical component. Nonetheless, the simplifications upon which they are based allow us to develop an understanding of the individual components that influence the spatial resolution in STEM imaging. In order to understand the electron probe as the result of the collective effect of all four factors mentioned above, the electron probe need to be described on the basis of wave optics rather than on purely geometrical grounds, which is beyond the scope of this chapter. However, it is important to note that the above considerations about the individual contributions to the STEM probe and their dependence on the illumination semi-angle qualitatively remain valid. For instance, it is still the spherical aberration and the diffraction limit that define the optimum STEM probe of a conventional scanning transmission electron microscope.

2. Electron scattering and collection

When a fine focused electron probe interacts with the specimen, the high-energy incident electrons are scattered to the different angle. Through collectors with different collection angle or different configures, different imaging contrast can be obtained corresponding to different modes, such as BF, ABF and ADF etc. (Pennycook, 2011).

An electron diffraction pattern consisting of a set of convergent beam discs can be obtained when a thin crystal is oriented along a principal zone axis. If a circular STEM detector is positioned in the transmission disc, the obtained STEM image after the probe scanning over the sample is a BF image. According to the principle of reciprocity mentioned in Section 4.2, the BF imaging is similar with HRTEM coherent imaging, and its contrast is sensitive to the sample thickness and the defocus values.

Instead of the circular detector, if an annular detector is chosen to collect the electrons in the transmission disc, ABF images will be obtained. Comparing with BF images, ABF images are also sensitive to the defocus value, the feature of coherent imaging, but not to the sample thickness, which will be specifically introduced in Section 4.5. Still using the annular detector, if we decrease the camera length so that electrons of the transmission disc are excluded, ADF images are obtained. In the case of HAADF imaging with inner collection angle at least more than 50 mrad, the image contrast is insensitive to both the sample thickness and the defocus value but mainly dependent on the atomic number, Z, which is mainly ascribed to the incoherent TDS electrons. HAADF imaging, therefore, is also named Z (atomic number) contrast imaging, which will be introduced in Section 4.4. In the next, the modes of BF imaging (coherent imaging) and ADF imaging (incoherent imaging in some cases) are selected as examples to briefly explain their imaging contrast.

The amplitude distribution of the incident electrons at the exit surface of the sample can be described by a wave function $\Psi(K)$, which is the Fourier transformation of the object wave function $\varphi(R)$ multiplied by the phase factor due to the aberration of condenser lens

$$\Psi(K) = F^{-1}(\varphi(R))e^{-i\chi(K)} \qquad (4.3.9)$$

where $\chi(K)$ for an uncorrected microscope is dominated by the coefficient of spherical aberration and defocus value,

$$\chi(K) = \pi\left(\Delta f \lambda K^2 + \frac{1}{2}C_s \lambda^3 K^4\right) \qquad (4.3.10)$$

After the Fourier transformation into the image plane, Eq. 4.3.9 becomes a convolution and we have

$$\psi(R) = \varphi(R) \otimes F^{-1}[e^{-i\chi(K)}] \qquad (4.3.11)$$

In the TEM case, the Fourier transformation of the phase changes due to aberrations, $F^{-1}[e^{-i\chi(K)}]$, is usually called contrast transfer function. While in the STEM case, it can be treated as the probe amplitude distribution, $p(R)$. After squaring of Eq. (4.3.11), the bright filed image intensity can be obtained as the square of a convolution

$$I_{BF}(R) = |\varphi(R) \otimes p(R)|^2 \qquad (4.3.12)$$

which is the reason that bright field phase contrast images can show positive or negative contrast depending on the phase of the transfer function.

While for the imaging of a STEM annular detector, it is necessary to assume that the annular detector collects all of the scattering to convert pure phase variation to intensity variations in the ADF image (Misell et al., 1974; Engel et al., 1974). In this case we obtain the fundamental equation for incoherent imaging

$$I(R) = |\varphi(R)|^2 \otimes |p(R)|^2 \qquad (4.3.13)$$

Sometimes, Eq. (4.3.13) is simplified to be

$$I(R) = O(R) \otimes P(R) \qquad (4.3.14)$$

That is, the intensity of incoherent imaging is described as a convolution of an object function $O(R)$ with a STEM probe intensity profile $P(R)$ (shown in Fig. 4.3.2), and the latter also can be regarded as the point spread function. The Fourier transform of the image will therefore be a product of the Fourier transformation of the object function and the Fourier transformation of the probe intensity. The latter is known as the optical transfer function and its typical form is shown in Fig. 4.3.3 (accelerating voltage 300 kV, C_s 1 mm, defocus -44.4 nm). Unlike the phase contrast transfer function (PCTF) for BF imaging, it shows no contrast reversals and decays monotonically as a function of spatial frequency.

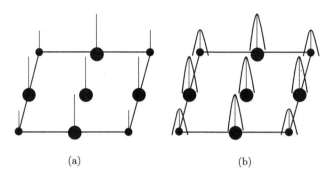

(a) (b)

Fig.4.3.2 (a) Ni$_3$Al $\langle 100 \rangle$ projected model with face-centered cubic structure and object functions represented by the weighted lines. (b) The experimental image interpreted as a convolution of the probe intensity and the object function.

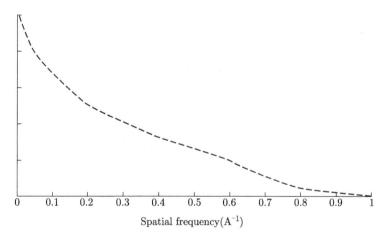

Fig.4.3.3 A typical optical transfer function for incoherent imaging in STEM with accelerating voltage 300 kV, C_s 1 mm, defocus −44.4 nm (Nellist, 2011).

4.3.2 Ronchigram

The electron "Ronchigram" (James and Browning, 1999), or "shadow image" is one of the most useful ways of characterizing and optimizing the probe. This is because the intensity, formed at the diffraction plane, varies considerably with angle, and this variation is a very sensitive function of lens aberrations and defocus (Cowley, 1986). When the excitation of each illumination electron optical component is slightly changed, very small misalignments become apparent by translations in the pattern that depart from circular symmetry. Furthermore, the presence or absence of interference fringes in the pattern indicates the amount of incoherent probe broadening due to instabilities and the effect of a finite source size. Fig. 4.3.4 shows schematically the ray diagram for Ronchigram formation. The probe remains stationary and the post-specimen intensity is recorded as a function of angle by a CCD camera or equivalent device.

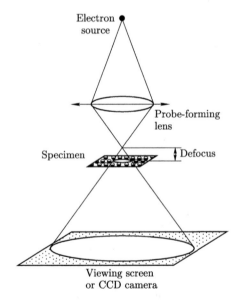

Fig.4.3.4   A ray diagram showing formation of an electron Ronchigram (James and Browning, 1999).

Experimentally, to observe the Ronchigram, apertures are removed after the specimen and a large convergence angle (> 100 mrad) is selected by inserting the largest condenser aperture. The Ronchigram can then be directly observed on the microscope phosphor screen or on a TV-rate CCD camera positioned beneath the phosphor screen. Camera length and positioning are controlled with the projector lenses and shift coils.

Typical Ronchigrams at the amorphous edge of a specimen are shown in Fig. 4.3.5. At large defocus as in Fig. 4.3.5(a) and (d), the electron cross-over is at a

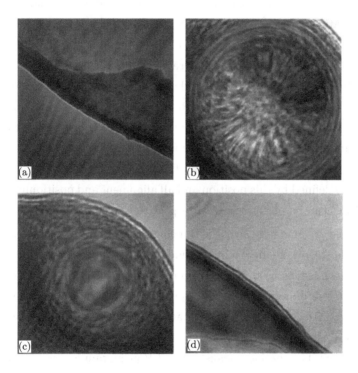

Fig.4.3.5 Experimental electron Ronchigrams of a thin amorphous carbon layer (James and Browning, 1999). (a) Large underfocus-rays at all angles cross the optic axis after the specimen and a shadow image of the specimen edge is seen. (b) Small underfocus-low angle rays cross the optic axis after the specimen. High-angle rays cross before the specimen, because of the effect of spherical aberration. The shadow image therefore changes in magnification as a function of angle and critical angles occur where there is infinite radial and azimuthal magnification. Departures from circular symmetry indicate the presence of astigmatism. (c) Gaussian focus-the lowest angle rays cross the axis at the specimen; higher angle rays cross before it, due to the effect of spherical aberration. (d) Overfocus-rays at all angles cross the axis before the specimen and a shadow image of the specimen edge is visible.

relatively large distance from the specimen, along the optic axis, and a projection image is observed. Due to the opposite sign of defocus values, the projection images in Figs. 4.3.5(a) and (d) are reversed: the sample is at the top right in (a) but at the bottom left in (d). As Gaussian focus is approached, an angular dependence to the magnification emerges as shown in Fig. 4.3.5(c), due to lens aberrations and the manner in which they change the phase of the electron beam. At slight underfocus (see Fig. 4.3.5(b)), the azimuthal and radial circles of infinite magnification can be seen. These are the angles at which defocus and spherical aberration effectively cancel and they are characteristic of Ronchigrams from a round, probe-forming lens. Axial astigmatism can be very accurately corrected by exciting the stigmator coils so that these Ronchigram features are circularly symmetric. As the beam is focused, the central, low angles display the highest magnification. The coma free axis is clearly defined by this position and all alignment and positioning of detectors and apertures can be performed with respect to this spot.

The prime advantage of using a Ronchigram is that the coma-free axis is directly visible. In other alignment methods, the current or voltage center of the objective lens must be used as the reference and this is not always sufficiently accurate. Next, the illumination beam alignment can be very accurately checked by wobbling first the condenser lens excitation and then the microscope high tension. If there is a misalignment of the beam between condenser and objective lenses, there will be a periodic translation of Ronchigram features as the wobbling takes place. This can be corrected using the condenser alignment coils (CTEM bright tilt) so that the features only oscillate in and out symmetrically about the coma-free axis.

The probe has now been aligned with respect to the coma-free axis. Control of its intensity distribution is now dependent on the exact illumination lens settings and the size of the STEM condenser aperture that is subsequently inserted to exclude aberrated beams at high angles. Fig. 4.3.6 shows Ronchigrams from a thin region of Si $\langle 110 \rangle$ at slight defocus. Diffraction effects are clearly presented in the pattern and lattice fringes are observed if the probe coherence is great enough. It is the movement of these fringes across the relevant STEM detectors, as the probe is scanned, that gives rise to image contrast. At high angles in the Ronchigram, the fringes are distorted mainly because of spherical aberration of the condenser lens. This effect is lucid in Fig. 4.3.7(a), a Ronchigram of silicon ⟨111⟩. Fringes correspond to the 0.192 nm {220} planar spacing. Their distortion, a function of angle from the center, is circularly symmetric, as expected when spherical aberration is the only aberration of significance. At low angles the hexagonal arrangement of the lattice planes is more obvious from the fringe pattern. Analysis of such Ronchigrams, and derivation of electron optical parameters from them, has been carried out by Lin and Cowley (1986). Also, at large defocus, when a shadow image at low magnification is visible, the crystal Kikuchi lines are seen (Fig. 4.3.7(b)). Since the coma-free axis position is already known, it is simple to adjust the specimen tilt so that the desired zone axis is aligned precisely for the sub-micron specimen area that is of interest.

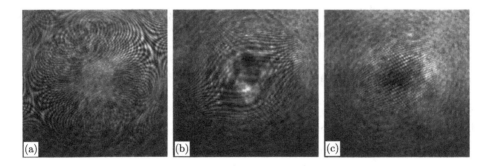

Fig.4.3.6 Ronchigrams of a thin region of silicon ⟨110⟩ showing diffraction effects and fringes arising from the specimen periodicities (James and Browning, 1999). Visibility of the characteristic fringes depends on precise tilting of the specimen and the amount of probe coherence in a direction perpendicular to the relevant crystal lattice plane. (a) Small underfocus-lattice fringes are visible near the Ronchigram center and they become heavily distorted further out in angle. The distortion is due to phase changes introduced by the lens spherical aberration. (b) Near Scherzer focus-the central fringes become large and wide. Their area corresponds to the entire overlap region between zero-order and relevant diffracted beams. (c) Slight overfocus-fringes are visible with size and spacing that decreases with increasing angle from the Ronchigram center.

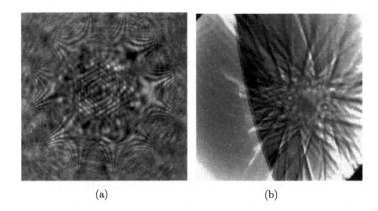

(a)   (b)

Fig.4.3.7 (a) Ronchigram of a thin region of silicon ⟨111⟩. Interference gives rise to fringes corresponding to the 0.192 nm {220} periodicity. Spherical aberration of the probe-forming optics causes significant distortions of the hexagonal pattern away from the center. (b) Ronchigram at the edge of a silicon ⟨110⟩ crystal showing Kikuchi lines when the probe is well overfocused. James and Browning, 1999.

4.4 HAADF imaging

In general, HAADF imaging is thought to be mainly dependent on the accumulated atomic number of the column, being ascribed to the incoherent imaging, and contrary to high-resolution TEM imaging. Its contrast is not sensitive to small changes of defocus values and specimen thickness. In reality, however, the intensity distribution of the electron probe changes with the variation of defocus value. In crystal materials, channeling effect plays an important role in the distribution of incident electrons when transmitting along the zone axis. Thus, in some cases defocus and sample thickness as well as Debye Waller factors and other factors take an effect on the image contrast of HAADF and then HAADF imaging cannot intuitively interpreted in terms of the atomic numbers. In this section, the origin of Z dependence as to HAADF imaging will be introduced at first, and then the other influencing factors on HAADF imaging such as defocus, thickness and Debye Waller factors will be specified, respectively.

4.4.1 Z dependence

1. Principle of Z-contrast imaging

According to the Theory of Bloch Waves, disruptions in the atom periodicity, including displacement disorder and chemical disorder, cause scattering of the Bloch wave states. Chemical disorder is negligible in HAADF imaging because its contribution decreases with Δk. On the other hand, diffuse scattering from displacement disorder increases with Δk as $1 - e^{-(\Delta k)^2 \langle U^2 \rangle}$, where $\langle u^2 \rangle$ is the mean-squared displacement during thermal motion of the atoms and scales linearly with temperature, T. Differences in atomic size disorder can also make a contribution to the HAADF image, and HAADF contrast could perhaps be used to measure this type of disorder. In the present section, however, we discuss the thermal contribution to the high-angle electron scattering, thermal diffusion scattering (TDS).

When $(\Delta k)^2 \langle u^2 \rangle \gg 1$, the Debye–Waller factor, $e^{-(\Delta k)^2 \langle u^2 \rangle} \ll 1$, strongly suppresses the intensity of Bragg peaks, i.e. the coherent scattering, justifying the assumption of incoherent imaging in HAADF measurements. Even remaining some coherent scattering, after integration over a large angular range, the effects of coherence would be fatherly suppressed. Moreover, when the probe size is smaller than the spacing between aligned atomic columns in a crystal, the atom columns are illuminated sequentially as the probe is scanned over the specimen. Then each electron is often considered as confined laterally to one atomic column (called channeling effect and to be introduced in Section 4.4.3), so the image is not affected by coherent interference between different columns. Above factors cause the HAADF imaging to be an incoherent imaging, different from the HRTEM imaging, coherent imaging.

Owing to the incoherence of the scattering, HAADF image contrast is independent of some wave interference issues involving the structure factor of the unit cell, the presence of forbidden diffractions, or some defects. The interpretation of the image is almost intuitive, that is, can be interpreted as scattering from individual atoms without phase relationships between them.

The individual atoms have their own form factors, and these must be considered when accounting for the intensity of the high-angle scattering. Combining this factor with the thermal diffuse intensity mentioned above, $1 - e^{-(\Delta k)^2 \langle u^2 \rangle}$, the intensity of the high-angle incoherent scattering depends on Δk as

$$I_{HAADF} = |f_{at}(\Delta k)|^2 [1 - e^{-(\Delta k)^2 \langle u^2 \rangle}] \quad (4.4.1)$$

The atomic form factor for electrons, f_{at}, approaches the limit of Rutherford scattering at large Δk. In this case,

$$I_{HAADF} = \frac{4Z^2}{a_0^2 \Delta k^4} [1 - e^{-(\Delta k)^2 \langle u^2 \rangle}] \quad (4.4.2)$$

where a_0 is the Bohr radius. The thermal diffuse intensity, the factor in the square braces in Eq. (4.4.2), approaches 1 for large Δk so that

$$I_{HAADF} = \frac{4Z^2}{a_0^2 \Delta k^4} \quad (4.4.3)$$

For the characteristic feature of HAADF is the dependence of atomic number, Z, HAADF imaging is also called "Z-contrast imaging". In general, HAADF images are usually formed by collecting elastically scattered electrons with the inner collection angle over 50 mrad, and the HAADF image intensity is thought to be proportional to $Z^{1.7}$. Thus, as to the HAADF imaging, Eq. (4.3.14) can be rewritten, given the sample is composed of the N atoms

$$I_{HAADF}(r) = \left[\sum_{i=1}^{N} Z_i^{1.7} \delta(r - r_i) \right] \otimes P(r) \quad (4.4.4)$$

Comparing with ABF imaging with its intensity roughly proportional to the $Z^{1/3}$ in some cases (to be introduced in Section 4.5), HAADF imaging is more suitable to observe heavy atoms rather than light atoms, especially surrounded by heavy atoms.

Finally, it should be mentioned that the high-angle scattering is nearly elastic. Owing to its large Δk, however, it does involve "multiphonon scattering", where energy from the high-energy electron is used to create multiple phonons (quanta of vibrations) in the sample, so that not all of the HAADF signal is elastic in origin. Inelastic scattering can make minor contributions to the HAADF image, at least for elements of low Z, and this contribution is also incoherent.

2. Application

Similar to the high-resolution TEM imaging, high-resolution HAADF imaging can offer the crystal structure information at the atomic scale besides the compositional information. In the following, one example (Ge et al., 2012) will be shown that through a single HAADF image both compositional interfacial width and order (Ni_3Al phases)–disorder (Ni phases) interfacial width are determined at the same time in one kind of nickel based superalloy, in which interfaces play an important role in determining the mechanical properties,.

Fig. 4.4.1(a) is a typical HAADF image of the as-cast superalloy in [001] direction with the cuboidal γ' phase (Ni_3Al with the space group $Pm-3m$) precipitated in the γ matrix (Ni with the space group $Fm-3m$), and the thickness of the sample is about 30 nm determined by the relative log-ratio method on electron energy loss spectra (Egerton, 2011). Analysis of EDS shows that less heavy atoms partition into the γ' phase, which makes it display with low contrast in the HAADF image, vice versa (Ge et al., 2010; 2011). Figs. 4.4.1(b) and (c) are raw atomic resolution HAADF images of γ and γ' phases, respectively, with the corresponding diffractograms inset on the bottom right. Some dots with higher contrast compar-

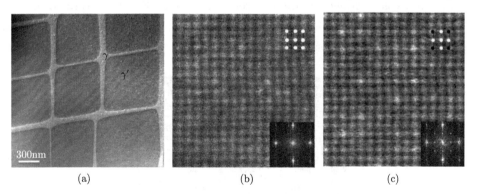

(a) (b) (c)

Fig.4.4.1 (a) Low-magnification HAADF image of the superalloy and (b) and (c) high resolution HAADF images of γ and γ' phases, respectively, with their corresponding diffractograms inset on the bottom right. The projected structure models of γ and γ' are insets on the top right in (b) and (c), respectively. Black dots correspond to Al atoms and white dots Ni atoms.

ing others can be observed clearly especially in Fig. 4.4.1(c), which corresponds to the heavy atoms added into superalloys for solid solution hardening such as Mo, Ta, W and Re et al.. According to the projected structural models inset top right (black dots represent Al atoms, white dots Ni atoms) in Figs. 4.4.1(b) and (c), it can be concluded that heavy atoms distribute randomly in γ phases, while in γ' phases they are preferentially located in Al site, which is in agreement with the results of the atom probe tomography (Blavette et al., 2000). This kind of ordering of the distribution of heavy atoms makes the difference between the adjacent {002} planes in γ' phases, i.e., Ni planes and (50% Ni+50% Al(or heavy atoms)) planes alternates in $\langle 001 \rangle$ direction, while in γ phases there is no obvious difference between the adjacent {002} planes, which can be used to distinguish γ and γ' phases.

Fig. 4.4.2 is a raw [001] high-resolution HAADF image of interfaces, and the difference of the image contrast between γ and γ' phases is obvious. The averaged intensity profile across the interface corresponding to the area denoted by a rectangle shown in Fig. 4.4.2 is plotted Fig. 4.4.3(a). The lower background intensity in the left side, γ' phases, can be observed, which origins from lower concentration of heavy atoms as mentioned above, vice versa. Thus, from Fig. 4.4.3(a) the transition width of chemical composition in the superalloy can be determined to be about 2.2 nm, 6 atomic {001} planes.

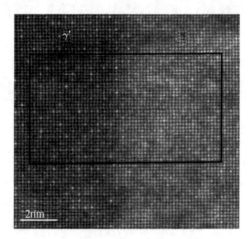

Fig.4.4.2   Raw high-resolution HAADF image of the interface.

In Fig. 4.4.3(a), an additional feature should be noted that the image intensity keeps almost constant in the right side, while it fluctuates in the left side, which is due to the alternation of Ni planes and (50% Ni+50% Al) planes in the ordered γ' phases. To show the variety of the image intensity clearly, the intensity ratio of each atomic column to its adjacent column was made as shown in Fig. 4.4.3(b). The ratio in the right side, the disordered γ phase, remains almost constant, i.e. 1, while in the left side, the ordered γ' phase, the ratio alternates between about 0.9 and 1.1. Thus, from Fig. 4.4.3(b) the interfacial width of Ni and Ni_3Al can be determined to be about 2.2 nm, same as the compositional width. Besides the equality of two kinds of interfacial width, comparing Fig. 4.4.3(a) with Fig. 4.4.3(b), it also can be found that the transition area of the chemical composition across the interfaces of γ/γ' is exactly the same as that from the ordered phase to the disordered phase as denoted by two dotted lines.

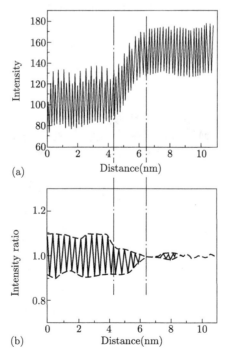

Fig.4.4.3 (a) Averaged intensity profile across interfaces corresponding to the area denoted by a rectangle in Fig. 4.4.2 and (b) intensity ratio of each atomic column to adjacent column in (a).

4.4.2 Effect of focus

As discussed in Section 4.1, electron probe is the Fourier transformation of the condenser lens aberration so that the probe, accurately speaking, is not like a delta function but have an intensity distribution with the subsidiary peaks symmetrically located on the side of the central peak of probe as shown in Fig. 4.4.4, the intensity profiles with series of defocus values as well as the corresponding simulated HAADF images of Si in [110] direction. From the simulated images in Fig. 4.4.4 it can be seen that image contrast changes not too much with the defocus from −500 to −900 Å, which is different from the TEM imaging, but with further deviation of defocus value the HAADF image contrast deteriorate (with defocus −300 Å) and even not reflect the structure any more (−1100 Å). This phenomenon can be interpreted by the probe intensity profiles as shown in Fig. 4.4.4 that with the variation of focus not only the FWHM of the central peak (usually defined as the resolution) changes but also the intensity of the subsidiary peaks. In the following, another example will be given that it is just due to the influence of the subsidiary peak that fictitious features could appear in the HAADF image (Pennycook et al., 1995).

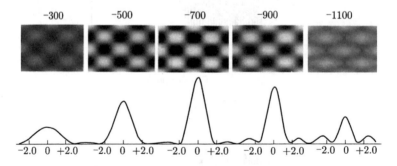

Fig.4.4.4 Simulated defocus series for Si [110] with corresponding probe intensity profiles (100 kV, C_s=1.3 mm, optimum objective aperture semiangle 10.3 mrad), giving a minimum probe size of 0.22 nm at the optimum Scherzer focus of −69.3 nm. Rennycook et al., 1995.

Fig. 4.4.5 shows the through-focal HAADF STEM images of $SrTiO_3$ in [001] direction. The intensity of oxygen columns, as highlighted by the arrowheads, seems to be enhanced in the image at $\Delta f = -70$ nm, which is above the Scherzer focus ($\Delta f = -50$ nm) of the lens ($C_s = 1.0$ mm). Both are similar with thickness 20 and 60 nm, indicating that that feature in image contrast is not sensitive to the thickness (Yamazaki et al., 2001).

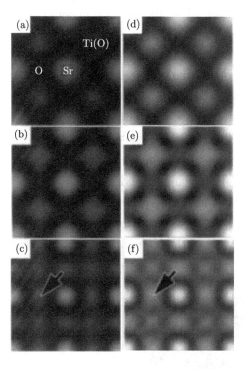

Fig.4.4.5 HAADF image of [001] SrTiO$_3$ of 20 (a), (b), (c) and 60 nm thickness (d), (e), (f), calculated with probes at defocus. of $\Delta f = -30, -50, -70$ nm, respectively. Arrow indicates artificial spots. Yamazaki et al., 2001.

The wave fields, which are formed in the SrTiO$_3$ crystal by incident beams probed into an oxygen column (at P in Fig. 4.4.6), were calculated and shown for defocus of –50 and –70 nm in Fig. 4.4.6(a) and (b), respectively, which disclose the intensity distribution of the wave fields along the depth (effect of sample thickness on HAADF imaging will be detailedly introduced in next section). At $\Delta f = -50$ nm, the wave packet channels almost exclusively along the oxygen columns as shown in Fig. 4.4.6(a), and the subsidiary peak of the probe is weaker and does not fall into the neighboring Sr and Ti columns. Because the high-angle TDS cross section of oxygen is much lower than that of titanium and strontium, originating from the $Z^{1.7}$ dependence of HAADF imaging, there is no contrast at O columns in images shown in Fig. 4.4.5(b) and (e). On the other hand, when the probe at $\Delta f = -70$ nm is located at the position of an O column P in Fig. 4.4.6(c), its subsidiary peak coincides with the surrounding Sr columns A and B, and Ti columns C and D, as illustrated using double-headed arrow stretching between Fig. 4.4.6(c) and (d), and the wave packet formed by the subsidiary peak channels along these Sr and Ti columns (see Fig. 4.4.6(b)). The HAADF detector thereby counts much TDS electrons from Sr and Ti, although the probe is located at the position of O column. Consequently, bright spots appear at the O column positions as if the O columns were visible (see Fig. 4.4.5(c) and (f)).

Above example shows the influence of focus on HAADF image contrast due to the contribution of the subsidiary peaks. Besides this, according to the imaging principle in the light optics, different focus corresponds to the different imaging plane, so with different focus the information of the different plane of sample along the beam direction can be obtained if there is a sufficient depth resolution.

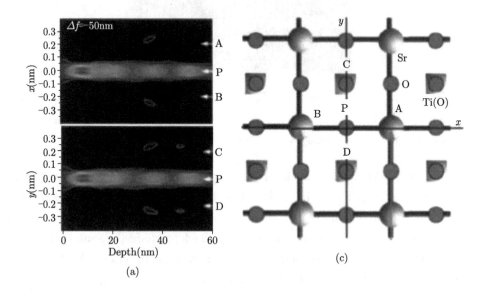

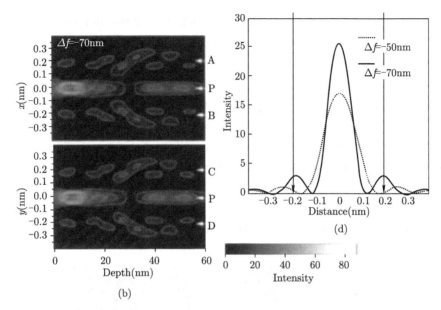

Fig.4.4.6 (a) and (b) Intensity distribution of the wave field along the depth of A-D columns and P in [001] SrTiO$_3$, calculated for probes focus on P at $\Delta f = -50, -70$ nm, respectively. (c) Atomic column positions of the SrTiO$_3$, indicating A-D and P in (a) and (b). (d) Calculated probe functions with a semiangle of $\alpha = 12$ mrad at $\Delta f = -50, -70$ nm, respectively. Yamazaki et al., 2001.

In the absence of other aberrations, it is well known from the optics literature (Born and Wolf, 1980) that the intensity at the center of the probe along the beam direction, z-axis, can be described as

$$I(z) = \frac{4\lambda^2}{\pi^2 z^2 \alpha^4} \left[\sin\left(\frac{\pi z \alpha^2}{2\lambda} \right) \right]^2 \quad (4.4.5)$$

where α is semi-convergent angle, so that the FWHM of the probe along the beam direction, the depth resolution, is given by

$$\delta_{\text{STEM}} = \frac{1.77\lambda}{\alpha^2} \quad (4.4.6)$$

which means the depth resolution of the HAADF is inversely proportional to the square of the semi-convergent angle.

For 300 kV aberration-corrected STEM with semiangle of convergence 22 mrad, the depth resolution is about 7.3 nm (Borisevich et al., 2006) and for scanning confocal electron microscope (Nellist et al., 2008), it is further bettered by about one third, so that this kind of sectioning technique is feasible to image in three dimension. Although for conventional STEM, the sectioning technique is not practical because the semiangle of convergence is usually no more than 10 mrad (depth resolution is about 35 nm for 300 kV, $C_s = 1.2$ mm) due to the existence of the spherical aberration and image contrast can still be changed with different defoci. Fig. 4.4.7(a) is a HAADF image almost under the optimum condition on some kind of crept superalloy such that the crystal lattice can be observed besides the matrix γ and precipitate γ' phases. While in Fig. 4.4.7(b) far from the Scherzer focus condition, the lattice fringes disappear but bright dots, heavy atom clusters, show up instead.

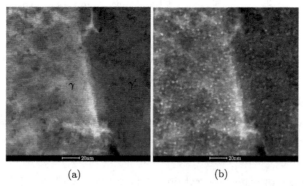

(a) (b)

Fig.4.4.7 HAADF images of some crept nickel superalloy with different focus value: (a) and (b) near and far from the Scherzer focus condition, respectively.

In all, defocus value of the condenser lens have effects on the STEM probe profile as well as the imaging plane, which will influence the contrast of HAADF image like the atomic number. And due to the enhanced depth resolution of aberration-corrected STEM microscopes according to Eq. (4.4.6), more attention should be drawn to interpret image contrast.

4.4.3 Effect of thickness-channeling/dechanneling effect

As mentioned in Section 4.3.1, the STEM image intensity can be treated as a convolution of object function with the probe intensity profile. In that case, the entire sample is thought to be at the same focus and then the object function is convoluted with single probe intensity profile. But in reality, the intensity profile changes along the beam direction when the probe transmits through the materials, especially in the zone axis of the crystals as mentioned in Fig. 4.4.6, the probe channeling along the atomic column, which is so-called channeling effect. In this section, the influence of thickness and channeling effect are introduced at first. Then one example as to the image contrast of dopant atoms in crystals is given to show the influence of the channeling effect on HAADF image contrast. At last, some examples of dechanneling effect are shown, which is useful in the practical observation.

Considering the effect of thickness, the HAADF intensity described by the Eq. (4.3.14) should be restated mathematically. The differential contribution to the HAADF-STEM image from the layer of atoms at a depth z is simply denoted

$$\frac{dI(r,z)}{dz} = O(r,z) \otimes P(r,z) \qquad (4.4.7)$$

The final image intensity $I(r)$ for a sample thickness t is

$$I(r) = \int_0^t \frac{dI(r,z)}{dz} dz = \int_0^t O(r,z) \otimes (r,z) dz \qquad (4.4.8)$$

The initial probe wave function $P(r,0)$ is readily calculable usings the Fourier transformation of the condenser lens aberration. If the probe stayed in this form, it would be easy to calculate an image from Eq. (4.4.8), but that is not what happens. Instead, in a zone-axis oriented crystalline specimen, the probe strongly channels and changes along the atomic columns (Howie, 1966), which, in essence, is because the atom cores present a more attractive positive potential than the interstitial regions between atoms.

A deeper explanation of channeling can be formulated by analogy to light transmission down optical fibers. Rewritten Snell's law (http://en.wikipedia.org/wiki/Snell's_law) in terms of the wavelength, λ,

$$\frac{\lambda_1}{\lambda_2} = \frac{\sin\theta_1}{\sin\theta_2} \qquad (4.4.9)$$

Owing to the attractive potential of atomic nucleus the kinetic energy of electrons will be larger when electrons are close to atoms, so the wave vector in the columns of atom core is largest. That is, the wavelength of the electron in the columns of atom cores, λ_2, is shorter than the wavelength in the interstitial regions, λ_1. If the electrons fly along the direction parallel to the atomic column, the incidence angles in the atomic columns are nearly 90° as shown in Fig. 4.4.8, then the glancing angle ϕ is nearly zero. Because $\lambda_2 < \lambda_1$ in Fig. 4.4.8, from Eq. (4.4.9) $\sin\theta_2 < \sin\theta_1$. The critical condition shown in Fig. 4.4.8 has $\sin\theta_1 = 1$. In this case it is impossible for θ_2 to become larger while satisfying Snell's Law because $\sin\theta_1$ cannot exceed 1. What happens at glancing angles shallower than the critical angle ϕ_{crit} is total reflection at the interface. That is, the electron bounces back into the column of atom cores, which leads to the channeling effect.

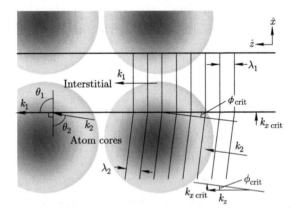

Fig.4.4.8 Electron wave functions in a column of atoms, and in the interstitial region between atoms (Fultz and Howe, 2008). The critical condition is shown, where the wave vector in the interstitial region is at 90° with respect to the interface normal.

On the other hand, however, because crystals have a high density of atomic nuclei and electrons, the incident electrons with high probability suffer a high angle Rutherford scattering or energy loss in collision with other electrons and leave the channel. This is "dechanneling" process or "tunneling" process, which is prominent at the area with the local atomic arrangement or relaxation, such as dislocation and interface. Tunneling will be severe for the critical glancing angle ϕ_{crit}, because the electron wave function is a constant through the interstitial region and therefore appears with full amplitude in the next column of atoms. Tunneling is suppressed if the glancing angle ϕ is smaller, but it will occur even when $\phi = 0$.

Therefore, in practice, the critical angle is an important consideration in forming narrow probe beams with high-quality objective lenses. For lenses with small aberrations, larger convergent angles, α, are preferred for probe formation. A larger α allows a larger range of Δk, and hence a smaller width of the probe beam, improving lateral resolution, as well as the smaller depth of field, improving the depth resolution (see Eq. (20)). This strategy works well until α exceeds ϕ_{crit}, when only some of the electrons are channeled effectively, and many electrons are tunneled. The lateral resolution is sustained by the channeled electrons, but there is a background "noise" caused by the electrons that do not channel.

When crystals are in the zone axis and the probe channels along the column, its intensity changes quasi-periodically at the same time, which will make things complicated. In the next, it will be demonstrated by the interpretation of impurity atom contrast as an example.

At first the channeling effect was simulated as shown in Fig. 4.4.9 by means of a plane-wave multislice simulation. The initial probe produced with accelerating voltage 200 kV, C_s 1.0 mm, convergent angle 10 mrad, defocus 45.0 nm, is placed on one side of a Si $\langle 110 \rangle$ dumbbell, and we follow the probe intensity as it propagates along the atomic column. As shown in Fig. 4.4.9(a), the probe quickly becomes much more intense directly on the atomic column, reaching a maximum intensity around $z = 10$ nm. The probe then dechannels somewhat, spreading intensity away from the atomic column, so that the on-column intensity is a minimum at $z = 22$ nm. The probe then rechannels, but some intensity has spread to the adjacent atomic column.

From Fig. 4.4.9 it can be seen that the HAADF image intensity of an impurity in crystals depends on its depth. For samples of thickness less than the first channeling intensity maximum, an impurity at the bottom of the sample will have higher image intensity than an impurity at the top. Therefore, if the sample is thinner than the first channeling maximum, and the likelihood of having more than one impurity in a column is small (low impurity concentration and a thin sample), we might be able to determine the depth of an impurity in the column from its image intensity.

If the sample is thicker than the first channeling maximum, then intensity as a

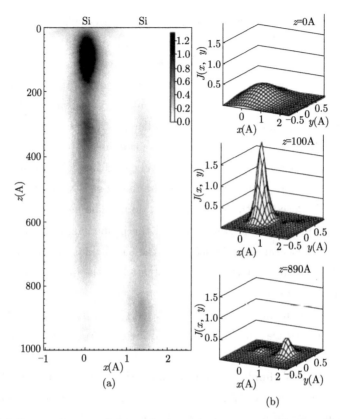

Fig.4.4.9 (a) Grey-scale map of the probe intensity propagating down a Si ⟨110⟩ atomic column. x is position across the dumbbell, z is position along the column. The Si atomic columns are at $x = 0$ and 0.135 nm. The probe starts exactly on the left-hand side of the Si dumbbell at $x = 0$: Probe conditions are accelerating voltage 200 kV, $C_s = 1.0$ mm, 10 mrad aperture, and 45.0 nm defocus. (b) Surface plots of probe intensity versus x and y at the indicated z. Voyles et al., 2003.

function of depth is no longer single-valued and no quantitive information about the individual column occupancies or depth of impurities can be obtained from the intensity. However, if the information we seek is only the average impurity concentration (which is one of the relevant quantities for integrated circuits, for example) the channeling contribution averages out for thin samples.

At even larger thickness, we see that the probe intensity becomes greater on the neighboring atomic column than the column under that beam. The image of an impurity near the bottom of such a sample will therefore appear in the wrong atomic column. But the average dopant concentration still could be determined from such an image. These concerns emphasize the continued necessity of using thin (generally less than 10 nm thick) samples, even when very high spatial resolution imaging, such as with a spherical-aberration corrected STEM, is available.

It is interesting that an aberration-corrected probe should not always be the probe of choice for detection of single dopant atom inside the crystal, although the depth resolution is improved for the aberration-corrected lens as indicated by Eq. (4.4.6). From the probe intensity of Si along [110] direction as the function of propagation depth, as shown in Fig. 4.4.10, if a dopant atom is located at 4–5 nm below beam entry surface, an aberration-corrected probe will clearly have an advantage. While a dopant atom located 10 nm from the entrance surface has about higher visibility when imaged using non-corrected 0.17 nm probe instead of aberration-corrected 0.08 nm probe.

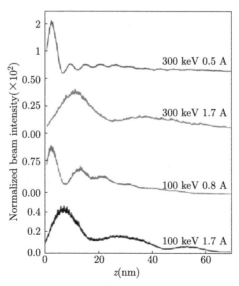

Fig.4.4.10 Probe intensity in Si along [110] direction as the function of propagation depth (Mittal and Mkhoyan, 2011). Four intensity variations are compared with different accelerating voltages and probe sizes.

The same method can be applied to study the imaging of interstitial impurities as follows. In some structures and orientations, such as Si ⟨110⟩, an interstitial impurity may sit between the projections of the atomic columns. If the impurity is off-column, the probe wave function is not enhanced by channeling but may even be depleted. Fig. 4.4.11 shows what happens to a probe placed exactly between

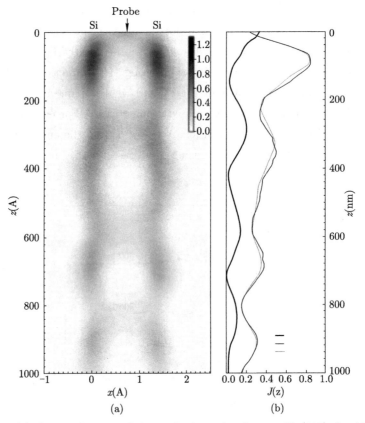

Fig.4.4.11 (a) Grey-scale map of the probe intensity down a Si ⟨110⟩ dumbbell when the probe starts exactly between the two atomic columns at $x = 68$ pm. (b) The probe intensity within 3 pm of each atomic column, and the initial x and y position of the probe as a function of z. As in Fig. (4.4.9), the probe is strongly channeled onto the atomic columns, leaving almost no intensity below the initial probe position at $z = 14.5$ and 43 nm. Voyles et al., 2003.

the two atomic columns in a Si $\langle 110 \rangle$ dumbbell. For parameters same as in Fig. 4.4.9, the probe is quickly channeled onto the two adjacent columns, leaving little intensity at the initial probe positions between the columns for $z > 4$ nm. This effect is reduced with a smaller probe (Voyles et al., 2003), and significantly less evident at the actual Si interstitial position, which is farther away from any of the $\langle 110 \rangle$ atomic columns, but the general principle remains: image contrast for off-column impurities will not be enhanced due to channeling. At best the contrast will be depth independent. At worst, off-column impurities near the bottom of samples that are not very thin may be invisible. Therefore, if we want to image interstitial impurities, we could work along a zone axis that puts interstitials on-column.

On the other hand, tilting the crystals to deviate high-symmetry zone axes can avoid the strong probe channeling, which can also be used to observe interstitial impurities by HAADF imaging. Fig. 4.4.12 is a tilt series of high resolution ADF

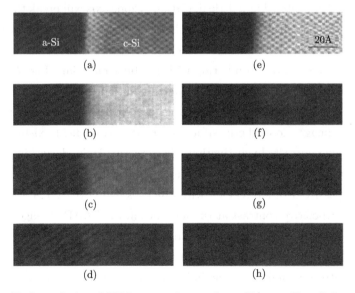

Fig.4.4.12 High-resolution ADF images of amorphous-Si/crystalline-Si interface taken at various tilt positions (Yu et al., 2008). The column on the left is the LAADF images and the column on the right is the HAADF images. From top to the bottom, each pair is taken simultaneously with the sample tilt away from the [110] zone axis at an angle of $0°, 1°, 2°$ and $4°$, respectively.

images of interface between amorphous Si and crystalline Si. The column on the left is the low angle ADF (LAADF) images and the column on the right the HAADF images. From top to bottom, each pair is taken simultaneously with the sample tilted away from the [110] zone axis at an angle of $0°, 1°, 2°$ and $4°$, respectively. Fig. 4.4.12 shows that the HAADF signal decreases much faster than the LAADF signal upon tilting. The rapid decay of the HAADF signals reflects the fact that channeling electrons along aligned atomic columns contribute significantly to the HAADF images, but little to the LAADF images. Therefore, through tilting the crystals the channeling effect can be avoided and then the interstitial dopant atom can be observed by the HAADF imaging. In all, to observe the interstitial impurities two methods can be selected: to put the impurities along some atomic columns so that channeling effect works or to make the crystal deviated from the zone axis, dechanneling effect works.

Fig. 4.4.12 indicates that a little deviation of zone axis will break the satisfaction of the Channeling effect. Besides this, if there exists local lattice distortion, for instance dislocation or interface, channeling effect can also break out, which will lead to abnormal contrast comparing with the bulk materials. Fig. 4.4.13 shows a LAADF image and a HAADF image taken simultaneously from the interface of amorphous Si and crystalline Si (Voyles et al., 2003.). There is a bright band in the LAADF image along the interface where the strain field exists, indicating a positive strain contrast. In distinction, the interface looks darker in the HAADF image, so that the contrast due to the strain field is negative in the HAADF image. Images from other regions with thicknesses above 200 Å show a similar phenomenon: positive contrast at the interface in the LAADF image and negative contrast in the HAADF image. The abnormal contrast at the interface can be understood by the dechanneling effect. Electrons which should have been scattered into high scattering angle as in the bulk materials are scattered into the low angle due to the lattice distortion of the interface, leading to the bright contrast of LAADF images and lower contrast of HAADF ones.

In a word, when electron beam transmits through the crystals it prefers piping along the atomic column, so the area along the principal axis or plane will show brighter contrast than other area in the HAADF images, while the defect such as

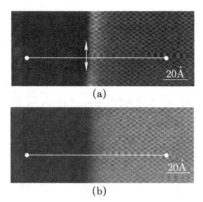

Fig.4.4.13 Experimental ADF images of amorphous Si/crystalline Si interface taken simultaneously at ∼35 nm thick by two detectors (Yu et al., 2004). (a) LAADF (20–64 mrad), (b) HAADF (64–200 mrad), The two-headed arrow indicate the location of the last visible atomic plane on the c-Si side. (200 kV JEOL 2010F TEM/STEM).

dislocations and interfaces show darker contrast. Moreover, the intensity of electron probe, when it transmits along the zone axis, changes quasi-periodically with the different depth, so dopant atoms located with different depth may display different HAADF contrast. These two factors should be taken into consideration as to the interpretation of HAADF images.

At the end of this section, two more examples (see Fig. 4.4.14, ADF images of a certain kind of crept superalloys) are given as to the influence of STEM image contrast by channeling effect and dechanneling effect, respectively, which is useful in experiments. In Fig. 4.4.14(a), a HAADF image, besides the high and striped contrast from top to bottom corresponding to the γ phases due to the partition of more heavy elements, there is octopus-shape bright contrast in the center of the image. This abnormal contrast can be explained by channeling effect. Due to the existence of distortion in the crystal lamina only the circular area as shown in the image is in zone axis, while in the area of octopus claw some crystal plane is parallel the electron beam. This kind of image contrast is similar like the dark field (DF) diffraction contrast, which can be used for guidance of aligning of the crystals in the STEM mode. Analogously, in Fig. 4.4.14(b), the medium angle ADF (MAADF) image, besides the striped contrast of γ phases, there are abnormal white contrast denoted by a red arrow, from the top left to the bottom right, which is actually originated from dislocations due to the dechanneling effect. Different with the DF imaging technique, all dislocations can be observed at the same time with bright contrast in STEM LAADF or MAADF imaging even if in the zone axis, which is useful to estimate the dislocation density of samples.

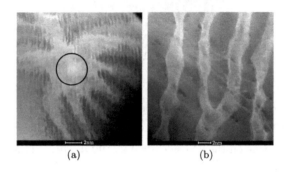

Fig.4.4.14 Images of a certain kind of crept superalloy in (a) HAADF mode and (b) MAADF mode, respectively. The area denoted by the circle in (a) with high contrast is in the zone axis, and the white line in (b), one of which denoted by a red arrow, is a dislocation.

4.4.4 Effect of Debye-Waller factor

According to Eq. (4.4.1), it is obvious that the HAADF image intensity is related with the Debye-Waller factor, $e^{-(\Delta k)^2 \langle u^2 \rangle}$. Specifically speaking, a large thermal displacement $\langle u^2 \rangle$, corresponding to a small Debye-Waller factor, lead to the increase in the high-angle scattering probability and thus increase of HAADF image contrast. The effect of Debye-Waller factor in simulation has been reported recently for several different materials (Ishizuka, 2002; Haruta et al., 2009; Findlay et al., 2009; Blom, 2012). In the experiment, Haruta et al. (2009) has observed the abnormal contrast that the contrast of Sn column appears brighter than the La columns, although the atomic number of La ($Z = 57$) is larger than that of Sn ($Z = 50$), which was interpreted by Debye-Waller factor that atom Sn with smaller Debye–Waller factor produced greater HAADF signal than atom La.

In this section, another example (LeBeau et al., 2009) will be given to show that to determine the atomic structure precisely, which is dependent on achieving the excellent agreement between the HAADF image contrast and the simulation, more factors, such as Debye Waller factor and space incoherence, should be taken into consideration besides the atomic number and thickness.

To avoid strain and nonstoichiometry, the study uses a single crystal of lead tungstate ($PbWO_4$), which contains two cations with relatively large Z ($Z_{Pb} = 82$ and $Z_W = 74$). The experimentally measured Debye-Waller factors were used for the atom displacements in the frozen phonon image simulations accounting for TDS. Fig. 4.4.15 shows a simulated HAADF intensity line scan along [001] across pairs of Pb and W columns with and without TDS. The term "with thermal diffuse scattering" refers to a frozen phonon calculation averaged over several configurations of the atoms (enough to obtain a reasonable convergence). By "without thermal diffuse scattering" we mean a simulation in the frozen phonon model us-

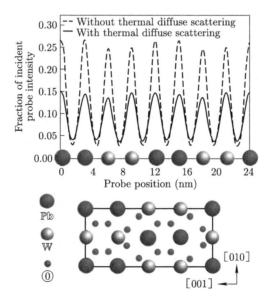

Fig.4.4.15 (Top row) Simulated intensity line scan along [001] in a [100] projection for a 17.5-nm-thick $PbWO_4$ sample without thermal diffuse scattering (dashed line) and with thermal diffuse scattering (solid line). A [100] projection of the $PbWO_4$ is shown in the bottom row. LeBeau et al., 2009.

ing a single configuration with no displacements of the atoms. With TDS, the overall intensity scattered onto the HAADF detector decreases and the contrast is reduced by 19%. Both are caused by a significant decrease in the atomic column intensities due to thermal vibrations. Without TDS, the Pb columns would have appeared brighter than the W columns, as expected because of their larger Z. With TDS, however, the Pb and W columns have almost identical intensities. They alternate slightly in intensity (depending on the thickness) because of the different Debye-Waller factors for each column.

In addition to correctly accounting for Debye-Waller factors in the simulations, accurate measurements of the thickness are essential for comparisons between theory and experiment. In particular, strongly scattering crystals, such as $PbWO_4$, are much more sensitive to errors in the experimental thickness determination than crystals with smaller Z. Comparing with EELS, the position-averaged convergent beam electron diffraction (PACBED) patterns yield more precise and highly accurate thickness values for it is very sensitive to small changes in thickness. Although PACBED relies on comparison with pattern simulations, further validation of its accuracy comes from analyzing image intensities as a function of thickness determined by PACBED. Excellent agreement can be obtained by comparing mean HAADF image intensities in simulation and experiments as the function of thickness determined by PACBED. But the maximum atom column intensities (the intensity at which 1.5% of the total number of image pixels were above the absolute maximum image signal) still differ by a factor of 1.2–1.3 if simulations do not take into account the cumulative effects of a finite source size, sample vibration, or any other sources contributing to spatial incoherence, and a similar mismatch is observed for the background signal (the largest intensity for which 1.5% of all image pixel values fell below the absolute minimum image signal). To model the influence of spatial incoherence, the simulated images were convolved with a Gaussian function (Klenov et al., 2007; Nellist and Rodenburg, 1994). Fig. 4.4.16 shows that after convolution with a Gaussian with a 0.115 nm FWHM, excellent agreement between experiments and simulations is obtained, and the match between experiments and simulations is within about 5% (mean error).

The above mentioned results fatherly point to potential challenges in quantita-

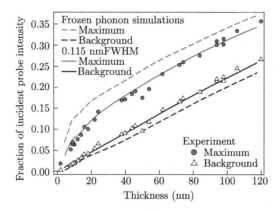

Fig.4.4.16 Comparison of experimental (symbols) and simulated (line) image signals as a function of thickness. Both the maximum signal and minimum (background) signal are shown. The dashed and solid lines represent simulations without and with the effects of spatial incoherence taken into account by convolution with a 0.115 nm FHWM Gaussian, respectively. LeBeau et al., 2009.

tive HAADF. In particular, comparisons with simulations for crystals with heavy elements are much more sensitive to errors in foil thickness measurements than those for more weakly scattering materials. A second challenge is knowledge of the Debye-Waller factor of each atom. Although Debye-Waller factors can be measured for bulk materials, this is not the case at interfaces or near defects. A large contrast mismatch between simulations and experiments may take place due to neglecting TDS in the simulations and/or the unknown magnitude of the Debye-Waller factor in epitaxially strained films. Most importantly, because of the great sensitivity to thermal vibrations and channeling effects, image simulations are required to fully understand the image contrast, contrary to common perception that HAADF-STEM images are intuitively interpretable in terms of the atomic numbers present.

4.5 ABF imaging

Light elements play key roles in a wide range of materials and devices. For example, oxygen atoms in materials critically affect the properties of oxide dielectrics, ferroelectrics, and superconductors. Much effort has been made to directly observe light elements inside materials using TEM. But light atoms scatter electrons weakly, making them difficult to detect. Recently, Jia and co-workers (2004; 2005) used aberration corrected TEM with negative spherical aberration to image oxygen concentrations in complex oxides and grain boundaries, but such imaging method usually requires very thin specimens (usually < 10 nm) for reliable interpretation. HAADF imaging allows direct image interpretation over a wide thickness range, but the signal scales strongly with atomic number: when heavy elements are present, light elements are barely visible. STEM EELS permits elemental analysis down to atomic level (Muller et al., 2008), but the signal-to-noise ratio is still low and detailed preprocessing is required. There remains a critical need for a real time imaging mode to robustly image light elements over a wide thickness range for materials science and device engineering applications.

A candidate technique was recently introduced by Okunishi et al., who showed images using an annular detector located within the bright field region, in which both light and heavy atom columns were simultaneously visible. By analogy to ADF imaging, this was called ABF imaging.

4.5.1 Basic principle

Based on the reciprocity principle, as mentioned in Section 4.2, the ABF imaging mode could be equally realized in HCI TEM. PCTF description is used to give more fundamental, straightforward insights into why ABF-STEM is able to provide enhanced phase contrast in terms of the improved lens properties. Image formation of the phase contrast relies on a phase transfer of the scattered wave by the objective lens, which is described by the lens transfer function $-i\chi(q)$, where $\chi(q)$ is a wavefront aberration function of the scattering vector q. PCTF is given as an imaginary part of the lens transfer function, and hence it results in $\chi(q)$ in the case of axial illumination. For off-axis HCI, the tilt-incident wave K as well as the scattered wave k is affected by the lens aberrations, and therefore the corresponding PCTF, $L(q)$, is written as (Rose, 1974)

$$\frac{\lambda_1}{\lambda_2} = \frac{\sin\theta_1}{\sin\theta_2}$$

$$L(q) = \int_{\theta_c} \int_{\phi} \sin(\chi(k) - \chi(k)) d\phi d\theta_c \qquad (4.5.1)$$

where ϕ represents an azimuth around the optical axis, θ_c represents a cone angle ranging from $\theta_c^{\max}$ to $\theta_c^{\min}$ and the scattering vector q is defined as $k - k$. To simplify the equation, the third-order C_s is set to be zero, and now the fifth-order spherical aberration C_5 dominates the phase transfer of the lens. Therefore, with fixed θ_c, PCTF is given as

$$L(u, \theta_c) = \oint \sin\left[\frac{\pi}{3} C_2 \lambda^5 \{((u + \theta_c \cos \phi)^2 + (\theta_c \sin \phi)^2)^3 - \theta_c^6\}\right] d\phi \qquad (4.5.2)$$

where u presents the magnitude of q, and λ is the wavelength of electrons. The HCI conditions are tuned according to Eq. (4.5.1) and (4.5.2), and the well-optimized PCTF is obtained with 11 mrad $\leqslant \theta_c \leqslant$ 22 mrad. Information transfer now remarkably extends up to 22.5 nm^{-1}, which is far beyond from that of the typical axial illumination ($\sim$8 nm^{-1}) and corresponds to the real-space correlation length of 44.4 pm. It is noted that the entire shape of the HCI PCTF curve, like a gently sloped hill as shown in Fig. 4.5.1, may work effectively for increasing the visibility of the weak-scattering light atoms, given the condition that the phases of the wave are almost equivalently transferred over wide high-frequency ranges. In the following, the basic imaging principle of ABF will be fatherly introduced with the help of image simulation. Similarities are found as to ABF imaging with HAADF imaging: Insensitivity to sample thickness and Z dependence in some cases.

Fig. 4.5.2 shows an ABF and a BF defocus-thickness map for SrTiO$_3$ viewed along the [011] direction. The main feature of the ABF map is that there is a band of images centered around zero defocus and stretching out overall thickness values, in which the form of the ABF image does not change very much. Moreover, the form of these images directly represents the column arrangement of the specimen and resembles the experimental images. Thus, for suitable defocus value, ABF imaging shares with HAADF imaging the property of being relatively insensitive to specimen thickness. The range of defocus values defining this band is fairly narrow, extending from around –4 to 4 nm. The first effect of increasing the deviation of the defocus value from zero is a reduction in contrast, though as one goes further out the qualitative appearance of the images also changes.

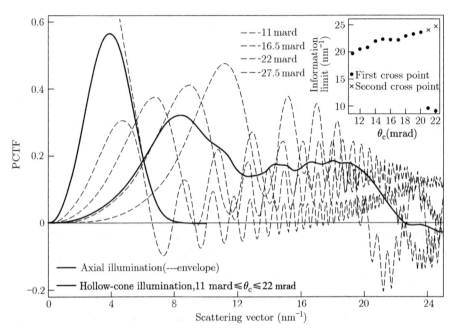

Fig.4.5.1 HCL PCTF calculated on the basis of Eq. (4.5.1) and (4.5.2) with $\lambda = 2.5$ pm and $C_5 = 1.5$ mm. Each of the dotted curves represents PCTF with the fixed θ_c (Eq. (4.5.2), and its corresponding first-cross point where the curve first becomes zero (second-cross means vice versa) is plotted for each, as shown at the upper right. Note that the first-cross points occur in the low-frequency region less than 10 nm^{-1} with θ_c values larger than 20 mrad, around which the θ_c^{max} may well be optimized. The solid red curve shows the HCI PCTF integrated over $\theta_c^{min} \sim \theta_c^{max}$. PCTF with axial illumination is shown by the solid blue curve, calculated with the representative aberration-corrected TEM parameters (C_s=−40 μm, defocus = 9 nm, C_c =1.4 mm), and the corresponding damping envelope derived with the energy spread of the beam $\Delta E = 0.3$ eV is shown by the dashed blue line. This PCTF is shown by inverted values for comparison with the present HCI PCTF. Ishikawa et al., 2011.

The robustness of the visibility and interpretability of both light and heavy atom columns in the ABF image with respect to thickness is better appreciated when compared with the more usual BF images. Dramatic changes in pattern and contrast can be observed as a function of both thickness and defocus in a BF

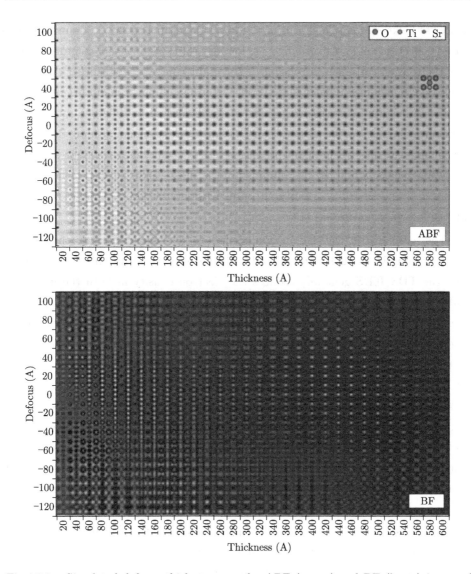

Fig.4.5.2 Simulated defocus-thickness map for ABF (upper) and BF (lower) image of SrTiO3 viewed along the [011] orientation. The ABF images are simulated with the detector ranges 11–22 mrad, while for the BF imaging a 0–4 mrad detector was assumed. Finite source size has not been taken into account in either map. Findlay et al., 2010.

defocus-thickness map as shown in the lower panel in Fig. 4.5.2, assuming a 0–4 mrad collection angle. At zero defocus, for very thin specimens all columns, including the oxygen columns, are visible with bright contrast. For thicker specimens, the Sr columns quickly disappear. Both Ti and O columns remain bright, but the degree of brightness varies with thickness and does so at different rates for the different columns. So, for instance, for a 50 nm thick specimen at 0 nm defocus the Sr columns are dark with a faint bright halo, the O columns are perceptible only as very faint peaks, while the Ti columns are the most visible peaks. Such an image would be impossible to correctly interpret without the aid of simulations relying on the known structure.

The main reason why BF imaging is not robust with respect to thickness as ABF imaging is thought to be related with the coherent s-state interference effect and the TDS. TDS generally reduces the electron density in both the inner and outer areas of the bright field region. In the outer area of the bright field region, this reduction reinforces the reduction given by the coherent s-state interference effect-both effects combine to produce an absorptive ABF signal. However, in the inner area of the bright field region this reduction opposes the increase given by the coherent s-state interference effect——whether the contrast in a small bright field detector is positive or absorptive depends on which effect dominates for each column. We see this explicitly in the BF defocus-thickness map in the lower panel in Fig. 4.5.2: for thicknesses beyond 20 nm or so the Sr/O columns have dark contrast (TDS dominates) while the Ti and O columns have bright contrast (the s-state effect dominates), though the amount of bright contrast oscillates with thickness and does so at different rates for the different elements (because the s-state oscillations are atomic species dependent). The resulting image is visually rather confusing. Thus, coherent BF imaging is not nearly as robust with respect to thickness as ABF imaging and HAADF imaging.

Another key feature of HAADF imaging, as mentioned in last section, is the well-known $Z^{1.7}$ scaling. In ABF imaging, in reality, some similar trend can also be observed in some cases. The on-column signal is plotted, averaged over the thickness region 30–60 nm (averaged to remove some of the remnant oscillatory behavior in this region) as a function of atomic number. It might have been expected that the role of absorption would ensure that this value decreased monotonically with increasing atomic number. This is true for the convergent angle $\alpha = 32$ mrad and $16 \leqslant \theta_c \leqslant 32$ mrad parameters as shown in Fig. 4.5.3(b), but in Fig. 4.5.3(a), for parameters $\alpha = 22$ mrad and $11 \leqslant \theta_c \leqslant 22$ mrad, it is seen that the signal starts to decrease with increasing Z but then increases lightly beyond $Z \approx 30$. A partial explanation for this can be found by plotting the mod-square coupling into the s-state by probe wavefunction as a function of atomic number, as shown in the dashed line on these plots. For $\alpha = 22$ mrad, it is found that at $Z \approx 30$ the coupling into the s-states reaches a maximum, and beyond this Z value decreases again.

The trend in Fig. 4.5.3(b) is sufficiently monotonic that we may consider approximately the Z scaling. For conceptual simplicity, rather than matching to this

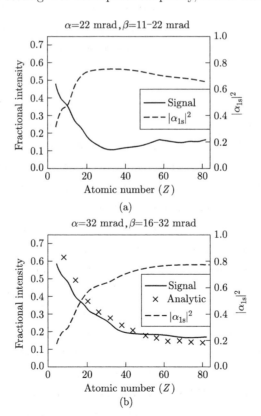

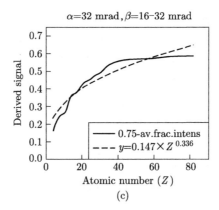

Fig.4.5.3 Average on-column ABF signal over the thickness range 30–60 nm as a function of atomic number for (a) $\alpha = 22$ mrad and $11 \leqslant \theta_c \leqslant 22$ mrad, and (b) $\alpha = 32$ mrad and $16 \leqslant \theta_c \leqslant 32$ mrad. The right vertical axis shows the mod-square coupling amplitude to the s-state. (c) Plot of the specimen free signal (0.75) minus the averaged signal in (b), along with an approximate fit of this signal based on a simple parametric model. Findlay et al., 2010.

plot, let us instead consider the "signal" as the depth of the troughs relative to the specimen free limit. This produces the solid line plot in Fig. 4.5.3(c), for which we can now search for a simple parametric model. Fitting to the power law relation aZ^b, it is found that $a = 0.147$ and $b = 0.336$ to give the best fit (using nonlinear least squares optimization based on the Levenberg–Marquardt algorithm, fitting to a sampling of Z in steps of 2 between 4 and 30 and in steps of 4 between 30 and 82). This is shown with the dashed plot in Fig. 4.5.3(c). As a crude rule of thumb, $Z^{1/3}$ is a loosely tolerable description here, though it must be borne in mind that a fixed probe and detector aperture geometry are assumed as well as along-column spacing and Debye–Waller factor. The non-monotonic form of Fig. 4.5.3(a) is evidence enough that varying these parameters can change the Z-dependence. The weak Z dependence is a mixed blessing. While it is this feature which means that light and heavy atom columns are visible simultaneously, cf. the latter swamping the former as is the case in HAADF, it also means that even under ideal conditions the technique will struggle to discriminate elements of similar atomic number. For atomic species discrimination, simultaneous HAADF imaging is the more useful tool.

4.5.2 Other limitation comparing with HAADF

With smaller spacing between the columns, it is not only the "background" signal which might change in specimens. Anstis, Cai and Cockayne (2003) have shown that the s-state channeling model, the basis for the qualitative understanding of the ABF image contrast from light columns, breaks down when the intercolumn spacing becomes small enough that the s-states of adjacent columns overlap. It will lead to an important limitation of the ABF technique: intercolumn spacing cannot reliably be measured if the columns are quite close together. It should be noted, though, that the same is true of HAADF, albeit to a smaller degree. Also, it is important to appreciate that the visibility of the contrast from either column is not significantly affected (cf. the HAADF case in which the contribution from the heavy columns due to probe spreading make it impossible to identify the light column location for the finer spacing considered here).

Another limitation of ABF indicated by the simulation is that ABF may not be a reliable tool for quantitatively assessing the degree of distortion because there is appreciable overlap for the different degrees of static displacement. While for HAADF imaging, there is greater discrimination between the signal resulting from the different degrees of static disorder, so HAADF imaging may be the better tool for trying to quantitatively assess the degree of distortion. Analogously, the fluctuation of ABF signal are found in the presence of increasing vacancies, which means that this technique is not well suited to quantifying the vacancy concentration. But because the fidelity of the HAADF signal wanes more rapidly with disorder and vacancies than the ABF signal, it is the less reliable technique for clearly detecting the presence of more distorted columns. On the contrary, robustness of ABF imaging allows us to locate the columns reliably even in the presence of a significant number of vacancies.

Another pertinent and interesting issue is the effect of specimen tilt on the form and contrast of ABF images. The ABF images are seen to be more sensitive to tilt with [011] $SrTiO_3$ as the example, with significant distortions in the form of the images by a 12 mrad (~0.69°) tilt, especially for the thicker specimen. While the HAADF images retain their qualitative form out to the same tilt, though the tilt may notably reduce the absolute signal strength.

4.5.3 Applications

As mentioned above, the main advantage of ABF imaging is to observe light elements, especially surrounded by heavy atoms. In this subsection, two examples will be given: firstly, the lightest element hydrogen is observed for the first time in YH_2 compound (Ishikawa et al., 2011); secondly, the ABF imaging technique is used to study the phase separation during the charging processing in $LiFePO_4$ (Gu et al., 2011), one of the most promising energy storage materials.

The YH_2 compound has a fluorite-type structure, for which the individual yttrium-and hydrogen-atom columns distinguishably appear along the [010] projection as shown in Fig. 4.5.4(a). Using the ABF condition with $11 \ll \theta_c \ll 22$ mrad set up in accordance with the optimized HCI condition, the hydrogen-atom columns in a crystalline compound YH_2 was successfully imaged. Faint but distinct dark-dot contrast can be recognized at the hydrogen-atom column positions in the ABF image (Fig. 4.5.4(b)), the validity of which is well demonstrated by the averaged intensity profiles across the yttrium and hydrogen sites. It is noted that no significant intensity is observed at the relevant positions in the BF image (Fig. 4.5.4(c)) or the ADF image (Fig. 4.5.4(d)). It is therefore concluded that, only when the sufficiently extended PCTF is realized by HCI/ABF conditions, the hydrogen atoms can be successfully detected by phase-contrast imaging that reveals atoms as dark dots within a WPO approximation. According to the log-ratio method using EELS, specimen thickness in the observed region (Fig. 4.5.4(b)) was estimated to be approximately 8 ± 2 nm, which is thin enough to apply a WPO approximation. Image simulations based on the multislice method are carried out with the estimated thickness of 8 nm, reproducing fairly well all the observed features, ABF, BF and ADF STEM images, as inset in each of the images of Figs. 4.5.4(b)–(d). It is also confirmed by the simulation of ABF imaging that no significant intensity occurs at the hydrogen sites when the hydrogen atoms are removed from the structure (that is, the hypothetic fluorite YH_2 structure where all the hydrogen sites are vacant). This strongly supports the validity that the observed intensity is indeed originated from the hydrogen atoms themselves, not by the imaging artefacts of phase contrast.

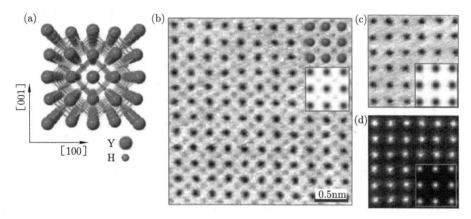

Fig.4.5.4 ABF, BF and ADF–STEM images of the crystalline compound YH$_2$. (a) Crystal structure of YH$_2$ viewed from the [010] direction. (b)–(d), ABF, BF and ADF images obtained with the detector ranges 11–22 mrad, 0–22 mrad and 70–150 mrad, respectively. Simulated images are inset in images (b)–(d), and the YH$_2$ unit-cell projection is overlaid in (b). Ishikawa et al., 2011.

Due to the great power of resolving light atoms, ABF imaging has been utilized to observe lithium element to study the working mechanism of lithium ion batteries. Here, direct observation of Li staging in partially delithiated LiFePO$_4$ is given as an example (Gu et al., 2011). In the ABF image of the pristine LiFePO$_4$ as shown in Fig. 4.5.5(a), the lithium sites are in the middle of two oxygen sites, and marked by yellow circles. In the partially delithiated materials Li$_{0.5}$FePO$_4$ part of the lithium remains in the lattice after charging at every other row as labeled by yellow circles in Fig. 4.5.5(c), and the Li extraction sites are indicated by red arrows. The high resolution atomic image shows clearly that a first order lithium staging structure exists for the partially delithiated Li$_{1-x}$FePO$_4$ ($x \approx 0.5$) samples, which has never been observed by other technique, and challenges previously proposed LiFePO$_4$/FePO$_4$ two-phase separation mechanisms. Obviously, a clear structural evolution picture of other important lithium-containing electrodes could also be disclosed now at atomic resolution based on the ABF technique, which is indeed essential to understand Li-storage mechanisms in important energy storage materials and other materials containing small and light elements.

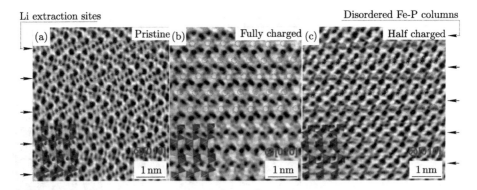

Fig.4.5.5 ABF micrographs showing Li ions of partially delithiated LiFePO$_4$ at every other row. (a) Pristine material with the atomic structure of LiFePO$_4$ shown as inset; (b) fully charged state with the atomic structure of FePO$_4$ shown for comparison; and (c) half charged state showing the Li staging. Note that Li sites are marked by yellow circles and the delithiated sites are marked by orange circles. Gu et al., 2011.

4.6 Scanning Moiré fringe imaging

Atomic resolution HAADF images are usually recorded at a magnification of 15 M as shown in Fig. 4.6.1(a) or more. From the same area, two other HAADF image were taken, Figs. 4.6.1(b) and (c), with the magnification of 5.1 M and 1.3 M, respectively. Surprisingly, the contrast periodic in (b) is not four times as large as that in (c) but almost same as that. Undoubtedly, the lattice-like contrast of Fig. 4.6.1(c) is an artifact. The lattice spacing from (c) was measured to be 0.77 nm, while the real interplanar distance of this alloy should be about 0.18 nm. Detailed study at different magnifications indicates that this artifact contrast is from scanning interference between the electron beam and atomic lattice (Su and Zhu, 2010). This lattice spacing was observed to be a function of the scanning spacing and the angle between the scanning and the specimen lattice, and independent of the setting of the Digital Micrograph software or the dwelling time per pixel. To distinguish this imaging method with the Moiré fringe resulting from two overlapping crystal lattices which may also appear in STEM imaging, these fringes are named as Scanning Moiré Fringes (SMF).

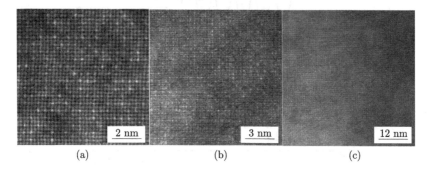

Fig.4.6.1 (a)–(c) HAADF images of some superalloy along ⟨001⟩ zone axis taken with Titan 80–300 at magnifications 14.5 M, 5.1 M and 1.3 M from the same area, respectively.

A geometrical model is shown in Fig. 4.6.2 to illustrate the origin of the artificial lattice. In scanning, a periodic grating lattice is created as shown in Fig. 4.6.2(a). When the scan spacing (d_s) is close to the spacing of the specimen lattice (d_l), the probe would interfere periodically with arranged atom columns in the sample, as depicted in Fig. 4.6.2(b). Therefore, the STEM image records the feature resulting from the interference of d_s-d_l. This process can be viewed as one wave modulated by another wave with a close frequency as a special case of frequency beating. Sinusoidal function $\sin(2\pi f x)$ is used to denote the spatial waves with frequencies f, and f_s is the spatial frequency of the electron probe, and f_1 is the spatial frequency of the specimen's lattice. Their sum is

$$\sin(2\pi f_s x) + \sin(2\pi f_1 x) = 2\cos\left(2\pi \frac{f_s - f_1}{2} x\right) \sin\left(2\pi \frac{f_s + f_1}{2} x\right) \quad (4.6.1)$$

The beating is an envelope function of the sum, deduced as $2\cos\left(2\pi \frac{f_s - f_1}{2} x\right)$, as shown in Fig. 4.6.2(c), with the frequency $\frac{f_s - f_1}{2}$, which is smaller than either scanning or lattice frequency. It is worth noting that this model is purely based on geometric consideration of the scan and specimen lattice only, and does not affect the physical process of electron–sample interaction. Accordingly, the interference HAADF image still is a Z-contrast image.

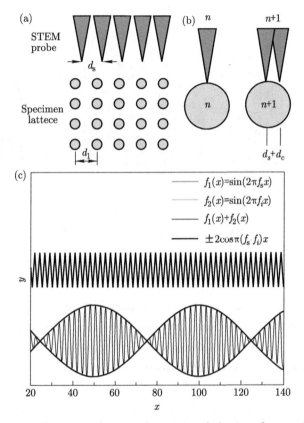

Fig.4.6.2 Schematic diagrams showing the origin of the interference image. (a) The general view of the scanning grating and the specimen lattice. (b) The interference between neighboring scanning probes and the atomic columns. (c) The beating of two waves with frequencies f_s and f_1. Here, f_s is assumed to be 5% higher than f_1 and the amplitudes of two functions are identical. Su and Zhu, 2010.

One advantage of the SMF method is that it can be used for observation at a lower magnification and in some cases atomic lattice can also be imaged, as shown in Fig. 4.6.1(c) and Fig. 4.6.3(a), a HAADF image of interface of $Ba_{0.3}Sr_{0.7}TiO_3$

(BST) and SrTiO$_3$ (STO) along the $\langle 100 \rangle$ zone axis of STO. At such lower magnifications, the specimen suffers less damage from beam irradiation, and also less carbon contamination is produced, both of which are challenging issues in acquiring high-quality STEM images at high magnification especially for beam-sensitive samples.

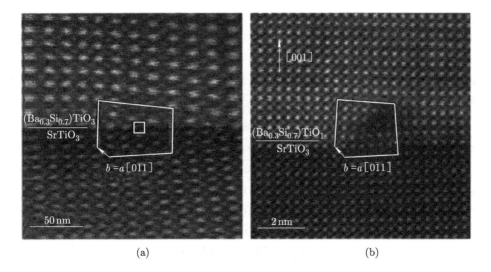

Fig.4.6.3 (a) A SMF image at the magnification of 500 k, showing the enlarged image of a misfit dislocation at the interface of BST film and STO substrate. (b) A HAADF image taken at the magnification of 15 M on the green square region shown in (a), where in a Burger vector of $b = a[0\bar{1}1]$ is indicated by a Burger circuit. Su and Zha, 2010.

One can also use SMF to observe dislocations at coherent interfaces at lower magnification as shown in Fig. 4.6.3(a) with a dislocation-like distortion observed. It was acquired with the scanning direction, x, along the interface. In Fig. 4.6.3(a), we can even determine the "Burgers vector" of a misfit dislocation, $\boldsymbol{a}[0\bar{1}1]$, at the interface. The HAADF image taken at the same area at a magnification 15 M×, Fig. 4.6.3(b), clearly verifies the existence of such a dislocation with $\boldsymbol{a}[0\bar{1}1]$ Burgers vector. Therefore, using this method to image dislocations with a large field of view gives the convenience in measuring the intensity of dislocations.

4.7 Application on micro-area analysis

In the above sections, different imaging modes in STEM have been introduced, such as HAADF and ABF, which is suitable to image heavy atoms and light atoms, respectively. Besides these imaging, another important application of STEM is analysis on micro area by combination with other analytical signal as mentioned in Section 4.1, for instance, EELS, EDS and NBD, etc.. The spectroscopy imaging will be specifically introduced in Chapter 5. For the completeness, however, in this Section, one example is still given as to the combination between STEM and the analytical methods, EDS and NBD, to achieve the compositional and structural information of the micro area.

Fig. 4.7.1 shows a HAADF image of a nickel-based alloy after creep test. Comparing with the image of the as-cast alloy as shown in Fig. 4.4.1, variations in image contrast can be observed in the γ phase and the γ'/γ' interface shows brighter contrast. To study the redistribution of the major constituents during the creep tests, combination of STEM and EDS are used to analyze the area denoted by a black rectangle in Fig. 4.7.1(a). From element mapping images as shown in Fig. 4.7.1(b), enrichment was found as to Cr, Co, Mo, W and Re, γ forming elements, in the γ phase adjacent to the γ/γ' interface. And among heavy elements W and Re were obviously enriched, which can be clearly seen in their distributions along the direction perpendicular to the interface, shown in Fig. 4.7.1(c). That is why interface shows brighter contrast than other area in Fig. 4.7.1(a). Moreover, in the area of γ phases with low contrast in Fig. 4.7.1(a) few W and Re as well as Cr, Co and Mo has been found but more Al, Ni and Ta, γ' forming elements, instead. It suggests that these areas may change into γ' phases from γ phases due to the element redistribution during the creep test. To prove this hypothesis two ways can be used, atomic resolution HAADF imaging like Fig. 4.4.1 and NBD. Diffraction patterns can be recorded by the axial CCD camera simultaneously with HAADF imaging and EDS collection. One pixel in the HAADF image or EDS mapping corresponds to one diffraction pattern. The NBD pattern corresponding to the area of γ phases with low contrast is shown in Fig. 4.7.1(d). Ordered γ' phase is identified, which confirms the results of EDS mapping.

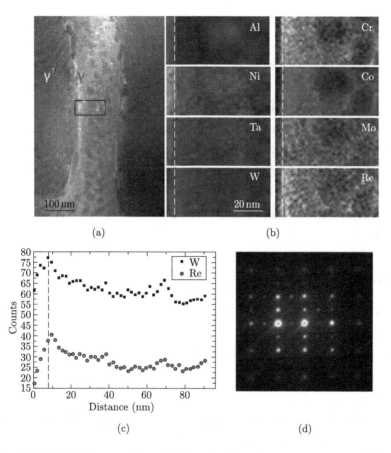

Fig.4.7.1 (a) HAADF image of a nickel-based alloy after the creep test. (b) Element mapping images of major constituents corresponding to areas denoted by a black rectangle in (a). (c) Distribution of element Re and W along the direction perpendicular to the interfaces. Dashed lines schematically indicate the γ/γ' interfaces. (d) Diffraction pattern corresponding to the low-contrast area in γ phases in (a).

4.8 Discussion and conclusion

For conventional STEM aberrations of the condenser lens play an important role in determining the point spread of the probe and thus limit the resolution. The invention of the aberration corrector in the end of last century, however, increased the resolution greatly, and in the present equipped with double aberration-correctors, the TEAM 0.5 microscope is capable of producing images with 50 pm resolution (http://ncem.lbl.gov/TEAM-project/index.html). Thus, resolution may not be a problem concerned by electron microscopist anymore. But on the other hand, several factors as mentioned in this chapter, such as channeling effect, Debye-Waller factor and focus, et al. influence the HAADF image contrast and make it deviation from the Z-contrast dependence, especially in aberration-corrected microscopes. Therefore, to fully understand the image contrast and obtain the quantative results, the HAADF image simulations are required, contrary to common perception that HAADF-STEM images are intuitively interpretable in terms of the atomic numbers.

References

Anstis, G.R., Cai, D. Q., Cockayne, D.J.H., Limitations on the s-state approach to the interpretation of sub-angstrom resolution electron microscope images and microanalysis. Ultramicroscopy, 2003 (94): 309–327.

Ardenne, M.v., Das Elektronen-rastermikroskop, theoretische grundlagen. ZPhy, 1938 (109): 553–572.

Ardenne, M.v., Das Elektronen-rastermikroskop. Praktische Ausführung, Zeitschrift für Technische Physik, 1938 (19): 407–416.

Blavette, D., Cadel, E., Deconihout B., The Role of the Atom Probe in the Study of Nickel-Based Superalloys. Mater. Charact., 2000 (44): 133–157.

Blom, D.A., Multislice frozen phonon high angle annular dark-field image simulation study of Mo–V–Nb–Te–O complex oxidation catalyst "M1". Ultramicroscopy, 2012 (112): 69–75.

Borisevich, A.Y., Lupini, A.R., Pennycook S.J., Depth sectioning with the aberration-corrected scanning transmission electron microscope. Proc. Natl. Acad. Sci. U. S. A., 2006 (103): 3044–3048.

Born, M., Wolf, E. Principles of Optics. Oxford: Pergamon Press, 1980.

Chu, M.W., Liou, S.C., Chang, C.P., Choa, F.S., Chen, C.H., Emergent Chemical Mapping at Atomic-Column Resolution by Energy-Dispersive X-Ray Spectroscopy in an Aberration-Corrected Electron Microscope. Phys. Rev. Lett., 2010 (104): 196101.

Cowley, J.M., Electron diffraction phenomena observed with a high resolution STEM instrument. JEMT, **3** (1986): 25–44.

Cowley, J.M., Image contrast in a transmission scanning electron microscope. ApPhL, **15** (1969) 58–59.

Crewe, A.V., High resolution scanning microscopy of biological specimens. Philosophical Transactions of the Royal Society of London Series B-Biological Sciences, 1971 (261): 61.

Crewe, A.V., Scanning electron microscopes-is high resolution possible. Science, 1966 (154): 729.

Crewe, A.V., Isaacson, M., Johnson, D., Electron energy loss spectra of nucleic acid bases. Nature, 1971 (231): 262.

Crewe, A.V., Salzman, D.B., ON THE OPTIMUM RESOLUTION FOR A CORRECTED STEM. Ultramicroscopy, 1982 (9): 373–377.

Crewe, A.V., Wall, J., A Scanning microscope with 5 a resolution, JMBio, 1970 (48): 375.

Crewe, A.V., Wall, J., Langmore J. Visibility of single atoms. Science, 1970 (168): 1338.

Crewe, A.V., Wall, J., Welter, L.M., A high-resolution scanning transmission electron microscope. JAP, 1968 (39): 5861–5868.

D'Alfonso, A.J., Freitag, B., Klenov, D., Allen, L.J. Atomic-resolution chemical mapping using energy-dispersive x-ray spectroscopy. Phys. Rev. B, 2010 (81): 100101.

Egerton, R.F., Electron Energy-Loss Spectroscopy in the Electron Microscope. Third ed., Springer, 2011.

Engel, A., Wiggins, J.W., Woodruff, D.C., Comparison of calculated images generated by 6 modes of transmission electron-microscopy. JAP, 1974 (45): 2739–2747.

Erni, R., Aberration-Corrected Imaging in Transmission Electron Microscopy: An Introduction. London: ICP/Imperial College Press, 2010.

Findlay, S.D., Shibata, N., Ikuhara, Y., What atomic resolution annular dark field imaging can tell us about gold nanoparticles on TiO2TiO2 (110). Ultramicroscopy, 2009 (109): 1435–1446.

Findlay, S.D., Shibata, N., Sawada, H., Okunishi, E., Kondo, Y., Ikuhara, Y. Dynamics of annular bright field imaging in scanning transmission electron microscopy. Ultramicroscopy, 2010 (110): 903–923.

Findlay, S.D., Shibata, N., Sawada, H., Okunishi, E., Kondo, Y., Yamamoto, T., Ikuhara, Y., Robust atomic resolution imaging of light elements using scanning transmission electron microscopy. ApPhL, 2009 (95): 191913.

Fultz, B., Howe, J.M., Transmission Electron Microscopy and Diffractometry of Materials, 3rd ed. New York: Springer, 2008.

Ge, B., Luo, Y., Li, J., Zhu, J., Study of γ/γ' Interfaces in Nickel-Based Single-Crystal Superalloys by Scanning Transmission Electron Microscopy. Metall. Mater. Trans. A, 2011 (42): 548–552.

Ge, B., Luo, Y., Li, J., Zhu, J., Tang, D., Gui, Z., Study of γ/γ' interfacial width in a nickel-based superalloy by scanning transmission electron microscopy.PMagL, 2012 (92): 541–546.

Ge, B.H., Luo, Y.S., Li, J.R., Zhu, J., Distribution of rhenium in a single crystal nickel-based superalloy. Scripta Mater., 2010 (63): 969–972.

Gu, L., Zhu, C., Li, H., Yu, Y., Li, C., Tsukimoto, S., Maier, J., Ikuhara, Y. Direct observation of lithium staging in partially delithiated $LiFePO_4$ at atomic resolution. J. Am. Chem. Soc., 2011 (133): 4661–4663.

Hanssen, K.J., Contrast transfer of electron-microscope with partial coherent illumination .a. ring condensor, Optik, 1971 (33): 166–&.

Haruta, M., Kurata, H., Komatsu, H., Shimakawa, Y., Isoda, S. Effects of electron channeling in HAADF-STEM intensity in La2CuSnO6. Ultramicroscopy, 2009 (109): 361–367.

Howie, A., Diffraction channelling of fast electrons and positrons in crystals. PMag, 1966 (14): 223–237.

Inada, H., Su, D., Egerton, R.F., Konno, M., Wu, L., Ciston, J., Wall, J., Zhu, Y., Atomic imaging using secondary electrons in a scanning transmission electron microscope: Experimental observations and possible mechanisms, Ultramicroscopy, 2010.

Ishikawa, R., Okunishi, E., Sawada, H., Kondo, Y., Hosokawa, F., Abe, E., Direct imaging of hydrogen-atom columns in a crystal by annular bright-field electron microscopy. Nat Mater, 2011 (10): 278–281.

Ishizuka, K., A practical approach for STEM image simulation based on the FFT multislice method. Ultramicroscopy, 2002 (90): 71–83.

James, E.M., Browning, N.D., Practical aspects of atomic resolution imaging and analysis in STEM. Ultramicroscopy, 1999 (78): 125–139.

Jia, C.L., Thus, A., Urban, K., Atomic-scale analysis of the oxygen configuration at a $SrTiO_3$ dislocation core. Phys. Rev. Lett., 2005 (95).

Jia, C.L., Urban, K., Atomic-resolution measurement of oxygen concentration in oxide materials. Science, 2004 (303): 2001–2004.

Kimoto, K., Asaka, T., Nagai, T., Saito, M., Matsui, Y., Ishizuka, K., Element-selective imaging of atomic columns in a crystal using STEM and EELS. Nature, 2007 (450): 702–704.

Klenov, D.O., Findlay, S.D., Allen, L.J., Stemmer, S., Influence of orientation on the contrast of high-angle annular dark-field images of silicon. Phys. Rev. B, 2007 (76): 014111.

Komoda, T., Electron microscopic observation of crystal lattices on level with atomic dimension. Jpn. J. Appl. Phys., 1966 (5): 603–607.

LeBeau, J.M., Findlay, S.D., Wang, X., Jacobson, A.J., Allen, L.J., Stemmer, S., High-angle scattering of fast electrons from crystals containing heavy elements: Simulation and experiment. Phys. Rev. B, 2009 (79): 214110.

Lin, J.A., Cowley, J.M., Calibration of the operating parameters for an HB5 stem instrument. Ultramicroscopy, 19 (1986): 31–42.

Liu, J., Scanning transmission electron microscopy of nanoparticles, in: Z. Wang (Ed.) Characterization of nanophase materials, 2000: 81–132.

Mathews, W.W., The Use of Hollow-Cone Illumination for Increasing Image Contrast in Microscopy. Trans. Am. Microsc. Soc., 1953 (42): 190–195.

Misell, D.L., Stroke, G.W., Halioua, M., Coherent and incoherent imaging in scanning-transmission electron-microscope. Journal of Physics D-Applied Physics, 1974 (7): L113–L117.

Mitome, M., Takayanagi, K., Tanishiro, Y., Improvement of resolution by convergent-beam illumination in surface profile images of high resolution transmission electron microscopy. Ultramicroscopy, 1990 (33): 255–260.

Mittal, A., Mkhoyan, K.A., Limits in detecting an individual dopant atom embedded in a crystal. Ultramicroscopy, 2011 (111): 1101–1110.

Muller, D.A., Kourkoutis, L.F., Murfitt, M., Song, J.H., Hwang, H.Y., Silcox, J., Dellby N., Krivanek O.L., Atomic-scale chemical imaging of composition and bonding by aberration-corrected microscopy, Science, 2008 (319): 1073–1076.

Nellist, P.D., The Principles of STEM Imaging, in: S.J., Pennycook, P.D., Nellist (Eds.) Scanning Transmission Electron Microscopy: imaging and analysis, Springer New York, 2011: 92.

Nellist, P.D., The Principles of STEM Imaging, in: S.J., Pennycook, P.D., Nellist (Eds.) Scanning Transmission Electron Microscopy: imaging and analysis, Springer New York, 2011: 91–116.

Nellist, P.D., Cosgriff, E.C., Behan, G., Kirkland, A.I., Imaging Modes for Scanning Confocal Electron Microscopy in a Double Aberration-Corrected Transmission Electron Microscope. Microsc. Microanal., 2008 (14): 82–88.

Nellist, P.D., Rodenburg, J.M., Beyond the conventional information limit: the relevant coherence function. Ultramicroscopy, 1994 (54): 61–74.

Okunishi, E., Ishikawa, I., Sawada, H., Hosokawa, F., Hori, M., Kondo, Y. Visualization of Light Elements at Ultrahigh Resolution by STEM Annular Bright Field Microscopy. Microsc. Microanal., 2009 (15): 164–165.

Pennycook, S.J., A Scan Through the History of STEM, in: S.J. Pennycook, P.D. Nellist (Eds.) Scanning Transmission Electron Microscopy: imaging and analysis, Springer New York, 2011: 1–90.

Pennycook, S.J., Seeing the atoms more clearly: STEM imaging from the Crewe era to today. Ultramicroscopy, 2012.

Pennycook, S.J., Jesson, D.E., Chisholm, M.F., Browning, N.D., McGibbon, A.J., McGibbon, M.M., Z-Contrast Imaging in the Scanning Transmission Electron Microscope. Microsc. Microanal., 1995 (1): 231–251.

Rose, H., Nonstandard imaging methods in electron-microscopy. Ultramicroscopy, 1977 (2): 251–267.

Rose, H., Phase-contrast in scanning-transmission electron-microscopy. Optik, 1974

(39): 416–436.

Su, D., Zhu, Y.M., Scanning moire fringe imaging by scanning transmission electron microscopy. Ultramicroscopy, 2010 (110): 229–233.

Voyles, P.M., Grazul, J.L., Muller, D.A., Imaging individual atoms inside crystals with ADF-STEM. Ultramicroscopy, 2003 (96): 251–273.

Yamazaki, T., Kawasaki, M., Watanabe, K., Hashimoto, I., Shiojiri, M., Artificial bright spots in atomic-resolution high-angle annular dark field STEM images. J. Electron Microsc., 2001 (50): 517–521.

Yu, Z., Muller, D.A., Silcox, J., Effects of specimen tilt in ADF-STEM imaging of a-Si/c-Si interfaces. Ultramicroscopy, 2008 (108): 494–501.

Yu, Z., Muller, D.A., Silcox, J., Study of strain fields at a-Si/c-Si interface. JAP, 2004 (95): 3362–3371.

Zeitler, E., Thomson, M.G.R., Scanning transmission electron microscopy. 2, Optik, 1970 (31): 359.

Zeitler, E., Thomson M.G.R., Scanning transmission electron microscopy .1, Optik, 1970 (31): 258.

http://en.wikipedia.org/wiki/Snell's_law

http://ncem.lbl.gov/TEAM-project/index.html

5
Spectroscopy

Zhihua Zhang, Yonghai Yue and Jiaqing He

5.1 Introduction

Analyzing the composition, microstructure and electronic structure of materials at the atomic/nanoscale is key issues for material science, which provides the information necessary for correlating the microstructures and the measured chemical and physical properties of a solid specimen. Energy-dispersive X-ray spectroscopy (EDS) and Electron energy-loss spectroscopy (EELS) are two important techniques for determining the structure and/or chemical composition of a solid. Benefited from the development of electron microscope, the application of EDS and EELS is growing rapidly worldwide. Combined with HRTEM, diffraction, and other TEM-related techniques, EDS and EELS can be used to characterize the relationship among crystal structure, chemistry and physical functions, especially in nanomaterial research.

EDS technique use the generated X-rays when the electron beam strikes a TEM specimen to detect nearly all the elements in the periodic table. EDS is accomplished by X-ray spectrometry, which takes advantage of modern semiconductor technology, also it comprises a detector interfaced to signal-processing electronics and a computer-controlled multi-channel analyzer display. EELS is the analysis of the energy distribution of electrons that have interacted inelastically with the specimen. There are three aspects can be revealed by EELS: the electronic structure of the specimen atoms, the nearest-neighbor distributions and their bonding, the dielectric response. The measurement of transmitted electrons' energy spectrum was first created by Ruthemann (1941). Now, two types of commercial spectrometers are manufactured: the magnetic prism spectrometer (Gatan) and the omega filter (Zeiss). The analytical pouer of EELS is iucreasing with the production of an elemental map (by selecting an appropriate ionization edge and correcting for the pre-edge background) This allows elemental segregation to be imaged in a semi-quantitative manner.

Originally, EDS detectors are sensitive to relatively heavy elements while EELS is more suitable to analyze the light elements. With the development of EDS detectors, wavelength-dispersive spectrometers with parallel-recording detectors are available for the TEM and can detect elements down to Li (Terauchi et al., 2010a) with an energy resolution of typically 1eV but relatively low solid angle (Terauchi et al., 2010b). These developments make EDS competitive to EELS for the detection of light elements in a TEM specimen. and there still exist some differences between EDS and EELS. First, for the same incident beam current, the count rate of characteristic X-rays (detectable by EDS) is less than that of core-loss electrons (detectable by EELS), the spectrometer collection efficiency of EELS is higher than EDS. However, the EELS background is generally higher than that of EDS and the characteristic features of EELS are edges but not peaks, this makes the features in EELS less visible than the corresponding peak in the X-ray spectrum. Second, EELS offers slightly better spatial resolution than X-ray emission spectroscopy because the volume of specimen which gives rise to the energy-loss signal can be limited by means of an angle-limiting aperture. Spatial resolution is a major factor determining the minimum detectable mass of an element. However, for EELS, very thin specimen is needed, if the specimen is too thick, plural scattering will greatly increase the background and make the electron energy loss edges invisible. In contrast, EDS can tolerate thicker specimens (up to a few hundred nanometers). Third, the quantification of EELS don't need to involve in the use of standards. The signal-intensity ratios only depend on the physics of the primary excitation and are largely independent of the spectrometer. For EDS, the relative intensities of the peaks depend on the properties of the detector. If the detector parameters are not precisely known, the factors needed by quantification cannot be accurately calculated. Forth, EELS offers greater elemental sensitivity for certain specimens and also offers possibility of additional information, including an estimation of the local thickness of a TEM specimen and information about its crystallographic and electronic structure. However, in order to have a deep understanding on the information provided by EELS, the physics underlying has to be involved. To summarize, there are advantages and disadvantages of these two techniques, the operator can choose one alternatively during the study.

The development of EDS and EELS is remarkable. duving these years, the improvements are achieved partly by TEM instrument, such as the successful correction of spherical aberration and chromatic aberration of electron lenses, allowing sub-angstrom spatial resolution in TEM and STEM images. With an aberration-corrected STEM and efficient detectors, elemental maps can show individual atomic columns in crystal, using either EDS or EELS techniques (Watanabe et al., 2010a; b). Some new techniques, such as magnetic (linear and chiral-dichroic) measurements are also developed (Schattschneider et al., 2006). Computer software has been improved in data processing, all of these improvements have ensured EDS and EELS techniques to solve real material problems, while not mainly a playground for physicists.

There are some good textbooks and review papers about the EDS and EELS, therefore in this chapter we will summarize the recently development in this research field. For the details of the history and principles of spectroscopy, the readers are recommended to read the literatures of Williams and Carter (1996) and Egerton (1996).

5.2 Principle of EDS and EELS

Zhihua Zhang

5.2.1 The physics of EDS and EELS

In TEM, we can take interaction of incident electrons and atoms in the specimen as scattering. When the incident electrons transfer an appreciable amount of energy to the specimen, that is inelastic scattering. EDS and EELS are both correlated with the inelastic scattering. Inelastic scattering occurs as a result of Coulomb interaction between the incident electron and the atomic electrons in the specimen.

Some inelastic processes can be understood as the excitation of a single atomic electron to a higher energy level (Fig. 5.2.1). The initial-state lies typically some hundreds or thousands of electron volts below the Fermi level of the solid and the final-state exist only above the Fermi level. So the inner-shell electron can make the excitation only if it absorbs an amount of energy similar to or greater than its original binding energy. The energy comes from the incident electrons during the interaction. The incident electrons lose an equal amount of energy and transfer the energy to the atom in the specimen. After losing a characteristic amount of energy, the incident electrons transmitted through the specimen and are directed into a high-resolution electron spectrometer, which separates the electrons according to their kinetic energy and produces an EELS showing the number of electrons (scattered intensity) as a function of their decrease in kinetic energy. So EELS deals directly the primary process of electron excitation, while EDS deal with the de-excitation process. After the interaction, the target atom is left in a highly excited (or ionized) state, an outer-shell electron (or an inner-shell electron of lower binding energy) undergoes a downward transition to the vacant "core hole" and some of the excess energy is liberated as electro-magnetic radiation (X-rays). X-ray photons generated enter the EDS detector, where the charge pulse is proportional to the energy of the incoming X-ray. Thus the EDS spectra is a histogram of X-ray inten-

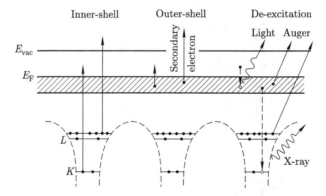

Fig.5.2.1 EDS is related with the de-excitation process of photon and electron emission on the right; EELS is related with the primary process of inner-and outer-shell excitation (shown on the left).

sity versus energy and consists of several approximately Gaussian-shaped peaks. The characteristic peaks are due to X-rays emitted when ionized atoms return to the ground state, therefore they contain the elemental information in the analyzed volume.

5.2.2 Spectroscopic character of EDS and EELS

Typical EDS spectra obtained from TEM are shown in Figs. 5.2.2(c) and (d), which are plots of X-ray counts (imprecisely termed "intensity") versus X-ray energy. The characteristic of EDS Spectra appear as Gaussian shaped peaks superimposed on

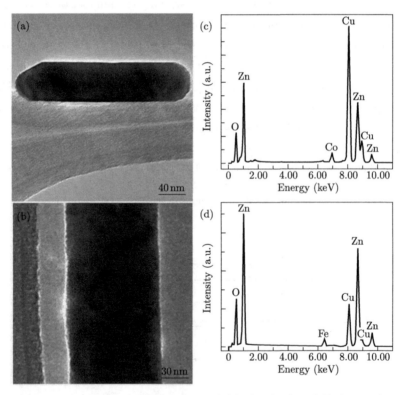

Fig.5.2.2 Low magnification TEM images of (a) the Co-doped ZnO sample; and (b) the Fe-doped ZnO sample; (c) and (d) typical EDS results obtained from Co-doped and Fe-doped ZnO sample.

a background of bremsstrahlung X-rays. By choosing different analysis region in Figs. 5.2.2(a) and (b), the elemental composition at the nanoscale can be identified. Quantification of the spectrum involves removal of the background and determination of the true relative intensities of the characteristic peaks. EDS is easy to use and the operator can quickly interpret the readout within a very short time by routines software.

Compared with EDS, EELS spectrum is more complicated. A typical EELS spectrum recorded from a thin specimen over a range of about 1000 eV is shown in Fig. 5.2.3. The spectrum are plots of electron counts (intensity) versus electron loss energy and can be divided mainly into two parts, that is low-loss region (0∼40 eV) and core-loss region (above 40 eV). The low-loss spectrum (also named as valence electron energy loss spectrum, VEELS) includes zero-loss peak and peaks corresponding to the inelastic scattering from outer-shell electrons. Zero-loss peak represents electrons that are transmitted without suffering measurable energy loss. Outer-shell electrons can undergo single-electron excitation, expressed as a series of peaks in low-loss spectrum, providing the information of joint density of states (JDOS) near the band gap in an insulator or semiconductor. Outer-shell inelastic scattering may also involve an oscillation of the valence electron density, known as

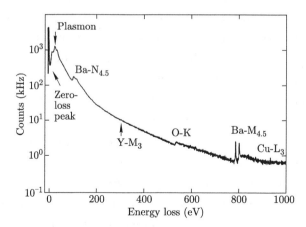

Fig.5.2.3 Typical electron energy loss spectrum recorded from a thin specimen over a range of about 1000 eV.

a plasma resonance. The excitation can also be described in terms of the creation of a pseudo-particle, the plasmon. Plasmon excitation and single-electron excitation represent alternative modes of outer-shell inelastic scattering. The core-loss spectra locate at higher energy loss and take the form of edges rather than peaks. The intensity of core-loss edges rises rapidly and occurs at the ionization threshold, whose energy-loss coordinate is approximately the binding energy of the corresponding atomic shell. So the ionization edges present in an energy-loss spectrum reveal which elements are present within the specimen. The fine structure in the low-loss peaks and the ionization edges reflect the crystallographic or energy band structure of the specimen, which will be discussed in the following parts.

5.2.3 Basic operational mode

To produce the EDS shown in Fig. 5.2.2, a TEM or a STEM image have to be obtained, by which the interested specimen area can be chosen. In TEM mode, EDS spectrum was recorded at a particular point on the specimen or (more precisely) integrated over a circular region defined by an incident electron beam. In STEM mode, more information can be obtained by choosing different operational mode. EDS spectrum can be taken by producing a stationary probe on the point feature of the analysis region, the composition information to the point thus can be obtained. If the EDS spectrum collection was carried out through a line-scan using a STEM, the composition corresponding to each point can be obtained and the element distribution along this line can be analyzed. Element distribution image can be obtained by acquiring an EDS spectrum at each pixel as a STEM probe is rastered over the specimen. Examples of EDS collected at different operational mode using STEM are given in the following sections.

EELS spectrum collection can also be carried out using a conventional TEM or a STEM. The point spectrum, the line-scan, and the spectrum image similar to EDS can be obtained. If the EELS spectrum was obtained using a TEM, the performance of the EELS system (energy resolution, collection efficiency, and spatial resolution of analysis) is affected by the properties of the TEM imaging lenses and the way in which they are operated (image or diffraction mode). Good energy resolution requires that an electron-beam crossover of small diameter be placed at the spectrometer object plane. In practice, this crossover is either a low-magnification image of the specimen or a portion of its diffraction pattern (just the central beam, if a bright-field objective aperture is inserted). Since the spectrometer in turn images this crossover onto the EELS detector or the energy-selecting slit, what is actually recorded represents a convolution of the energy-loss spectrum with the diffraction pattern or image of the specimen, sometimes called spectrum diffraction of spectrum image mixing. If the TEM is operated in image mode, with an image of the specimen of magnification M on the viewing screen, the spectrometer is said to be diffraction coupled because the projector lens crossover then contains a small diffraction pattern of the specimen. The angular range of scattering allowed into the spectrometer (the collection semi-angle β) is controlled by varying the size of the objective lens aperture. The region of specimen giving rise to the energy-loss spectrum is determined by a spectrometer entrance aperture and corresponds to a portion of the image close to the center of the TEM viewing screen (before the screen if lifted to allow electrons through to the spectrometer). If the TEM is operated in diffraction mode, with a diffraction pattern of camera length L at the viewing screen, the spectrometer is image coupled because the projector crossover contains an image of the illuminated area of the specimen. The area of specimen being analyzed is determined by the electron-beam diameter at the specimen or else by a selected area diffraction aperture, if this aperture is inserted to define a smaller area.

5.2.4 Qualitative and quantitative analysis of EDS

Before the quantitative microanalysis of the EDS spectrum, qualitative analysis should be carried out firstly. Qualitative analysis requires that every peak in the EDS spectrum be identified unambiguously and statistically.

The first and most important step in qualitative analysis is to acquire a spectrum across the widest possible energy range, so that all the characteristic peaks present in the spectrum. Then, the computer system can be used to run an automatic identification on the peaks in the spectrum. If the spectrum is simple, containing a few well-separated peaks, this automatic step may be all that is required. However, if the spectrum contains many peaks, and particularly if peak overlap is occurring, misidentification may occur during such an "autosearch" routine. Manual indexing is thus necessary. There is a well-established procedure we can follow as discussed in detail by Williams and Carter (1996). Here a short introduction was given. First, the most intense peak can be labeled based on the automatic identify step. Then repeat the exercise, until all the major peaks are accounted for. For some special cases, it is tricky to conclude if the small peaks appeared in EDS. The spectra can be reacquired at a much lower dead time and at a different accelerating voltage or specimen orientation to see if the small peaks still exist. In some cases, high energy resolution is necessary to dissolve overlap peaks.

The best operating conditions of TEM for qualitative analysis of EDS should maximize the X-ray count rate and with the minimum of artifacts. The spectrum should be obtained from a large area of the specimen firstly and then from smaller areas. The highest operating voltage should be used to get the most X-ray counts in the specimen. Pick a portion of the specimen that is single phase in the area of interest and is well away from strong diffraction conditions, so as to minimize crystallographic effects and coherent bremsstrahlung, etc.

The quantification routines of EDS is based on cliff-Lorimer equation

$$\frac{C_A}{C_B} = k_{AB}\frac{I_A}{I_B} \qquad (5.2.1)$$

where C_A and C_B are weight percent of each element in a binary system, I_A and I_B are the measured characteristic EDS peak intensities and k_{AB} is the Cliff-Lorimer factor. Generally, the intensity of K_α lines is used to perform the quantitative analysis. And the EDS should be acquired with enough counts, the specimen should be close to 0° tilt (away from a strong two-beam dynamical-diffraction condition) and should be oriented with the thin portion of the wedge facing the detector to minimize spurious effects and X-ray absorption.

To quantify an EDS spectrum, the measuring of the peak intensities is the first step, the background subtraction and peak integration are involved. The second step is the determination of a value for the k_{AB} factor. These steps are accomplished by several available software routines in the modern EDS system. The background refers to the intensity under the characteristic peaks generated by the "bremsstrahlung" or "braking radiation" process as the beam electrons interact with the coulomb field of the nuclei in the specimen. The best approach to background subtraction depends on whether the region of interest is in low-energy regime, and if the characteristic peaks are close together or isolated. Window methods (Goldstein et al., 1986) are simple and are applicable to isolated characteristic peaks and on a linear portion of the background. More sophisticated mathematical approaches (Goldstein et al., 1986) are most useful for multi-element spectra and/or those containing peaks below ∼1.5 keV. The method that gives the most reproducible results should be chosen. There are two ways to determine k factors (Williams, 1987): experimental determination using standards and calculation from first principles, The former is slow and laborious but gives the most accurate values, the latter is quick and painless but the results are less reliable. The values of the peak intensities and the correct value of the k factor are inserted into the Cliff-Lorimer ratio equation, the weight percent of each element can be obtained.

5.2.5 Analysis of EELS

Discussed as above, the fine structure in the low-loss EELS and the core-loss ionization edges reflect the crystallographic or energy band structure of the specimen. The specimen thickness, optical functions and the JDOS can be obtained from low-loss EELS. An example is given to clarify the steps to quantify low-loss EELS (Zhang et al., 2008). Fig. 5.2.4 is the loss function of a potential optical material $YBa_3B_9O_{18}$ and is derived from the low-loss EELS. To obtain such information, two steps should be performed: Zero-loss extraction and single-scattering deconvolution. Zero-loss extraction is one of the most important steps for the quantification of low-loss EELS, as emphasized by many people (Ryen et al., 1998; Hu and Jones, 2005) because the high-energy tail covers features of the low-loss region. A number of different models have been used to simulate the shape of the zero-loss peak (Bangert et al., 1997; Batson et al., 1986; Raether, 1980). After the zero-loss peak is extracted appropriately, the plural scattering can be removed by Fourier-log deconvolution method (Egerton 1996). Then the band gap of the materials can be determined by fitting the edge-on slope of the loss function. Alternatively, fitting the edge-on slope of the loss function near the band gap using the $(E - E_g)^{0.5}$ for direct band gap and $(E - E_g)^{1.5}$ for indirect band gap material serves as a more precise method (Rafferty and Brown 1998). The $(E - E_g)^{1.5}$ curve generates a best fit for the experimental loss function (fitting result was shown as the dashed line in Fig. 5.2.4), leading to the determination of the band gap as ~6.3 eV for YBa$_3$B$_9$O$_{18}$ and the nature of the materials band gap is indirect.

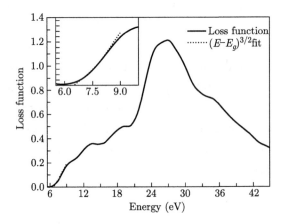

Fig.5.2.4 The loss function of a potential optical material YBa$_3$B$_9$O$_{18}$, which is derived from the low-loss EELS.

Above the band gap, there are five well-resolved peaks in the loss function, located at ~9.4, ~13, ~19.2, ~26.2, and ~34.6 eV, respectively. The dominant peak at 26.2 eV can be assigned to the bulk plasmon oscillation. The highly damped bulk plasmon peak is characterized by its large excitation cross sections for fast electrons and its energy is related to valence electron density. A brief estimation of the collective plasmon excitation with the Drude formula (Lazar et al., 2003) using the free-electron approximation can also give the assignment. Other peaks should originate from the single electron excitation from the valence band (VB) to the empty density of states (DOS) in the conduction band (CB), and their profiles are expected to have direct correlation with the JDOS between occupied and unoccupied states in the energy bands (Benthem et al., 2001).

If the electronic structures of the YBa$_3$B$_9$O$_{18}$ were calculated, the individual transitions related with the EELS can be assigned. The origin of each interband transition observed in the EELS can be determined by comparing their energetics to the calculated total DOS (TDOS). The interband transition strength J_{CV} is defined as the convolution of the valence and CB DOS weighted by the dipole selection rule matrix elements and can be directly compared to the JDOS. Such a quantity can be deduced from the equation

$$J_{CV}(E) = (m_0^2/e^2\hbar^2)(E^2/8\pi^2)[\varepsilon_2(E) + i\varepsilon_1(E)] \qquad (5.2.2)$$

where the dielectric function $\varepsilon(\omega)$ is obtained from the material loss function $\text{Im}(-[1/\varepsilon(\omega)])$ via the Kramers-Kronig analysis. For computational convenience, the prefactor $m_0^2 e^{-2}\hbar^{-2}$ was taken to be unity so that the $J_{CV}(E)$ spectra plotted have a unit of eV^2 instead of the unit of density. By using simple algebraic transformations, other optical parameters can be obtained.

As discussed above, Low-loss spectra provide the information of the electronic structures (through the volume plasmons and single-electron excitation) and the optical properties. The thickness of the specimen can also be determined from the low-loss spectra as discussed in general introduction of EELS (Egerton, 2010). If the spectrometer system offers sub-electron volt energy resolution, the fine structure of low-loss spectra is more easily distinguished, making a monochromated TEM a preferred option. Many studies related with plasmons (including the bulk plasmon and surface plasmon) are carried out recently due to the application of monochromators and the combination of modern STEM-EELS (Nelayah et al., 2007; Gass et al., 2006).

For core-loss spectrum, the elemental analysis can be performed and the structural information can be obtained from the energy loss near edge spectrum (ELNES). The ELNES of an ionization edge represents approximately a local densities of states at the atom giving rise to the edge, and thus can serves as a coordination fingerprint when applied to minerals and organic complexes. The fine structure changes with the orientation of the anisotropic specimen relative to the incident beam (Jiang et al., 2002). The atom location by channeling enhanced microanalysis (ALCHEMI) method is based on planar channeling, can be used to determine the doping site of the transitional metal elements in the dilute magnetic semiconductor matrix (Zhang et al., 2009). The changes in the threshold energy of an ionization edge between different atomic environments is named as chemical shift, which can provide information about the charge state and atomic bonding in a solid. A radial distribution function (RDF) specifying inter-atomic distances relative to a particular element can be derived by Fourier analysis of the extended fine structure (EXELFS), starting about 50 eV beyond an ionization edge threshold. The L_2 and L_3 edges of transition metals are characterized by white-line peaks at the ionization threshold. The energy separation of these peaks reflects the spin-orbit splitting of the initial states of the transition and their prominence (relative to the higher E continuum and to each other) varies with atomic number Z. As a new applications of EELS, spin-state measurements have been performed by energy-loss magnetic chiral dichroism (EMCD). The feasibility of EMCD measurement was firstly proposed in 2003 (Hébert and Schattschneider, 2003), and experimentally demonstrated in 2006 in single crystalline Fe thin film epitaxially grown on a GaAs substrate (Schattschneider et al., 2006). The equivalence of the momentum transfer q in EELS to the linearly polarization vector ε in X-ray absorption spectroscopy (XAS) (Stöhr, 1999) allows the magnetic circular dichroism to be measured in the absence of spin polarized electron beam. Two requirements must be satisfied to enable the EMCD measurement. Firstly, two chiral positions for spectra acquisition must be generated. This can be achieved by tilting the specimen to a two-beam condition, as illustrated in Fig. 5.2.5. The "+" and the "−" position in the diffraction pattern are lined up with a mirror symmetry with respect to the g reciprocal lattice vector. They are equivalent to the result generated by the left and the right circular light in X-ray magnetic circular dichroism (Stöhr, 1999). EELS acquired at "+" and "−" thus correspond to opposite chirality conditions due to the opposite sign of momentum transfer (q) at "+" and "−" positions, respectively. Secondly, magnetization of the specimen must have its net magnetic moment aligned to a direction being parallel/antiparallel to the incident electron beam, to achieve a maximum dichroism (Stöhr, 1999). The initial magnetization of the samples can be easily achieved with the TEM objective lens pole piece, which gives approximately 2T magnetic field, being large enough to saturate the magnetic materials. Since the first work about EMCD in 2006, many methods (Verbeeck et al., 2010; Stöger-Pollach et al., 2011) have been tried to improve the signal/noise ratio.

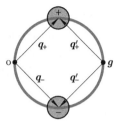

Fig.5.2.5 Two chiral positions of a two-beam condition in the diffraction pattern for EMCE spectra acquisition.

Energy-filter TEM (EFTEM) makes the TEM having the ability to display two-dimensional distributions of specific elements (also known as elemental mapping). Spectrum-imaging software makes it easy to obtain an EFTEM image. At least two images have to be recorded before and after the ionization edge during the experiment. The simplest procedure is to subtract the pre-edge and post-edge images. This two-window procedure works well enough for edges with high jump ratio but is unsatisfactory when quantitative results are required. Another simple procedure involves dividing the post-edge and pre-edge images, yielding a jump-ratio image that is largely insensitive to variations in specimen thickness and diffracting conditions. Three-window method (Crozier, 1995) is more appropriate to the quantitative elemental mapping. However, three-window modeling produces a noisier image than the two-window methods, so a good strategy is to first acquire a jump-ratio image from the area of interest, requiring a relatively short exposure time. If the results are encouraging, the three-window method can then be used to obtain a more quantitative elemental map. With the development of aberration corrected TEM, atomic resolution images have been obtained (Muller et al., 2008).

5.2.6 Artifact of EDS and EELS

Although the operating conditions are optimized for both EDS and EELS, however some artifacts are still introduced into the spectrum. For example, "escape" peaks and "internal fluorescence" peaks are appeared during the EDS signal-detection course. "Sum" peaks may occur during the EDS signal processing. For EELS, the relative low energy resolution makes the signal near to the band gap invisible, and some approximations have to be adopted to derive the true information. All the individual diodes will differ slightly in their response to the incident electron beam, therefore there will be a channel-to-channel gain variation in intensity. Gathering many spectra and superimposing them can bring the readout noise problem. There may be the problem of incomplete readout of the display. The operator should be aware of all the artifacts in the EDS and EELS spectra.

5.3 EDS+TEM and EDS+STEM

Yonghai Yue

5.3.1 Why we combine TEM/STEM with EDS

As we known, TEM just gives us 2D projected images of thin 3D specimens and you may not identify the nature of different specimens simply from a series of 2D images taken by a TEM imaging system even though you have become an excellent microscopist. This is accomplished by X-ray spectrometry, which is one way to transform the TEM into a far more powerful instrument, called an analytical electron microscope (AEM). After EDS was induced into a TEM, you can easily identify different samples with there different spectra, it will be possible to obtain information details in a matter of minutes or even seconds. On the other hand, taking the advantage of the high spatial resolution of TEM, elemental analysis focused on microscopic region became possible. With the development of aberration-corrected STEM and efficient detectors, combined with EDS or EELS, elemental maps showing individual atomic columns in a crystal are now feasible. TEM/STEM combined with EDS will form a more powerful tool in materials' characterization.

5.3.2 EDS configuration in TEM

Conventional EDS systems include a sensitive X-ray detector, a liquid nitrogen dewar for cooling, and software to collect and analyze energy spectra. The detector is mounted in the sample chamber of the main instrument at the end of a long arm, which is itself cooled by liquid nitrogen. Today, TEMs constitute arguably the most efficient and versatile tools for the characterization of materials. EDS was available as an option on many TEMs and even more widespread on other electron beam instruments such as SEMs. Take the advantage of the high spatial resolution of TEM, elemental analysis focused on microscopic region became possible. The modern energy-dispersive X-ray spectrometer, when fitted to a transmission electron microscope (Fig. 5.3.1), is capable of detecting the characteristic X-rays of all the elements present in the sample which are of greater atomic number than sodium. Fig. 5.3.1(a) shows the configuration of conventional EDS cooled by liquid nitrogen in TEM. Fig. 5.3.1(b) shows the cross section of a Si(Li) detector of the EDS. In order to compensate the impurities of commercial Si, the intrinsic zone in Si is generated by the diffusion of Li atoms. When X-rays deposit energy in the semiconductor, electrons are transferred from the valence band to the conduction band, creating electron-hole pairs. Since characteristic X-rays typically have energies well above 1 keV, thousands of electron-hole pairs can be generated by a single X-ray. The number of electrons or holes created is directly proportional to the energy of the X-ray photon. It is difficult to confirm all the X-ray energy can be converted to electron-hole pairs, but it is enough to collect sufficient signal to distinguish most elements in the periodic table. After the creation of electrons and holes, the electron detectors will separate them by an internal reverse bias across a very narrow p-n junction.

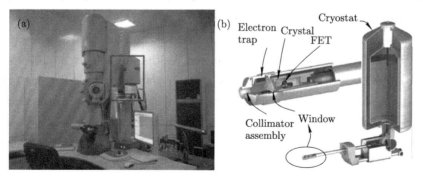

Fig.5.3.1 Image shows the configuration of EDS in TEM. (a) Image of Tecnai F20 with EDS; (b) Cross section of a Si(Li) detector.

The incoming X-ray photons will produce pulses in Si(Li) detector, then the detector will give output pulses proportional to the X-ray energy. These pulses are sorted in a multi-channel analyzer to form a histogram or X-ray spectrum. Every photon arriving at the EDS detector is counted, regardless of its energy, so that this serial device has many of the advantages of a parallel detection system. The performance is limited, however, by the dead-time of the detector and by the maximum count rate which can be processed. This limit may be reached in a channel other than that of interest, thereby limiting the count rate in the channel of interest. Excitations in the silicon detector itself may also give rise to spurious peaks. An electron-probe-forming lens is required for high spatial resolution. This requirement, together with the side-entry stage needed for EDS (and possible mechanical instabilities produced by the X-ray detector and its liquid-nitrogen dewar) may conflict with the requirements for HREM work. Through the use of the symmetrical condenser-objective lens, together with a side-entry stage, however, some impressive design compromises have been achieved. Lighter elements may be detected using special windowless detectors or ultra-thin window detectors (Williams and Joy, 1984).

5.3.3 Energy resolution

Referring to resolution, we can remember several different "resolutions", such as time resolution, spatial resolution and so on, Here, for EDS, we must talk about its energy resolution which means the ability to identify elements by distinguishing their spectral peaks at different energies. The natural line width of the emitted X-rays is only a few eV but the measured widths are usually $\gg$ 100 eV. The electronic noise in the EDS system is a major source of the difference between the practical and theoretical energy resolutions and the width of the electronic noise is described as the "point-spread function" of the detector. The poor energy resolution of EDS (typically $\sim 135 \pm 10$ eV) is a major limitation of any kind of semiconductor detectors, we need to examine this concept more closely. There will be energy peak overlaps among different elements, particularly those corresponding to X-rays generated by emission from different energy-level shells (K, L and M) in different elements.

Excellent energy resolution specifically in the low energy range is essential for ambitious tasks in nano-analysis. For silicon-based detectors, energy resolution typically changes with line energy. The energy resolution is commonly measured on peaks recorded from Mn K_α radiation and expressed in terms of the full width at half maximum (FWHM). Energy resolutions for detectors are usually specified in accordance with ISO 15632:2002, where the energy resolution of the K_α peaks of manganese (Mn), carbon (C) and fluorine (F) must be stated at 1000 or 2500 cps. The smaller the value, the better the energy resolution of the detector.

5.3.4 Development of EDS

The silicon drift detector (SDD) technology is a type of energy dispersive solid state detector which was first commercially introduced by Bruker. Silicon drift detectors (SDDs) have been gaining in popularity over the past few years due to improvements in peak stability over time and when subjected to a wide range of beam currents. It utilizes a special drift field structure to guide charges produced by absorbed X-rays to an extremely small readout anode (As shown in Fig. 5.3.2). The structure is optimized for each type of SDD-smaller droplet (Fig. 5.3.2A) and larger round detector (Fig. 5.3.2B) for efficient charge transportation. The obvious advantages of SDDs over Si(Li) detectors include the fact that they are liquid nitrogen (LN$_2$) free, have no moving parts and also have the ability to process large numbers of counts without flooding the detector. A new generation of large area, high resolution SDDs is now available for SEMs and TEMs and these detectors are available in sizes up to 80 mm^2 for a single sensor.

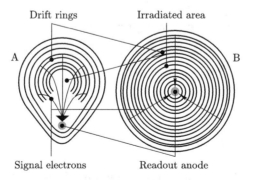

Fig.5.3.2 Sketch map for the special drift field structure of SDD.

Recently, a new type of EDS system named XFlash® 6 have been marketed by Bruker. Fig. 5.3.3 shows the image taken from XFlash® 6T for TEM. It uses an on-chip amplifier to achieve energy resolutions suitable for EDS element analysis with only moderate cooling, achieved with thermoelectric Peltier coolers. The drift fields, the low capacitance of the readout anode, and the integrated FET ensure the excellent energy resolution and high speed of Bruker's detectors. It also achieves the largest solid angle detectors for TEM and S/TEM offer minimum mechanical and electromagnetic interference. They provide optimum take-off angle and avoid the necessity of sample tilt. Fig. 5.3.4 shows the best energy resolution range (BERR) of different detectors, we can see XFlash® 6 combines the best available detector technology with an optimized and well adapted hybrid signal processor.

Fig.5.3.3   XFlash® 6T for TEM.

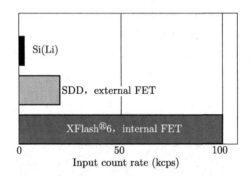

Fig.5.3.4   Best energy resolution range (BERR) of different detectors.

5.3.5 TEM sample effect

In TEM, most samples are thin foils or small particles, so it is therefore sufficient to discuss the count rate N of characteristic X-ray lines from thin films. Certain corrections (such as Atomic Number Correction, Absorption Correction, Fluorescence Correction) of the formula will be necessary for X-ray emission from thin foils. The readers are recommended to read the book of Reimer and Kohl (2008). Electron waves in single crystals are Bloch waves with nodes and antinodes at the nuclear sites depending on the crystal orientation. So the influence of crystal orientation should also be considered. In TEM, the tilt can be determined exactly from the location of Kikuchi lines in a simultaneously recorded diffraction pattern. This method can also be expanded to electron energy-loss spectroscopy by measuring the EELS signal intensity at ionization edges (Tafto and Krivanek, 1982; Tafto and Lehmpfuhl, 1982).

An important problem for X-ray microanalysis (XRMA) of thin films in TEM is the unwanted contribution from continuous and characteristic quanta generated not in the irradiated area but anywhere in the whole specimen and specimen cartridge by electrons scattered at diaphragms above and below the specimen and from X-ray fluorescence due to X-ray quanta generated in the column. Additional diaphragms have to be inserted at suitable levels to absorb scattered electrons and X-rays, whereas the objective diaphragm should be removed during XRMA. The number of unwanted quanta can be further reduced by constructing the specimen holder from light elements such as Be, Al, or high strength carbon (Bentley et al., 1979).

5.3.6 EDS+TEM

Many commercial systems have been developed to acquire spectra and images, and many of these in terms of the file formats used within the computer system. Next, we will take the conventional EDS based on Si(Li) detectors as a typical exam to show how it works combined with TEM. Fig. 5.3.5 shows the sample geometry for analysis with an EDX detector, the position of the EDS detector is above the specimen to minimize the background signal in the spectrum. The sample is typically tilted to an angle between 15° and 30° towards the detector to minimize absorption of low-energy X-rays within the sample. Therefore, even sample regions not hit by the electron beam or parts from the pole pieces near the sample can contribute to EDX spectra due to the generation of additional X-rays, which may be due to the excitation of atoms by the X-rays and secondary electrons generated in the illuminated volume. Thus, particular care has to be taken to account for a possible excitation of X-rays in neighboring areas in quantitative EDX analysis.

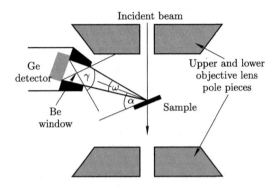

Fig.5.3.5 Sample geometry for analysis with an EDX detector (Petrova, 2006).

EDS is well suited for the detection of heavy elements. The EDX spectra can be obtained in the following formats:
- Point measurement: This is the spectra obtained by measuring the X-rays in a very small area (down to 1 nm diameter with reasonable count rate is possible) of the sample.
- Line scan: This is the spectra obtained by positioning the small electron probe sequentially on points along a line. For each point position an EDX spectrum is obtained.
- Area scan/map: This is a spectra obtained by positioning the small electron probe sequentially on points on a two-dimensional grid to form a map of the area.

When used in "spot" mode, a user can acquire a full elemental spectrum in only a few seconds; for line scan mode, we can use this mode to analysis the chemical composition along a preset line, which can easily detect the composition distribution on the interface or boundary. Area scan mode is a more powerful mode which can easily offer a whole map of composition distribution of your materials, combined with STEM mode, you can got more details about the material.

Fig. 5.3.6 shows a typical example to examine the chemical composition of Cu-Zn-Se polycrystalline film both in the interior of its constituent grains and at the grain boundaries between them (Chen et al., 2014).

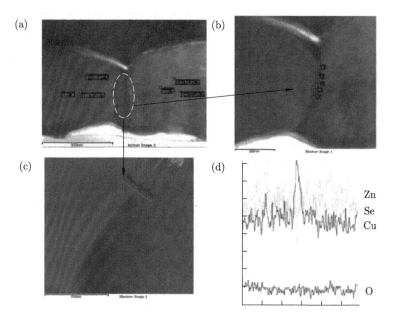

Fig.5.3.6 Ways of TEM/EDS analysis used: (a) EDS point analyses within two neighboring grains, (b) EDS point analyses along the grain boundary, (c) EDS scanning across the boundary, and (d) the magnified view of the resultant profiles in (c).

5.3.7 EDS+STEM

STEM was first developed by A. Crewe and co-workers at the University of Chicago (Crewe, 1980) who used it to obtain the first electron microscope images of individual atoms in 1970 (Crewe et al., 1970). Unlike a TEM image, STEM operates as a SEM, obtain an image by scanning a fine probe over a sample. The first image of isolated heavy atoms on a thin carbon film was obtained at 30 kV.

There are three main imaging modes in STEM distinguished via different detectors.
- Bright-field STEM: use a small axial detector;
- Dark-field STEM: use an annular detector with a small inner cutoff, collecting elastic scattering to form an image;

- High-angle annular dark-field (HAADF) Z-contrast mode: use a larger inner and outer detector cutoffs to reduce the contribution of elastic Bragg scattering and increase the contribution of localized thermal diffuse scattering (TDS).

During the literature, the last two modes are often not distinguished, we usually use "ADF-STEM" to define them. Since ADF-STEM images show a strong dependence on atomic number, so coupled with EDS, we can get more details comparing with TEM plus EDS. Fig. 5.3.7 shows the sketch map of EDS coupled with STEM. For a STEM DF image, the scattered electrons fall onto the ADF detector. This gives rise to a fundamental difference between the TEM and STEM DF modes. DF TEM images are usually formed by permitting only a fraction of the scattered electrons to enter the objective aperture. STEM images are formed by collecting most of the scattered electrons on the ADF detector. Therefore, STEM ADF images are less noisy than DF TEM images. Because lenses are not used to form the STEM image (although they are used to form the probe), the ADF images do not suffer aberrations, as would the equivalent off-axis TEM DF image.

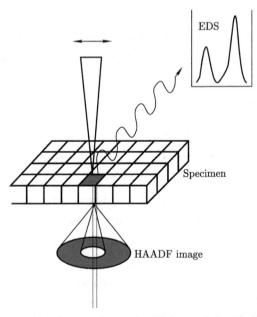

Fig.5.3.7 Sketch map shows the EDS coupled with STEM.

Since TEM or STEM is used, the method can be applied to areas as small as a few hundred angstroms in diameter, while the detection sensitivity is limited by that of the EDS system to about 0.1 atomic percent. Elements which are neighbors in the periodic table can normally be readily distinguished.

Using ABF-STEM coupled with EDS, we can easily get the chemical distribution information of the sample. In order to investigate the structural stability of S/(Carbon Nano-Tube@MicroPorous Carbon), Guo and Wan (2012) employed TEM characterizations on the S/(CNT@MPC) cathode after 200 cycles at 0.1 °C in the carbonate electrolyte. Fig. 5.3.8 shows an ABF-STEM image taken from the MPC region which shows that short-chain-like sulfur molecules including S_{2-4} are

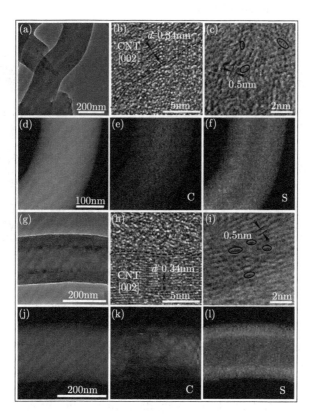

Fig.5.3.8 Structural characterization of S/(CNTMPC) before and after 200 cycles at 0.1 °C.

well dispersed in the micropores of MPC. Coupled with elemental mapping which reveals a uniform distribution of sulfur in the MPC layer.

Over the past few years, at least two high voltage HREM machines have been built with a resolution close to one angstrom, while image processing techniques and aberration correctors in both the TEM and STEM mode have brought the same resolution (or better) to medium voltage machines. STEM, with its fine probe and multitude of detectors for spatially-resolved spectroscopy, is the ideal instrument for the study of inorganic nanostructures. With the newly developed aberration-corrected STEM technique (Huang et al., 2011; Gu et al., 2011; He et al., 2011; Shao-Horn et al., 2003) equipped with ADF and HADDF detectors. Fig. 5.3.9 shows the spinel structure of LTO.

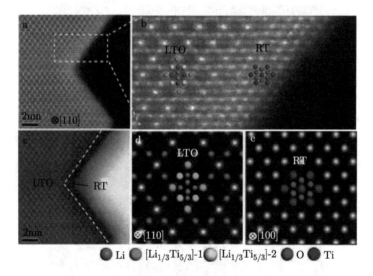

Fig.5.3.9 STEM images of LTO-RT-600: (a, b) HAADF and (c) corresponding ABF images of LTO-RT-600 NSs. The simulated HAADF images of (d) LTO projected from [110] direction and (e) rutile-TiO$_2$ projected from [100] direction. The insets show the arrangements of atoms.

EDS coupled with STEM can easily distinguish different composition because it is easy for STEM to distinguish light and heavy atoms from the difference of contrast. Fig. 5.3.10 shows the image of CuO-coated multi wall carbon nanotubes with Co-catalyst particles analyzed with the XFlash® 6T |30 (Bruker), the spot size can be down to 1.5 nm, beam current is 400 pA. The measurement was performed with two different acquisition times of 3 min.

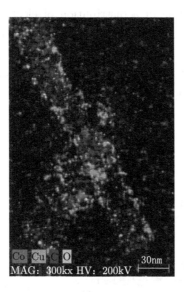

Fig.5.3.10 CuO-coated multi wall carbon nanotubes with Co-catalyst particles analyzed with the XFlash® 6T |30. (Herrmann, S. and Wachtler, T.)

5.4 EELS-TEM

Jiaqing He

Since the EEL spectra contain large amounts of information from specimen as stated in details in the section above, analytic electron microscopes have almost been equipped with spectrometer to acquire the EEL spectra in TEM or STEM mode. On the other hand, energy filtering devices have also been equipped to the electron microscopy to display the feature of EELS through imaging, such as energy filtering TEM (EFTEM) or STEM spectrum imaging (SI).

Generally, the commercial EEL spectrometer is Gatan PEELS, and is always equipped below the viewing screen (post-column), as illustrated in Fig. 5.4.1

(Williams and Carter, 2009) The project crossover acts as the object point or plane, and the electrons passing through the selection entrance aperture are deflected by the magnet prism through the drift tube, and forms a spectra in the dispersion plane, consisting of a distribution of electron intensity versus energy loss. Electrons with different energy (i.e. different energy loss) will be deflected in the different point in the dispersion plane. Finally, a parallel acquisition system (unlike serial detection) gathers the information from the dispersion plane and forms the EEL spectra through CCD camera.

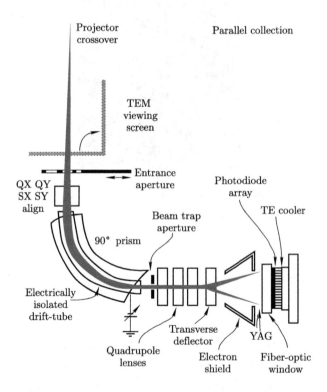

Fig.5.4.1   Schematic illustration of Gatan PEELS.

In a conventional TEM, the EELS can be obtained in image and diffraction modes, the ray paths of two modes are shown in Fig. 5.4.2 (Egerton, 2011).

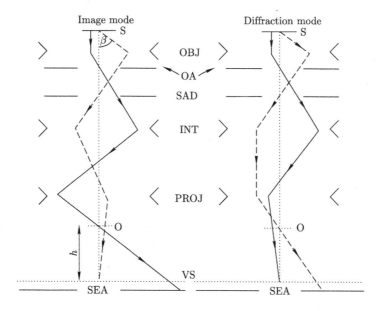

Fig.5.4.2 Ray paths of TEM image mode(left) and diffraction mode(right).

5.4.1 Image mode

If the TEM is operated in *image mode* (also diffraction coupled mode for the project lens crossover contains a small diffraction pattern), with an image of the specimen of magnification M on the viewing screen. The size of this diffraction pattern is represented by a camera length: $L_o = h/M$, and can be as small as 1 μm.

The collection semi-angle β is controlled by varying the size of the objective lens aperture. The region of specimen giving rise to the energy-loss spectrum is determined by a spectrometer entrance aperture (SEA) and corresponds to a portion of the image close to the center of the TEM viewing screen (before the screen is lifted to allow electrons through to the spectrometer).

More precisely, the diameter of analysis is $d = 2R/M'$, where R is the SEA radius and $M' = M(h'/h)$ is the image magnification at the SEA plane, h' being height of the projector lens crossover relative to the SEA.

This mode is always used for rapid analysis.

5.4.2 Diffraction mode

If the TEM is operated in *diffraction mode* (also image coupled mode for the projector lens crossover contains an image of specimen), with a diffraction pattern of camera length L at the viewing screen. The image magnification at O is $M_o = h/L$ and is typically of the order of 1. Unless the objective aperture limits it to a smaller value, the collection semi-angle is $\beta = R/L'$, where $L' = L(h'/h)$ is the camera length at the SEA plane. To ensure that the SEA is centered on the optic axis, TEM diffraction shift controls have to be adjusted for maximum intensity of some sharp spectral feature. Alternatively, these controls can be used to select any desired region of the diffraction pattern for energy analysis. The area of specimen being analyzed is determined by the electron-beam diameter at the specimen or else by selected area diffraction (SAD) aperture, if this aperture is inserted to define a smaller area.

This mode is always combined with STEM.

5.4.3 Comparisons

Keast (2001) has summarized the differences between the three modes as Table 5.4.1 (STEM mode to be discussed in next section).

Table 5.4.1 Important parameters and considerations for the collection of EEL spectrum under different modes of microscope operations.

	TEM diffraction coupled	TEM image coupled	STEM
Incident illumination	parallel	parallel	Converged probe
Object plane of spectrometer	Diffraction pattern	image	specimen
Selection of	Spectrometer	Selected area	Electron

	TEM diffraction coupled	TEM image coupled	Continued STEM
sample area	entrance aperture and magnification	aperture or electron beam	beam
Collection defined by	Objective aperture	Spectrometer entrance aperture and camera length (without objective aperture)	Collector aperture
Effect of chromatic aberration	Error in area of analysis	small	small
Effect of spherical aberration	Small for small α and β	Small	Incident beam only

Because of the importance of the object plane and the collection angle and resolution for electron optics, next we summarize the three items in different microscopy operations for EELS.

1. The Object Plane

As with any lens, the spectrometer takes electrons emanating from a point in an object plane and brings them back to a point in the image (dispersion) plane. The object plane of the spectrometer or filter depends on the detail of the machine you are using, as listed in Table 5.4.2. Obviously, for all modes except dedicated STEM with no post-specimen lens, the project lens crossover, namely the back-focal plane, acts as the object plane. In different modes, there are different things (diffraction pattern or image) in the back-focal plane. For the dedicated STEM without any post-specimen lens, the object plane is the specimen itself.

Table 5.4.2 The object plane in different microscope operations

Mode	Object plane
Dedicated STEM with no post-specimen lens	The plane of the specimen
TEM/STEM, or DSTEM with post-specimen lens	The back-focal plane of the projector lens (post-column)

2. Collection Angle

The collection angle always determines what extent of electrons we can obtain after interactions with specimen. If it is too large, too much noise or unexpected information will be collected, and on the other hand, too small, the gathered electrons will be too small to make a right decision.

Usually, first of all, we should know the scattering angle of typical electrons (e.g. scattered by inner shell electrons). The typical energy-loss electrons (producing plasmon, interband transition, phonons and ionizations) have its characteristic scattering angle. We should adjust the operational collection angle to gather large extent of electrons that we desire. Table 5.4.3 lists the typical characteristic scattering angle and energy loss.

Table 5.4.3 Typical electron energy-loss interactions and its scattering angles

Type of electron scattering	Angular range/mrad	Energy loss/eV
Unscattered	0	0
Elastic scattering	10~100	~0
Phonons	10~100	< 1
Interband transition	1~2	< 10
Plasmon	< 1	5~30
Ionization	10~100	10~1000

The collection angle in different microscopy operations is determined by different conditions (e.g. objective aperture, spectrometer entrance aperture) and can be calculated as listed in Table 5.4.4. For details of deduction, we can read Chapter 37 of Williams (2009). For the TEM image mode, if no objective aperture is inserted, the collection angle will be too large to gather all the electrons that will lack the sincerity of information from EELS and thus, this mode is always used for rapid analysis.

Table 5.4.4 Calculations of collection angle β under microscope operations

Mode	Collection angle β
DSTEM with no post-specimen lens	$\approx d/2h$ (d-diameter of the entrance aperture, h-distance from the specimen to the aperture)
TEM-image mode	(no objective apertures) too large ($> \sim 100$ mrad) $\approx d/2f$ (d-diameter of the objective-aperture, f-the focal length of the objective lens)
TEM/STEM diffraction mode	(D-distance between the project crossover and the recording plane, D_A-distance from the crossover to the entrance aperture, d-diameter of the entrance aperture, L-camera length)

3. Resolution

Egerton has summarized the spatial resolution and energy resolution in EELS in different modes in his book (2011), and main conclusions are listed below.

Only taken chromatic aberrations into account, in the TEM image mode, analysis area determined by the selection aperture (e.g. spectrometer entrance aperture or objective aperture) cause a greater error in the analysis area; while for analysis area determined by the electron beam (see the formula below, TEM diffraction or STEM mode), the error of analysis area is small.

$$\Delta r_c \approx C_c \alpha \frac{\Delta E_0}{E_0} \tag{5.4.1}$$

where C_c is the coefficient of chromatic aberrations, α is the convergence angle, ΔE_0 is the energy spread of electron beam.

The energy resolution can be affected by the energy spread of the incident electrons before reaching specimen, the broadening by the spectrometer's optics, and the spatial resolution of the electron detector.

$$(\Delta E)^2 \approx (\Delta E_0)^2 + (\Delta E_{so})^2 + (s/D)^2 \qquad (5.4.2)$$

where D is the spectrometer dispersion.

For the double-focusing spectrometer, the image formed is a convolution of the EELS and the DP or the image of specimen. The second term can be explicitly expressed as below:

$$(\Delta E_{so})^2 \approx (M_x d_0/D)^2 + (C_n \gamma^n/D)^2 \qquad (5.4.3)$$

Where M_x is the magnification of an ideal spectrometer, d_0 is the diameter of the object, C_n is the n-th order aberrations, γ is the divergence semi-angle of the beam entering the spectrometer.

For the operational mode without SEA, the broadening can be optimized by adjusting the lens through changing the divergence semi-angle of the beam γ and the diameter of analysis area. But for modes with SEA, the divergence semi-angle of the beam entering the spectrometer is restricted ($M'd/2 = R$ in image mode, $\beta M' = R$ in diffraction mode). But for lower magnification can be adjusted by reduce β in image mode or analysis area diameter in diffraction mode.

5.4.4 EFTEM

Except for spectra featuring the information of specimen, we can also obtain images to display these feature spectra by using the imaging or diffraction capabilities of TEM. An image-forming spectrometer is used as a filter that accepts energy losses within a specified range, giving an energy-selected image or diffraction pattern. First, let's introduce the two commercially energy filters: GIF and Omega Filter (Williams and Carter, 2009; Egerton et al., 2011).

1. Gatan Image Filter

The GIF for energy filtering is nearly the same for spectroscopy, but with energy selection slit and the post-spectrometer optics, see Fig. 5.4.3(a). The energy selection slit is placed on the dispersion plane (EELS) to select a specified energy range, and the post-spectrometer optics is equipped after the energy selection slit as shown in Fig. 5.4.3(a), aiming to magnify the energy filtering images and correct the imaging defects (e.g. second order aberrations).

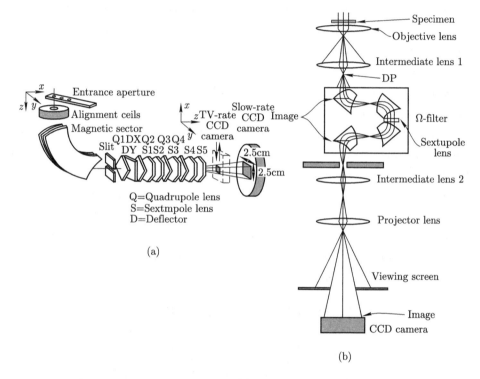

Fig.5.4.3 schematic illustrations of GIF (a) and Omega (b) energy filtering imaging (Williams and carter, 2009; Egerton et al., 2011).

2. Omega Filter

The omega filter was specifically designed to obtain energy filtering images. The omega filter is integrated into the TEM and sits between the intermediate and the projector lenses and consist of a set of magnetic prism arranged in an Ω shape, see Fig. 5.4.3(b). The recording CCD detectors only receive electrons that have come through the filter. So all images or DPs only consists of electrons of a specific selected energy. So in the Ω filter, information we get is always in the filtered mode. In TEM image mode, we usually project an image into the prism which is focused on a DP in the back focal plane of the intermediate lens, and thus, the entrance aperture to the spectrometer selects an area of specimen and β is governed by the objective aperture. Electrons following a particular path through the spectrometer can be selected by the post spectrometer slit. Therefore, only electrons of a given energy range, determined by the slit width, are used to form the image projected onto the TEM CCD.

Using these two filtering devices, we can obtain energy filtering images formed by electrons with a specified energy.

Energy-filtering TEM can filtered out inelastically scattered electrons that produce background "fog" to improve contrast in images or DPs, and more importantly, can create elemental maps by forming images with inelastically electrons.

By operating the spectrometer in a diffraction-coupled mode and adjusting the spectrometer excitation or accelerating voltage so that the zero-loss peak passes through the energy-selecting slit, Egerton obtained a zero-loss image which was produced with greater contrast and/or resolution than the normal (unfiltered) image (Egerton, 1976), as shown in Fig. 5.4.4.

Fig.5.4.4   (a) Unfiltered and (b) zero-loss micrographs of a 40 nm epitaxial gold Film. Energy filtering increased the crystallographic contrast by a factor of 2.

Dieterele et al. (2008) used EFTEM method of elemental mapping, Fig. 5.4.5, to analyze the evolution of the local chemical composition and element distribution on the nanoscale of nanocrystalline $La_{0.5}Sr_{0.5}CoO_{3-\delta}$ (LSC) thin films deposited with Sr under different conditions.

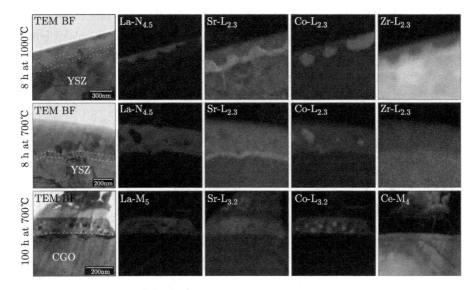

Fig.5.4.5 EFTEM of chemical compositional mapping of $La_{0.5}Sr_{0.5}CoO_{3-\delta}$.

5.5 EELS-STEM and applications

Jiaqing He

5.5.1 Basic theory of EELS-STEM

Except for TEM mode, EELS can be also obtained in STEM mode. In EELS-STEM mode, electrons of high angle scattering will be collected by an annular detector to form the atomic number contrast to reveal the compositional distribution; on the other hand, electrons of low angle scattering (almost inelastic) will come through the EEL spectrometer to form energy loss spectra to show the electronic structures (Keast et al., 2001; Egerton, 1976). Meanwhile, for sake of the atomic resolution of the STEM probe (even higher resolution with spherical aberration corrected), the EEL spectrum can be even with atomic resolution, and thus we can obtain detailed information from area in the sub-angstrom scale other than the average consequence from a large analysis area, which will be of great advantage to investigate the interface. Batson (1993) employed this mode to detailed characterize the Si-SiO$_2$ interface, in Fig. 5.5.1(b), showing a strong Si^{2+} signal at the interface. Not only the atomic sites and distribution can be observed, but also the details of bonding and electronic structures can be resolved. Muller et al. (1993) has also mapped sp^2 and sp^3 states at the interface using STEM-EELS method at sub-nanometer scale.

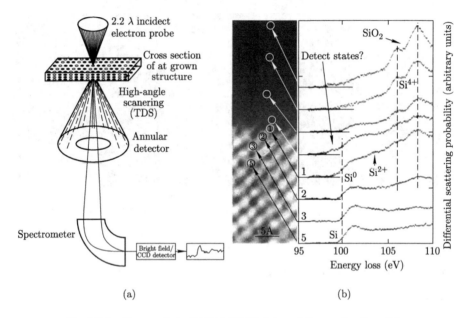

Fig.5.5.1 Ray path in STEM-EELS (a) and its application (b).

In EELS-STEM, the object plane and the collection angle has been listed in above section, and the analysis area is determined by the electron beam; the spatial resolution is determined by the electron beam diameter (no selected area aperture) as in TEM diffraction mode, which can also be affected by the chromatic and spherical aberrations, for dedicated STEM, independent of energy loss. For STEM with post lens, the spatial resolution remains the same; the energy resolution can be compared to that in TEM diffraction mode: the effect of size of electron beam can be negligible if with a slight defocus, and the aberrations term by spectrometer can be minimized by the appropriate spectrometer design, and if keep the magnification small with the order of 10, the total energy resolution can be as good as 0.5 eV (Scheinfein and Isaacson, 1984).

While in STEM mode, for sake of nano probe forming and the atomic resolution, we can also scan over an area of interest and captures an EEL spectrum at every pixel. This is term spectrum imaging, unlike EFTEM, which is the image formed by a specified energy-loss electrons, as Fig. 5.5.2 shows (Verbeeck et al., 2004).

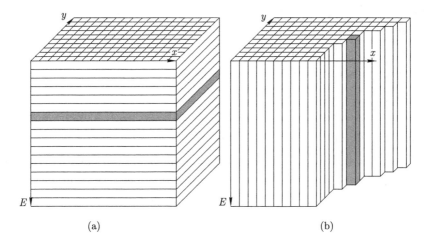

Fig.5.5.2 Differences between (a) EFTEM and (b) STEM

Meanwhile, we can also select a specified energy to create compositional maps or even store a full spectrum at each pixel. Thus, we can obtain a three dimensional data cube, sampled by taking a large set of energy filtered images with a small energy slit (Verbeeck et al., 2004). Each image is taken at a slightly different energy loss, sampling the energy loss axis in a certain region of interest.

5.5.2 Applications of EELS

If we want to make full use of the EELS, firstly we should pre-processing such as background subtraction and removing the plural scattering effect, which has stated in details in Egerton's book (2011). This section will not cover this part, and we only list some important applications in characterization of physical and chemical information.

For sake of complexity in EELS, different regions can tell us different information from the specimen and instruments. Table 5.5.1 below gives us an overview of its application (Colliex, 1996).

Table 5.5.1 Different types of information accessible through a detailed analysis of an EEL spectrum.

Spectrum domain	Information accessible	Required processing technique	Field of application
Whole spectrum, ZLP	Thickness, total inelastic scattering	Unsaturated zero-loss peak	Very general
Low-loss region	Average electron density	Plasmon-line properties	Metallic alloys
Low-loss region	Interband transition/joint density of states	Kramers-Kronig transformation, calculation of dielectric constants	Optical and transport properties, comparison with VUV spectra, intergranular Van der Waals forces
Low-loss region	Interface/surface properties	Study of the interface plasmon modes	Interface and boundary structure and chemistry
Core-loss region	Qualitative and quantitative chemical analysis	Core-edge weight	Nanoanalysis in any type of materials
Core-loss region	Site-symmetry, bonding type and bond length	Core-ELNES, comparison with fingerprints, molecular orbital, multiple scattering or band structure calculation	Site-selected valence state, charge transfer, bonding and structure environment

Continued

Spectrum domain	Information accessible	Required processing technique	Field of application
Core-loss region	Radial distribution function	Core-EXELFS	Site-selected crystal coordinate
Core-loss region	Density of holes in local sites	White-line intensities	Electron configuration in intermetallic, insulators, superconductors, magnetic properties

The EELS has always been made use to qualitative and quantitative analysis of elements, which has been discussed in section 5.2. Here we list some applications in sample thickness determination and electronic structures identifications using different regions of the spectra.

1. Thickness measurement

Generally, we can use the log-ratio method to measure the sample thickness:

$$t = \lambda_p(\beta) \ln(I_T/I_0) \qquad (5.5.1)$$

where t-specimen thickness, λ_p-plasmons' mean free path of inelastic scattering, I_T-intensity of electron beam, I_0-intensity of zero loss peak (Egerton and Cheng, 1987).

If we know the value of mean free path, then we can calculate the specimen thickness. Meanwhile, Oikawa et al. employed this method inversely to make use of the equal thickness fringes to obtain the mean free path as shown in Fig. 5.5.3.

2. Chemical shift

When two atoms form the ionized bonds, the anode loses an electron and thus becomes more attractive to electrons nearby, making the electronic orbit lower into the deeper level; and in turn, the electronic orbit of anion will be in the higher

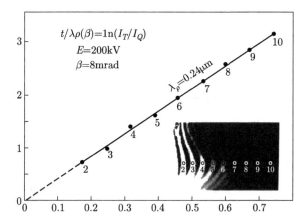

Fig.5.5.3 Sample thickness determination using log-ratio method.

level. This level change will generate a slight shift in the EEL spectrum, and this shift is defined "chemical shift".

Kimoto et al. (1997) measured the chemical shift of Si $L_{2,3}$, L_1 and K edge in SiO_2/Si_3N_4 deposition at spatially resolved EELS-TEM.

Otten et al. (1985) also applied EELS to determine the oxidation state of transition metal Ti, Mn and Fe in minerals, and simultaneously determined the chemical shifts as shown in Fig. 5.5.4. Obviously, the more electrons are lost from a single atom or the higher valence the atom is, the more bound electrons are to the atom, and thus the more energy incident electrons loss.

3. Band gap

Low-loss region contains information from not only plasmons but also interband transition. For semiconductors, the least energy for interband transition should be the band gap. If the EELS have a good enough energy resolution, we can obtain the threshold of interband transition in low-loss region. Through the low-loss region in the EELS with spatial resolution of sub-angstrom and energy resolution of (0.5∼ 0.2 eV) can be advantageous over optical spectroscopy that lacks in spatial resolution in micron scale, but a slightly better energy resolution about 1 meV.

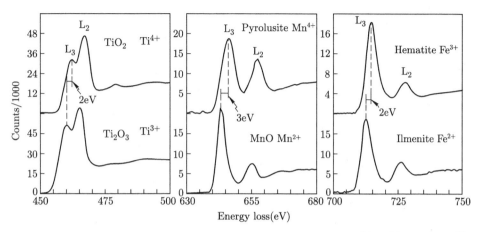

Fig.5.5.4 Chemical shifts as a function of oxidation state of Ti^{3+}-Ti^{4+}(2eV), Mn^{3+}-Mn^{4+}(3eV), Fe^{2+}-Fe^{3+} (2eV).

Rafferty et al. (2000) measured the band gap of MgO cube with the electron beam along the [100] and [110] direction through deconvolution and background subtraction to remove the tail of ZLP in the EEL spectrum in STEM with resolution of 0.22 eV and 0.5 nm, and proved the former should be more effective to remove the ZLP tail effect, and the threshold in the low-loss region indicates the band gap, in Fig. 5.5.5.

Kimoto et al. (2005) has used a monochromator-equipped TEM to improve the energy resolution and to obtain the EEL spectrum of silicon, and finally observe the onset 1.1 eV of the low-loss region without any numerical calculations, Fig. 5.5.6, which is perfect match with a known indirect band gap of silicon, and also found the tail of ZLP intensity placed more important than the FWHM of ZLP to evaluate the band gap.

4. Core-hole effect

In metals, the hole generated by electron excitation in the inner shell will be screened by surrounding electrons, while in nonmetals, such screen effect will be weak and the interaction between the holes and excited electrons should be taken into consideration.

Duscher et al. (2001) has demonstrated only taking delocalized core-hole ef-

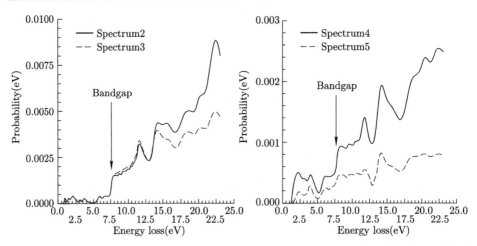

Fig.5.5.5 Band gap spectra taken with the cube aligned along the [100] (left) and [110] (right) directions. (solid line-deconvolution, dashed line-subtraction).

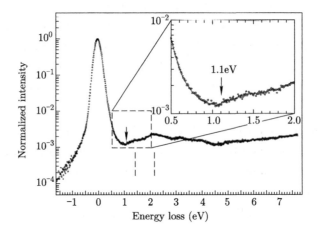

Fig.5.5.6 The band-gap measurement of silicon using the monochromator.

fects that makes results of ab initio fully self-consist LDA calculations match more precisely with the experimental ELNES in Si, SiO_2, MgO, SiC.

Nufer et al. (2001) compared the differences of ELNES of the Al-K and Al-L_1 edges in Al_2O_3 by experimental results and calculations with an without core-hole effects, and they found their uniformity can be attributed to the differently effective screening of the core-hole. In Fig. 5.5.7(a), Al-L_1 edge (blue) is scaled to the theoretical data at the maximum of the PDOS calculated without core-hole simulation; while in Fig. 5.5.7(b), the Al-K (blue) aligned and scaled to the theoretical data at the maximum of the PDOS with core-hole. They concluded that the localization of the initial state is a key factor to observe the core-hole effect in the ELNES because 1s electrons are more delocalized than 2s electrons.

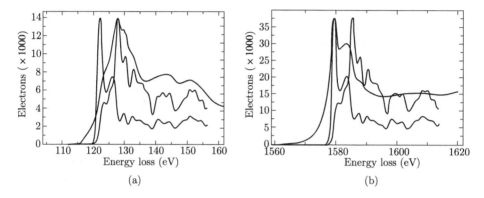

Fig.5.5.7 Comparison of the unoccupied Al p-PDOS with(green) and without(red) a core-hole and ELNES(blue) at the Al-L_1 (left) and Al-K (right) edge in α-Al_2O_3(bulk).

5. Charge transfer

For rare earth and transition metals, there are too many unfilled 3d states and accordingly there are peaks that satisfy the dipole selection rule (p-d), and this phenomena is "white lines" (for white in X-ray absorption spectra), and we can measure the intensity of such peaks to calculate the occupancy of 3d electrons.

Murakami et al. (1998) made use of white lines in ELNES to verify martensitic transformation. In Fig. 5.5.8, $L_{2,3}$ correspond to the transition $2p_{1/2} - 3d$ and $2p_{3/2} - 3d$, and we can find the intensity of Ti $L_{2,3}$ became slightly weaker and accordingly, the intensity of Ni $L_{2,3}$ turned a bit higher, and it was concluded that the filled states density of 3d in Ti became smaller, and in turn, that of Ni turned larger.

Yang et al. (2008) also used aberration corrected STEM-EELS and directly found the hole transfer from the rocksalt CoO to hexagonal CoO_2 layers. Layer

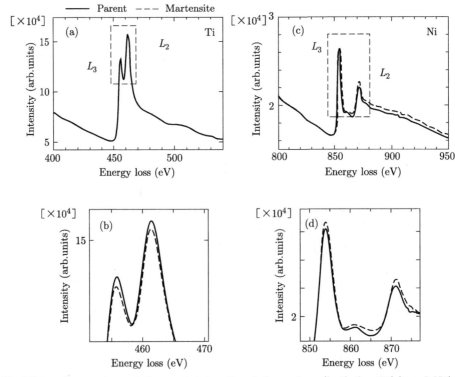

Fig.5.5.8 Comparison of the peak intensity of $L_{2,3}$ edges (both for Ti(a) and Ni(c)) between the parent phase(solid line) and the martensite (dashed line). Enclosed pictures fir the dotted area in(a) and (c) were represented in (b) and (d), respectively

measure charge transfer in $Ca_3Co_4O_9$. Tan et al. (2011) calculated the Mn and Fe $L_{2,3}$ ratio to determine the oxidation state.

Yang and Zhu (2000) applied the ratio of Fe $L_{2,3}$ to quantitatively measure the charge transfer in nanocrystalline-amorphous $Fe_{73.5}Cu_1Mo_3Si_{13.5}B_9$ alloy, as shown in Fig. 5.5.9. The 3d occupancies were determined with the normalized white-line intensities in both crystalline and amorphous phases by the empirical method $I'_{3d} = 10.8\ (1{-}0.10n_{3d})$ (Pearson et al. 1993). The enhancements of normalized white-line intensities in crystalline phase compared to amorphous phase indicate a depletion of about $0.25{\pm}0.06$ electron/atom from the outer d states in amorphous phase during amorphous-crystalline transformation.

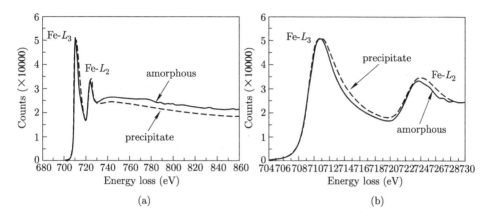

Fig.5.5.9 (a) Background-subtracted EEL spectra of the iron $L_{2,3}$ edges taken from α-(Fe,Si) precipitate and adjacent amorphous phase. (b) A redisplay of the Fe L_2 edge region for illustration the enhancement of the white line. The enhancements of $L_{2,3}$ white line spectrum obtained from crystalline compared to amorphous indicating that some electrons have left Fe 3d states upon crystallization.

6. Radial distribution function

EXELFS can reveal the local atomic environment electrons with excess energy after ionization will lose energy greater than 50 eV and radiate a wave and this wave will suffer single scattering which can be strongly affected by the surrounding atomic arrangement.

Ito et al. (1999) has applied this method to experimentally measure the short range order in $Pd_{30}Ni_{50}P_{20}$ bulk metallic glass in TEM-EELS, as shown in Fig. 5.5.10. The procedure should be the following steps:

- Conversion of $\chi(E)$ from energy to k-space $\chi(k)$;
- Isolation of the oscillatory component of the EXELFS($\chi(E)$) by using n-spline fitting over the k range of interest.
- Correction for k-dependence of backscattering by multiplying with k^n where $n = 1, 2$ or 3.
- Truncation of $\chi(k) * k^n$ by multiplying with a window function.
- Fourier transform of $X(k) * k^n$ to give a raw RDF (phase shift uncorrected),

$$FT(\chi(k) * k^n) = |RDF|.$$

- Correction for phase shifts to convert $|RDF|$ peak positions into interatomic distances.

Alamgir et al. (2003) has also used this method to investigate the short range order in other bulk metallic glasses of $Pd_xNi_{80-x}P_{20}$ system.

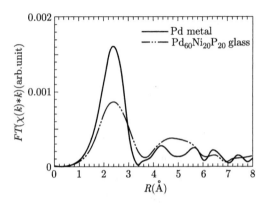

Fig.5.5.10   Fourier transform of $(\chi(k) * k)$ obtained experimentally from Pd and $Pd_{30}Ni_{50}P_{20}$ bulk metallic glasses(without phase shift correction)

7. Composition detection in situ test

EELS, combined with in situ test method, we can detect the composition change during in situ test which will provide us more details on the structure evolution. Han et al. (2007) use EELS to study the structure evolution process of Si nanowires under super plasticity. Fig. 5.5.11 shows a series of EELS spectra taken from a Si NW prior to and after large-strain plastic deformation. The images in Fig. 5.5.11 were taken along the same line across the NW at a time interval of 1 h. The Si L_{2-3} edge characters of amorphous Si and SiOx from the literature (Schulmeister and Mader 2003) are also shown in Figs. 5.5.11(e) and (f), respectively, for comparison. This confirms that sp3 bonding is preserved in the plastic-deformation-induced disordering of Si structure. This work is a typical example shows the powerful function of EELS+TEM.

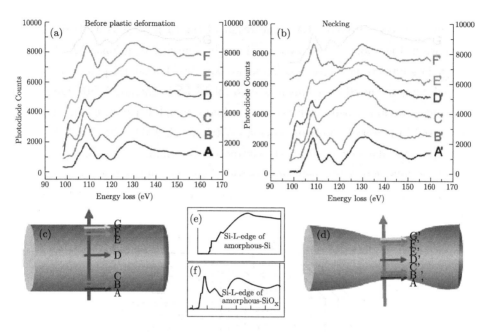

Fig.5.5.11 Series of EELS spectra taken from a Si NW prior to and after large-strain plastic deformation. Both series of EELS spectra were taken across the NW width. (a) A series of spectra taken from the Si NW prior to plastic deformation. (b) A series corresponding to the NW with a superplastic necking feature. (c), (d) Schemes of the locations at which the spectra were taken for (a) and (b), respectively. (e), (f) Reference (Schulmeister and Mader 2003) EELS spectra for the Si L edge of Si and amorphous SiOx, respectively. (g), (h) Line scan EDS spectra of elemental oxygen and silicon for the Si NW before and after neck formation. The surface oxygen layer thickness was measured as the distance from the starting position A or A' to the first peak position of oxygen in both oxygen spectra.

5.6 Spectrum imaging

Yonghai Yue

5.6.1 Spectrum imaging

As the term implies, spectrum imaging (SI) collects a full spectrum at every pixel in the digital image, so you can only do this in STEM mode although there are analog versions in energy-filtered TEM. The result of the SI process is a 3D data cube, as shown in Fig. 5.6.1, where, $\Delta x, \Delta y$ show the spatial dimensions, ΔE shows the energy loss dimension, vertical columns are spectra, horizontal planes are energy filtered images. The SI term was first used for EELS in the late 1980s. Only much later did SI became feasible for X-ray mapping although it is now common enough to be used in materials problem solving (Wittig et al. 2003). Fig. 5.6.2 shows the EDS spectrum data cube (Bruker). Extracting information like this from the SI data cube tells you how much more powerful the SI approach is compared with spot or line profile analyses. Now it is possible to collect EELS and EDS simultaneously.

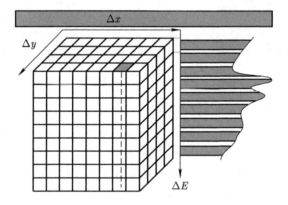

Fig.5.6.1   EELS spectrum schematic.

Spectrum imaging is collection, visualization and processing of detailed spectroscopy information to creat images or line profiles. Typically elemental information is extracted, and any signal can be used to generate image contrast. On a historical note, you should be aware that SI methods have been practiced in other fields, such as radio astronomy for several decades and, indeed, Legge and Hammond took the output of their EDS and WDS spectrometers and synchronized the detectors' output pulses with the position of the beam 30 years ago. We can go back to the spectrum imaging at any time if we have stored the data cube, we can recheck the data, re-do any analysis, search for other spectral features, look at different images at different energies and yet always have the original image and spectra at your disposal.

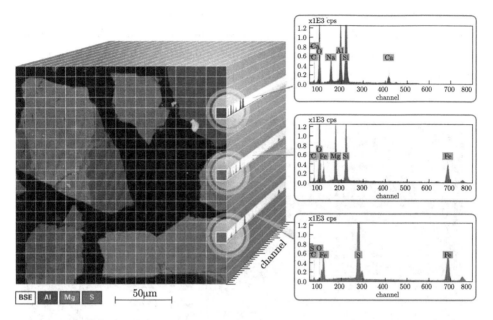

Fig.5.6.2 The hypermap data cube contains a spectrum for each pixel.

5.6.2 Basic spectrum image processing

After you got the spectrum imaging data, you will find that you can slice and dice the data cube in many ways which will reproduce all the other methods of analysis that we've described. We can select a single pixel in the x-y plane or a set of individual pixels based on the spots you want to analyses. Likewise, we can select a line of pixels in the image, effectively slicing the cube along the x-z, y-z, or some combination of these directions and thus produce spectrum-line profiles. But we should remember to remove the background and integrate the signal at each point to generate image.

5.6.3 EFTEM acquisition

As mentioned before, after the energy filter, inelastically scattered electrons which produce background "fog" will be filtered and the contrast of the images will be improved, the elemental map was created with no inelastically electrons. Then EFTEM can acquire an image containing a narrow range of energies, the spectrum image data cube is filled one energy plane at a time, as shown in the sketch map in Fig. 5.6.3. Fig. 5.6.4 demonstrate an example which shows the zero-loss image. See more details in transmission electron microscopy edited by Williams and Carter (1996).

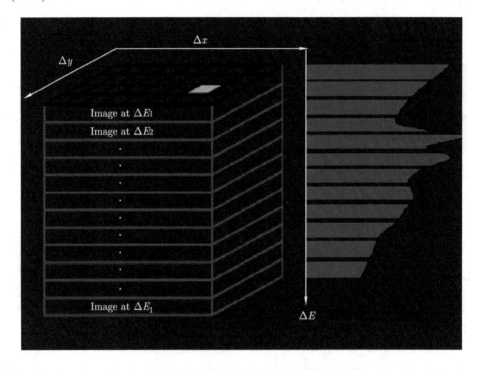

Fig.5.6.3   Sketch map for EFTEM

5.6.4 Artifacts and instabilities

Artifacts and instabilities cause errors when projecting spectra or energy planes from spectrum image. There are many factors which can introduce errors to the spectrum image such as current drift, energy drift, and spatial drift and so on, which have been shown in Fig. 5.6.5. So it will be important to do corresponding correction to guarantee the real spectrum image.

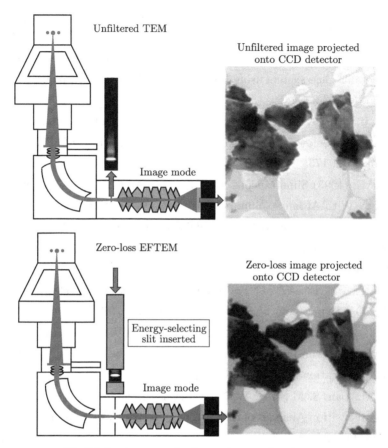

Fig.5.6.4 An example shows the zero-loss image.

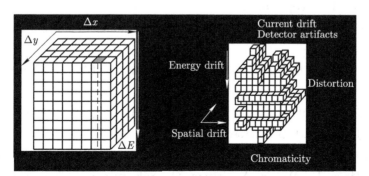

Fig.5.6.5 Artifacts and instabilities which can introduce errors to the spectrum image.

References

Alamgir, F. M., Jain, H., Williams, D. B. and Schwarz, R. B. (2003) The structure of a metallic glass system using EXELFS and EXAFS as complementary probes. Micron, **34**, 433.

Bangert, U., Harvey, A.J., Keyse, R. (1997) Assessment of electron energy-loss spectroscopy below 5 eV in semiconductor materials in a VG STEM. Ultramicroscopy, **68**, 173.

Batson, P.E. (1993) Simultaneous Stem Imaging And Electron-Energy-Loss Spectroscopy With Atomic-Column Sensitivity. Nature, **366**, 727.

Batson, P. E., Kavanagh, K. L., Woodall, J. M., Mayer, J. W. (1986) Electron energy-loss scattering near a single misfit dislocation at the GaAs/GaInAs interface. Phys. Rev. Lett. **57**, 2729.

Benthem, K. van, Elsässer, C. and Frence, R. H. (2001) Bulk electronic structure of $SrTiO_3$: Experiment and theory. J. Appl. Phys. **90**, 6156.

Bentley, J., Zaluzec, N. J., Kenik, E. A., Carpenter, R.W. (1979) Optimization of an analytical electron microscope for X-ray microanalysis: instrumental problems. Oak Ridge National Lab., TN (USA).

Chen,T. M., Lan, S. M., Uen, W. Y., Yang, T. N., Chang, K. J. (2014) Structure and Composition Analysis of the Cu-Zn-Se Ternary Compounds by TEM/EDS. J. Crys. Growth. **388**, 87.

Colliex, C., (1996) New Trends in STEM-Based Nano-EELS Analysis. Journal of Electron Microscopy, **45**, 44.

Crewe, A., Wall, J. and Langmore, J. (1970) Visibility of single atoms. Science, **168**, 1338.

Crewe, A. (1980) The Physics of the high resolution STEM. Rep. Prog. Phys, **43**, 621.

Crozier, P. A. (1995) Quantitative elemental mapping of materials by energy-?ltered imaging. Ultramicroscopy, **58**, 157.

Dieterle, L., Bach, D., Schneider, R., Stömer, H., Gerthsen, D., Guntow, U., Ivers-Tiffée, E., Weber, A., Peters, C., Yokokawa H. (2008) Structural and chemical properties of nanocrystalline $La_{0.5}Sr_{0.5}CoO_3$-delta layers on yttria-stabilized zirconia analyzed by transmission electron microscopy. Journal of Materials Science, **43**, 3135.

Duscher, G., Buczko, R., Pennycook, S. J., Pantelides, S. T. (2001) Core-hole effects on energy-loss near-edge structure. Ultramicroscopy, **86**, 355.

Egerton, R.F. (1976) Inelastic-scattering and energy filtering in transmission electron-microscope. Philosophical Magazine, **34**, 49.

Egerton, R.F. (2011) Electron Energy-Loss Spectroscopy in the Electron Microscopy. New York: Springer.

Egerton, R.F. and Cheng, S.C. (1987) Measurement of local thickness by electron energy-loss spectroscopy. Ultramicroscopy, **21**, 231.

Egerton, R.F. (1996) Electron Energy Loss Spectroscopy in the Electron Microscope. 2nd edition, Plenum Press.

Gass, M. H., Koziol, K., Windle, A. H. and Midgley, P. A. (2006) Four-dimensional spectral tomography of carbonaceous nanocomposites. Nano. Lett. **6**, 376.

Goldstein, J. I., Williams, D. B. and Cliff, G. (1986) Principles of analytical electron microscopy. New York, Plenum Press.

Grovenor, C. R. M., Batson, P. E., Smith, D. A. and Wong, C. (1985) As segregation to grain boundaries in Si. Philos. Mag. A. **50**, 409.

Gu, L., Zhu, C. B., Li, H., Yu, Y., Li, C. L., Tsukimoto, S., Maier, J., Ikuhara, Y. (2011) Direct observation of lithium staging in partially delithiated LiFePO$_4$ at atomic resolution, J. Am. Chem. Soc,**133**, 4661.

Hébert, C. and Schattschneider, P. (2003) A proposal for dichroic experiments in the electron microscope. Ultramicroscopy, **96**, 463.

Han, X. D., Zheng, K., Zhang, Y. F., Zhang, X. N., Zhang, Z. and Wang, Z. L. (2007) Low-Temperature In Situ Large-Strain Plasticity of Silicon Nanowires, Adv. Mater, **1**, 8.

He, X., Gu, L., Zhu, C., Yu, Y., Li, C., Hu, Y. S., Li, H., Tsukimoto, S., Maier, J., Ikuhara, Y. (2011) Direct Imaging of Lithium Ions Using Aberration-Corrected Annular-Bright-Field Scanning Transmission Electron Microscopy and Associated Contrast Mechanisms. Mater. Express, **1**, 43.

Hu, K. X., Jones, I. P. (2005) Low electron energy-loss spectroscopy study of the electronic structure of matrix and S19, S13 boundaries in SrTiO3. J. Phys. D:

Appl. Phys. **38**, 183.

Huang, R., Hitosugi, T., Findlay, S. D., Fisher, C. A. J., Ikuhara, Y. H., Moriwake, H., Oki, H., Ikuhara, Y. (2011) Real-time direct observation of Li in $LiCoO_2$ cathode material. Appl. Phys. Lett, **98**, 051913.

Ito, Y., Alamgir, F. M., Jain, H., Williams D. B. and Schwarz, R. B. (1999) EXELFS of metallic glasses. Bulk Metallic Glasses. Symposium, **31**, 6.

Jiang, N., Jiang, B., Spence, J.C. H., Yu, R. C., Li, S. C. and Jin, C. Q. (2002) Anisotropic excitons in MgB2 from orientation-dependent electron-energy-loss spectroscopy. Phys. Rev. B. **66**, 172502.

Keast, V. J., Scott, A. J., Brydson, R., Williams, D. B. and Bruley, J. (2001) Electron energy-loss near-edge structure-a tool for the investigation of electronic structure on the nanometre scale. Journal of Microscopy, **203**, 135.

Kimoto, K., Sekiguchi, T. and Aoyama, T. (1997) Chemical shift mapping of Si L and K edges using spatially resolved EELS and energy–filtering TEM. Journal of Electron Microscopy, **46**, 369.

Kimoto, K., Kothleitner, G., Grogger, W., Matsui, Y.(2005) Ferdinand Hofer Advantages of a monochromator for bandgap measurements using electron energy-loss spectroscopy. Micron., **36**, 185.

Lazar, S., Botton, G., Wu, M.-Y., Tichelaar, F. and Zandbergen, H. (2003) Materials science applications of HREELS in near edge structure analysis and low-energy loss spectroscopy. Ultramicroscopy, **96**, 535.

Muller, D. A, Tzou, Y., Ray, R., Silcox, J. (1993) Mapping SP2 and SP3 states of carbon at subnanometer spatial resolution. Nature, **366**, 725.

Muller, D. A., Kourkoutis, L. F., Murfitt, M., Song, J. H., Hwang, H. Y., Silcox, J, Dellby, N. and Krivanek, O. L. (2008) Atomic-scale chemical imaging of composition and bonding by aberration-corrected microscopy. Science, **319**, 1073.

Murakami, Y., Shindo, D., Otsuka, K. And Oikawa, T. (1998) Electronic structure changes associated with a martensitic transformation in a $Ti_{50}Ni_{48}Fe_2$ alloy studied by electron energy-loss spectroscopy. Journal of Electron Microscopy, **47**, 301.

Nelayah, J., Kociak, M., Stephan, O., Garcia de Abajo, F. J., Tence, M., Henrard,

L., Taverna, D., Pastoriza-Santos, I., Liz-Marzan, L. M., and Colliex, C. (2007) Mapping surface plasmons on a singlemetallic nanoparticle.metallic nanoparticle. Nat. Phys. **3**, 348.

Nufer, S., Gemming, T., Elsäser, C., Kötlmeier, S., Rüle, K. (2001) Core-hole effect in the ELNES of α-Al$_2$O$_3$: experiment and theory. Ultramicroscopy, **86**, 339.

Otten, M.T., Miner, B., Rask, J.H., Buseck P. R. (1985) The determination of Ti, Mn and Fe oxidation states in minerals by electron energy-loss spectroscopy. Ultramicroscopy, **18**, 285.

Pearson, D.H., Ahn, C. C. and Fultz, B. (1993) White lines and d-electron occupancies for the 3d and 4d transition metals. Phys. Rev. B. **47**, 8471.

Petrova, R.V. (2006) Quantitative high-angle annular dark field scanning transmission electron microscopy for materials science, Ph.D dissertation, University of Central Florida, Orlando.

Raether, H. (1980) Excitation of Plasmons and Interband Transitions by Electrons. Berlin, Springer press.

Rafferty, B., Pennycook, S. J. and Brown, L. M. (2000) Zero loss peak deconvolution for bandgap EEL spectra. Journal of Electron Microscopy, **49**, 517.

Rafferty, B. and Brown, L. M. (1998) Direct and indirect transitions in the region of the band gap using electron-energy-loss spectroscopy. Phys. Rev. B. **58**, 10326.

Ruthemann, G. (1941). Diskrete Energieverluste schneller Elektronen in Festkörpern. Naturwissenschaften, **29**, 648.

Ryen, L., Wang, X., Helmersson, U., Olsson, E. (1998) Determination of the complex dielectric function of epitaxial SrTiO$_3$ films using transmission electron energy-loss spectroscopy. J. Appl. Phys. **85**, 2828.

Schattschneider, P., Rubino, S., Hébert, C., Rusz, J., Kuneš, J., Novák, P., Carlino, E., Fabrizioli, M. Panaccione, G. and Rossi, G. (2006) Detection of magnetic circular dichroism using a transmission electron microscope. Nature, **441**, 486.

Scheinfein, M. and Isaacson, M. (1984) Design and Performance of 2nd Order Corrected Spectrometers for Use with the Scanning-Transmission Electron-Microscope. Scanning Electron Microscopy, pp. 1681–1696.

Schulmeister, K. and Mader. W. (2003) TEM investigation on the structure of amorphous silicon monoxide. Journal of Non-Crystalline Solids, **320**, 143.

Shao-Horn, Y., Croguennec, L., Delmas, C., Nelson, E. C., O'Keefe, M. A. (2003) Atomic resolution of lithium ions in $LiCoO_2$. Nat. Mater. **2**, 464.

Stöger-Pollach, M., Treiber, C. D., Resch, G. P., Keays, D. A. and Ennen, I. (2011) Real space maps of magnetic properties in Magnetospirillum magnetotacticum using EMCD. Micron, **42**, 461.

Stöhr, J. (1999) Exploring the microscopic origin of magnetic anisotropies with X-ray magnetic circular dichroism (XMCD) spectroscopy. Journal of Magnetism and Magnetic Materials, **200**, 470.

Tafto J., Krivanek O. L. (1982) Site-specific valence determination by EELS. Phys. Rev. Lett, **48**, 560.

Tafto J., Lehmpfuhl G. (1982) Direction dependence in EELS from single crystals. Ultramicroscopy, **7**, 287.

Tan, H., Turner, T., Yücelen, E., Verbeeck, J. and Van Tendeloo, G. (2011) 2D Atomic Mapping of Oxidation States in Transition Metal Oxides by Scanning Transmission Electron Microscopy and Electron Energy-Loss Spectroscopy. Phys. Rev. Lett. **107**, 107602.

Terauchi, M., Koike, M., Kukushima, K., and Kimura, A. (2010b) Development of wavelength-dispersive soft x-ray emission spectrometers for transmission electron microscopes–an introduction of valence electron spectroscopy for transmission electron microscopy. J. Electron Microsc, **59**, 251.

Terauchi, M., Takahashi, H., Handa, N., Murano, T., Koike, M., Kawachi, T., Imazono, T., Koeda, M., Nagano, T., Sasai, H., Oue, Y., Onezawa, Z., and Kuramoto, S. (2010a) Li K-emission measurements using a newly developed SXES-TEM instrument. Microsc. Micronal, **16**, 1308.

Verbeeck, J., Tian, H. and Schattschneider, P. (2010) Production and application of electron vortex beams. Nature, **467**, 301.

Verbeeck, J., Van Dyck, D. and Van Tendeloo, G. (2004) Energy-filtered transmission electron microscopy: an overview. Spectrochimica Acta Part B: Atomic Spectroscopy, **59**, 1529.

Watanabe, M., Kanno, M., and Akunishi, E. (2010b) Atomic-level chemical analysis by EELS and XEDS in aberration-corrected scanning transmission electron microscopy. JEOL News. **45**, 8.

Watanabe, M., Okunishi, E., and Aoki, T. (2010a) Atomic-level chemical analysis by EELS and XEDS in aberration-corrected scanning transmission electron microscopy. Microsc. Microanal, 2010, **16**, 66.

Williams D.B. and Carter C. B. (2009) Transmission Electron Microscopy: a textbook for materials science. New York: Springer.

Williams D. B. (1987) Practical analytical electron microscopy in materials science, 2nd edition, New Jersey, Philips Electron Optics Publishing Group.

Williams D. B. and Carter C. B. (1996) Transmission electron microscopy: A textbook for materials science. New York and London, Plenum Press, Chapter 34.

Williams, D. B., and Carter, C. B. (1996) Transmission electron microscopy. Springer. Wittig, J. E., Al-Sharaba, J. F., Doerner, M., Bian, X. P., Bentley, J. and Evans, N. D. (2003) Influence of Microstructure on the Chemical Inhomogeneities in Nanostructured Longitudinal Magnetic Recording Media, Scr. Mater. **48**, 943.

Xin, S., Gu, L., Zhao, N. H., Yin, Y. Y., Zhou, L. J., Guo, Y. G. and Wan, L. J. (2012) Smaller Sulfur Molecules Promise Better Lithium-Sulfur Batteries. J. Am. Chem. Soc, **134**, 1850.

Yang, G., Ramasse, Q and Klie, R. F. (2008) Direct measurement of charge transfer in thermoelectric $Ca_3Co_4O_9$. Phys. Rev. B, **78**, 153109.

Yang, G.Y. and Zhu, J. (2000) 3D occupancy determination from Fe $L_{2,3}$ electron energy-loss spectra of nanocrystalline-amorphous $Fe_{73.5}Cu_1Mo_3Si_{13.5}B_9$ alloy. Journal of Magnetism and Magnetic Materials, **220**, 65.

Zhang, Z. H., Wang, X. F., Xu, J. B., Muller, S., Ronning, C. and Li, Q. (2009) Evidence of intrinsic ferromagnetism in individual dilute magnetic semiconducting nanostructures. Nature Nanotechnology, **4**, 523.

Zhang, Z. H., Yang, J. J., He, M., Wang, X. F. and Li, Q. (2008) Electronic structure of a potential optical crystal $YBa_3B_9O_{18}$: Experiment and theory. Appl. Phys. Lett. **92**, 171903.

6
Aberration Corrected Transmission Electron Microscopy and Its Applications

Lin Gu

In Chapter 3, it has shown that TEM is very powerful for the study of microstructures of materials, which helps to deepen understanding of the relationship between the microstructures and properties. With the diversification and complication of materials there is an urgent need for observing microstructures at atomic scale. During the past decade, the development of aberration correction technology has improved the spatial resolution of a TEM to sub-angstrom and the energy resolution to 0.1 eV. In this chapter we will firstly introduce the theory and methods of aberration correction in TEM. Then some new applications of aberration corrected TEM in materials science will be exampled.

6.1 Basics of aberration correction

6.1.1 Aberration

Similar to an optical imaging system, an ideal TEM imaging system does not exist. Aberration of an electron microscope refers to the difference between an image formed by an actual electron-optical system and that formed by an ideal electron-optical system. In general, aberration theory of electron lens is completely derived from optics. Here we briefly review some knowledge about optics.

When a light ray passes through an optical system, an object point is imaged into an image point. If we only considers the light rays nearest to the axis of a lens, without regard to the quadratic term and the high-order term of the distance to the axis or included angle of light ray and axis, the theory that satisfies the above conditions is commonly called Gaussian optics or ideal imaging. Otherwise, aberration will occur (Zhu and Ye, 2010).

In an optical system, aberration can be divided into two categories: one is generated by monochromatic light, called monochromatic aberration, which includes spherical aberration, coma, astigmatism, curvature of field and distortion; the other is generated by polychromatic light, called chromatic aberration considering that the optical lens has different refractive indexes for lights with different wave lengths.

Similarly, in electron microscope, main factors influencing its spatial resolution also include spherical aberration, astigmatism and chromatic aberration. We will briefly introduce several main aberrations influencing electron microscope and some methods to correct them below. For clarity, it will be assumed that other aberrations do not exist (but actually all aberrations exist) when discussing one of them.

1. Spherical aberration

Spherical aberration is caused by different capabilities of central zone and peripheral zone of electromagnetic lens to converge electrons. Off-axial electrons are more refracted when passing the lens compared with paraxial electrons, so that they do not converge on one point, as shown in Fig. 6.1.1, and a round spot emerges in image plane. The calculation shows that transverse diffusion Δr is proportional to cube of convergence angle θ and the proportionality coefficient is called coefficient of spherical aberration C_s, which can be expressed as: $C_s = \dfrac{\Delta r}{\theta^3}$.

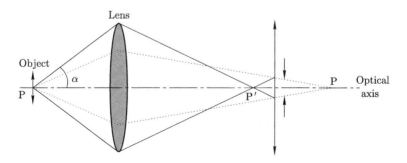

Fig.6.1.1　Schematic of the spherical aberration of a lens.

In an optical system, we can eliminate its spherical aberration with concave lens (concave lens and convex lens have spherical aberration with opposite symbols). As for electron microscope, we can adjust its spherical aberration by use of "concave lens" — multilevel field unit, and we call such a device as a spherical aberration corrector. Spherical aberration correction is currently at leading edge of electron microscopy and has greatly improved resolution of an electron microscope down to sub-angstrom level.

2. Astigmatism

Astigmatism here refers to axis astigmatism and it happens in case of asymmetry of a magnetic field of magnetic lens. As shown in Fig. 6.1.2, the asymmetry makes magnetic lens have different focusing capabilities in two mutually perpendicular planes. In case of focusing in the tangential plane (Plane A), focusing may not has happened in the plane perpendicular to it-meridional plane (Plane B); while in case of focusing in the meridional plane, over focusing begins in tangential plane. Thus image of an object point in any image plane is not a point any more. For instance, an electron beam running in Plane A is focused at point P_A while that running in Plane B is focused at point P_B, so the image of a round object becomes diffused elliptical shape. In practice, this type of astigmatism can be corrected using an electromagnetic astigmatism stigmator and this method has been widely used in various models of electron microscopes.

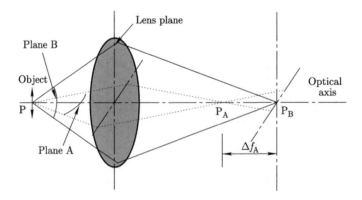

Fig.6.1.2 Schematic of the astigmatism.

3. Chromatic aberration

Aberration resulting from non-monochromaticity of electrons is called chromatic aberration, which is caused by the difference in electron energy. Magnetic lens has different capabilities to focus electrons with different energy, so the image of an object point becomes a round spot in Gaussian image plane, as shown in Fig. 6.1.3 below.

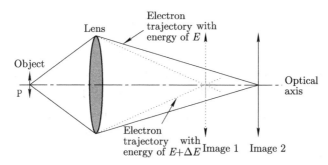

Fig.6.1.3 Schematic of the chromatic aberration.

Chromatic aberration of an electron microscope is mainly attributed to two aspects, one is unstable accelerating voltage; the other is interaction between electron beam and specimen, leading to electronic energy loss when transmitting the specimen. After correcting spherical aberration, information resolution limit is only subject to diffusion of electron wave energy. To obtain a better image, the next step is to eliminate chromatic aberration and current strategy is to correct chromatic aberration using a monochromator.

Description of the above aberrations only involves the simplest cases, while actual aberration theory is more complex. Generally speaking, based on Gaussian optical formula, if a sin function expansion including cubic term of angle is substituted in an relevant formula of the refraction law, the formula generated will be different from Gaussian optical formula in as many as 5 aspects, which were firstly conducted by Seidel and called Seidelsum. Each item corresponds to a kind of aberration, including spherical aberration, coma, astigmatism, curvature of field and distortion. Such aberrations are obtained through expansion of sin function to cube of angle, so they are commonly called the third-order aberration. By parity of reasoning, there are fifth-order aberration, seventh-order aberration, etc.. However, algebraic expression of these aberrations is very complex. Because of its nonsignificant influence and very dedicated application, it is usually unknown. With in-depth study on aberration-corrected electron microscopy, influences of higher-order aberrations emerge after elimination (or reduction) of the third-order aberration (Zhu and Ye, 2010).

6.1.2 Development of spherical aberration corrector

Since Knoll and Ruska invented electron microscope in 1932 (Knoll and Ruska, 1932), improving resolution of electron microscope has always been an important objective pursued by researchers. In 1936, Scherzer (Scherzer, 1936) proposed that the spherical aberration of a rotationally symmetric electromagnetic lens was unlikely to be eliminated through design of field distribution itself. This is called Scherzer Theorem.

To eliminate the spherical aberration of an electron lens, at least one of the following three conditions must be dropped: 1) Electron-optical system is rotationally symmetric; 2) lens field is static; 3) space charge does not exist in optical path. Afterwards, Scherzer had been working on the issues concerning correction of spherical aberration of electron lens (round lens for short). He put forward in 1947 (Scherzer, 1947) that the most promising approach was to abandon the rotational symmetry and to introduce multi-pole field unit in optical path, and the latter was not rotationally symmetric. While some scholars proposed other ideas later on, such as abandoning the condition of static field and introducing alternating field lens; or introducing space charge in optical path. However, introduction of multi-pole field unit has always been the mainstream of aberration correction. One year later, Dennis Gabor (Gabor, 1948) put forward that two main axial aberrations, chromatic aberration and spherical aberration, could be eliminated through combination of fourth-order and eighth-order magnetic lenses.

As suggested by Scherzer, electrostatic corrector was used, including two cylindrical lenses, a rotationally symmetric lens and three octupole lenses, therefore the structure was very complex. In 1949, Seeliger (Seeliger, 1949; Seeliger, 1951) built and tested the system according to Scherzer's suggestion. During test, Seeliger found that clear image could be obtained after eliminating aberration; and spherical aberration could be corrected through appropriate adjustment of octupole lens. However, actual resolution capability of electron microscope could not be improved due to power instability rather than spherical aberration of objective lens at that time. In 1954, Mollenstedt came up with another approach. He adjusted illuminating aperture angle to 20 mrad and increased spherical aberration to such a level to blur the image so as to make spherical aberration become limiting factor of resolution (Mollenstedt and Hubig, 1954). In case of corrector of Seeliger, resolution was increased by 7 times and image contrast was improved greatly.

To make the corrector applicable to electron microscope with higher accelerating voltage, Deltrap advanced spherical aberration corrector by use of electromagnetic quadrupole lens and octupole lens in 1964 (Deltrap et al., 1964). Afterwards, Rose presented electromagnetic corrector available for correcting spherical aberration and chromatic aberration at the same time in 1971 (Rose, 1971; Rose, 1971). All these correctors are composed of multiple units, many of which can influence paraxial trajectory in objective lens, and such units require accurate adjustment and have to maintain particularly high stability. Due to such technical difficulties, the stringent requirements could not be met at that time. Consequently, various efforts to improve resolution of a TEM through correction of spherical aberration had failed in more than 40 years. Up to the late 1970s and the early 1980s, Koops (Koops et al., 1977) and Hely (Hely, 1982) put forward again theoretically that electromagnetic multi-pole corrector also could be used to correct axial chromatic aberration. Before 1995, due to low vacuum technology of instruments as well as short life and low brightness of electron source, the development of electron microscope developed slowly in this period and electron microscope generally remained on the level of low acceleration voltage and low resolution (~ 10 nm). The first successful practice was conducted in 1995, Zach and Haider (Zach and Haider, 1995) succeeded in correcting spherical aberration and chromatic aberration of an objective lens simultaneously in a dedicated low-voltage SEM and had improved resolution from 5 nm to 1.8 nm under 1 kV.

An ideal converging lens would image a point in the object to a corresponding point in the image. In reality, as a result of aberrations, the image is broadened into a point-spread disk. Fig. 6.1.4 shows the case of spherical aberration (Urban, 2008). Point spread arises from the refraction power of a real lens increasing with the incident angle of the beams entering the lens. As a result, electrons are scattered in specimen and there is certain distance between Gaussian image plane that is defined by low-angle and high-angle beams converging point. Focal length changes markedly with current of electromagnetic coils. Such defocusing also induces point spread, which is regarded as spherical aberration. Although these two represent the most substantial reasons for spherical aberration, there are many other reasons, not only broadening of point, but also an angular distortion of the point spread disk

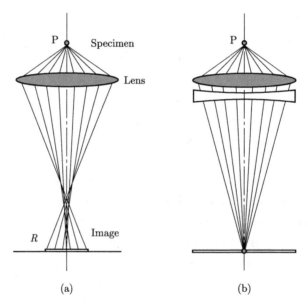

(a) (b)

Fig.6.1.4 (a) Schematic illustration of spherical aberration of a converging lens. The scattering power increases with the increasing angle (with respect to the optical axis) at which the electrons enter the lens. As a result, the focal length of the beams passing the lens at its periphery come to a focus, a distance in front of the Gaussian image plane that is defined by the paraxial, low-angle beams. The image of a point P of the specimen is broadened into a "point-spread disc" of radius R. (b) Spherical aberration is compensated by combining the converging lens with a suitable diverging lens. In electron optics, diverging lenses are realized by combinations of multipole lenses (Urban, 2008).

(Urban, 2008).

In practice, for correcting spherical aberration of a conventional lens, a diverging lens is added to compensate for the too-high refraction power of a high-angle scattered beam. In TEM and STEM, spherical aberration correction can be achieved using different systems, for example, a double-hexapole system (Urban, 2008). Correction is achieved due to the fact that the primary second-order aberrations of the first hexapole (a strong three fold astigmatism) are exactly compensated by the second hexapole element. Due to their nonlinear diffraction indices, the two hexapoles additionally induce a residual secondary, third-order spherical aberration which is rotationally symmetric with a negative sign, thus cancelling the positive spherical aberration of the objective lens.

The first equipment composed of a probe, an imaging aberration correctors and an energy filter was installed in Oxford in 2003 (Fig. 6.1.5(a), JEOL 200 kV 2200FS) (Kirkland and Meyer, 2004). Adjustment of the imaging corrector was

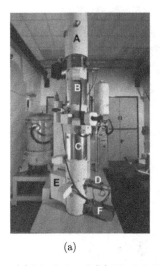
(a)

(b)

Fig.6.1.5 (a) Mark 1 and (b) Mark 2 200kV double aberration corrected TEM instruments installed at Oxford. In (a), the probe forming and imaging correctors are marked as B and C and the in-column energy filter is marked as E (A is the electron gun). In (b) the additional frame leading to improved mechanical stability is visible (Kirkland and Meyer, 2004).

achieved using a Zemlin tableau of diffractograms calculated from images of a thin amorphous area and recorded at several tilt azimuths with constant tilt magnitudes. These datasets provide a linear evaluation on defocus caused by angle introduction and on coefficient of wave aberration function resulting from two-fold astigmatism.

A recent instrument is shown in Fig. 6.1.5(b). Some special stabilization devices have been added to the microscope to reduce mechanical vibration so as to improve stabilities in both the high voltage and objective lens to reach values of typically ca. 5×10^{-7}. Some devices are featured by monochromatic electron-beam source and makes beam energy dispersion reduced to ca. 0.1 eV, thus reducing effect generated due to temporal coherence and providing energy resolution of spectrum directly comparable with that achievable from cold field emission sources. By 2007, direct and discernible TEM resolutions can realize less than 80 pm under acceleration voltage between 200 and 300 kV (as shown in Fig. 6.1.6), limited to the effects of chromatic aberration and residual mechanical, acoustic and electrical instabilities.

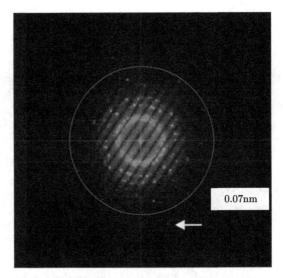

Fig.6.1.6 Youngs fringe diffractogram calculated from images taken from a sample of amorphous carbon film supported gold particle using the Oxford Mark 2 aberration corrected instrument (JEOL 2200MCO) operated at 200 kV. Clear information transfer extending to 0.07 nm is visible with addition reflections (marked) from the gold particles visible at 0.05 nm (Kirkland and Meyer, 2004).

The further developments will focus on correction of higher-order coefficients in the wave aberration and chromatic aberration of through the incorporation of further mechanical and acoustic isolation and more complex geometric correction (Rose, 2005).

6.1.3 Optimum corrected imaging conditions

As having been mentioned, phase-contrast imaging conditions for a HRTEM are defined by the balance of phase shifts between positive third-order spherical aberration C_3 and negative defocus C_1. Under the well-known Scherzer condition it can be given by (Scherzer, 1947):

$$C_1 = -\lambda^{1/2} C_3^{1/2} \qquad (6.1.1)$$

This leads to an interpretable resolution limit of

$$d = 0.625 \lambda^{3/4} C_3^{3/4} \qquad (6.1.2)$$

where λ is the wavelength of the incident electrons. For an aberration-corrected TEM, two contingent conditions are defined. The first one is balance between the positive fifth-order spherical aberration, negative third-order spherical aberration and defocus (it should be noted that this will lead to the optimal negative C_3 and positive C_1, which are defined as follows),

$$C_3 = -2.88 \lambda^{1/3} C_5^{2/3} \qquad (6.1.3)$$
$$C_1 = 1.56 \lambda^{1/3} C_5^{1/3} \qquad (6.1.4)$$

In turn this defines limit of interpretable resolution given by C_5,

$$d = 0.625 \lambda^{5/6} C_5^{1/6} \qquad (6.1.5)$$

A contingent optical condition has been proposed in the position of limited chromatic aberration. Initial case can be defined in setting of such a variable chromatic aberration to guarantee compliance of the phase-contrast transfer function with temporal coherent information limit. Under these conditions, suitable values

C_1, C_3 and C_5 are given by

$$C_1 = 1.7\pi\Delta \tag{6.1.6}$$

$$C_3 = -3.4\frac{(\pi\Delta)^2}{\lambda^2} \tag{6.1.7}$$

$$C_5 = 1.3\frac{(\pi\Delta)^3}{\lambda^2} \tag{6.1.8}$$

$$\Delta = C_c\left\{\left(\frac{\Delta V}{V}\right)^2 + 4\left(\frac{\Delta I}{I}\right)^2 + \left(\frac{\Delta E}{E}\right)\right\}^{1/2} \tag{6.1.9}$$

where C_c is the chromatic coefficient, V the accelerating voltage, ΔV the fluctuation in the accelerating voltage, I the current in the probe-forming lens, ΔI the fluctuation of the lens current, E the energy of the electron beam and ΔE the spread in energy of the beam. The optimal phase-contrast imaging is defined above. However, pure amplitude contrast may be caused because C_3 and C_1 may be zeroed (Lentzen et al., 2002). Under follow-up conditions, phase-contrast transfer function has a zero value while amplitude transfer function is the greatest value. Importantly, the imaging pattern is inapplicable in uncorrected conventional TEM.

6.1.4 Indirect compensation and exit wave reconstruction

Several methods have been developed for experimental determination of axial spherical aberration coefficient and measurement the first-order spherical aberration based on known multiple beam tilting geometry. These provide evaluation on coefficient of wave aberration function, namely, $W(\theta, \varphi)$. Here it is expanded to the third order and expressed in the form of polarity to make the radial and the azimuthal dependence of coefficients apparent. Such measurements can be used in the retrieval of phase and coefficient of the sample exit wave function in both the linear and nonlinear imaging models(Kirkland and Myer, 2004).

$$W(\theta, \phi) = |A_0|\theta\cos(\phi - \phi_{11})$$
$$+ \frac{1}{2}|A_1|\theta^2\cos 2(\phi - \phi_{22}) + \frac{1}{2}C_1\theta^2$$

$$+ \frac{1}{3}|A_2|\theta^3 \cos 3(\phi - \phi_{33}) + \frac{1}{3}|B_2|\theta^3 \cos(\phi - \phi_{31})$$
$$+ \frac{1}{4}|A_3|\theta^4 \cos 3(\phi - \phi_{44}) + \frac{1}{4}|S_3|\theta^4 \cos 2(\phi - \phi_{42}) + \frac{1}{4}C_3\theta^4 + \ldots$$
$$\tag{6.1.10}$$

Such a method has been proved to improve the resolutions of uncorrected microscopes, and complements direct spherical aberration correction.

Direct correction has advantages that it may be a process of linear acquisition by using only a single image without subsequent acquisition. However, at present, correction of spherical aberration in TEM has only been expanded to the third order and the recorded data only include intensity. By contrast, indirect compensation and reconstruction of an exit wave function of a complex sample to compensate any order of spherical aberration coefficient is feasible in theory that is limited only by measurement accuracy. Its disadvantage lies in that offline technology requires a subsequent acquisition process and a record of multiple imaging data in a series of exposing radiation.

Practically, indirect and direct methods can be combined. For a series of focus geometry, parallel illuminating system is unnecessary due to elimination of axial coma aberration induced by tilting. Therefore, the current density in a sample is maintained when the emitter current is reduced and light converges, so that a reduced energy transmission is provided and information limit is improved. For a series data of tilting angle, reduction of axial coma aberration led in less crucial focusing conditions by angle results, which make a given tilting angle and multi-polar tilting magnitude possible without introduction of any change of focal distance. Intrinsic compensation of such high-order spherical aberration up to the fifth order is also possible.

6.1.5 Quadrupole-octupole correctors

Quadrupole-octupole correctors compensate both chromatic and spherical aberration in essence while a hexapole corrector can only be used to eliminate the spherical aberration. A design of a quadrupole magnetic lens is shown in the Fig. 6.1.7 (Y. Liao, 2007). The optical axis is inward perpendicular to paper along the direction of the electron beam propagation, the blue lines indicate the directions of the magnetic field and the red lines indicate the directions of Lorentz forces acting on the electron. As shown in Fig. 6.1.7, the quadrupole magnetic lens generates simple focusing effect in one azimuthal direction while generating defocusing effect in the perpendicular direction at the same time. Therefore, the quadrupole can be used to focus electron beams in linear direction (called line focus).

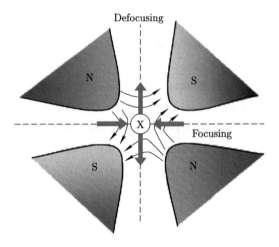

Fig.6.1.7   A quadrupole design (Y. Liao, 2007).

A design of an octupole magnetic lens is shown in Fig. 6.1.8 (Y.Liao, 2007). The optical axis is inward perpendicular to paper along the directions of the electron beam propagation, the blue curves indicate the directions of the magnetic field lines and the red and purple red lines indicate the direction of the Lorentz forces of the electron. An octupole can be used to adjust quadratic distortion of electron beam. In addition, according to Laplace's equation (Y.Liao, 2007),

$$\phi(r,\theta) = r^N \{p_N \cos(N\theta) + q_N \sin(N\theta)\} \quad (6.1.11)$$

where the r and θ are polar coordinates. We can understand the field in the system of octupole, which can be given through the following equation:

$$\phi(r,\theta) = r^3 \{p_3 \cos(3\theta) + q_3 \sin(3\theta)\} \quad (6.1.12)$$

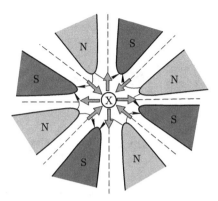

Fig.6.1.8 An octupole design for spherical aberration (C_s) correction. The plum arrows show focusing function, while the red arrows show defocusing function (Y. Liao, 2007).

That is to say, it changes with r^3, directly represents third-order aberrations and increases with distance of optical axis. However, the third-order spherical aberration (occurring to spiral lens) cannot be canceled in case that the paraxial optical system is spherically symmetric due to lack of cylindrical symmetry system. In other words, an octupole magnetic lens always generates parasitic aberration because it has cubic radial dependence and fourfold azimuthal symmetry (fourfold astigmatism).

The current quadrupole-octupole correctors essentially consist of 4 groups of quadrupoles and at least 3 groups of octupoles magnetic lenses, which are used for correcting third-order axial spherical aberration. The third quadrupole-octupole refers to coma-free plane of an objective lens. Fig. 6.1.9 shows an example of such setup (Y. Liao, 2007). A round beam accompanied by a positive spherical aberration for the first group of quadrupole-octupole magnetic lens (Q1-O1), generates elliptical elongation transversally and is accompanied by a negative spherical aberration in O3. In the system, if octupole correctors are turned off, the overall effect of the four groups of quadrupole magnetic lens (Q1, Q2, Q3 and Q4) will be a group of round lenses. Due to functions of Q1 and Q3, the QO system is used to correct spherical aberrations of corresponding x axis and y axis.

However, just as mentioned above, an octupole correction always results in parasitic aberrations. Therefore, the second octupole (O2) acting on the round beam with an opposite sign to O1, and O3 is used to correct parasitic imperfection of Q1 and Q3 in the QO corrector.

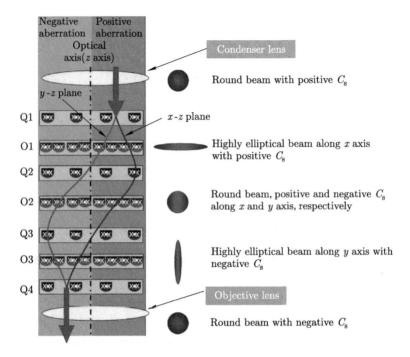

Fig.6.1.9 Schematic illustration of a quadrupole-octupole (QO) corrector consisting of four quadrupoles and three octopoles. In addition, the grey background on the left represents negative spherical aberration for the schematic electron trajectories, while the red background on the right represents positive spherical aberration (Y. Liao, 2007).

As a discussion case of parasitic aberration, the electron beam distortions composed of the first order and the third-order and two-fold and four-fold symmetry are introduced into the Nion quadruple – an octupole third-order corrector. Therefore, the parasitic aberrations not only depend on fundamental symmetries of manufactured TEM systems, but also can limit the performance of a corrector in any order and azimuthal symmetry essentially. In the quadruple-octupole magnetic lens correction system, parasitic aberrations generally comprising hexapole-like fourth-order forms, owns one-fold, three-fold and five-fold symmetries as indicated in Fig. 6.1.10.

By use of an aberration corrector composed of electrostatic magnetic quadrupoles and octupoles, an actual reduction of the spatial resolution limit by a factor

Fig.6.1.10 Aberration coefficients of $C_{4,1,a}$, $C_{4,3,a}$, and $C_{4,5,a}$. (Y. Liao, 2007)

of approximately three was first achieved by Zach and Haider in a low-voltage SEM in 1995. However, due to the large field (off-axis) aberration of the corrector, it was not used for image correction in any TEM.

Fig. 6.1.11 shows a TEM system composed of a C_3/C_5 corrector. It can be

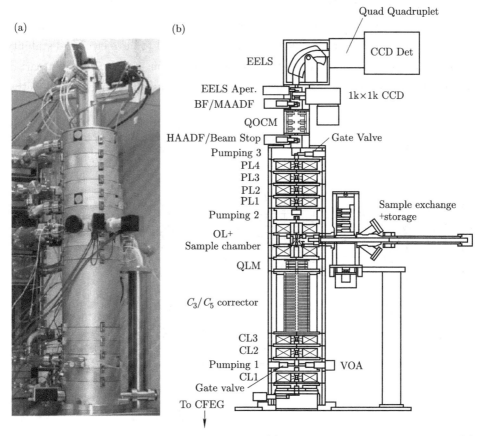

Fig.6.1.11 A TEM column with a corrector (a) and its schematic cross-section (b) (Krivanek et al., 2008).

noted that this TEM has no phosphor screen and has its incident electrons emitted from the bottom and the detector is on the top which is different from the most commercial TEM systems.

The spherical aberration corrector from Nion Company is composed of seven components, namely, four quadrupoles (red part) and three octupoles (blue part), which can be installed in the middle of two collecting lenses at the bottom and one objective lens at the top of a VG Microscopes HB501 STEM (Dellby et al., 2001). The quadrupoles enable an electron beam to enter the octupoles in a shape of pencil-like cross section and afterwards the octupoles correct the spherical aberration generated in the objective lens. Fig. 6.1.12 shows that there is a net +1 in both the x-z and y-z planes. The corrector is preinstalled in the scan coils and under the objective lens. For such a device, it is crucial to have electrical stability of 0.2 ppm and good mechanical stability to its operation success. The right part of Fig 6.1.12 shows hardware structure of the corrector installed under the objective

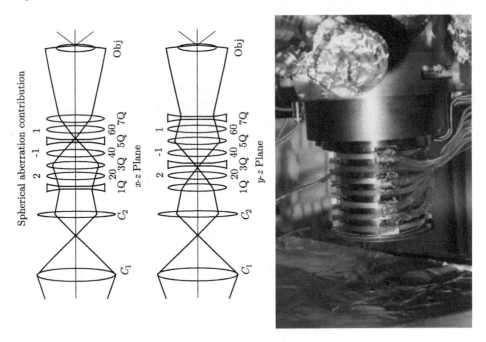

Fig.6.1.12 Hardware structure of corrector installed under objective lens.

lens.

Successful cases of spherical aberration correctors include those of quadrupole-octupole C_s/C_zc (spherical aberration and chromatic aberration) correctors of Zach, et al. for low-voltage SEM, hexapole spherical aberration correctors of Haider, et al. (Haider et al., 1995; Haider et al.,1998) for 200 kV TEM and quadrupole-octupole spherical aberration correctors of O.L. Krivanek, et al. in 100 kV STEM, as well as quadrupole-octupole spherical aberration corrector specifically designed by O.L. Krivanek, et al. of Nion Company for STEM in 1999 (Krivanek et al., 1999).

Fig. 6.1.13 shows schematic diagram of a spherical aberration corrector manufactured for dedicated STEM VG HB5. The spherical aberration corrector is installed between the second condenser lens and the scan coils and in front of the

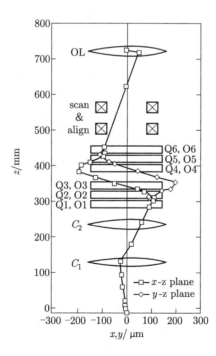

Fig.6.1.13 Principal electron-optical elements of the C_s-corrected STEM and its first-order electron trajectories (Krivanek, 1999).

VG objective lens. It consists of six identical stages, each of which has 12 poles and comprises a quadrupole and an octupole as well as winding coils on each pole. The quadrupole and octupole coils greatly determine the trajectory of an electron passing the corrector. These weak winding coils generate small multi-pole components and make parasitic aberration ineffective.

6.1.6 Hexapole correctors

Spherical aberration of objective lens can be compensated by using two groups of hexapoles. The principle of optics is: 1) the non-rotationally symmetric second-order aberration generated by the first group of hexapoles can be compensated by the second group of hexapoles; 2) for hexapoles having the power of nonlinear diffraction, they can generate rotationally symmetric third-order spherical aberration which is subordinate. But the symbol of the subordinate third-order spherical aberration coefficient is opposite to that of the spherical aberration coefficient of an objective lens, and correspondingly spherical aberration of objective lens can be compensated by application of appropriate incentive intensity (current). In practical application, a spherical aberration correction system for an objective lens is a nearly aplanatic objective lens system which is installed behind the objective lens and is composed of two groups of hexapoles and two groups of additional transfer doublet round lenses, as shown in Fig. 6.1.14 (Urban et al., 1999). This system not only can eliminate the spherical aberration of an objective lens, but also can eliminate the paraxial coma and astigmatism and additional axial aberration induced by alignment. To eliminate each radial isotropic component in paraxial coma and astigmatism, coma-and-astigmatism-free plane N_0 of an objective lens and the spherical aberration correction system must coincide, which can be realized through the first group of transfer-lens doublet D_1. To reduce each anisotropic component of azimuth angle in paraxial coma and astigmatism, current direction of the first group of transfer-lens doublet D_1 is designed to be opposite to current direction of objective lens. In this way, paraxial coma and astigmatism can be minimized. The second group of transfer-lens doublet D_2 is used to reflect the first hexapole electromagnetic group in the second hexapole electromagnetic group without any enlargement, so as to eliminate the second-order aberration of hexapole electromagnetic group completely. Meanwhile, spherical aberration of the system is also compensated by subordinate and negative third-order spherical aberration of electromagnetic lens.

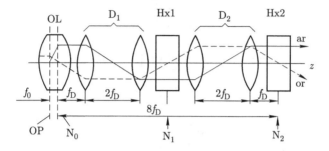

Fig.6.1.14 Scheme of an aberration-corrected objective lens system. OL denotes the objective lens (f_0 denotes the focal length of the objective lens), D_1 and D_2 are the two transfer round-lens doublets of focal length f_D, Hx1 and Hx2 mark the two hexapoles. OP is the object plane and N_0, N_1 and N_2 denote the position of the outer nodal planes of the doublets. These are also the coma-free planes. The path of an axial ray is denoted by "ar", the path of an off-axial ray is denoted by "or" (Urban et al., 1999).

The above spherical aberration correction system for an objective lens has been successfully installed in a 200 kV Philips CM 200 FEG ST TEM and is in operation successfully. After being installed with the spherical aberration correction system, the point resolution of this TEM is improved from the original 0.24 nm to 0.13 nm.

At present, design of a hexapole spherical aberration corrector involves two multipole systems only. Firstly, uncorrected remaining intrinsic aberration is the fifth-order sixfold astigmatism, which limits the available minimum probe dimension for the large aperture angles. The fifth-order spherical aberration of a hexapole correction system can be adjusted between positive and negative values, which makes six-fold astigmatism in design of a hexapole spherical aberration corrector be the only limit to the fifth-order axial aberration (Müller et al., 2006).

Design of a hexapole spherical aberration corrector for a STEM is shown in Fig. 6.15. In the middle of the last collecting lens and the objective lens, it is composed of two original hexapole elements (HP1 and HP2), five transfer lenses (TL11/TL12,

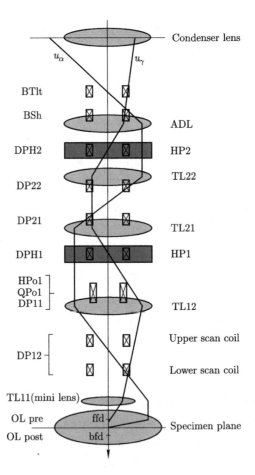

Fig.6.1.15 Schematic drawing of the optical design of a hexapole STEM corrector with transfer lenses (TL), adaptor lens (ADL), hexapole elements (HP), alignment deflectors (DP), beam tilt (BTlt) and beam shift (BSh) coils, and stigmators (QPol, HPol). In addition, the courses of the axial ray u_α and of a selected field ray u_γ are depicted (Müller et al., 2006).

TL21/TL22 and ADL), eight x/y alignment deflectors (DP11/DP12, DP21/DP22, DPH1/DPH2, BTlt and BSh), and two stigmators (QPol/HPol) for twofold and threefold astigmatism.

If a condenser mini lens is available, it can act as a replacement for the lower transfer lens TL11 and appropriately weighted DC offsets to the upper and lower scan coils emulate the deflector DP12.

The axial fundamental ray u_α and the selected field ray u_γ are described in Fig. 6.1.15. The selected field ray u_γ is chosen such that there is zero value in aperture plane with negligible coma aberration nearby front focal plane of the objective lens.

This stable primary hexapole filed results from two multipole elements, and each with six ferromagnetic pole pins. One hexapole element is assembled from pole pins, pole pieces, coils and yoke, as shown in the Fig. 6.1.16. In current electromagnetic design, center hole of multipole element has a diameter of 4 mm, pole piece is 30 mm long measured along the optical axis, and nickel-iron alloy is used to serve as pole pins to generate accurate regeneration magnetic field.

Fig.6.1.16 Single hexapole element of a STEM hexapole corrector having six ferromagnetic pole pins with pole pieces, coils, and yoke. The outer diameter of the yoke is 152 mm. The liner tube inside the hexapole bore is not shown in this picture (Müller et al., 2006).

6.2 Aberration corrected electron microscopy

The development of TEM itself is in close relation to the development of nanometer materials. With microminiaturization of electronic devices and rapid development of material science, nanotechnology, etc., there is an urgent need to observe material structure at atomic scale, so as to understand the relation between material structure and its property, and breakthrough of aberration correction enables TEM to achieve sub-atomic resolution, as shown in Fig. 6.2.1. (Muller, 2009). This section introduces new opportunity brought about to electron microscopy by aberration correction: 1) aberration-corrected HRTEM; 2) aberration-corrected electron holography; 3) aberration-corrected STEM; 4) atomic resolution chemical imaging using both EDX and EELS; 5) atomic-resolution electron tomography.

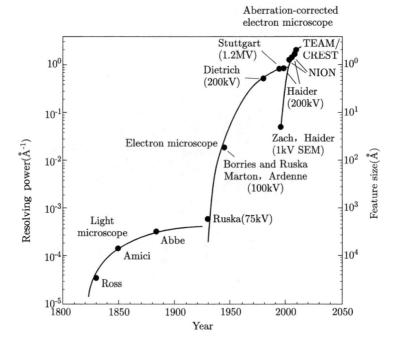

Fig.6.2.1 Hardware advances in imaging microscopies (Muller, 2009).

6.2.1 Aberration-corrected HRTEM

1. Introduction

As having been mentioned in Chapter 3, TEM has been an important bridge for human to know microworld. Knoll and Ruska built the first TEM in 1931. In later 1970s, HRTEM and AEM were built. Now HRTEM has become an indispensable comprehensive instrument for analysis on crystal structure and chemical composition. During observation of the world, no matter by naked eyes or by a microscope, spatial resolution is always the most important index for determination of image definition. Interface, dislocation, segregation, interstitial atom and other defect structures impose great influence on physical, chemical and electrical properties of materials and devices, thus, it is the research objective of high-resolution electron microscopy to obtain detailed information of atomic structure and point resolution of "sub-angstrom" is necessary for obtaining details of defected atomic structures. Resolution of HRTEM has broken through a new spatial scale of "1 Å" (1 Å=10^{-10} m, viz. one ten billionth meter) upon decades of effort by electron microscopy experts and technicians. The ability to observe microworld has been greatly expanded due to this important progress, enabling our visual world to directly touch microscopic atom world much deeper. It is necessary to develop a new-generation TEM equipped with components of quasi-monochromatic electron source, collecting lens system with dimension of electron beam spot less than 0.1 nm, objective lens spherical aberration corrector, aberration-less projection lens, energy-filtering imaging system, etc., for the purpose of achieving resolution of "sub-angstrom" in medium-voltage electron microscope.

To improve resolution of a traditional HRTEM, various theoretical methods and new technical means have been developed, and C_s-corrected electron microscopy and high-resolution microscopic image processing and image simulation therein have facilitated the development of sub-angstrom electron microscopy, and have gained great progress during the past decade. Remarkable achievement has been gained in C_s-corrected electron microscopy with respect to improving image quality. As having been mentioned, the resolution of a traditional TEM (Scherzer resolution) is defined as: $d_{\text{resolution}} = 0.66 C_s^{\frac{1}{4}} \lambda^{\frac{3}{4}}$, where λ is the wavelength of the incident electron and C_s is the spherical aberration coefficient of the lens. Thus the resolution can be improved through either rising the accelerating voltage of the electron microscope or/and reducing the spherical aberration of the objective lens. In history, ultrahigh-voltage HRTEM with electron energy up to several MeV have been built which indeed improved the resolution down to 1 Å. However, this kind of microscope is high purchase cost, and high maintenance cost, and can damage most of materials studied. During the past decade, improve resolution by manipulating spherical aberration coefficient (C_s) through developing new and practical spherical aberration correctors have gained great success. Basing on Rose theory, Zach and Haider succeeded in developing a spherical aberration and chromatic aberration corrector composed of quadrupole-octupole electromagnetic lens groups used for low-voltage SEM in 1995. In 1997, Haider, et al. developed a new-type spherical aberration corrector composed of two hexapole electromagnetic lenses (correctors) and two transmission double lenses (composed of two parallel lenses) applicable to a TEM for the first time, as shown in Fig. 6.2.2.a. In 1999, Krivanek, et al. from Nion Company developed a spherical aberration corrector composed of quadrupole-octupole electromagnetic lenses used for a 100 kV STEM, as shown in Fig. 6.2.2.b. The new-type spherical aberration corrector developed by Haider, et al. was firstly installed in a CM200 FEG TEM in Jülich in Germany in 1997 with the C_s value adjustable from conventional 1.23 mm to any value between +2 mm and −0.05 mm. But this kind of corrector can unfortunately only correct the spherical aberration and makes the chromatic aberration increased from 1.3 mm to 1.7 mm in the beginning. Nevertheless, the information resolution of the microscope has been increased from its original 0.24 nm to 0.13 nm. In addition, due to reduced contrast delocalization, phase contrast in image has been greatly improved. In particular, non-periodic structure like interface and defect can be provided very bright, sharp and clear image. In 2001, Oxford University of UK developed a twofold spherical aberration corrector which can correct spherical aberration of both condenser lens and objective lens on the basis of JEM 2010FEF TEM (Lentzen et al., 2002; Urban et al., 1999; Haider et al., 1998; Haguenau et al., 2003).

As shown in Fig. 6.2.2, the hexapole aplanatic system put forward by Rose includes three major parts:

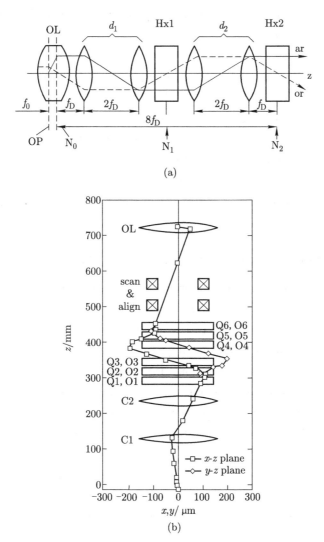

Fig.6.2.2 Diagram of spherical aberration corrector. (a) set at TEM; (b) set at STEM (Krivanek et al., 1999; Urban et al., 1999).

1) The first transfer lens system behind objective lens, which is composed of two round lenses;

2) Two hexapole field units;

3) The second transfer lens group between two hexapole field units, which is also composed of two round lenses.

Since later 1990s, high-resolution STEM began to become a main steam in TEM world. The development of STEM technology has gone through several important stages. Crewe, et al. invented the annular detector and introduced FEG gun in a STEM in 1970; Pennycook, et al. obtained HAADF (High-angle annular dark field) image from high-temperature superconducting layered material ($YBa_2Cu_3O_7$) in 1988. The image of the layered structure of atoms with different atomic numbers is directly given, and difference in contrasts for different atoms in the image can be properly explained by change of their atomic numbers (Smith et al., 2008; Smith et al., 2012).

2. Principle of spherical aberration corrected HRTEM

In practical, it is not enough to improve resolution just by incorporating the corrector itself. In addition, it is necessary to compensate parasitic second-order axial deviation, slanting spherical aberration and three-fold astigmatism which are in dependent relationship with spherical aberration, as well as to properly prevent nonspherical axial third-order aberration, star aberration and four-fold astigmatism. For this purpose, in the new instrument, the aberration coefficients are determined by means of an extended version of the diffractogram-tableau method proposed by Zemlin, et al. Amorphous diffractogram of specimen is recorded after correction change-over switch is off, and under focus and two-fold astigmatism induced by tilting the illumination is evaluated on-line. Correction of angle error start after correction change-over switch is on. A proper adjustment program is also put into operation at the same time, and this semi-automatic adjustment program needs 15 min. The difference between on-line and off-line correction change-over switch refers to Fig. 6.2.3.

Fig. 6.2.3(b) shows the correction of diffractogram and system in alignment. All diffractograms are generally in the same shape, indicating that spherical aberration of objective lens is almost eliminated. Even in case of tilting angle of electron beam at 30 mrad, residual axial aberration will only cause small under-focus and two-fold astigmatism.

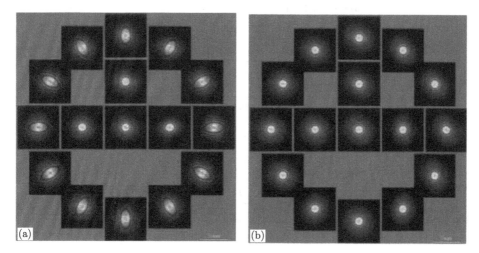

Fig.6.2.3 Diffractogram tableau of the microscope. (a) uncorrected, and (b), after correction for spherical aberration and proper alignment. The beam tilt angle is 10.8 mrad in both cases and the azimuthal angles vary between 0 and 2π in steps of $\pi/6$. The essentially identical shape of the diffractograms in (b) indicates the vanishing spherical aberration (Haider et al., 1998).

Fig. 6.2.4.shows comparison of PCTF: (a) TEM equipped with a spherical aberration corrector: $C_s = -30$ μm and $f = 10$ nm (over-focused); (b) traditional TEM: $C_s = 0.5$ mm and $f = -42.5$ nm (under-focused). The Scherzer resolution is about 0.2 nm. However, the TEM resolution can reach 0.1 nm or even smaller after equipped with the spherical aberration corrector. Fig. 6.2.5. displays a typical HRTEM image of the dumbbell structure with small black dots corresponding to Si atoms, and white dots to channels. Fig. 6.2.6. shows an image of a crystal along the [110] direction, which is similar to the dumbbell image with the point-to-point resolution of 89 pm.

Point-to-point resolution expressed by Scherzer formula in TEM image is (Scherzer, 1949):

$$\delta = 0.66 \sqrt[4]{C_s \lambda^3} \qquad (6.2.1)$$

where C_s is the 3rd-order spherical aberration coefficient and λ is the wavelength of the incident electrons.

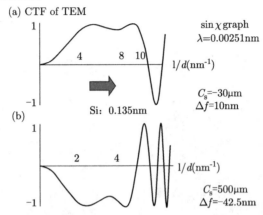

Fig.6.2.4 Phase contrast transfer functions for C_s-corrected (a) and ordinary (b) TEMs (Tanaka et al., 2008).

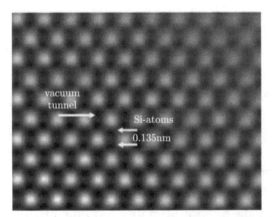

Fig.6.2.5 200 kV C_s-corrected TEM image of a silicon crystal observed along the [011] direction, $C_s = 0 \pm 1$ μm, slightly under-focused (Tanaka, 2008).

Resolution also depends on chromatic aberration damp which results from energy transmission and energy loss of the incident electron beam in specimen and chromatic aberration of the objective lens. It can be expressed as follows:

$$\delta_c = C_s \alpha \Delta \qquad (6.2.2)$$

$$\Delta = \sqrt{\left(\frac{\Delta E}{E}\right)^2 + \left(\frac{\Delta E_0}{E}\right)^2 + 4\left(\frac{\Delta I}{I}\right)^2} \qquad (6.2.3)$$

wherein, ΔE represents fluctuation of accelerating voltage; ΔE_0 energy loss of specimen; ΔI fluctuation of current of lens; the value α is a correction half-angle used in the objective lens.

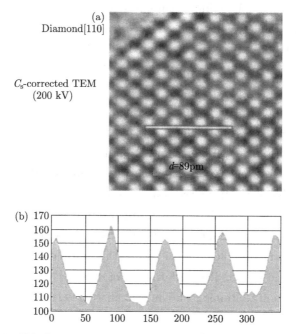

Fig.6.2.6 (a) 200 kV C_s-corrected TEM image of a diamond crystal along the [011] direction and (b) intensity profile along the line shown in (a). The image has a point-to-point resolution of 89 pm (Tanaka, 2008).

6.2.2 Aberration corrected electron holography

Spherical aberration-corrected TEM has greatly promoted the development of electron holography. It is interesting that the original purpose of electron holography is to eliminate spherical aberration of objective lens, so as to improve resolution (Lichte and Lehmann, 2008). Electron holography can be performed according to the following steps: its first step is to obtain phase contrast image through interference on imaging. The second step is to reconstruct image on the basis of the obtained phase contrast image, and conduct correction against spherical aberration through treatment by experienced method in the process, and aplanatic image can be obtained theoretically. Holography without objective lens was adopted in early electron holographic microscopy. However, since resolution is limited by spatial coherence, complete diffraction information cannot be received without an objective image plane, thus, the current common electron holography of off-axis image plane with objective lens is adopted. In spite of this, spherical aberration is still the major factor influencing resolution. Imaging quality generally depends on two factors: transverse resolution and signal resolution. Transverse resolution mainly depends on TEM recording holographic image, and is in relation to information limit of TEM, stripe resolution of holographic image, precision of spherical aberration correction, etc.. As for information limit, phase shift rather than image distortion caused by spherical aberration can be eliminated through selection of optimal defocusing without spherical aberration correction (Lichte, 1991). As for stripe resolution, it can be improved through adjustment of optimal position of the biprism. The signal resolution is determined mainly by the electron dose accumulated in one reconstructed pixel and by the contrast of the hologram fringes, as well as technical aspects like performance of detector (Lichte, 1996). Therefore, transverse resolution at atomic scale can be realized for electron holography without spherical aberration corrector. However, under atomic-scale resolution, information resolution cannot be improved by posterior method since point spread function is too large and due to attenuation electron beam for various reasons, contrast is lowered while influence of signal noise is strengthened. In particular to light atoms, displacement signal caused by scattering is too weak. So the resolution is still not satisfactory. Spherical aberration corrector can improve signal resolution of a single atom, with basic principle of improving utilization rate of coherent electron (Lichte et al., 2010). In addition, usage of spherical aberration corrector is good for precise correction of phase difference of holographic image in posterior method.

Principle of eliminating aberration of holographic test by posterior method is as below. Suppose the originally emitted objective wave is

$$o(\boldsymbol{r}) = a(\boldsymbol{r})\exp(i\varphi(\boldsymbol{r})) \qquad (6.2.4)$$

It arrives at CCD through various lenses, and image wave at this time is

6 Aberration Corrected Transmission Electron Microscopy ...

$$b(r) = A(r)\exp(i\Phi(r)) \tag{6.2.5}$$

a and A represent amplitudes of objective wave and image wave, respectively, and φ and Φ respectively the phases of objective wave and image wave. Supposing the amplitude of the reference wave is A_0, equation of superposition of objective wave and reference wave is:

$$I_{\text{hol}}(r) = I_0 + I_{\text{ima}}(r) + 2|\mu|A_0 A(r)\cos(2\pi q \cdot r + \Phi(r)) \tag{6.2.6}$$

$I_{\text{hol}}(r)$ represents light intensity of holographic image, I_0 and $I_{\text{ima}}(r)$, respectively, light intensities of reference wave and image wave, μ interference degree, indicating degree of consistence by distribution of intensity and energy of lighting system. In addition, in consideration of influence of non-elastic scattering electron, intensity of image wave can be expressed as below:

$$I_{\text{ima}}(r) = I_{\text{ima,el}}(r) + I_{\text{ima,inel}}(r) \tag{6.2.7}$$

$$I_{\text{ima,inel}}(r) = I_{\text{ima}}(r)P(r) \tag{6.2.8}$$

$I_{\text{ima,el}}(r)$ and $I_{\text{ima,inel}}(r)$ respectively represent light intensities contributed by elastic scattering and non-elastic scattering electrons, and $P(r)$ probability of generating non-elastic electron scattering.

$$A_{\text{el}}(r) = A(r)\sqrt{1-P(r)} \tag{6.2.9}$$

$$V_{\text{inel}}(r) = \sqrt{1-P(r)} \tag{6.2.10}$$

$A_{\text{el}}(r)$ represents scattering amplitude of elastic electron and $V_{\text{inel}}(r)$ attenuation factor of non-elastic scattering. Contrast ratio of stripes of holographic image can be expressed as below:

$$V = |\mu|V_{\text{inst}}V_{\text{MTF}}V_{\text{inel}} = \hat{V}V_{\text{inel}} \tag{6.2.11}$$

V_{inst} represents intrinsic contrast ratio, V_{MTF} transfer function factor from Mobulation of CCD, $\hat{V}$ from electron microscope, while V_{inel} from specimen. It shall be noted that bending and movement of stripes not only indicate intended phase displacement caused by specimen, but also indicate fake effect caused by electric charge on diaphragm, biprism and specimen holder, as well as distortion caused by projection lens, CCD, etc. Fourier Transform is conducted for image wave $b(r)$, empty-state reference is conducted to re-position, extract and correct man-made phase diagram, and then inverse Fourier transform is conducted to real space to obtain:

$$b_{rec}(r) = \hat{V} A_0 A_{el}(r) \exp(i\Phi(r)) \tag{6.2.12}$$

Conduct normalization for $\hat{V} A_0$ to obtain:

$$b_{rec}^n(r) = A_{el}(r) \exp(i\Phi(r)) \tag{6.2.13}$$

$b_{rec}^n(r)$ represents image wave obtained through inverse Fourier transform and A_0 the amplitude of the reference electron wave. Calculation of convolution of image wave and phase difference is conducted in accordance with transmission theory to re-construct original image. Supposing the TEM is in good coherence and lighting status, we can describe isoplanatic model by a wave transfer function (Born and Wolf, 1993)

$$wtf(q) = \exp(-i\chi(q)) \tag{6.2.14}$$

$\chi(q)$ represents phase difference function of the objective wave and it is in relation to phase difference coefficient and spatial frequency rather than to r of real space. If frequency spectrum of objective wave is $S_{Obj}(q) = FT[o(r)]$, frequency spectrum $S_{ibj}(q)$ of image wave can be obtained through transmission, then

$$b(r) = FT^{-1}[S_{ima}(q)] = FT[S_{Obj}(q) wtf(q)] \tag{6.2.15}$$

$\chi(q)$ represents additional phase caused by aberration during transmission, and can be expressed as:

$$\chi(q) = 2\pi k \left(\left(\frac{C_1}{2}\right) \left(\frac{q}{k}\right)^2 + \left(\frac{C_s}{2}\right) \left(\frac{q}{k}\right)^4 \right) \tag{6.2.16}$$

C_1 represents defocusing coefficient, C_s spherical aberration coefficient, and that's why optimal defocusing amount can eliminate spherical aberration. In consideration of interference of electron wave due to spatial coherence function E^{sc} and chromatic aberration function E^{chr}, $WTF(q) = E^{sc}(q) E^{chr}(q) \exp(-i\chi(q))$; wave transfer function is expressed by generated digital image plane during reconstruction; if precise spherical aberration coefficient is known, ultimately reconstructed objective wave will be

$$o_{rec}(r) = FT^{-1} \left[\frac{FT[b_{rec}^n(r)]}{WTF(q)} \right] \tag{6.2.17}$$

then aplanatic image is obtained.

It seems that geometric distortion no longer dominates due to "correction" by the above method. However, there are still some other influences of phase difference that limit precision of holographic experiment, mainly in transverse resolution and noise of reconstructed wave. As for transverse resolution, reachable resolution q_{max} is given by information limit q_{lim} of TEM, which is also in relation to E^{sc} and E^{chr}. As for noise limit, viz. whether a good transverse resolution image can be obtained under direct influence of phase detection limit. Phase detection limit refers to minimum phase difference of two distinguishable pixels detected under fringe contrast V and signal-to-noise ratio snr, and it can be expressed as $\delta\varphi_{lim} = \frac{snr}{V}\sqrt{\frac{2}{N_{res}}}$ where N_{res} represents number of electron received by each distinguishable pixel, which is in relation to coherent current density j_{coh} of reconstructed wave, and can be expressed as $N_{res} = 2\frac{j_{coh}t}{e(2q_{res})^2}$, wherein, t represents exposure time, e electron charge and q_{res} maximum wave numbers during reconstruction. $j_{coh} = -\ln(|\mu|)\frac{\varepsilon B_{ax}}{\pi(kw_{hol})^2}$, ε represents elliptical factor of light, B_{ax} brightness of electron source and k number of electron wave and w_{hol} represents width of holographic image. $w_{hol} \geq 4 \cdot psf = 2C_s\left(\frac{q_{max}}{k}\right)^3$, psf represents diameter of point spread function $PSF(r)$ which is expressed as $PSF(r) = FT^{-1}[WTF(q)]$, and they are all in relation to resolution q_{max} during optimal holographic focusing (Lichte et al., 1991). In consideration of attenuation $V = |\mu|V_{inst}V_{MTF}V_{inel}$ of fringe contrast V and detection quantum efficiency DQE (q) of camera, the ultimate phase detection limit can be expressed as:

$$\delta\varphi_{lim} = \frac{4\sqrt{\pi}snrC_s}{|\mu| \cdot V_{inel} \cdot V_{inst} \cdot V_{MTF}\sqrt{-\ln(|\mu|)\frac{B_{ax}}{ek^2}\varepsilon t DQE(q_c)}} \times \frac{q_{max}^4}{k^3} \quad (6.2.18)$$

q_{max} is taken for q_{res}. In posterior method, in case of correction by digital phase plane $\chi_{num}(q)$, it is necessary to know that precise aberration correction coefficient reaches $\delta\chi = |\chi(q) - \chi_{num}(q)| \leq \pi/6$; as for non-corrected TEM, all phase differences below quadrupole in relation to q shall be corrected, which means accurate information of nine correction coefficients is necessary, wherein, since actual tilting angle of electron beam cannot be obtained precisely, influence of C_s to other parameters is very large, such as ε.

Due to correction of spherical aberration corrector, $\chi(q)$ disappears and $\chi(q)$ gradient and psf are reduced greatly, thus improvement of holographic test is shown in the following aspects.

1) Information limit is improved. Transverse resolution finally depends on information limit, while information limit can only be eliminated by priori method rather than by posterior method. To spatial coherence function

$$E^{sc}(q) = \exp\left(-\pi^2 \frac{k^2 \alpha_{ill}^2}{\ln 2}(\mathrm{grand}\chi(q))^2\right) \qquad (6.2.19)$$

α_{ill} represents angle of emergent electron of electron source. When $\chi(q)$ gradient is zero, E^{sc} is 1. And to time coherence function E^{chr} can be obtained by correction coefficient C_c of chromatic aberration corrector.

2) Signal intensity is strengthened. Since imaging diaphragm is opened much wider, large-angle scattering electrons also contribute to image wave. Firstly, transverse resolution is improved, and signal of phase shift of atoms in image wave is also improved. As for phase shift of a single atom, if transverse resolution changes from 5 nm^{-1} to 10 nm^{-1}, 2 index factors are increased for phase shift of a single atom, 2 more index factors will be increased for phase shift of a single atom if transverse resolution changes from 10nm^{-1} to 20 nm^{-1}, and at last phase shift will reach an upper limit (Linck et al., 2006).

3) Signal resolution is enhanced. Width of holographic image shall meet $w_{hol} \geqslant 4 \cdot \mathrm{PSF} = 2C_s \left(\frac{q_{max}}{k}\right)^3$; as for traditional TEM, too many pixels are needed to reach atomic-scale resolution; if intensity scatters widely, signal resolution will be very low. After correction by spherical aberration corrector, psf attenuates to atomic scale, and width of holographic image can be selected as small as possible to be within view interesting us, such as a crystal boundary and its surrounding environment. In addition, holographic image becomes narrow, correspondingly coherent current is larger and exposure time can be reduced. In this way, attenuation of fringe contrast due to instability is also reduced.

4) Spherical aberration corrector enables more free adjustment of magnifying

power, equivalent focal distance and other optical parameters, thus improving flexibility of better distinguishing or using new optical path. Under intermediate magnification, lens of corrector can be used as long-focus objective lens, which can increase magnification multiple, e.g., in case of imaging for a magnetic specimen which shall undergo testing in field-free environment. The other benefit for holography of atomic-scale resolution is increase of magnification multiple from objective plane to the first image plane. In this way, too high voltage of biprism is unnecessary, thus reducing attenuation of fringe contrast due to instability of voltage.

5) Crystal orientation can be observed by tilting electron beam. The better the transverse resolution is, the stricter the requirement of electron beam for orientation of a crystal will be. But common mechanical method of tilting specimen is not precise enough. Tilting electron beam in traditional electron microscope cannot substitute tilting specimen for aberration parameters will be influenced. Small aberration will result in great change of result, while in C_s-corrected electron microscope, aberration due to tilt of electron beam is kept at a low value. Thus, it is available to use slight tilt of electron beam to substitute tilt of specimen in C_s-corrected electron microscope, and the method is effective in particular during real-time reconstructed equipment. Small transverse displacement can be compensated by changing of corresponding biprism.

6) Adjustment by posterior method will be more precise. Adjustment by posterior method shall meet $\delta\chi = |\chi(\boldsymbol{q}) - \chi_{\text{num}}(\boldsymbol{q})| \leqslant \frac{\pi}{6}$, otherwise, there will be large error during numerical correction. In conventional electron microscope, $\chi(\boldsymbol{q})$ changes greatly and is not easy to be met, while $\chi(\boldsymbol{q})$ of residual aberration in C_s-corrected electron microscope changes slightly and numerical artefacts are obviously reduced.

7) Higher-order aberrations like fifth-order spherical aberration, star spherical aberration and fourth-order spherical aberration which have become important in corrected system can be further corrected by fine adjustment through holographic posterior method. If spherical aberration corrector is not used, degree of freedom of adjustable parameters will be reduced, thus these higher-order aberrations will be difficult to be eliminated.

6.2.3 Aberration-corrected STEM

1. The development of STEM

Crewe and his colleagues laid a foundation for modern scanning TEM (STEM). They acquired monatomic resolution image for the first time and electron energy loss spectroscopy (EELS) at the same time by means of using cold field emission source and annular detector, thus making aberration corrector a fast development and generating sub-angstrom resolution, single light atom image, spectrogram of two-dimensional atomic resolution, etc. STEM designed by Manfred von Ardenne putting imaging lens before the samples instead of putting them behind the samples according to TEM design by Ruska. He realized only monitoring rather than focusing that transmitted electron needs to generate high resolution image, so STEM optical device can avoid chromatic aberration caused by energy loss arising from electron through samples. However, this is only the idealization in principle. Although he achieved 10 nm resolution image, he failed to use field emission source, so he gave up STEM design concept based on Ruska's TEM design (Pennycook, 2012).

Almost 30 years later, Crewe picked up STEM again. He realized the necessity of using high brightness cold field emission gun to achieve high electron beam intensity of small probe, and then the first instrument achieving resolution of 30 Å was born. Through a few years' development, resolution had reached 5 Å and then 2.5 Å (Pennycook, 2012). However, Pennycook working in Oak Ridge National Laboratory of America is the one who made the imaging reach the level of atomic resolution. He and his colleagues developed a complete set of method and theory for conducting Z-contrast imaging of atomic resolution for crystalline structure by means of a set of 100kV STEM with high resolution (Ye et al., 2003). This resolution achieved single atom image in electron microscope, which was formed by the ratio of elastic signal collected by annular to non-elastic signal collected by spectrograph. However, the ratio of cross section was about proportional to atomic number Z, so the acquired image was called Z-contrast image or HAADF image (Pennycook, 2012). There are two necessary conditions to obtain high resolution

Z-contrast image in experiment, i.e. high brightness electron beam spot of atomic scale and annular detector (Ye et al., 2003).

2. Advantages of STEM's aberration correction (Walther et al., 2013)

1) The improvement of spatial resolution of parallel beam's lighting and convergent beam's lighting means the smaller lattice fringe and crystalline information may be obtained further; meanwhile, upon given direction, lattice fringe in crystalline grain of small mixed crystal is more likely to be observed.

2) Displacement can be reduced, which prevents size of small crystalline grain and interface width from being disturbed by fringe effect during measurement and makes measuring results more reliable.

3) Under better signal to noise ratio condition, quantitative chemical analysis at thin region (because projection effect of overlapped particles shall be avoided.) shall be conducted by means of using EDX or EELS.

In the future, aberration correction of additional chromatic lens may be beneficial to further improving resolution of energy filter transmission electron microscopy (EFTEM), enabling larger energy window to be allowed during energy selection imaging (ESI) instead of long series of little energy range. The application of STEM with high resolution will benefit from the chromatic aberration and generate smaller beam diameter containing a certain proportion of total intensity, which decreases signals at the back of adjacent atomic columns on the contrary, thus local EELS analysis is more reliable.

3. Image modes

The smaller the probe is, the better the performance of STEM will be. Different selection of aperture and other parameters will lead to big "probe tail", which will decrease contrast of image and cause non-physical delocalization during microscope analysis. Until cold field emission source is used later, size of probe is mainly subject to the third-order spherical aberration ($C_s = C_{s3}$) of objective lens. Detector shall be optimized from two aspects, i.e. selecting correct diaphragm size of objective lens and defocusing with specific value of third-order spherical aberration as source demagnified sufficiently from the condenser lens. Astigmatism and defocusing shall be adjusted by naked eyes based on experience, but the situation has changed with the emergence of new-generation aberration corrector. Third-order spherical aberration correction is becoming controllable and higher-order aberration correction is becoming more important. A large collection of new aberrations to be controlled are also introduced in typical multiple-order corrector; resolution ratio is high enough for typical electron energy, so chromatic aberration will not be neglected any more. The size of aperture can be selected via adjusting the size of "flat" region in Ronchigram, but it is just a rough estimate, and part of flat region in Ronchigram will still influence the aberration error of image quality (Kirkland, 2011).

Scherzer has shown how to counteract big third-order spherical aberration with first-order defocusing and objective lens diaphragm with suitable size, which is only a part of correction but resolution ratio is extended immensely. Optimization conditions of BF-CTEM (bright field convergence) and ADF-STEM are different. Phase error of ADF through diaphragm of objective lens shall be zero, but small aberration of BF will generate phase difference of $\pi/2$. It is inevitable that higher-order aberration is failed to be correct although aberration corrector is used. Aberration corrector can control various low-order aberrations and partially counteract high-order aberration of objective lens angle within small range. No corrector is perfect, so error always exists in each process, and some small low-order residual aberrations always exist in final conditions of optimization corrector, whereas these aberrations are likely to become the limiting factor for aberration corrector (Krivanek et al., 2008; Intaraparsonk et al., 2008).

HAADF imaging performed in an aberration corrected STEM can easily be achieved at atomic resolution right now. However, interface sharpness of actually quantitative at atomic scale may be more complex. For example, it is not clear that how all atomic columns or signals of background HAADF influence measured sharpness or width of the single atomic layer. The capacity of HAADF to provide atomic scale information has improved obviously upon the development of aberration corrector. For example, Super STEM1 is the first set of field-emission gun STEM of UK100kV based on aberration correction, which can achieve spatial resolution of 1 Å.

Phase contrast imaging in STEM becomes totally different after the emergence of aberration correction. With the expansion of plane phase region, diaphragm of collector (be equivalent to condenser aperture of TEM) is expanded in order of magnitude, but most of current instruments are free of seriously damping information transmission, so phase contrast image with high resolution is easy to be obtained under sound signal to noise ratio condition. Fig. 6.2.7 shows the phase contrast image, octahedral rotation around the interface of $BiFeO_3/La_{0.7}Sr_{0.3}MnO_3$ is traced, and it can be seen from the image that several unit cells of octahedral rotation of $BiFeO_3$ thin film around the interface disappear (Borisevich et al., 2010). These results are generated with emergence of third-order aberration corrector, but the design of fifth-order corrector has already appeared and was achieved soon afterwards. Under the accelerating voltage of 300 kV, resolution of 0.63 Å was achieved in the [211] direction of GaN; soon after, resolution of 0.47 Å was achieved in the $\langle 114 \rangle$ direction of Ge. According to incoherent imaging of physics, limiting resolution is still subject to noise because balance is required between resolution and electron-beam current, but STEM succeeds in mastering the resolution of recorded image firmly. As for the application in spectroscopy, some spatial resolutions usually contribute to higher signal. Accelerating voltage needs to be considered, which becomes especially important for lighter elements having lower threshold value of collision damage energy, so it is more inclined to use lower accelerating voltage. Fig. 6.2.8. shows single layer BN which collects scattered electrons across range of 58 to 200 mrad as many as possible by means of using middle angle ADF detector. After elimination of noise and probe tail via Fourier filtering, image becomes clearer and it's easier to distinguish B and N which are impurity atoms in fact. O is replaced by N while C pair is replaced by BN of a unit, small holes will be formed on hexacyclic thin film of C, but the holes will be filled with moving carbon atoms very soon, and now Z-contrast image is visible to either light elements or heavy elements (Krivanek et al., 2010).

The most useful approach for light elements is annular bright field (ABF) and approximately half of diaphragm range formed by probe is covered by annular detector upon this mode. Just like what Rose put forward at the beginning, ABF is phase contrast image mode, so it has high sensitivity, but it is still featured by

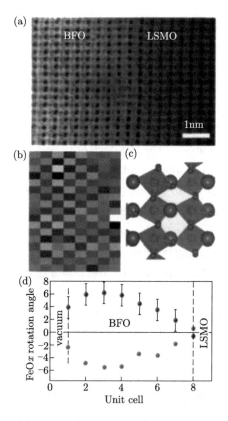

Fig.6.2.7 Oxygen octahedral rotation angles from BF STEM: (a) image of the 3.2 nm BFO/5 nm LSMO/STO ultrathin film, acquired simultaneously with the HAADF image, (b) corresponding 2D map of in-plane octahedral rotation angles in BFO showing checkerboard order (color scale on left); (c) BFO structure in rhombohedral (001) orientation showing the tilt pattern, (d) line profiles obtained from the map in (b) with two checkerboard "sublattices" added up separately: odd (top) and even (bottom); two bad points in the map corresponding to a hole in the sample are taken out of the averaging. The image in (a) was rescaled to correct for drift; the map in (b) was corrected for local Bi-Bi angle variations; error bars in (d) are equal to the standard deviation of local Bi-Bi angles (Borisevich et al., 2010).

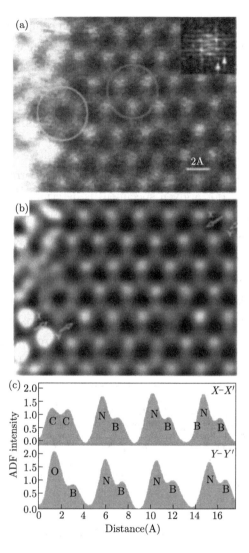

Fig.6.2.8 ADF-STEM image of monolayer BN. (a) as recorded. (b) Corrected for distortion, smoothed, and deconvolved to remove probe tail contributions to nearest neighbours. (c) Line profiles showing the image intensity (normalized to equal one for a single boronatom) as a function of position in image b along $X-X'$ and $Y-Y'$ (Krivanek et al., 2010).

incoherence because of high angle scope. Light atom column in heavy atom column can be imaged more easily by ABF image, so HAADF image greatly decreases the adaptation scope compared with ABF image. A result deserved to be concerned is the imaging of hydrogen atom column in hydride.

6.2.4 Atomic resolution chemical imaging using EDX and EELS

Working at the atomic level is facilitated by ultrasensitive energy-dispersive X-ray detectors in combination with C_s correction of the STEM probe. The signal strength in EDX spectroscopy (for a given specimen) is determined by three experimental STEM parameters: the accelerating voltage, the probe current, and the collection efficiency of the detector. For the atomic resolution chemical mapping, the first such results were published by D'Alfonso and co-workers (D'Alfonso et al., 2010; Allen et al., 2012) using a test sample of $\langle 001 \rangle$ SrTiO$_3$. Fig. 6.2.9(a) shows the HAADF line scan along $\langle 110 \rangle$ and Fig. 6.2.9(b) shows the simultaneously recorded EDX line scans. The EDX line scans shown in Fig. 6.2.9(b) are consistent with the HAADF image in Fig. 6.2.9(a). The Sr signal follows the reference HAADF line scan and this is to be expected since the HAADF contrast should be dominated by the Sr

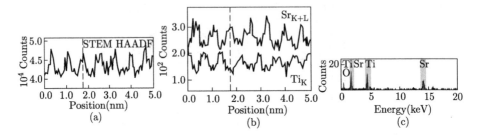

Fig.6.2.9 (a) Reference HAADF line scan of the SrTiO$_3$ sample along the $\langle 110 \rangle$ direction. The corresponding EDX line scans shown in (b) are for Ti and Sr and were constructed from the integrated shaded region of the X-ray spectrum indicated in (c), where a typical EDX spectrum is shown. The spectrum in (c) corresponds to the position indicated by the pink dashed lines in (a) and (b) (D' Alfonso et al., 2010).

columns. The Sr line scan in Fig. 6.2.9(b) was constructed from the sum of the Sr K-and L-shell ionization events. The corresponding Ti line scan only uses the Ti K-shell. The alternating Sr-Ti-Sr structure is clearly seen in the two line scans in Fig. 6.2.9(b). Fig. 6.2.9(c) shows a typical EDX spectrum that is instructive to notice that the count rates are quite low, so is the accompanying background. While the X-ray detector count statistics may be considered low, the peak-to-background (P/B) ratio is high. Therefore improved X-ray collection efficiency is ultimately necessary to maximize the signal for a sustainable electron dose if we are to obtain usable increases in the signal-to-noise ratio (S/N). For state-of-the-art EDX detectors P/B ratios between 60:1 and 100:1 are possible depending on the energy loss of the characteristic absorption edge in the EDX spectrum. We must stress that all of the data of Fig. 6.2.9. are obtained from a machine with no spherical aberration correction. It can be seen from the difference between Fig.6.2.9 and Fig.6.2.10 obviously. Fig.6.2.10 shows the effective scattering potentials for HAADF, Ti K-shell X-ray emission for EDX, and Sr K-plus L-shell X-ray emission. The HAADF potential is the most peaked and corresponds closely to the projected structure. It is more peaked around the Sr columns than on the Ti columns.

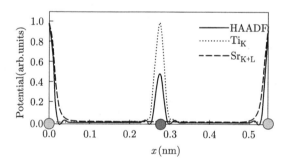

Fig.6.2.10 Effective Z-contrast inelastic scattering potential (denoted HAADF due to the high-angle annular dark field detector used in such imaging) for a detector spanning 60–160 mrad and effective ionization potentials for Ti K-shell X-ray emission and for the combined Sr K- and L-shell energy-dispersive X-ray (EDX) spectroscopy signals. The scattering potentials are for the $\langle 110 \rangle$ direction for a single unit cell of $SrTiO_3$ oriented along [001] and with the distance along the $\langle 110 \rangle$ direction denoted by x. The Sr and Ti sites are indicated by the blue and red circles, respectively. Each line plot has its maximum value normalized to unity (D' Alfonso et al., 2010).

Fig. 6.2.11 shows a crystalline specimen of GaAs along ⟨001⟩ zone axis and show chemical maps obtained for both K-shell and L-shell in Ga and As. These images

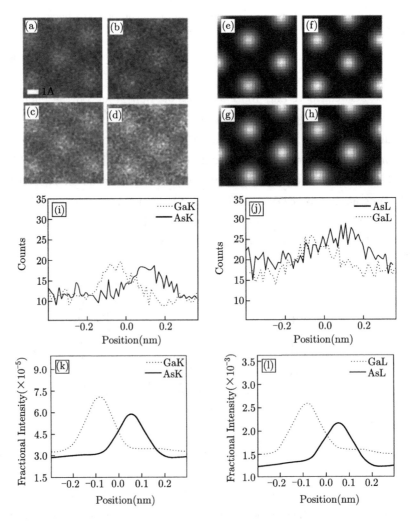

Fig.6.2.11 Experimental energy-dispersive X-ray (EDX) results for (a) the Ga K-shell, (b) the As K-shell, (c) the Ga L-shell, and (d) the As L-shell. The corresponding simulations are shown in (e), (f), (g), and (h). Line scans in the experimental data for (i) the K-shell and (j) the L-shell. The simulations corresponding to (i) and (j) are shown in (k) and (l), respectively (Allen et al., 2012).

are unfiltered, representing the raw intensities of the characteristic edges at each point in the map. The corresponding simulations are shown in Fig. 6.2.11(e)–(h). It should be noted that the predicted column positions in the simulations concur with the experiments. Line scans extracted from Fig.6.2.11(a)–(d) are shown in Fig. 6.2.11(i)–(j). As the data for each image was recorded simultaneously, probe channeling was the same for each data point. Not surprisingly, Fig. 6.2.11 (i)–(j) demonstrates a pronounced delocalization about the atomic site when images are constructed from the lower lying L-shell ionization event. Obviously evident in Fig. 6.2.11(i)–(l) is a difference in localization between the Ga and As signals. At the same time, we see a delocalized background. This delocalized background is attributed to the dechanneled thermally scattered electrons, as simulations using only the elastic channel predict a reduced background.

The authors have shown the advantages of atomic-resolution chemical mapping using EDX. New ultrasensitive EDX detectors installed on microscopes with C_s correction have made possible the routine collection of high-quality EDX maps with atomic resolution.

Main challenge for EELS is that chemical signal of inner electron spectrum of atomic-resolution weakens, typically including scattered electrons of 10^{-6}–10^{-3}. In recent years, progress in STEM equipment has allowed such technology to become the actual tool for analyzing materials. These progresses involve spherical aberration correction, improving spectrometer design, improving luminance of electron source and better stability of electron microscope, which will be explained in detail later.

STEM-EELS with high brightness electron source and spherical aberration correction is shown in Fig. 6.2.12. High-energy (100–300 keV) electrons formed by spherical aberration correction irradiate on sample surface area at atomic scale. Atomic-size electron beam passes the sample when raster scanning, which can generate spectrogram for each region of electron beam by recording electron energy loss spectrum. Data set of one spectrogram can reach 2–3 dimensions, including the energy spectrum from respective span to line baffle plate or beam position of one image. For EELS mapping, the nuclear electron of each atom shall be excited from each spectrum correspondingly to generate the atomic resolved mapping, but chemical mapping obtained by EDX refers to X-rays flow controlled by beam scanning. Generally, instrument can also be equipped with wide-angle ADF image detector to record ADF together with spectrogram (EDX map). The direct relation between the atomic-resolution contrast of ADF image and sample structure enable the role of ADF image determining electron beam position very useful, especially, contributive to explaining energy spectrum data.

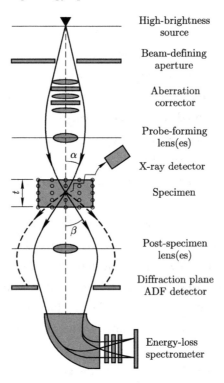

Fig.6.2.12 Schematic of the STEM setup used for atomic-resolution core-level spectroscopy, showing the major electron-optical elements (blue) and various detectors (orange). See the color plate (Dwyer, 2013).

6.2.5 Atomic-resolution electron tomography

Atomic-Resolution Electron Tomography (ET) refers to that acquiring the intuitional image of objects in 3D atomic arrangement status with atomic-resolution level. Aberration-corrected TEM and STEM provide a new opportunity to acquire atomic-resolution image of quantitative and directly interpretable data (Tanaka, 2008). 3D atomic image can be acquired through various approaches (Saghi and Midgley, 2012; Smith, 2012): certain amount of 2D images, 3D depth sectioning and tilting reconstruction, viz. record images at atomic-resolution level at each tilting angle and then reconstruct the 3D atomic images with computer.

1. 3D ET in aberration corrected TEM

When we desire to acquire ET for HRTEM with atomic-resolution, the sample is required to have similar weak phase object at different tilting angles; meanwhile, since HRTEM can only be acquired under proper zone axis, it only allows tilting in small angle range. The effective projection drawing acquired is seldom reconstructed relative to the conventional ET, so reconstruction method shall be improved correspondingly. Jinschek, et al. (Jinschek et al., 2008) developed a new reconstruction method, namely discrete tomography (DT). The crystal that the DT based on is assembled by discrete atomic layers, and it allows reconstructing the 3D crystal structure with less than 10 projection drawings even with noise and defects. The basic principle is: supposing crystal atoms are distributed in regular grid; the grid has some deviations due to defects, so it is occupied or in empty, and it is impossible to occupy more than 1 atom and fraction atoms in the same position. It acquires occupation atom number and linear equation respectively from horizontal, vertical, diagonal and back-diagonal directions; the corresponding value can only be in binary system (occupation of atom is shown with "1", and the vacancy is shown with "0"), thus the only occupation information is obtained. DT method has another advantage in that the measured value is directly related to atom number which is an integer value.

2. 3D ET in aberration corrected STEM

3D reconstruction image acquired with HAADF-STEM draws much attention than phase contrast due to its strength distribution image that can be directly explained. In one image mode, 3D depth sectioning can only be acquired under STEM; the basic principle is that scan the object layer by layer at different focal lengths by using scanning beam and finally gather information at all layers up for reconstruction. With the development of aberration correction technology, scanning beam can be reduced to sub-angstrom level under STEM at present to enable the scanning beam to have high lateral resolution and longitudinal atomic-resolution; another mode is series tilting, and discrete reconstruction technology can be used. However, knowing lattice periodicity required and vacancy was supposed as non-existed in early days, which limited the accurate reconstruction under 3D atomic-resolution. Scott, et al. (Scott et al., 2012) developed a 3D atomic-resolution reconstruction technique which is to know the advance information by using HAADF-STEM, viz. equally sloped tomography (EST). It replaces the conventional angle increment with equal slope increment method to collect a series of tilting pictures and acquire the data in consistent with that in reciprocal space with an iterative algorithm based on Fourier transform. Different from the DT, EST requires acquiring complete series tilting and eliminating projection on zone axis to avoid channel effect (Saghi et al., 2011). With EST, Scott, et al. obtained the 3D structure with nanoparticle of ~10 nm Au and resolution of 2.4 Å. Since EST requires acquiring complete series tilting, the sample is required keeping steady when being irradiated by electron beam. For some samples that are intolerant for radiation, limited tilting images can be obtained, so Compressed Sensing-Electron Tomography (CS-ET) is developed. 3D atomic-resolution reconstruction image can be successfully acquired with CS-ET although it has not been reported in relevant references. CS-ET takes advantage of the sparsity of the tomogram in a chosen transform domain and employs a convex optimization algorithm to find the sparsest solution in the domain, subject to consistency with the acquired data.

3. Challenges to atomic-resolution ET

Periodic structure is distributed in disperse, so 3D atomic-resolution image can only be reconstructed by utilizing the projection drawing less than the conventional ET. Generally speaking, the more the 2D image acquired, the higher the resolution of electron tomography reconstruction will be. It is applied to 3D atomic-resolution tomography of non-crystalline and polycrystalline materials; but for crystalline material, the principle is not applicable since atom structure can only be obtained in several main zone axes directions.

In terms of experiment, ET requires no irradiation from electron beam, structure change or kinetics absorption in image acquiring process. It has been able to acquire the atomic arrangement and distribution of ideal crystalline material with AR-ET, but there are still some challenges to 3D atomic-resolution tomography for crystals with defects. To solve the problem, it is necessary to improve the method and establish complex reconstructing algorithm.

6.3 Applications of aberration corrected electron microscopy

6.3.1 Direct imaging of light element by ABF-STEM

TEM is one of the important means to observe material microstructure under high spatial resolution. It has been an inevitable tendency for scientific and technological development at present to investigate the physicochemical property of materials with high spatial resolution (less than 0.1 nm) and to solve the specific problems in condensed matter physics. Due to various aberrations of objective lenses, especially the spherical aberration thereof, it is very difficult for the conventional transmission electron microscope to achieve sub-angstrom resolution. It often needs complex image processing and calculation to obtain improved high resolution. Meanwhile, TEM faces an another important technical challenge, namely how to directly image light element (such as hydrogen (H), lithium (Li), boron (B), carbon (C), nitrogen (N), oxygen (O), etc.). Because of their relatively small scattering cross-section, it is almost unable to directly identify the spatial position of light atom with the conventional imaging method. With the advent of spherical aberration corrector at end of the 20th century, spherical aberration corrected transmission electron microscopy was widely applied, which enabled to directly acquire structure information at the sub-angstrom level. In particular, the latest ABF-STEM makes it possible to directly observe light elements. This section will mainly present the application of ABF-STEM on direct observation of H atom and Li atom.

1. Direct observation of lithium at the atomic scale

As the positive material for lithium ion battery, the inherent poor electronic conductivity and ion transport of LiFePO$_4$ are the important factors affecting its performance. It can effectively improve the dynamics performance through methods like reducing particle size and coating carbon layer. The lithium ion batteries composed of the modified LiFePO$_4$ cathode have been applied to electric vehicle and stable energy storage device. However, there is debate on de-lithiation and/or lithiation mechanism of the LiFePO$_4$. Generally, the lithium insertion/extraction into/from LiFePO$_4$ is a two phase reaction, namely LiFePO$_4$ $\leftrightarrow$ Li$^+$ + FePO$_4$ + e$^-$. To better show the reaction mechanism of LiFePO$_4$, many models have been proposed, such as core-shell model (Padhi et al., 1997), domino-cascade (Delmas et al., 2008)and phase transformation wave model (Burch et al., 2008). We are unclear about the real process due to limited direct observation of Li ion.

Recently, Gu Lin, et al. (Gu et al., 2011) successfully realized the direct observation of Li ion in LiFePO$_4$ with the advanced ABF-STEM. Fig. 6.3.1(a) shows the schematic diagram of ABF imaging, the ABF image of pure phase LiFePO$_4$ in Fig. 6.3.1(b) directly reveals the lithium atom sites with corresponding structure schematic shown in the inset. Detailed contrast analyses by the line profile in Fig. 6.3.1(c) acquired at the box region in Fig. 6.3.1(b) confirmed the lithium contrast in comparison to that of oxygen.

Fig. 6.3.2 shows ABF images of LiFePO$_4$ with different electrochemical state viewed from the [110] zone axis. In comparison to the pristine state shown in Fig. 6.3.2 (a), in which the lithium sites are marked by yellow circles, the all lithium ions are extracted after full charge shown in Fig. 6.3.2(b), in which the delthiated sites are marked by orange circles. After partially charge ($\sim$0.5 mole of Li is extracted), part of the lithium remains in the lattice (as shown in Fig. 6.3.2 (c)) and preferably occupy every second layer along the b axis, thereby exhibiting unexpected staging phenomenon. It is the first time to observe the interleaving removal and even "staging" phenomenon of Li ion in partially charged LiFePO$_4$, which is not consistent with phase boundary migration, shell structure and other various reaction models proposed. It is of great significance to deeply understand the lithium storage mechanism and "staging Phenomenon" of the material.

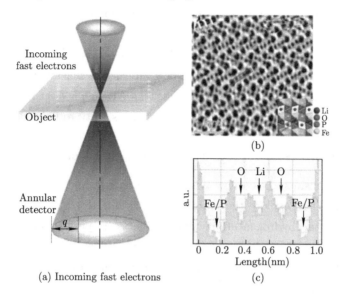

Fig.6.3.1 (a) The schematic diagram of ABF imaging geometry, (b)A demonstration of lithium sites within a LiFePO$_4$ crystal is shown in (b) with the corresponding line profile acquired at the box region shown in (c) to confirm the lithium contrast with respect to oxygen (Gu et al., 2011)

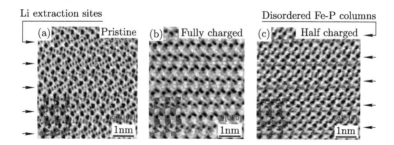

Fig.6.3.2 Initial image in LiFePO$_4$ [010] direction and completed charged and partially charged images are shot through the STEM-ABF technique (Gu et al., 2011).

Previously, Shao-Horn, Y, et al. observed the Li ion in LiCoO$_2$ through HRTEM (Shao-Horn et al., 2003), which, however, required complex operation and expensive calculation method (focal-series reconstruction). In their paper, the author states clearly that it is hard to observe the Li ion with the method and puts forth the possible substituting method, viz. realizing the observation of the Li ion with high resolution using high-voltage TEM (the energy is greater than 1MeV). At present, the ABF-STEM is used as the best alternative scheme, which realizes the direct observation of the Li ion. Meanwhile, the ABF-STEM ensures to be able to observe lighter element as hydrogen due to its sensibility to light element.

2. Direct atomic scale observation of the lightest element: hydrogen

Hydrogen is considered as a new type of clean energy, and lots of researches have been conducted to explore the material that is able to store hydrogen. Since arrangement of the hydrogen atom directly affects its performance, it is urgently needed to observe hydrogen atom. Thanks to the development of spherical aberration correction and ABF technique, the lightest atom H can be directly observed finally after oxygen, nitrogen, carbon, boron, lithium, etc. (Findlay et al., 2010; Oshima et al., 2010; Findlay et al., 2011). In the past several decades, electron microscope has witnessed great development, in particular to the appearance of spherical aberration correction of electromagnetic lens. Resolution of sub-angstrom level can be realized for the electron microscope at present, so it is more sensitive to light atom comparing with the HAADF image. TEM with spherical aberration correction successfully identifies the concentration of oxygen in oxides (Jia and Urban, 2004) and observes the detailed structure of single-layer border of graphene (Girit et al., 2009). HAADF technique realizes more remarkable observation, and it enables to observe nitrogen, boron in boron nitride (Krivanek et al., 2010); however, contrast of HAADF image exhibits a $Z^{1.7}$ dependence with respect to atomic number Z, and its sensibility to elements is related to the scattering power of the relevant atoms. The element that have small atomic number Z, such as hydrogen atom, it is impossible to be observed due to the weak scattering capacity. Contrast of images acquired with the newly developed ABF technique for detecting light atoms is based on phase contrast with wave interference, which requires the object (atoms) only to alter the phase of a wave (weak-phase object (Cowley, 1969),WPO), so it has great advantages on measurement of light element.

Abe, et al. observed hydrogen atom in YH$_2$ through the latest ABF technique on spherical aberration-corrected TEM (Ishikawa et al., 2011). It is extremely difficult to directly observe the H atom due to weak scattering power, the intolerance of irradiation and the instability of most substances containing hydrogen under electron microscope. To acquire the direct image of H atom, they selected YH$_2$, one of the most stable hydrides in thermodynamics, as the research object.

Fig. 6.3.3 (a) shows the structural diagram of crystal YH$_2$ along the [010] direction, its corresponding ABF, bright-field and ADF images are shown in Fig. 6.3.3 (b), (c) and (d), respectively. It can be found from comparison of (c) and (d) that it cannot observe hydrogen atom with bright-field and ADF under the same conditions, which embodies the superiority of ABF technique on observation of light element.

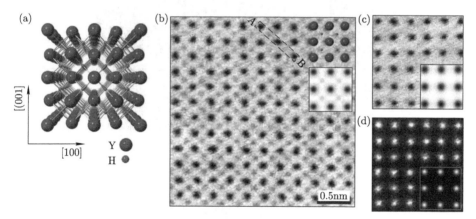

Fig.6.3.3 (a) The structural diagram of crystal YH$_2$ along the [010] direction, (b) ABF image, (c) bright-field image and (d) ADF image (Ishikawa et al., 2011).

In addition, the authors analyzed ABF images acquired along the [110] and [100] direction, and further confirm that ABF approach is reliable to observe hydrogen atom.

6.3.2 Applications of aberration correction high resolution TEM

The main advantage of spherical aberration correction is that structure-imaging artefacts due to contrast delocalization can to a great extent be avoided. These artefacts have turned out to be a major obstacle for the application of instruments with field emission guns in defect and interface studies (Thust et al., 1996). Contrast delocalization arises from the width of the aberration discs belonging to the individual diffracted electron waves whose diameter increases with the spherical aberration.

Fig. 6.3.4.a shows in a cross-sectional preparation, the interface of $CoSi_2$ grown epitaxially on Si (111) seen along ⟨110⟩ direction. The image was taken under Scherzer defocus conditions without aberration. At the interface, an approximately 2-nm-broad region of darker contrast can be seen. The width depends on the defocus value reach in a minimum at the Lichtede focus (Fig. 6.3.4(b)). If the structure is imaged in the aberration-corrected condition (Fig. 6.3.4(c)), the delocalization has

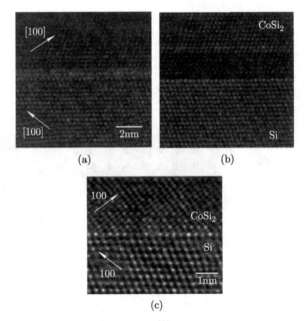

Fig.6.3.4 Structure images of an epitaxial Si (111)/$CoSi_2$ interface demonstrating the production of image artefacts by the effect of contrast delocalization due to spherical aberration. Images in (a) and (b) were taken in the uncorrected microscope at Scherzer defocus and Lichte defocus, respectively. (c) Image taken in the aberration-corrected state at Scherzer defocus does not show any delocalization ($C_s = 0.05$ mm) (Haider et al., 1998).

essentially disappeared and the interface is atomically sharp (Haider et al., 1998).

Spherical aberration correction improves the capability of elemental mapping, so it is able to calculate elemental mapping through the energy filtering image. The delocalization during energy loss can be calculated to 1.8 Å (Kohl and Rose, 1985), and is sufficient to allow atomic-scale resolution. Fig. 6.3.5 shows a HRTEM image of a cross-sectional sample of LaAlO$_3$ on LSAT ((LaAlO$_3$)$_{0.3}$(Sr$_2$AlTaO$_6$)$_{0.7}$). An ordered structure caused by O-deficiency can be seen in the LaCoO$_3$ film (the one pointed by arrow in Fig. 6.3.5.) (Hansteen et al., 1998).

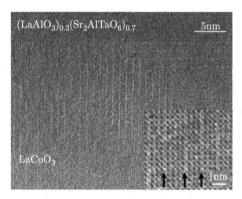

Fig.6.3.5 HRTEM image of LaAlO$_3$ on LSAT; the inset shows an order structure at a higher magnification (Kabius et al., 2009).

The same area shown in Fig. 6.3.5 has been used for C_c corrected elemental mapping. Fig. 6.3.6(b) is C_c corrected element maps of La, Fig. 6.3.6(a) shows the results of La recorded with an uncorrected Tecnai F20 (experiment parameter is unchanged: exposure time: 60s, energy loss for 860 eV of ~150 electrons/pixel) without installing spherical aberration corrector from the same sample. Fig. 6.3.7 shows the elemental maps acquired through line-scanning with ~1 nm wide lines. The apparent interface width measured from the uncorrected La map is > 2 nm while the C_c-corrected measurement gives an interface width of ~4 Å which is equivalent to the distance of La atoms in LaCoO$_3$. To sum up, the resolution of elemental mapping after C_c correction is only limited by noise and sample drifting, and further improvement in resolution by using higher beam currents and by minimizing the sample drift can be expected (Kabius et al., 2009).

Fig.6.3.6 Elemental maps of La: (a) uncorrected, (b) C_c corrected (Kabius et al., 2009).

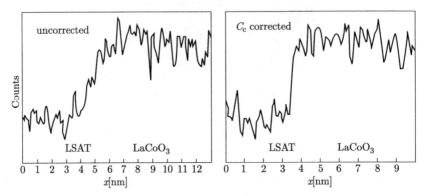

Fig.6.3.7 Line-scans derived from elemental maps of La across the interface LSAT/LaCoO$_3$ (Kabius et al., 2009).

It has been illustrated that application of spherical aberration correction in HRTEM brings the new possibility to improve the resolution; it improves the resolution not through the very high tension or reducing diameter of geometrical pole piece which could reducing spherical aberration of objective lens. Spherical aberration corrected transmission electron microscope operated at 200 kV prevents the sample from irradiation damage generated through atom collision under high tension. Besides, application of standard pole piece means that sample conversion and tilting will not be affected.

It requires to observe the materials microstructure more carefully with spherical aberration-corrected TEM. In quantitative study of the defects and surface in material, disadvantage of the spherical aberration-corrected TEM lies in that it requires reducing the contrast delocalization effect relying on C_s adjustment, and the advantage lies in that it is related to the application of image calculation technique. Fig. 6.3.5 contains the quantitative information prepared as computer language through electronic channel of the sample in quantum mechanics. Quantum mechanics image calculation and numerical wave function iteration technique are part of microscope practice at present.

It is incapable of exploring the application result of Bragg diffraction contrast imaging in expansion sampling after spherical aberration correction, such as crystal boundary and dislocation.

When the sample is a crystal, it forms HRTEM phase contrast image through the interference of incident wave and diffracted wave. Up to now, we have mostly used the linear relation formed through interference of incident wave and diffracted wave based on Fourier imaging principle. The point resolution of the conventional TEM is around 0.2 nm, and the nonlinear interference fringe, such as half-spacing lattice fringe, is non-interferential image contrast because the aberration damping of contrast transfer function makes the fringes not be images. After spherical aberration correction, resolution of TEM is better than 0.1 nm. It has become necessary to remove or decrease the nonlinear contrast for accurate image interpretation.

N. Tanaka, et al. has studied the process of forming HRTEM image after spherical aberration correction and has developed a new method, viz. comparing the two images acquired from spherical aberration-corrected TEM (Yamasaki et al., 2005). Fig. 6.3.8 shows an example of images before and after the subtraction. The contrast of the false image cannot correspond to the real atoms; the false contrast can be completely eliminated by image comparison. Furthermore, a newly revised image deconvolution method using a Wiener filter to recover the transmission of image information by the objective lens at lower spatial frequencies. The TEM image reflecting the real crystal structure were successfully acquire. It was also proved theoretically to the second order that the present subtraction method can even be applied to amorphous samples upon assuming the phase object approximation.

Another method of reducing the nonlinear image contrast is the adoption of negative C_s values, as proposed by the Jülich group (Jia et al., 2003). In this case,

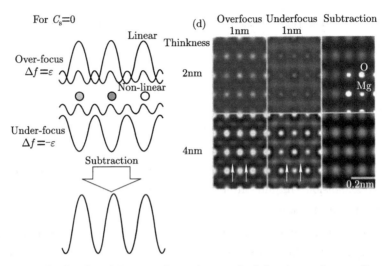

Fig.6.3.8　Newly developed image-subtraction method for decreasing nonlinear image contrast (Tanaka, 2008).

an ordinary linear term has a positive contrast (white contrast in a bright-field image), which is in agreement with white nonlinear image contrast.

Spherical aberration-corrected TEM has been adopted gradually in all laboratories; meanwhile, many scientists in electron microscopy cooperate with the international excellent laboratories to conduct the study on spherical aberration correction electron microscopy.

6.3.3　Applications of aberration correction STEM

Hexapole aberration corrector is used in TEM, and quadrupole or octupole corrector is used in STEM; upon some improvements, resolution of 1.36 Å of Si ⟨110⟩ dumbbell shape has been achieved with the accelerating voltage of 100 kV only (Dellby et al., 2001). Fig. 6.3.9 shows that single Bi atom is displayed as bright point in specific Si column, and the intensity of Bi atom depends on the location in Si column due to channel effect (Lupini and Dennycook, 2003). Resolution and sensitivity of EELS have been improved accordingly, which makes spectroscopy can distinguish La atom replaced separately in Ca column of $CaTiO_3$ (Varela et al., 2004). An amazing result is displayed from quasicrystal, Fig. 6.3.10 depicts the comparison of phase diagram of decagonal $Al_{72}Ni_{20}Co_8$ obtained by using HAADF detector and low angle ADF detector at the same time, HAADF image shows that 2 nm cluster in transition metal column is very clear, including the damage to symmetry center, which explains the origin of quasi-periodic tiling (Abe et al., 2003). However, in case of lower angle detector used in Penrose tiling mode, scattering intensity becomes more sensitive to amplitude of thermal vibration and specific position, and displayed bright point indicates the position of enhanced amplitude of thermal vibration. In case of a similar corrector being put into VG microscope HB603U, its resolution will reach sub-angstrom level. Fig. 6.3.11 shows the resolution of dumbbell shape Si ⟨112⟩ 0.78 Å (Nellist et al., 2004). At present, decagonal quasicrystal can display the position of each atom, including light element Al column which is hard to be distinguished by previous corrector. Fig. 6.3.12 shows the comparison of previous and current aberration correctors (Abe, 2012).

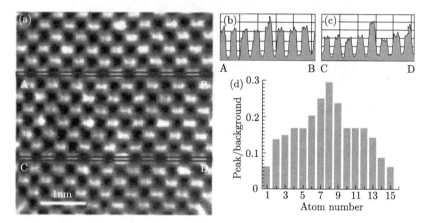

Fig.6.3.9 (a) HAADF image of Bi-doped Si [110]. (b) Intensity profile along line AB shows a Bi atom on the right hand column of a Si dumbbell. (c) Profile along line CD shows a Bi atom in each of the two column of dumbbell. (d) Histogram of Bi atom intensities (Lupini and Rennycook, 2003).

The new application of rare earth element dopant imaging with mechanical property improved is developed, especially in the Si_3N_4 ceramic field. It is also realized that the increase of diaphragm of objective lens not only increases the transverse resolution but also decreases the depth of field and makes it smaller than

Fig. 6.3.10 ADF images of decagonal $Al_{72}Ni_{20}Co_8$ obtained simultaneously with different detectors: HAADF detector, inner-angle ∼50 mrad (left), low angle ADF detector, inner-angle ∼30 mrad (right) (Abe, 2012).

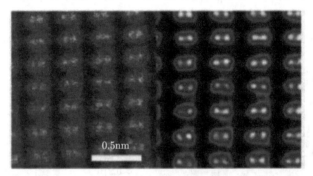

Fig. 6.3.11 Image of ⟨112⟩ Si recorded using a VG microscopes HB603U with Nion aberration corrector operating at 300 kV (left) and image after low-pass filtering and unwarping (right) (Nellist et al., 2004).

the typical sample thickness. Therefore, image can be got from different depths by changing focusing now. Image of single high-frequency atom can be obtained in dielectric layer of semiconductor with nanometer scale through above methods, and three-dimensional position and density of each high-frequency atom can be displayed. The slice with such thickness is very useful during analysis; however, probe focusing of specific depth is opposite to the trend of channel in electronic column, and the slice value of depth is related to relatively weak column. However, the defect of single Au point in Si nanowire is determined from focusing series directly; although small Au cluster is on the surface of nanowire, Au atom on the

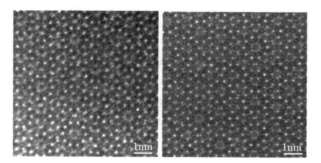

Fig. 6.3.12 Atomic-resolution HAADF images of the decagonal quasicrystal $Al_{72}Ni_{20}Co_8$ taken by a 300 kV VG Microscopes HB603U before (left) and after (right) installation of a Nion aberration corrector (Abe, 2012).

surface of nanowire center through focusing will become indistinct or cannot be seen. So the configuration for point defect inside can be observed.

The advantages of high resolution HAADF image are exhibited on study of the defect in materials, such as impurities in grain boundary, dislocation core structure, etc. Physical properties of material are influenced significantly by interface and crystal boundary. Impurities are prone to segregate in these defects, thus modifying the properties of material. Therefore, it is necessary to know about atomic-scale structure and composition of grain boundary. We take two examples to describe the application of HAADF imaging, which can provide important information that is unavailable through other approaches. Chisholm, et al. has studied Si crystal mixed in As, which tends to precipitate in the grain boundary (Chisholm et al., 1998). Fig. 6.3.13 depicts the HAADF image of Si in 23° tilt boundary which comprises a chain of dislocations (one edge dislocation and two mixed dislocations), and the intensity of atomic column of mixed dislocations is higher than those of other atomic columns by 20% or so, which signifies that As atom is accumulated in atomic columns of these dislocation core. They thought there were two As atoms among approximately 40 Si atoms in each atomic columns with respect of intensity. In addition, they thought As atom existed in grain boundary in form of As atom pair based on this experiment and in combination with theoretical calculation. The other example is the changes of grain boundary structure caused by precipitation of MgO crystal's impurities in grain boundary (Yan et al., 1998). The HAADF image

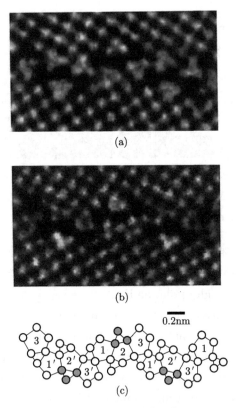

Fig. 6.3.13 (a) HAADF image of a symmetric 23° [001] tilt boundary in silicon. The dark features correspond to open channels formed in the center of the boundary dislocations in this ⟨001⟩ projection. (b) HAADF image of the same grain boundary after it has been doped with arsenic. The extra intensity seen at particular sites in the boundary indicates arsenic segregates to special sites in the boundary without otherwise changing the boundary structure. (c) A schematic of the projected atomic column positions obtained directly from the image (Chisholm et al., 1998).

of MgO crystal in crystal boundary is shown in Fig. 6.3.14, which directly displays that the structure of this crystal boundary is different from widely recognized atomic model of grain boundary put forward upon theoretical calculation by Harris, and is closer to the grain boundary model of densely arranged atoms proposed by Kingery. It also can be seen from the image that the intensity of some atoms in grain boundary is higher than that of others. The study of EELS in grain boundary shows that bright points is the result of calcium impurity (Ca, $Z = 20$) accumulating in these atomic columns. The theoretical calculation indicates the accumulation of impurities reduces the energy of such grain boundary.

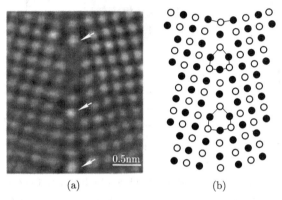

Fig. 6.3.14 (a) HAADF image of a symmetric 24° [001] tilt grain boundary in a MgO bicrystal. The white arrows highlight boundary columns with extra intensity. (b) Boundary structure model directly determined from the image. The open and closed circles indicate the location of the cations and anions on a single (001) plane (Yan et al., 1998).

As for the study of fine structure by EELS, as shown in Fig. 6.3.15, the anti-site defect of LiFePO$_4$, in special channel results in other channels being opened to Li ion transport, which is a very important phenomenon, for only retaining opened channels can enable battery to charge and discharge. An unexpected valence state is displayed in case of measuring $L_{2,3}$ peak ratio of iron with anti-site defect. If Fe replaces Li only, since its valence state is not +1 but exceeds +2, it will be higher than the valence state of native atom at the lattice site. Firstly, theoretical calculation reveals that there is an attraction between anti-site Fe (Fe$_{Li}$) and Li vacancy, which provides mechanism for Fe$_{Li}$. Although attraction is not enough to bind Li vacancy to Fe$_{Li}$, Li vacancy will be restricted in case of anti-site defect of accumulation through special channel. Li vacancy seldom continues to scatter but shuttles back and forth among Fe$_{Li}$ or always stops, thus resulting in reduction of net energy. The unusual energy-lowering mechanism is a feature of the one-dimensional nature of the Li diffusion paths (Lee et al., 2011).

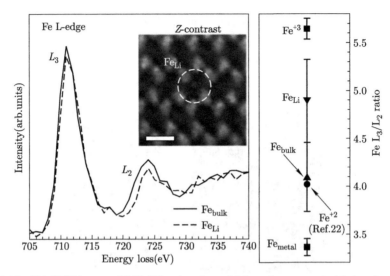

Fig. 6.3.15 HAADF image of LiFePO$_4$ showing a higher than expected intensity at some Li sites (red circle). EELS from such sites reveals the presence of Fe, and the Fe$_{Li}$ L$_{2,3}$ ratio is higher than that of Fe in the bulk lattice sites Fe$_{bulk}$. Scale bar is 0.5 nm (Lee et al., 2011).

The other development direction of the material study is to conduct three-dimensional reconstruction of surface and crystal structure of nanoscale crystal by means of HAADF image. Because HAADF image reflects atomic position of the crystal directly, HAADF image obtained from different projection directions can be used to achieve three-dimensional shape and structure of the crystal precisely. It will be greatly beneficial from the achievement of external shape of nanometer crystal to know about the property of nanometer materials.

6.3.4 Atomic-resolution measurement of oxygen vacancies concentration in oxide materials

1. Atomic-scale imaging of oxygen vacancy in SrTiO$_3$

Controlling and restricting oxygen vacancy plays an important role in applications of perovskite structure ferroelectric materials, voltage dependent resistor and field effect devices. In order to obtain the oxygen vacancy distribution in $SrTiO_3$, David A. Muller, et al. (Muller et al., 2004) adopted ADF-STEM approach and EELS, and demonstrated detection sensitivities of oxygen vacancy is one to four.

Electronic state information of O and Ti can be obtained, EELS and Ti/O ratio can be directly calculated by O-K peak and Ti-L peak intensity ratio, however such method is not applicable to the measurement of low oxygen vacancy concentration for the signal error generated is usually within several percentages. Therefore, David A. Muller, et al. sharpens EELS edge, which reflects the change of electronic structure. Fig. 6.3.16(a) is EELS spectrum of O-K edge, from which the peak pattern change under various oxygen vacancy concentrations can be observed. Its variation trend is related to cluster of crystal boundary in $SrTiO_3$ caused by oxygen vacancy. The nominal oxygen stoichiometric ratio is estimated by adopting Hall Effect and assuming two free carrier corresponding to one oxygen vacancy. Fig. 6.3.16(b) shows the more useful EELS of Ti-L edge obtained, which clearly indicates the variations of Ti^{3+} and Ti^{4+}, and that each oxygen vacancy has transferred two electrons to Ti 3d orbital. In this way, oxygen vacancy concentration of EELS can be obtained.

2. Atomic-resolution measurement of oxygen vacancy concentration of $BaTiO_3$ film

To obtain the oxygen vacancy information of $SrTiO_3$ through EELS has created the new path of studying vacancy and vacancy clustering in material through electron microscopy (Muller et al., 2004). However, the detailed occupation information of oxygen atom cannot be obtained via such method. For the purpose of getting the detailed occupation of oxygen atom, the resolution of TEM and S/N ratio image of shall be improved. C. L. Jia, et al. adopted negative spherical aberration imaging technique in aberration-corrected electron microscope to conduct atomic-resolution measurement on oxygen concentration of $\Sigma = 3$ {111} twin crystal plane in $BaTiO_3$ film. At the position of twin-plane boundary, about 68% is occupied by oxygen averagely with the rest vacancy (Jia and Urban, 2004).

The studied sample refers to $BaTiO_3$ film prepared by pulsed laser deposition method on $Pt/Ti/SiO_2/Si$ crystal under oxygen partial pressure of 0.1 mbar and

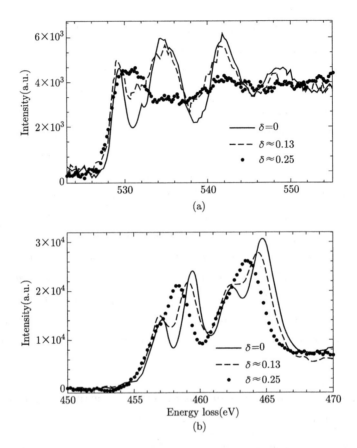

Fig. 6.3.16 Electron energy-loss spectra (EELS) for oxygen-deficient $SrTiO_{3-\delta}$, for $\delta < 0$, 0.13 and 0.25. (a) O-K edge, (b) The $Ti-L_{2,3}$ edge (Muller et al., 2004).

substrate temperature of 750 °C. The used electron microscope owns the electron-accelerating voltage of 200 kV and Philips CM200 field-emission gun electron microscope equipped with objective lens spherical aberration correction system adopts negative spherical aberration imaging technique, with $C_s = -40$ μm. Images are recorded by 1024 × 1024 CCD camera, with spatial resolution of about 0.02 nm/pixel.

Fig 6.3.17(A) shows the high-resolution image of $\Sigma = 3$ {111} twin boundary of $BaTiO_3$ [110] crystal, with "matrix" I, II and III referring to twin domains, respectively. Fig. 6.3.17(B) is the amplified image, from which we can see oxygen atom column lies between two Ti atom columns. Fig. 6.3.18 refers to "4" boundary in Fig. 6.3.17. The oxygen vacancy concentration estimated by measured intensity is between $0.4-0.7$. The average oxygen vacancy concentration obtained through

Fig. 6.3.17 (a) the high-resolution image of $\Sigma = 3$ {111} twin boundary plane of $BaTiO_3$ [110] crystal, (b) he amplified figure, from which we can see oxygen atom column lies between two Ti atom columns (Jia and Urban, 2004).

data analysis from all six boundaries is 0.68 ± 0.02, with standard deviation of 0.16.

The results obtained by the author mainly cover two meanings: First, HRTEM obtained through negative spherical aberration imaging technique can be used to measure oxygen vacancy concentration under atomic-resolution. By combining computer image simulation and quantitive intensity measurement, oxygen vacancy

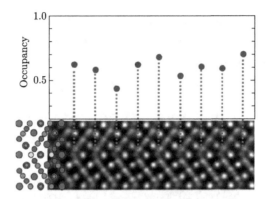

Fig. 6.3.18 Section of the twin boundary "4" (Fig. 6.3.17.) with the grain-boundary oxygen sites indicated (Jia and Urban, 2004).

in single lattice defect region can be measured. Such technique can be widely applied in defect study of oxide. Meanwhile, it can also be applied to other compounds rather than oxide only, including nitride and boride whose physical property is related to deviation of stoichiometric ratio. Secondly, it provides quantitive evidence for great reduction of oxygen vacancy. For example, 68% oxygen vacancy of $\Sigma = 3$ {111} twin boundaries in $BaTiO_3$ is occupied averagely, and the rest nearly one third is vacant.

3. Direct imaging oxygen vacancy concentration in $La_{0.5}Sr_{0.5}CoO_3$ film

Oxygen vacancy distribution and dynamics can directly control the performance of solid oxide fuel cell (SOFC), so the distribution of oxygen vacancy shall be obtained at the level of unit cell to obtain the working mechanism of cell. Young-Min Kim, et al. (Kim et al., 2012) developed the method of directly obtaining surface distribution map of oxygen vacancy of local unit cell via STEM.

Samples for experiment refer to $(La_{0.5}Sr_{0.5})CoO_{3-\delta}$ (LSCO) film grown on $NdGaO_3$ (NGO) and $(LaAlO_3)_{0.3}(Sr_2AlTaO_6)_{0.7}$ (LSAT) substrate. The reason for choosing such samples is their defect structure, electrochemical activity and the relation between electronic and ionic performances are very typical. Oxygen vacancy in LSCO is ordered, and its ordered phase is composed of ordered domains with different orientations of twin boundary and anti-phase boundary (Klie et al., 2001). The aberration-corrected STEM (HB603U with working voltage of 300 kV and Nion UltraSTEM with working voltage of 100 kV) is adopted.

Fig. 6.3.19 (a) and (b) refer to annular dark-field image (ADF) of LSCO film along $[110]_c$ (subscript "c" represents pseudo-cubic) orientation. Image contrast is proportion to $Z^{1.7}$, the brightest atom column refers to cation and the contribution of oxygen atom to contrast can be ignored. It can be observed from the image that structural feature of LSCO film is double unit cell, which is because of cation migration caused by oxygen vacancy ordering at (001) face. In ADF-STEM image, oxygen-deficient plane is darker than stoichiometric ratio one, so the overall contrast seems to be modulated. Such phenomenon is very common in material with oxygen vacancy ordering, such as cobaltite, ferrite and copper oxide (Inoue et al., 2010; Gazquez et al., 2011). Quantitive information of oxygen concentration in single layer cannot be obtained through observation of modulation contrast, therefore the author gets absolute stoichiometric ratio by adopting EELS and directly measuring lattice constant. Fig. 6.3.19(c) shows the EELS spectrum of LSCO/NGO film with vacancy order domain. O-K peak intensity reduces in non-stoichiometric ratio layer. Corresponding to the contrast variation of ADF image, the difference between O-K EELS peak integral intensity ratio of stoichiometric ratio layer and that of oxygen vacancy layer is $3.9 \pm 0.7\%$.

The existence of oxygen vacancy will cause structure change of perovskite oxide into brownmillerite structure which owns oxygen vacancy ordering in [100] pseudo-cubic orientation. For example, in LSCO, CoO_2 plane changes into oxygen vacancy plane CoO_{2-x}; when $x = 1$, Co cation is tetrahedral coordination in oxygen vacancy plane, while octahedral coordination in stoichiometric ratio plane (Fig. 6.3.20(b) and (c)). In the structural image of LSCO film on NGO obtained through aberration-corrected STEM, average space corresponding to Co ion is 0.30 ± 0.07 Å (Fig. 6.3.20.d), which is close to Co ion offset ~ 0.3 Å of brownmillerite structure in [100] orientation, thus LSCO oxygen depleted layer is completely tetrahedron and refers to $La_{0.5}Sr_{0.5}CoO_{2.5}$ in chemical correspondence. However, no Co ion pair is discovered in LSCO film grown on LSAT.

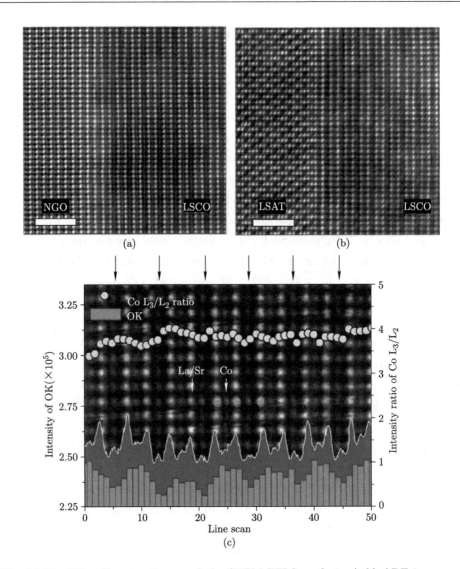

Fig. 6.3.19 Thin film structures and the STEM-EELS analysis. (a,b) ADF images of $[110]_c$-oriented LSCO thin films grown on different substrates, NGO (a) and LSAT (b). (c) A representative STEM-EELS result for a bulk region of the LSCO film grown on NGO substrate (Kim et al., 2012).

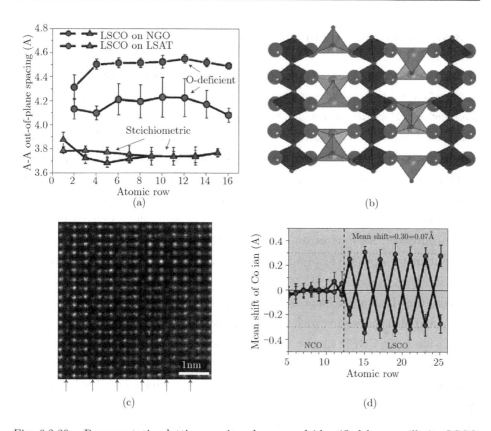

Fig. 6.3.20 Representative lattice spacing change and identified brownmillerite LSCO on NGO. (a) Interatomic spacing changes of A-site cations in the two respective systems, LSCO on NGO (red) and LSCO on LSAT (green), along the c axis. (b) Schematic of a brownmillerite structure $La_{0.5}Sr_{0.5}CoO_{2.5}$, where green spheres represent La/Sr atoms, blue spheres represent octahedral Co sites, orange spheres represent tetrahedral Co sites and small red spheres represent O atoms. (c) ADF STEM image of the [100] (b) (subscript (b) denotes brownmillerite)-oriented LSCO grown on NGO substrate showing the characteristic in-plane (vertical in the figure coordinates) pairwise shift of Co ions in the oxygen-depleted planes (solid arrows). The corresponding simulated image for the chemical composition of $La_{0.5}Sr_{0.5}CoO_{2.5}$ is given in the inset. (d) the error bars show the standard deviation with respect to averaging for each (vertical) atomic layer in the image (Kim et al., 2012).

6.3.5 Measuring strain distribution

Strain measurement is extremely important in semiconductor devices. For example, strained silicon technology can change the energy band structure, carrier mobility and device performance through processing technology, substrate growth and other methods. Currently, the relatively mature strain measurement methods with high sensitivity include X-ray method and micro-Raman spectroscopy method, but they are difficult for strain measurement in devices at nano-scale due to their low spatial resolution. On the contrary, TEM can meet such requirements due to its high spatial resolution. Electron microscopy methods of strain measurement mainly involve convergent-beam electron diffraction (CBED), high-resolution transmission electron microscopy (HRTEM) and dark-field holography based on geometric phase analysis (GPA) (Hytch et al., 2010). Therein, CBED method can analyze strain by measuring high-order Laue line in combination with the characteristics of small needle and high sensitivity, with the sensitivity accuracy as high as 2×10^{-4}. Moreover, wide-angle CBED can be used to get the distribution information of strain in samples by combining direct shadow image. However, considering elastic relaxation effect, the analysis of strain measurement is complicated (Houdellier et al., 2006). HRTEM is to analyze strain by directly measuring atom position. Warner, et al. analyzed the strain distribution in Z-shaped single-walled carbon nanotubes caused by stress bending by using spherical aberration-corrected HRTEM. When dark-field holography based on GPA is used to measure strain, strain distribution of the area about 200 nm can be obtained with satisfactory spatial resolution and strain analysis sensitivity. Its principle is that a beam of light of dark-field electron holography irradiates on complete strain-free crystal to be as the reference light, while the other beam irradiates on the object with strain to be measured as the object light. Both of them are superposed onto the screen through static biprism to generate coherence and form moire fringes, as shown in Fig. 6.3.21. (Hytch et al., 2010)

Coherent wave can be developed into $\psi(r) = \sum_g \psi_g(r) \exp\{2\pi i g \cdot r\}$ where r is on xy plane. And partial Fourier composition unit is $\psi_g(r) = a_g(r) \exp\{i\phi_g(r)\}$. The phase composition comprises of crystal field, applied magnetic field, material

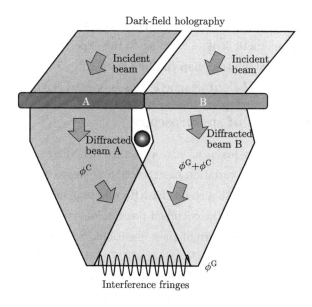

Fig. 6.3.21 Schematic diagram of dark field electron holography (Hytch et al., 2010).

average built-in potential and geometric phase: $\phi_g(\boldsymbol{r}) = \phi_g^C(\boldsymbol{r}) + \phi_g^M(\boldsymbol{r}) + \phi_g^E(\boldsymbol{r}) + \phi_g^G(\boldsymbol{r})$, among which the last one is also the phase difference between integrity region and strain region, coming from any partial displacement $\phi_g^G(\boldsymbol{r}) = -2\pi \boldsymbol{g} \cdot \boldsymbol{u}(\boldsymbol{r})$. Strain in x orientation is $\varepsilon_x = \partial \boldsymbol{u}(\boldsymbol{r})/\partial x = -(1/2\pi \boldsymbol{g})\partial \phi_g^G(\boldsymbol{r})/\partial x$. Thus, GPA theory points out that strain distribution can be obtained only by giving the phase difference between integrity region and strain region. But in order to get the strain distribution of large region, comparatively wider hologram should be obtained under Lorentz model. As the selected area aperture just equals to objective aperture which is too close to static biprism, so that traditional electron microscope cannot realize Lorentz model and electron holography simultaneously. Currently, dark-field hologrammetry under Lorentz model conducted on semiconductor device is mostly accomplished with the support of spherical aberration corrector. Under such model, objective lens of electron microscope is closed and spherical aberration corrector is opened to play the role of pseudo-Lorentz lens, so that the interferogram of holography can be obtained under small biprism voltage where that even the focal length of objective lens increases. By the way, aberration becomes very large under Lorentz model, esp. the aberration is difficult to be eliminated upon closing of class-VI electromagnetic lens. According to the experiment of Hytch, M. J., the S/N ratio of strain distribution map can be increased by 60% if aberration can be eliminated when class-VI electromagnetic lens is opened (Hytch et al., 2010).

6.3.6 Applications of spectroscopy

Mapping based on energy-loss near-edge structure (ELNES) surpasses the chemical signals extracted and the variation of extraction at near edge structure acts as a function of beam position. As a result, ELNES mapping embodies the atomic scale resolution, utilizing not only the occupied position of given atom, but also the specific intrinsic electronic structure of these positions. Such technology is extremely beneficial to understand electronic structure of materials. Besides, such method comes from stricter signal/noise (S/N) ratio, so the extracted energy loss of energy spectrum variation is required to be lower than 1 eV or smaller (Varela et al., 2005; Fitting et al., 2006; Kourkoutis et al., 2006).

ELNES mapping close to atomic-resolution should be dated back to the work of Batson, et al. (Batson, 1993) who obtained the information of Si valence state of Si/SiO_2 interface by changing $SiL_{2,3}$ near edge structure. Recently, Atomic-resolution ELNES has explained a series of problems connecting 2D chemical mapping in next-generation equipment (Varela et al., 2009; Tan et al., 2011).

Fig. 6.3.22 shows the results of Tan, et al. (Tan et al., 2011) who demonstrated the 2D mapping of Mn in Mn_3O_4. Mapping involves valence state trajectory in Mn fine structure. Under such cases, the splitting of Mn L_2 and L_3 white lines must be solved, which is also the reason of using monochromator. Variation of Mn $L_{2,3}$ fine structure is extracted by reference spectrum of multiple linear least square fit, with reference spectrum obtained through the average period. Variation of Mn $L_{2,3}$ fine structure at atomic scale is attributed to different Mn valence states in compounds, thus Mn^{2+} and Mn^{3+} valence states of predicted unit cell is displayed in Fig. 6.3.22. Moreover, Fig. 6.3.22 shows the valence state maps simulation, which can well correspond to experiment image. Such simulation is extremely important for understanding the provided mixing reference spectrum, that is to say, reference spectrum can include some mixing spectrum rather than demonstrating relatively pure Mn^{2+} and Mn^{3+} due to channel effect and delocalization effect of electron beam.

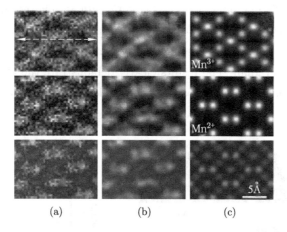

Fig. 6.3.22 Atomic-resolution mapping of Mn oxidation states in Mn_3O_4 along the [100] zone axis. (a) As-extracted spectral weights for Mn^{2+} and Mn^{3+}. A color map is displayed at the bottom (Red=Mn^{3+}, Green=Mn^{2+}). (b) Low-pass filtered data. (c) Simulated maps of the Mn^{3+} and Mn^{2+} signals (Tan et al., 2011).

Chemical mapping refers to be based on EDX spectrum. Therefore, an apparent advantage of STEM-EDX over STEM-EELS in the incoherent imaging regime is the efficiency of EDX detections of deep inner-shell ionizations. Another contribution of EDX surpassing EELS is that maps can connect with atomic structure in a simple way, thus leading to the effectiveness of object function description in excitation process. What is necessary is that the description of object function effectiveness shall be applied in the determined EELS-based mapping with personal factor displayable, different from EDX-based mapping. On the other hand, EDX lacks accurate structure and relatively low detection efficiency so that it cannot analyze electronic structure. Finally, all chemical signals can actually be obtained through 100 mrad semiangle by EELS. Just on the contrary, characteristic X-ray is excited from all orientations of sample, so an obvious proportion, viz. 4πsr, shall be covered by the detector for obtaining more effective EDX.

EDX mapping of atomic resolution has been elaborated by Chu, et al. (Chu et al., 2010), D' Alfonso, et al. (D' Alfonso et al., 2010), and Watanabe, et al. (Itakura et al., 2013). Fig. 6.3.23 shows the results by Chu, et al. (Chu et al., 2010), who described the EDX mapping of In$_{0.53}$Ga$_{0.47}$As [110] atomic resolution by adopting the operating instrument of 200 kV spherical aberration-corrected TEM/STEM with EDX, which is in effect collecting X-rays emitted from the sample over a solid angle of ∼0.13 steradian. The results clearly demonstrate the atomic structure of [110] crystal orientation, with positions of atomic columns in dumbbell layout and mixing In-Ga atomic columns clearly observed.

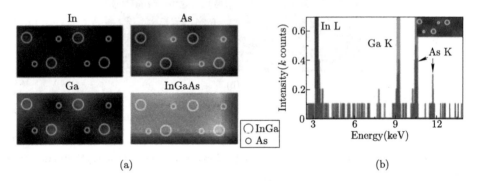

Fig. 6.3.23 (a) The STEM-EDX chemical mapping results for In, Ga, As, and the overlay of them. Open white circles, the atomic-column positions derived from the corresponding HAADF structural imaging in the inset of (b). Note that the missing In chemical contrast in the right-bottom InGa column is replenished by the Ga chemical contrast in the bottom left panel. (b) The integrated EDX spectrum and the associated spectral ranges exploited for the mapping in (a) (Chu et al., 2010).

In addition, STEM-EELS equipped with spherical aberration correction can better distinguish the interparticle size effect.

LiMnPO$_4$ is an alternative cathode material for lithium ion battery because of its high voltage (4.1 V), high heat stability and nontoxicity. The volume ratio of LiMnPO$_4$ below 40 nm is 136 mAh · g^{-1} at 5 C, and 120 mAh · g^{-1} at 10 C. It is indicated from the EELS analysis that the diffused distance of Li ion from the interior to the surface of initial particle has reduced by about 20 nm.

Meanwhile, the high spatial resolution spectrogram normally acquired by spherical aberration correction photology in the STEM is very necessary for understanding the physical property of transition metal oxide, such as manganite. The physical property of these perovskites is closely related to the occupation state of partly filled 3d orbit (i.e. the state of oxygen). The structure change of $L_{2,3}$ edge of Mn, and K edge of O is in linear relation with the normal valence state of Mn in the sample; such linear relation is extracted from the spectrogram of atomic-resolution and applied to the nominal oxidation state. Through some unexpected oscillation on the surface of Mn valence state table acquired in these spectrograms, and combining experiment and density functional calculation as well as motive power scattering simulation, the detailed explanation of these spectrograms is acquired. In $LaMnO_3$ (LMO), two kinds of non-equivalent O places can be distinguished by this method.

Transition metal L edges result in excitations of 2p electrons into empty bound states or the continuum. Thus, these edges show two characteristics features or white lines originating from transitions from the spin orbit split $2p_{3/2}$ and $2p_{1/2}$. For Mn, the L_3 line can be found around 644 eV and the L_2 line around 655 eV. Fig. 6.3.24(a) shows the Mn $L_{2,3}$ edges corresponding to $La_xCa_{1-x}MnO_3$ in case of different x, which are obtained in large region. Fig. 6.3.24(b) is the O-K edge and Mn-L edge demonstrated by EELS, in which LMO (black line) is around 530 eV and $CaMnO_3$ (CMO) (red line) is around 640 eV, and obvious displacement has occurred to the spectral line. Such phenomenon can coincide with the previous report well. Fig. 6.3.24(c) is a kind of Mn nominal oxidation state extracted, drawn and paired from the $L_{2,3}$ ratio of a series of $La_xCa_{1-x}MnO_3$ samples, from which we can obviously observe that the L_3/L_2 value reduces with the increase of oxidation state. The measured $L_{2,3}$ value in LMO and CMO compounds is consistent with the +3 and +4 valence of oxidation state in Mn compound. Fig. 6.3.24(d) shows a linear relation.

These methods can easily be applied to one-dimensional EELS line scans and to two-dimensional EELS spectrum images in order to extract relevant information on electronic properties. Fig. 6.3.25(a) is the Z-contrast image of $CaMnO_3$ sample, again from a very thin region near the CMO particle edge (thickness estimated around 2 – 3 nm). The CMO pseudocubic unit cell has been marked for clarity: a green circle for the Ca column and blue-yellow circles for Mn-O columns. Yellow

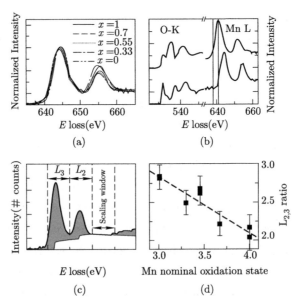

Fig. 6.3.24 (a) Mn $L_{2,3}$ edges for a series of $La_xCa_{1-x}MnO_3$ compounds with $x = 1$ (black), $x = 0.7$ (dark yellow), $x = 0.55$ (green), $x = 0.33$ (blue), and $x = 0$ (red). These spectra have been acquired from wide sample areas (by defocusing the electron beam or by averaging a number of spectra from different locations). The energy scale is nominal, and has been shifted and the intensity normalized so the L_3 lines match. (b) EEL spectra showing the O-K edge around 530 eV and the Mn L edge around 640 eV for LMO (black) and CMO (red). The spectra have been displaced vertically for clarity. (c) Sketch showing a generic Mn $L_{2,3}$ edge (black line) and the approximate position of the windows used for integration of the L_3 and L_2 line intensities, and also the window used to scale the Hartree-Slater cross-section step function (red line). After scaling and subtraction of this function, the remaining signals under the L_3 and L_2 lines (shaded) are integrated, and their ratio is calculated. (d) Dependence of the $L_{2,3}$ ratio with the formal oxidation state for a series of LCMO compounds. The red dashed line represents a linear fit to the data (Tan et al., 2011).

crosses show the O columns in between. Fig. 6.3.25(b) shows the simultaneously acquired ADF signal (Z contrast), clearly showing the CMO crystal lattice Mn-O columns appear bright, Ca columns dimmer. Spatial drift effects on this data set are observable, but minor. Fig. 6.3.25(c) is the signal picture of ADF image, which is acquired by the rate of 0.7 s/pixel. Fig. 6.3.25(d) is a two-dimensional Ca $L_{2,3}$ picture, while Fig. 6.3.25(e) shows the profile across a region equivalent to the one in Fig. 6.3.25.(b). Fig. 6.3.25(f) and 6.3.25(g) show the Mn $L_{2,3}$ image and its profile (averaged in the same way), while Fig. 6.3.25(h) and 6.3.25(i) show the equivalent O-K image and corresponding profile. While both the Mn-O column and the surrounding O columns contain the same number of O atoms, the signal from

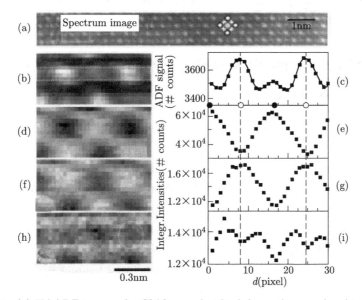

Fig. 6.3.25 (a) HAADF image of a CMO sample tilted down the pseudocubic [100] zone axis. The cubic unit cell has been marked. A green rectangle highlights the approximate window for spectrum image acquisition. (b) ADF signal acquired simultaneously with the spectrum image in (a). (c) Averaged profile along the region marked with a yellow rectangle in (b). (d), (f), and (h) Ca $L_{2,3}$, Mn $L_{2,3}$, and O K images, respectively. (e), (g), and (i) averaged profiles of (d), (f), and (h), respectively, using a window equivalent to the one depicted in (b). A white dotted circle shows the position of the volcano rim in (h). All EELS images have been generated by integrating a 35 eV wide window after background subtraction using a power-law fit from the EELS data after PCA. Green (blue-yellow) circles mark the approximate positions of Ca (Mn-O) columns along the scan (Tan et al., 2011).

the Mn-O column is reduced due to the electron probe being scattered beyond the EELS detector by the heavier Mn atoms.

In conclusion, after the spherical aberration correction is applied to the EELS, the distribution of elements in the samples can be observed more clearly, and the relation between each atom can be reflected more accurately. Therefore, after the application of this technology, the structure of samples and the influence of physical property can be well explained. In a word, with the continuous perfection and maturity of this technology, powerful support and guarantee will be provided for understanding the substance more clearly.

6.3.7 Studying quantum scale plasma oscillation of metal nanoparticles

Research on the plasma oscillation of metal nanoparticles has important significance in fields such as catalyst, biology and quantum optics, etc., and the performance has important application in superfine detection of biomolecule and gas, solar energy, etc (Anker et al., 2008; Bingham et al., 2010). Although the plasma properties of spherical particles with the diameter of more than 10 nm have been clearly investigated, the plasma properties of quantum scale system are hard to explain clearly due to the heterogeneity of overall measurement adopted by weak optical scattering and metallic bond. These difficulties hamper the control over the plasma properties of quantum scale particles in the process of real observation or design, especially for catalyst. Spherical aberration corrected STEM plays a key role in revealing the microstructure at the atomic scale, it can not only assist us to observe the morphology with high resolution, but also analyze the composition and chemical property of the samples via electron energy loss spectrum. Jonathan, et al. measured the different size of silver nanoparticle from 2 nm to 12 nm by using the spherical aberration corrected TEM, in combination with the EELS method under STEM mode (Scholl et al., 2012), which is characterized by directly conducting energy spectrum analysis on the morphology image of extremely tiny particles. They measured both the particle edges where the surface particle oscillation can be observed and the central position where the block mode can be observed. They found that when the electron beam directly approached the particle edge, it would cause the collective oscillation of the electron on the surface, i.e. localized surface plasma oscillation, and the energy was related to the dielectric constant, volume and shape of materials as well as the surrounding environment. When the electron beam passes through the center of the particle and causes collective oscillation of all electrons, the energy were only related to dielectric constant of materials, as shown in Fig. 6.3.26.

The measured results of different diameters are shown in Fig. 6.3.27. It can be known from Fig. 6.3.27 that with the decreasing diameter, blue shift will appear in the energy of surface plasma oscillation. According to the predication of classical Mie Theory, only 0.03 eV blue shift will appear when the diameter decreases from 11 nm to 1.7 nm, but the actual measured blue shift is about 0.5 eV. In addition, blue shift also appears when the diameter of collective oscillation is below 10 nm. All these are related to the qualitative change after the dielectric function is reduced to quantum scale. Afterwards, they built a quantized model and calculated the consistent results by first-principles. Such model can be applied to predict the plasma oscillation of other metals only by changing the parameters. So EELS of quantum plasma oscillation system shall make important contributions to forming new recognition of electronic mechanism, which shall be attributed to the

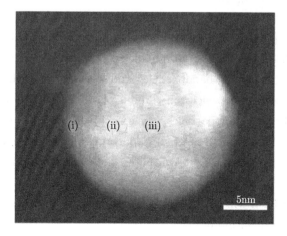

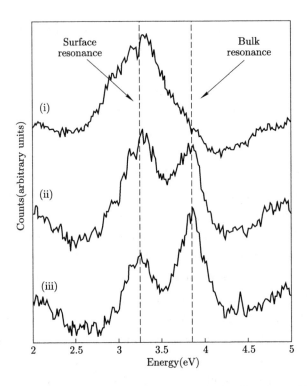

Fig. 6.3.26 Energy loss spectrum of Ag particles with the diameter of 20nm from edge to the center (Scholl et al., 2012).

development of spherical aberration-corrected TEM with atomic scale resolution, and STEM-EELS. Baston, et al. also made research on the stress mechanism and physical restriction of equi-nanoparticles and plasma via EELS (Batson et al., 2012). They found that fast electrons produce both attraction and repulsive force when passing through nanoparticles. Such force can be compared with the force of non-chemical bond, which plays an important role in studying the mechanical behavior of nanoparticles, etc. It will be an indispensable tool for later research on quantum scale plasma oscillation property, and a great number of in situ EELS experiments of STEM will be applied to the properties of metal and semiconductor plasma oscillation quantum dot.

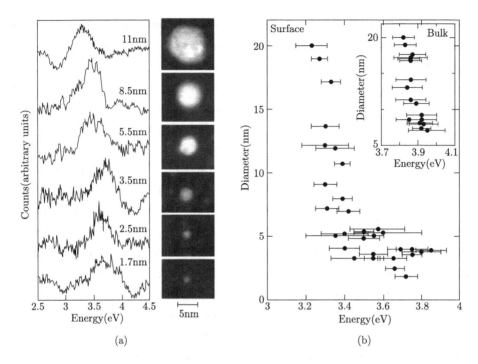

Fig. 6.3.27 (a) shows the plasma excitation energy loss spectrum of small localized surface of various diameters and the corresponding morphology image, while (b) refers to the energy of surface plasma oscillation of various diameters. The inserted figure refers to the corresponding energy of collective oscillation (with the error caused by fitting and auxiliary routine taken into consideration) (Scholl et al., 2012).

6.3.8 Studying the domain wall in ferroelectric thin films at atomic scale

Ferroelectric thin films have important applications in many fields, particularly possessing great potential in electronic industry and electron-optical devices. Since the performance of these devices is critically dependent on the ferroelectric materials' polarization switch strength and stability, it makes senses to study the polarization switching of these ferroelectric materials. Polarization switching is usually induced by an applied electric field. At present, 71° domain, 90° domain, 109° domain and

180° domain have been discovered in bulk materials, and the structure of domain wall has also been studied. For example, Streiffer, et al. studied the 90° domain structure of $PbTiO_3$ and $BaTiO_3$ through lattice fringe image under an electron microscope and estimated the thickness of domain wall upon the situation of defect, but failed to directly observe the atomic structure of domain wall (Streiffer et al., 2002; Fong et al., 2004). Some other methods like enabling Bragg diffraction in dark field or observing contrast stripes at domain boundary through the electron microscope as well as latter lattice fringe imaging all failed in the study at the atomic level. However the new technology on spherical aberration-corrected TEM developed recently makes this study possible. Jia Chunlin, et al. (Jia et al., 2011) observed the structure of 180° domain of $PbZr_{0.2}Ti_{0.8}O_3/SrTiO_3$ thin film structure and found that its ferroelectricity is generated due to the spontaneous polarization arising from the displacement of oxygen anion and other cation centers. They mainly studied transverse domain wall structure, vertical domain wall structure and mixed type domain wall structure, revealing the arrangement characteristics of electric dipole in these three domain walls; that is, the index of vertical domain wall plane was parallel to polarization vector and dipoles were arrayed head to head or tail to tail, charged; while the index of transverse domain wall plane was perpendicular to the polarization vector and dipoles were coupled head to tail, uncharged; the mixed domain wall was the combination of these two domain walls. They found that the width of vertical domain wall was the magnitude of single lattice cell and the width of transverse domain wall reached the magnitudes of 10 lattice cells, which was explained that the thicker domain wall could disperse the energy arising from compensation charge so that it can be stabilized; by establishing an appropriate model, they got the results in line with the ones obtained in experiment. Later, they also directly observed continuous rotation of electric dipoles from Flux-Closure domain structure via spherical aberration-corrected TEM with atomic resolution, shown as Fig. 6.3.28. (Jia et al., 2011). This directly verified the theory that Flux-Closure structure can be formed from local electric dipoles without compensation of surface charges.

By setting the extension of 90° domain to be very narrow, the first-principles simulation calculations were in excellent agreement with the experimental results.

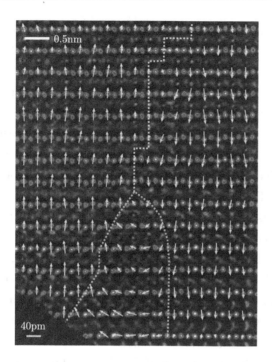

Fig. 6.3.28 Map of the atomic displacement vectors of PbZr$_{0.2}$Ti$_{0.8}$O$_3$/SrTiO$_3$ thin films. The displacement of the Zr-Ti atoms (arrows) from the center of the oxygen octahedra is shown here. The length of the arrows represents the modulus of the displacements. There is a continuous rotation of the dipole directions by 90° domain from down (right) to up (left), forming Flux-Closure structure (Jia et al., 2011).

The spherical aberration-corrected TEM with atomic resolution plays an important role in such experiments. It not only can distinguish the atomic species in different regions unit cell by unit cell, but also can quantitatively probe the position of the atoms. Therefore, this technology can be used to further study the details of the domain structure of this surface and interface, such as the electric domain of heterostructures or vortex-like structures in ferroelectric materials. Since these structures are very sensitive to the balance of the interface, even slight difference of interfacial property will largely affect the behavior of non-polarization areas.

Another method is to use in situ HAADF-STEM imaging technology to study dynamic ferroelectric switching. Christopher, et al. studied ⟨111⟩ direction of $BiFeO_3/La_{0.7}Sr_{0.3}MnO_3$ bilayer by use of this technology. (Nelson et al., 2011) They induced the switch of polarization by applying external electric field to the samples along the index direction of thin film plane and observed that 71° domain easily occurred during the polarization conversion of 180° domain, and the conversion of magnetic domain would be hindered when reaching the interface, particularly at the interface of heterostructures, because there would be many defects in the interface and the conversion of interface domain would require a higher negative voltage. However, for the domain conversion of non-heterostructure, no heavy atom defects were observed from thickness measurement, three-dimensional imaging and many other methods, and these barrier layers could break through by applying high bias; therefore, even though the position of oxygen atom was not directly visible in HAADF, these also could prove the theory that the ordered plane arising from oxygen vacancy results in the domain pinning. They also found that the introduction of electric filed would lead to a big displacement from the conversion at the tip, in particular the formation of interface domain. So far, the kinetics reasons for the appearance of 71° domain during the conversion of 180° domain have not determined the interface structure's impact on domain conversion can be further studied. Therefore, spherical aberration-corrected TEM and STEM will continue to play an important role in this area.

6.3.9 Structure characterization of hematite

Jingyue Liu, et al. (Liu et al., 2012) used hydrothermal method to synthesize a hematite with morphology controllable and pseudo-cubic boundary of $\{10\bar{1}2\}$ plane. In this typical synthesis experiment, the precursor or surfactant composition Fe $(NO_3)_3 \cdot 9H_2O$/poly (N-vinyl-2-pyrrolidone) (PVP) is dissolved in N, N-dimethyl formamide (DMF). The final solution is heated to 180 °C in the high-pressure furnace with different time. The method of imaging by HAADF spherical aberration STEM with a resolution of 0.08 nm is used to detect the synthetic samples.

Fig. 6.3.29 shows the spherical aberration-corrected HAADF image of precursor solution. Small bright spots represent the projection of independent Fe or superposed Fe atoms along the directions of incident electron beams. This chain structure, decorated with bright Fe atoms, most likely represents PVP molecules.

Fig. 6.3.29 Sub-angstrom resolution HAADF image of the precursor molecules dispersed onto carbon support (Liu et al., 2012).

Fig. 6.3.30 shows a series of HAADF images of hydrothermal synthesis method

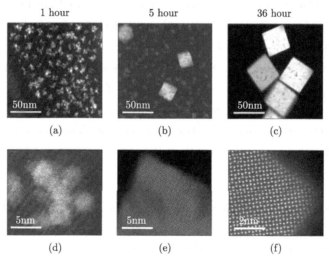

Fig. 6.3.30 Low (top) and high (bottom) magnification HAADF images of samples taken at different periods of time during the solvothermal synthesis process. The hematite pseudo-cubes are bounded by the $\{10\bar{1}2\}$ facets. The final pseudo-cubes possess almost atomically flat surfaces. The gray, patchy contrast in (c) and (e) suggests that channels, voids or probably polymer complexes may be present within the pseudo-cubes.(d) and (f) were obtained with the incident beam along the $[20\bar{2}1]$ zone axis, clearly revealing the atomic arrangement of the $\{10\bar{1}2\}$ surfaces of the synthesized hematite pseudo-cubes (Liu et al., 2012).

at different time periods. By heating for an hour in the high-pressure furnace at 180 °C, many small micro-crystals (Fig. 6.3.30a and b) will be aggregated to form the crystals with diameter of about 2 − 4 nm. Fe atoms or small clusters are also visible in Fig. 6.3.30b. After heating for 5 h, large hematite grains are formed with the diameter of about 20 − 30 nm. These crystals have a lot of small planes and pseudo-cubic morphology. This gray inharmonic contrast in pseudo-cubic indicates that open channel and polymer complexes may exist in the crystal. The pseudo-cubic plane (Fig. 6.3.30d) is a relatively flat; some islands are also suppressed; and some channels can be clearly visible on $\{10\bar{1}2\}$ planes. The diameter of smaller micro-crystals (Fig. 6.3.30c) or the crystals in the solution is about 5 − 10 nm. Approximately 36h later, the pseudo-cubic hematite crystals form the ones with diameter of 40 − 50 nm. At the stage after synthesis process, a large pseudo-cubic hematite is formed from a majority of small clusters and single Fe atoms. The pseudo-cubic planes are almost atomically smooth (Fig. 6.3.30f), but channels and vacancy structures are still visible.

Based on the analysis of the STEM image of various samples at different synthesis stages, they proposed that growth mechanism of pseudo-cubic hematite consists of the following simultaneous processes: 1) the aggregation of grains, 2) rapid diffusion of surfaces and interfaces, 3) Oswald ripening mechanism and growth level by level. The detailed analysis of growth process and the corresponding experimental data will be discussed.

6.3.10 Atomic-resolution imaging of catalyst nanoparticles

1. Aberration-corrected (S)TEM imaging

Catalysts are often used under certain atmosphere, while it is always difficult to directly observe gas molecules on the surface of catalyst particles in various reaction conditions. Recently, Yoshida, et al. (Yoshida et al., 2012) reported the imaging of CO molecules adsorbed on the gold nanoparticles (GNP) by use of aberration-corrected environmental TEM. This method can be applied to the understanding of catalyst reaction mechanism.

The CeO_2-supported gold nanoparticle is used as catalyst, which shows high catalytic activity to the CO oxidation at room temperature. FEI Titan ETEM with an accelerating voltage of 300 keV and equipped with an aberration corrector and special environmental device is used. Fig. 6.3.31 are high-resolution TEM images of gold nanoparticles obtained under the best defocus condition at 300 keV. A dark spot represents an Au atom column. The image analysis shows that {100} plane of Au nanoparticles has reconstructed surface under catalytic reaction condition. In the reconstructed surface, the atoms form hexagonal lattices on the outermost

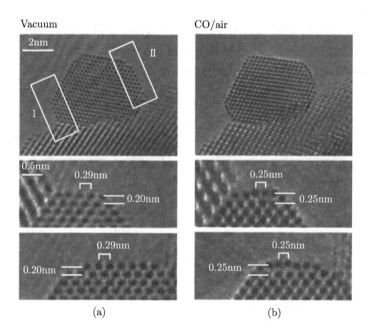

Fig. 6.3.31 Surface reconstruction of Au (100) plane under catalytic reaction conditions; (a) CeO_2–supported GNP high-resolution TEM in vacuum; (b) GNP high-resolution TEM in reaction atmosphere (with 1% of CO by volume in air). There are enlarged views below (a) and (b); in vacuum, the distance between the outermost layer and the second outermost layer of {100} plane is 0.2 nm, which is the same with the {200} planes spacing of bulk gold; the atomic distance near the outermost layer is 0.29 nm, which is the same with {100} plane spacing of bulk gold. In reaction atmosphere, both Au atomic distance and plane spacing are 0.25 nm, which is the result of the surface reconstruction of Au {100} plane (Yoshida et al., 2012).

layer and slightly distorted square lattices on the second outermost layer.

To improve the image contrast of light atoms like carbon and oxygen, the authors used low-energy electron microscopy to observe catalyst samples; they reduced the accelerating voltage down to 80 keV. A and B in Fig. 6.3.32 show the images of {100} plane of GNP obtained in vacuum and reaction atmosphere respectively, in which the GNP contrast is reduced and the contrast of carbon and oxygen is increased. Some unusual contrast variations (Fig. 6.3.32b) occurred in the reconstructed surface of GNPs samples in the reaction atmosphere; the simulated images (Fig. 6.3.32d) based the CO adsorption model (Fig. 6.3.32e and f) was obtained from first-principle calculation, which accords with the image obtained from experiment (Fig. 6.3.32.c) very well.

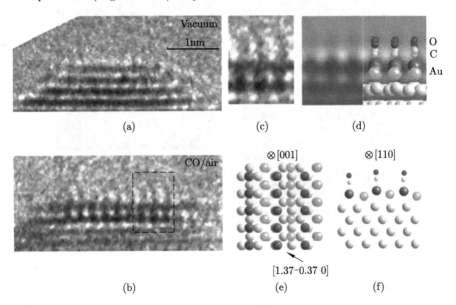

Fig. 6.3.32 Aberration-corrected ETEM images of Au{100}-hex reconstructed surface in (a) a vacuum and (b) a reaction environment. (c) The observed image in the rectangular region in (b) at higher magnification.(d) A simulated image based on an energetically favorable model. The model in (e) plan view along the [001] direction and (f) cross-sectional view along the [110] direction of gold (Yoshida et al., 2012).

The previous theoretical study (Mavrikakis et al., 2000; Molina and Hammer,

2004) showed that CO molecules were adsorbed more firmly than oxygen molecules on non-reconstructed GNP surface and CO molecules were considered to have a stronger adsorption in low coordination area than in surface area, such as steps, edges and corners. However, it was indicated that CO molecules were prone to be adsorbed on {100} surface of GNP combined with aberration-corrected environmental TEM images based on first-principles calculations.

Aberration-corrected STEM has important applications in the study of catalyst nanoparticles because it can directly image atoms and the contrast is proportional to the atomic number. For example, Sanchez, et al. (Sanchez et al., 2009) used JEOL-2010F electron microscopy with spherical aberration corrector to provide structural characterizations of Pt-Pd and Pd-Pt core-shell nanoparticles at atomic resolution with the diameter of electron beam spot up to 0.5 nm.

Fig. 6.3.33 is the HAADF-STEM image and FT image of Pt(core)-Pd(shell) nanoparticles, with zone axis being [001] and [1$\bar{1}$0] respectively, which is the same with the polyhedron with the most stable Pt(core)-Pd(shell) particles predicted by the theory. Fig. 6.3.34 is the model of atomic structure of Pt(core)-Pd(shell) particles corresponding to Fig. 6.3.33. The core-shell structure of samples can be obtained from experimental and simulated images.

The Fig. 6.3.35 shows the spherical aberration-corrected HAADF-STEM image of Pt/ZnO catalyst precursor (Liu, 2012). This sample is prepared by absorption-decomposition method; highly diluted H_2PtCl_6 (pH5.5 to pH6.8) is used with absorption time kept below 10 min and with ZnO nanowires and nanobelts as supports. The image shows: 1) the formation of particles and small clusters in early period of preparation process; 2) the existence of many independent Pt atoms, with some preferably gathering in a special location of ZnO supports seemingly. Furthermore, atomic arrangement of Pt clusters surfaces is recorded; some Pt atoms have a relatively high contrast. The detailed examination of ZnO nanobelts supports show that salt solution is completely converted to various orientations of micro-crystals in some region of single-crystal ZnO nanobelts during short decomposition-deposition preparation process. However, the modification of the substrate does not occur in ZnO nanowires, indicating different surface reactions. The adsorption of Pt is dependent on the surface of supports, the form of salt solution and pH value. The

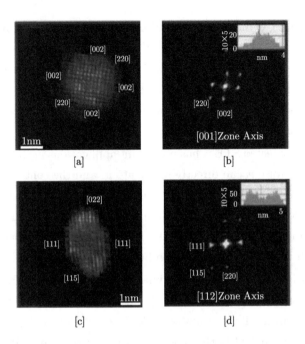

Fig. 6.3.33 The HAADF-STEM image and FT image of Pt(core)-Pd(shell) nanoparticles, with zone axis being [001] and [1$\bar{1}$0] respectively (Sanchez et al., 2009).

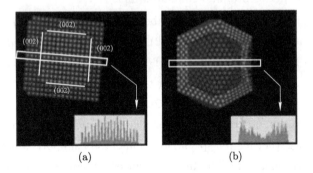

Fig. 6.3.34 The model of atomic structure of Pt(core)-Pd(shell) particles corresponding to Fig. 6.3.36 (Sanchez et al., 2009).

related observations about the preparation process of catalyst facilitate the understanding of the synthesis mechanism of catalyst at atomic scale.

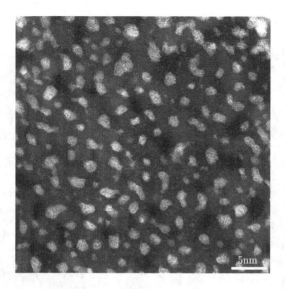

Fig. 6.3.35 The spherical aberration-corrected HAADF-STEM image of Pt/ZnO catalyst precursor (Liu, 2012).

2. Chemical and energy band information obtained from EDX and EELS analysis

EDX is to excite inner electrons by incident electrons for producing characteristic X-rays; since each element has corresponding characteristic X-rays, the composition of samples can be quantitatively analyzed through EDX, which has become the most fundamental, reliable and important electron microscopy analysis method. The EELS is related to the energy loss arising from the scattering of electrons and different atoms in samples. For some light elements, the EELS has better resolution.

For the same nano-catalyst materials of Pt-Pd core-shell structures, Khanal, et al. (Khanal et al., 2012) used JEOL JEM-ARM 200F electron microscopy equipped with a spherical aberration corrector, for EDX collection by use of the beam spot with the diameter of 0.13 nm and the current of 86 pA, to distinguish Pt and Pd. Fig. 6.3.36 shows EDX images of Pt-Pd core-shell structures of three different morphologies, where the Pt-L energy signals can be obtained from the core and Pd-L energy signals can be obtained from the shell; in this way, Pt and Pd are distinguished.

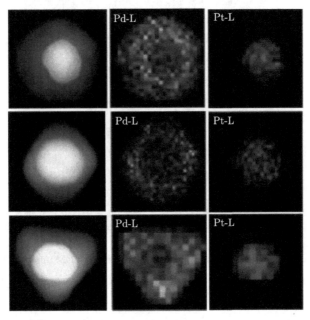

Fig. 6.3.36   EDX mapping of all three polyhedral core-shell nanoparticles (Khanal et al., 2012).

Wang, et al. (Wang et al., 2009) studied the oxygen reduction on Pt nano-catalyst particles (Pd or Pd_3Co) nanoparticles by use of C_s-STEM and EELS, as shown in Fig. 6.3.37. Meantime, nanoparticles were confirmed to have expected design structures by use of C_s-STEM intensity distribution and element distribution diagram. The catalytic activity of synthesized new nano-catalyst was improved by 5 to 9 times. Combined with DFT simulated structure, the catalytic activity was ultimately improved due to lattice mismatch caused by the collapse of (111) plane.

3. Three-dimensional imaging of nano-catalyst particles

Classic nano-catalyst is formed by overlapping nanoparticles on nano-porous support materials. The reaction activity of catalyst particles is not only subject to the size and shape distribution but also the pore diameter distribution, permeability of support material as well as catalyst distribution on support materials. Simple 2D

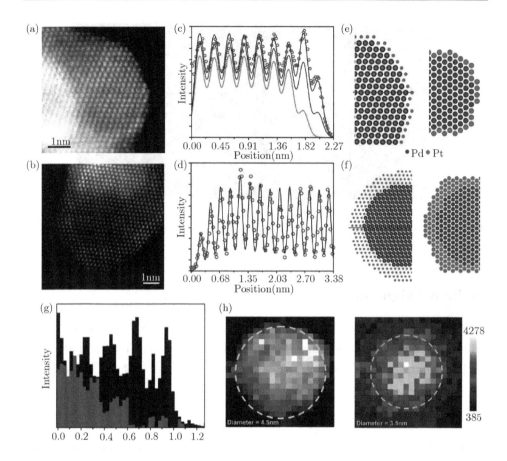

Fig. 6.3.37 Pt-Pd nanoparticles characterized using C_s-STEM and EELS (Wang et al., 2009).

projection mainly shows the catalyst's mean density at projection area, while the shape and pore diameter of the catalyst are very hard to be obtained (Kubel et al., 2005). Van Aert, et al. (Van Aert et al., 2011) studied Ag nanoparticles on Al substrate by use of atomic-resolution HAADF-STEM three-dimensional imaging. 3D structure of nanoparticles can be available from a series of pictures. The excellent match between the 3D reconstruction and the experimental data are illustrated in the Fig. 6.3.38.

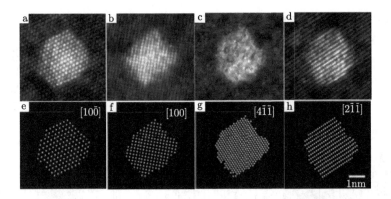

Fig. 6.3.38 Comparison of experimental images with projected 3D reconstructions (Van Aert et al., 2011).

Acknowledgments

It is a pleasure to acknowledge the help of many students Nijie Zhao, Xiao Chen, Min He, Jinan Shi, Licong Peng, Dongdong Xiao, Zhenzhong Yang and Shanming Li in preparing the manuscript.

References

A. I. Kirkland, R. R. Meyer, "Indirect" high-resolution transmission electron microscopy: Aberration measurement and wavefunction reconstruction, Microscopy and Microanalysis 2004, **10** (4): 401–413.

A. I. Kirkland, S. Haigh and L. Y. Chang, Aberration corrected TEM: current status and future prospects, Journal of Physics: Conference Series, 2008, **126**: 012034.

A. J. D'Alfonso, B. Freitag, D. Klenov and L. J. Allen, Atomic-resolution chemical mapping using energy-dispersive x-ray spectroscopy, Physical Review B, 2010, **81** (10): 4.

A. K. Padhi, K. S. Nanjundaswamy and J. B. Goodenough, Phospho-olivines as positive-electrode materials for rechargeable lithium batteries, Journal of the Electrochemical Society, 1997, **144** (4): 1188–1194.

A. R. Lupini, S. J. Pennycook, Localization in elastic and inelastic scattering, Ultramicroscopy, 2003, **96** (3–4): 313–322.

A. Thust, W. M. J. Coene, M. O. deBeeck and D. VanDyck, Focal-series reconstruction in HRTEM: Simulation studies on non-periodic objects, Ultramicroscopy, 1996, **64** (1–4): 211–230.

A. Y. Borisevich, H. J. Chang, M. Huijben, M. P. Oxley, S. Okamoto, M. K. Niranjan, J. D. Burton, E. Y. Tsymbal, Y. H. Chu, P. Yu, R. Ramesh, S. V. Kalinin and S. J. Pennycook, Suppression of Octahedral Tilts and Associated Changes in Electronic Properties at Epitaxial Oxide Heterostructure Interfaces, Physical Review Letters, 2010, **105** (8): 4.

B. Kabius, P. Hartel, M. Haider, H. Muller, S. Uhlemann, U. Loebau, J. Zach and H. Rose, First application of Cc-corrected imaging for high-resolution and energy-filtered TEM, J Electron Microsc (Tokyo), 2009, **58** (3): 147–155.

C. Delmas, M. Maccario, L. Croguennec, F. Le Cras and F. Weill, Lithium deintercalation in $LiFePO_4$ nanoparticles via a domino-cascade model, Nat Mater, 2008, **7** (8): 665–671.

C. Dwyer, Atomic-Resolution Core-Level Spectroscopy in the Scanning Transmission Electron Microscope, 2013, **175**: 145–199.

C. Kubel, A. Voigt, R. Schoenmakers, M. Otten, D. Su, T. C. Lee, A. Carlsson and J. Bradley, Recent advances in electron tomography: TEM and HAADF-STEM tomography for materials science and semiconductor applications, Microsc Microanal, 2005, **11** (5): 378–400.

C. L. Jia, K. Urban, Atomic-resolution measurement of oxygen concentration in oxide materials, Science, 2004, **303** (5666): 2001–2004.

C. L. Jia, K. W. Urban, M. Alexe, D. Hesse and I. Vrejoiu, Direct Observation of Continuous Electric Dipole Rotation in Flux-Closure Domains in Ferroelectric $Pb(Zr,Ti)O_3$, Science, 2011, **331** (6023): 1420–1423.

C. L. Jia, M. Lentzen and K. Urban, Atomic-resolution imaging of oxygen in perovskite ceramics, Science, 2003, **299** (5608): 870–873.

C. O. Girit, J. C. Meyer, R. Erni, M. D. Rossell, C. Kisielowski, L. Yang, C. -H. Park, M. F. Crommie, M. L. Cohen, S. G. Louie and A. Zettl, Graphene at the Edge: Stability and Dynamics, Science, 2009, **323** (5922): 1705–1708.

C. T. Nelson, P. Gao, J. R. Jokisaari, C. Heikes, C. Adamo, A. Melville, S. H. Baek, C. M. Folkman, B. Winchester, Y. J. Gu, Y. M. Liu, K. Zhang, E. G. Wang, J. Y. Li, L. Q. Chen, C. B. Eom, D. G. Schlom and X. Q. Pan, Domain Dynamics During Ferroelectric Switching, Science, 2011, **334** (6058): 968–971.

D. A. Muller, N. Nakagawa, A. Ohtomo, J. L. Grazul and H. Y. Hwang, Atomic-scale imaging of nanoengineered oxygen vacancy profiles in $SrTiO_3$, Nature, 2004, **430** (7000): 657–661.

D. A. Muller, Structure and bonding at the atomic scale by scanning transmission electron microscopy, Nature Materials, 2009, **8** (4): 263–270.

D. Burch, G. Singh, G. Ceder and M. Z. Bazant, Phase-transformation wave dynamics in $LiFePO_4$, Theory, Modeling and Numerical Simulation of Multi-Physics Materials Behavior, 2008, **139**: 95–100.

D. D. Fong, G. B. Stephenson, S. K. Streiffer, J. A. Eastman, O. Auciello, P. H. Fuoss and C. Thompson, Ferroelectricity in ultrathin perovskite films, Science, 2004, **304** (5677): 1650–1653.

D. Gabor, A New Microscopic Principle, Nature, 1948, **161** (4098): 777–778.

D. J. Smith, Development of aberration-corrected electron microscopy, Microscopy and Microanalysis, 2008, **14** (1): 2–15.

D. J. Smith, Progress and problems for atomic-resolution electron microscopy, Micron, 2012, **43** (4): 504–508.

E. Abe, Electron microscopy of quasicrystals-where are the atoms? Chemical Society Reviews, 2012, **41** (20): 6787–6798.

E. Abe, S. J. Pennycook and A. P. Tsai, Direct observation of a local thermal vibration anomaly in a quasicrystal, Nature, 2003, **421** (6921): 347–350.

E. J. Kirkland, On the optimum probe in aberration corrected ADF-STEM, Ultramicroscopy, 2011, **111** (11): 1523–1530.

F. Haguenau, P. W. Hawkes, J. L. Hutchison, B. Satiat-Jeunemaitre, G. T. Simon and D. B. Williams, Key events in the history of electron microscopy, Microscopy and Microanalysis, 2003, **9** (2): 96–138.

F. Houdellier, C. Roucau, L. Clement, J. L. Rouviere and M. J. Casanove, Quantitative analysis of HOLZ line splitting in CBED patterns of epitaxially strained layers, Ultramicroscopy, 2006, **106** (10): 951–959.

G. Mollenstedt, W. Hubig. Substandzdifferenzierung Im Elektronen-emissionsmikroskop Elektronenauslosung Durch Schragen Atomstrahlbeschuss, Optik, 1954, **11** (11): 528–539.

H. Hely, Test of An Improved Corrected Electron-Microscope. 2, Optik, 1982, **60** (4): 353–370.

H. Kohl, H. Rose, Theory of Image Formation by Inelastically Scattered Electrons in the Electron Microscope, 1985, **65**: 173–227.

H. Koops, G. Kuck and O. Scherzer, Test of An Electron-Optical Achromator, Optik, 1977, **48** (2): 225–236.

H. Lichte, Electron holography: Optimum position of the biprism in the electron microscope, Ultramicroscopy, 1996, **64** (1–4): 79–86.

H. Lichte, M. Lehmann, Electron Holography-Basics and Applications, Reports on Progress in Physics, 2008, **71** (1).

H. Lichte, M. Linck, D. Geiger and M. Lehmann, Aberration Correction and Electron Holography, Microscopy and Microanalysis, 2010, **16** (4): 434–440.

H. Lichte, Optimum Focus for Taking Electron Holograms, Ultramicroscopy, 1991, **38** (1): 13–22.

H. Müller, S. Uhlemann, P. Hartel and M. Haider, Advancing the Hexapole Cs-Corrector for the Scanning Transmission Electron Microscope, Microscopy and Microanalysis, 2006, **12** (6): 442.

H. Rose, Aplanatic Electron-Lenses, Optik, 1971, **34** (3): 285–289.

H. Rose, Properties of Spherically Corrected Achromatic Electron-Lenses, Optik 1971, **33** (1): 1–5.

H. Rose, Prospects for aberration-free electron microscopy, Ultramicroscopy, 2005, **103** (1): 1–6.

H. Tan, S. Turner, E. Yücelen, J. Verbeeck and G. Van Tendeloo, 2D Atomic Mapping of Oxidation States in Transition Metal Oxides by Scanning Transmission Electron Microscopy and Electron Energy-Loss Spectroscopy, Physical Review Letters, 2011, **107** (10).

H. Yoshida, Y. Kuwauchi, J. R. Jinschek, K. Sun, S. Tanaka, M. Kohyama, S. Shimada, M. Haruta and S. Takeda, Visualizing gas molecules interacting with supported nanoparticulate catalysts at reaction conditions, Science, 2012, **335**

(6066): 317–319.

J. A. Scholl, A. L. Koh and J. A. Dionne, Quantum plasmon resonances of individual metallic nanoparticles, Nature, 2012, **483** (7390): 421–468.

J. Gazquez, W. Luo, M. P. Oxley, M. Prange, M. A. Torija, M. Sharma, C. Leighton, S. T. Pantelides, S. J. Pennycook and M. Varela, Atomic-resolution imaging of spin-state superlattices in nanopockets within cobaltite thin films, Nano Lett, 2011, **11** (3): 973–976.

J. H. M. Deltrap, Correction of Spherical Aberration by Means of Nonrotational Symmetrical Lenses, Journal of Applied Physics, 1964, **35** (10): 3095.

J. M. Bingham, J. N. Anker, L. E. Kreno and R. P. Van Duyne, Gas Sensing with High-Resolution Localized Surface Plasmon Resonance Spectroscopy, Journal of the American Chemical Society, 2010, **132** (49): 17358–17359.

J. M. Cowley, Image Contrast in a Transmission Scanning Electron Microscope, Applied Physics Letters, 1969, **15** (2): 58.

J. N. Anker, W. P. Hall, O. Lyandres, N. C. Shah, J. Zhao and R. P. Van Duyne, Biosensing with plasmonic nanosensors, Nature Materials, 2008, **7** (6): 442–453.

J. R. Jinschek, K. J. Batenburg, H. A. Calderon, R. Kilaas, V. Radmilovic and C. Kisielowski, 3-D reconstruction of the atomic positions in a simulated gold nanocrystal based on discrete tomography: prospects of atomic resolution electron tomography, Ultramicroscopy, 2008, **108** (6): 589–604.

J. W. Lee, W. Zhou, J. C. Idrobo, S. J. Pennycook and S. T. Pantelides, Vacancy-Driven Anisotropic Defect Distribution in the Battery-Cathode Material $LiFePO_4$, Physical Review Letters, 2011, **107** (8): 5.

J. X. Wang, H. Inada, L. J. Wu, Y. M. Zhu, Y. M. Choi, P. Liu, W. P. Zhou and R. R. Adzic, Oxygen Reduction on Well-Defined Core-Shell Nanocatalysts: Particle Size, Facet, and Pt Shell Thickness Effects, Journal of the American Chemical Society, 2009, **131** (47): 17298–17302.

J. Y. Liu, Cui.L.F, Miao.S, Aberration-corrected STEM Investigation of The Growth Mechanism of Hematite Pseudo-cubic Nanocrystals, Microsc. Microanal. 2012, **18**.

J. Y. Liu, The Role of Aberration-corrected STEM in Developing Supported Catalysts, Microsc. Microanal. 2012, **18**: 2.

J. Yamasaki, T. Kawai and N. Tanaka, A simple method for minimizing non-linear image contrast in spherical aberration-corrected HRTEM, Journal of Electron Microscopy, 2005, **54** (3): 209–214.

J. Zach, M. Haider, Correction of Spherical and Chromatic Aberration in A Low-Voltage SEM, Optik, 1995, **98** (3): 112–118.

K. Urban, B. Kabius, M. Haider and H. Rose, A way to higher resolution: spherical-aberration correction in a 200 kV transmission electron microscope, Journal of Electron Microscopy, 1999, **48** (6): 821–826.

K. W. Urban, Studying atomic structures by aberration-corrected transmission electron microscopy, Science, 2008, **321** (5888): 506–510.

L. F. Kourkoutis, Y. Hotta, T. Susaki, H. Y. Hwang and D. A. Muller, Nanometer scale electronic reconstruction at the interface between $LaVO_3$ and $LaVO_4$, Physical Review Letters, 2006, **97** (25): 4.

L. Fitting, S. Thiel, A. Schmehl, J. Mannhart and D. A. Muller, Subtleties in ADF imaging and spatially resolved EELS: A case study of low-angle twist boundaries in $SrTiO_3$, Ultramicroscopy **106** (11–12) 2006: 1053–1061.

L. Gu, C. Zhu, H. Li, Y. Yu, C. Li, S. Tsukimoto, J. Maier and Y. Ikuhara, Direct observation of lithium staging in partially delithiated LiFePO4 at atomic resolution, J Am Chem Soc, 2011, **133** (13): 4661–4663.

L. J. Allen, A. J. D'Alfonso, B. Freitag and D. O. Klenov, Chemical mapping at atomic resolution using energy-dispersive x-ray spectroscopy, MRS Bulletin, 2012, **37** (01): 47–52.

L. M. Molina, B. Hammer, Theoretical study of CO oxidation on Au nanoparticles supported by MgO(100), Physical Review B, 2004, **69** (15): 22.

M. Born, E. Wolf, Principles of Optics, 1993.

M. C. Scott, C. C. Chen, M. Mecklenburg, C. Zhu, R. Xu, P. Ercius, U. Dahmen, B. C. Regan and J. W. Miao, Electron tomography at 2.4-angstrom resolution, Nature, 2012, **483** (7390): 444–491.

M. F. Chisholm, A. Maiti, S. J. Pennycook and S. T. Pantelides, Atomic Configurations and Energetics of Arsenic Impurities in a Silicon Grain Boundary, Physical Review Letters, 1998, **81** (1): 132–135.

M. Haider, G. Braunshausen and E. Schwan, Correction of The Spherical-Aberration of A 200-kV TEM by Means of A Hexapole-Corrector, Optik, 1995, **99** (4): 167–179.

M. Haider, H. Rose, S. Uhlemann, B. Kabius and K. Urban, Towards 0.1 nm resolution with the first spherically corrected transmission electron microscope, Journal of Electron Microscopy, 1998, **47** (5): 395–405.

M. Haider, H. Rose, S. Uhlemann, E. Schwan, B. Kabius and K. Urban, A Spherical-Aberration-Corrected 200 kV Transmission Electron Microscope, Ultramicroscopy, 1998, **75** (1): 53–60.

M. Haider, S. Uhlemann, E. Schwan, H. Rose, B. Kabius and K. Urban, Electron microscopy image enhanced, Nature, 1998, **392** (6678): 768–769.

M. Itakura, N. Watanabe, M. Nishida, T. Daio and S. Matsumura, Atomic-Resolution X-ray Energy-Dispersive Spectroscopy Chemical Mapping of Substitutional Dy Atoms in a High-Coercivity Neodymium Magnet, Japanese Journal of Applied Physics, 2013, **52** (5): 4.

M. J. Hytch, E. Snoeck and R. Kilaas, Quantitative measurement of displacement and strain fields from HREM micrographs, Ultramicroscopy, 1998, **74** (3): 131–146.

M. J. Hytch, F. Houdellier, F. Hue and E. Snoeck, Dark-field electron holography for the mapping of strain in nanostructures: correcting artefacts and aberrations, Electron Microscopy and Analysis Group Conference 2009, 2010, **241**.

M. Knoll, E. Ruska. The Electron Microscope, Zeitschrift Fur Physik, 1932, **78** (5–6): 318–339.

M. Lentzen, B. Jahnen, C. L. Jia, A. Thust, K. Tillmann and K. Urban, High-resolution imaging with an aberration-corrected transmission electron microscope, Ultramicroscopy, 2002, **92** (3-4): 233–242.

M. Linck, H. Lichte and M. Lehmann, Off-Axis Electron Holography: Materials Analysis at Atomic Resolution, International Journal of Materials Research, 2006, **97** (7): 890–898.

M. Mavrikakis, P. Stoltze and J. K. Norskov, Making gold less noble, Catalysis Letters, 2000, **64** (2–4): 101–106.

M. Varela, A. R. Lupini, K. van Benthem, A. Y. Borisevich, M. F. Chisholm, N. Shibata, E. Abe and S. J. Pennycook, Materials characterization in the aberration-corrected scanning transmission electron microscope, Annual Review of Materials Research, 2005, **35**: 539–569.

M. Varela, M. Oxley, W. Luo, J. Tao, M. Watanabe, A. Lupini, S. Pantelides and S. Pennycook, Atomic-resolution imaging of oxidation states in manganites, Physical Review B, 2009, **79** (8).

M. Varela, S. D. Findlay, A. R. Lupini, H. M. Christen, A. Y. Borisevich, N. Dellby, O. L. Krivanek, P. D. Nellist, M. P. Oxley, L. J. Allen and S. J. Pennycook, Spectroscopic imaging of single atoms within a bulk solid, Physical Review Letters, 2004, **92** (9): 4.

M. W. Chu, S. C. Liou, C. P. Chang, F. S. Choa and C. H. Chen, Emergent Chemical Mapping at Atomic-Column Resolution by Energy-Dispersive X-Ray Spectroscopy in an Aberration-Corrected Electron Microscope, Physical Review Letters, 2010, **104** (19).

N. Dellby, O. L. Krivanek, P. D. Nellist, P. E. Batson and A. R. Lupini, Progress in Aberration-Corrected Scanning Transmission Electron Microscopy, Journal of Electron Microscopy, 2001, **50** (3): 177–185.

N. Tanaka, Present status and future prospects of spherical aberration corrected TEM/STEM for study of nanomaterials, Science and Technology of Advanced Materials, 2008, **9** (1): 014111.

O. H. Hansteen, H. Fjellvag and B. C. Hauback, Crystal structure, thermal and magnetic properties of $La_3Co_3O_8$. Phase relations for $LaCoO_3$-delta (0.00 <= delta <= 0.50) at 673 K, Journal of Materials Chemistry, 1998, **8** (9): 2081–2088.

O. L. Krivanek, G. J. Corbin, N. Dellby, B. F. Elston, R. J. Keyse, M. F. Murfitt, C. S. Own, Z. S. Szilagyi and J. W. Woodruff, An electron microscope for the aberration-corrected era, Ultramicroscopy, 2008, **108** (3): 179–195.

O. L. Krivanek, M. F. Chisholm, V. Nicolosi, T. J. Pennycook, G. J. Corbin, N. Dellby, M. F. Murfitt, C. S. Own, Z. S. Szilagyi, M. P. Oxley, S. T. Pantelides and S. J. Pennycook, Atom-by-atom structural and chemical analysis by annular dark-field electron microscopy, Nature, 2010, **464** (7288): 571–574.

O. L. Krivanek, N. Dellby and A. R. Lupini, Towards sub-angstrom electron beams, Ultramicroscopy, 1999, **78** (1-4): 1–11.

O. Scherzer, Spharische Und Chromatische Korrektur Von Elektronen-Linsen, Optik, 1947, **2** (2): 114–132.

O. Scherzer, The Theoretical Resolution Limit of The Electron Microscope, Journal of Applied Physics, 1949, **20** (1): 20–29.

O. Scherzer. The weak electrical single lens lowest spherical aberration, Zeitschrift Fur Physik, 1936, **101** (1): 23–26.

P. D. Nellist, M. F. Chisholm, N. Dellby, O. L. Krivanek, M. F. Murfitt, Z. S. Szilagyi, A. R. Lupini, A. Borisevich, W. H. Sides and S. J. Pennycook, Direct sub-angstrom imaging of a crystal lattice, Science, 2004, **305** (5691): 1741.

P. E. Batson, A. Reyes-Coronado, R. G. Barrera, A. Rivacoba, P. M. Echenique and J. Aizpurua, Nanoparticle movement: Plasmonic forces and physical constraints, Ultramicroscopy, 2012, **123**: 50–58.

P. E. Batson, Simultaneous STEM Imaging and Electron-Energy-Loss Spectroscopy with Atomic-column Sensitivity, Nature, 1993, **366** (6457): 727–728.

R. F. Klie, Y. Ito, S. Stemmer and N. S. Browning, Observation of oxygen vacancy ordering and segregation in Perovskite oxides, Ultramicroscopy, 2001, **86** (3-4): 289–302.

R. Ishikawa, E. Okunishi, H. Sawada, Y. Kondo, F. Hosokawa and E. Abe, Direct imaging of hydrogen-atom columns in a crystal by annular bright-field electron microscopy, Nat Mater, 2011, **10** (4): 278–281.

R. Seeliger, Die Spharische Korrenktur Von Elektronenlinse Mittels Nicht-Rotationssymmetrischer Abbildungselemente, Optik, 1951, **8** (7): 311–317.

R. Seeliger, Versuche Zur Spharischen Korrektur Von Elektronenlinsen Mittels Nicht Rotationssymmetrischer Abbldungselemente, Optik, 1949, **5** (8–9): 490–496.

S. D. Findlay, N. Shibata, H. Sawada, E. Okunishi, Y. Kondo and Y. Ikuhara, Dynamics of annular bright field imaging in scanning transmission electron microscopy, Ultramicroscopy, 2010, **110** (7): 903–923.

S. D. Findlay, S. Azuma, N. Shibata, E. Okunishi and Y. Ikuhara, Direct oxygen imaging within a ceramic interface, with some observations upon the dark contrast at the grain boundary, Ultramicroscopy, 2011, **111** (4): 285–289.

S. I. Sanchez, M. W. Small, J. M. Zuo and R. G. Nuzzo, Structural Characterization of Pt-Pd and Pd-Pt Core-Shell Nanoclusters at Atomic Resolution, Journal of the American Chemical Society, 2009, **131** (24): 8683–8689.

S. Inoue, M. Kawai, N. Ichikawa, H. Kageyama, W. Paulus and Y. Shimakawa, Anisotropic oxygen diffusion at low temperature in perovskite-structure iron oxides, Nat Chem, 2010, **2** (3): 213–217.

S. J. Pennycook, Scanning transmission electron microscopy: Seeing the atoms more clearly, Mrs Bulletin, 2012, **37** (10): 943–951.

S. K. Streiffer, J. A. Eastman, D. D. Fong, C. Thompson, A. Munkholm, M. V. R. Murty, O. Auciello, G. R. Bai and G. B. Stephenson, Observation of nanoscale 180 degrees stripe domains in ferroelectric $PbTiO_3$ thin films, Physical Review Letters, 2002, **89** (6): 4.

S. Khanal, G. Casillas, J. J. Velazquez-Salazar, A. Ponce and M. Jose-Yacaman, Atomic resolution imaging of polyhedral PtPd core-shell nanoparticles by Cs-corrected STEM, J Phys Chem C Nanomater Interfaces, 2012, **116** (44): 23596–23602.

S. Van Aert, K. J. Batenburg, M. D. Rossell, R. Erni and G. Van Tendeloo, Three-dimensional atomic imaging of crystalline nanoparticles, Nature, 2011, **470** (7334): 374–377. T. WaltherI. M. Ross, Aberration Corrected High-Resolution Transmission and Scanning Transmission Electron Microscopy of Thin Perovskite Layers, Physics Procedia, 2013, **40**: 49–55.

V. Intaraprasonk, H. L. Xin and D. A. Muller, Analytic derivation of optimal imaging conditions for incoherent imaging in aberration-corrected electron microscopes, Ultramicroscopy, 2008, **108** (11): 1454–1466.

W. Y. M. Ye. H.Q., Progress in Transmission Electron Microscopy, 2003.

Y. Liao, Practical Electron Microscopy and Database, 2007.

Y. M. Kim, J. He, M. D. Biegalski, H. Ambaye, V. Lauter, H. M. Christen, S. T. Pantelides, S. J. Pennycook, S. V. Kalinin and A. Y. Borisevich, Probing oxygen vacancy concentration and homogeneity in solid-oxide fuel-cell cathode materials on the subunit-cell level, Nat Mater, 2012, **11** (10): 888–894.

Y. Oshima, H. Sawada, F. Hosokawa, E. Okunishi, T. Kaneyama, Y. Kondo, S. Niitaka, H. Takagi, Y. Tanishiro and K. Takayanagi, Direct imaging of lithium atoms in LiV_2O_4 by spherical aberration-corrected electron microscopy, Journal of Electron Microscopy, 2010, **59** (6): 457–461.

Y. Shao-Horn, L. Croguennec, C. Delmas, E. C. Nelson and M. A. O'Keefe, Atomic resolution of lithium ions in LiCoO$_2$, Nat Mater, 2003, **2** (7): 464–467.

Y. Yan, M. F. Chisholm, G. Duscher, A. Maiti, S. J. Pennycook and S. T. Pantelides, Impurity-Induced Structural Transformation of a MgO Grain Boundary, Physical Review Letters, 1998, **81** (17): 3675–3678.

Z. Saghi, D. J. Holland, R. Leary, A. Falqui, G. Bertoni, A. J. Sederman, L. F. Gladden and P. A. Midgley, Three-dimensional morphology of iron oxide nanoparticles with reactive concave surfaces. A compressed sensing-electron tomography (CS-ET) approach, Nano Lett, 2011, **11** (11): 4666–4673.

Z. Saghi, P. A. Midgley, Electron Tomography in the (S) TEM: From Nanoscale Morphological Analysis to 3D Atomic Imaging, Annual Review of Materials Research, 2012, **42** (1): 59–79.

Zhu J, Ye H. Insight for microstructure research of materials, ACTA Metallurgica Sinica, 2010, **46** (11): 15.

http://iamdn.rutgers.edu/?q=node/1168

7
In Situ TEM: Theory and Applications

Kun Zheng, Yihua Gao, Xuedong Bai and Renchao Che

7.1 In situ TEM observation of deformation-induced structural evolution at atomic resolution for strained materials

Kun Zheng, Ze Zhang and Xiaodong Han

7.1.1 Introduction

Transmission electron microscope (TEM) is one of the most powerful techniques to obtain the microstructure of materials. Equipped with energy dispersive X-ray (EDX) and electron energy loss spectroscopy (EELS), not only crystallographic structures but also chemical information at nano-scale, atomic scale and even sub-angstrom scale can be obtained. It helped human find the quasicrystal (Shechtman et al., 1984) and open the door of nano-world (Iijima, 1991). Except for these static structural studies, observing dynamic processes at high resolution is very important for development of nanoscience and nanotechnology. In situ TEM observations have been carried out since the 1960s (Hirsch et al., 1967). Nowadays, using TEM to monitor the dynamic processes and understand the physical mechanisms has been one of the interesting research fields.

The in situ TEM studies are benefited from the development of the techniques and instruments including laboratory-made and commercially available systems. Generally, in situ TEM is used to observe/monitor/record the dynamic responses and microstructural evolution of specimen resulting from external stimuli including heating, electricity, mechanical property etc. To apply these external stimuli, different types of TEM holders capable of straining, lasing, heating/cooling, electric are available commercially or are fabricated in the laboratory. During the past few years, many reviews have been provided for in situ TEM (Zhu et al., 2005; Han et al., 2007; Howe et al., 2008; Zhu and Li, 2010; Kiener and Minor, 2011; Golberg et al., 2012; Petkov, 2013; Wang et al., 2013). Here, we just introduce a part of in situ TEM about structural evolution observation at atomic scale for strained materials.

The atomic-scale visualization of structural evolution of a strained material is very important for further understanding the mechanisms of brittle-ductile transition, unusual plastic deformation, large-plasticity deformation and designing novel structures, new materials and applications. Owing to the atomic-scale pictures, the unusual dislocation initiation, partial-full dislocation transition, crystalline-amorphous transition have been disclosed. Let us introduce the experimental techniques based on TEM and the unusual phenomena.

7.1.2 Experimental techniques for in situ deformation in TEM

Up to now, nanoindentation and atomic force microscopy techniques are the most commonly used to perform in situ deformation in TEM. The Nanofactory TEM scanning tunneling microscopy (STM)/atomic force microscopy (AFM) holders (Agrait et al., 1993) and Hysitron picoindenter are typical devices for deformation testing (Shan et al., 2008). The precise movement in three dimensions can be controlled by the piezo-driven tube. The fixed specimen between the movable terminal and fixed one can be compressed or stretched and the structural evolution can be recorded real-time by the charge-coupled device (CCD) system. For Nanofactory TEM-AFM holder and Hysitron picoindenter, the force-displacement curves can be obtained simultaneously, while for TEM-STM holder, the electrical properties of specimen can be measured. The micro fabricated microelectromechanical systems (MEMs) fitted within a TEM specimen arm is another important strategy (Haque and Saif, 2002; Peng et al., 2008; Guo et al., 2011). These devices are usually based on silicon technology fabricated by lithography and etching techniques. Except for these commercial devices, other several in-lab-developed techniques and devices have also been used in studies of the structural evolution and strained physics of nanomaterials (Huang, 2005; Tang, 2010; Lin et al., 2011).

The above commercial holders or MEMs devices have very precise force-displacement feedback system which can provide quantitative stress-strain data along with revealing the structural evolution of strained materials. There are numerous studies showing valuable quantitative insight into deformation mechanisms of nanomaterials (Mompioua et al., 2006; Minor et al., 2006; Legros et al., 2008; Kiener et al., 2011; Wei et al., 2010). However, due to the limited chamber of TEM and the deformation-loading mode of holders or MEMs, these devices sacrifice with double-tilting capability, which makes it difficult to obtain the images conditions of high resolution TEM. As it is well known, direct atomic-scale imaging of dislocations, stacking faults, interface structures need the specific crystallographic orientation. Therefore, it is very necessary and important for keeping the double-tilting capable in order to achieve atomic-scale structural evolution information during the straining process.

Several groups have developed some methods or techniques to achieve this goal. Han group provided a method which could be used to deform the nanowires (NWs) in a commercial TEM grid with a colloidal/carbon thin film (Han et al., 2007). In this method, there was a key point that some precracks were made with the thin film. NWs were dispersed randomly with this supporting film. By using electron beam irradiation on this supporting film, the cracked parts would be induced to curve. NWs near the crack would be bent along with the deformation of supporting film. If you are lucky, a single NW bridging two broken parts of supporting film could be stretched. During this process, a particular crystallographic orientation can be achieved to obtain the atomic-scale images. However, this method is not controllable for the strain rate and deformation mode.

Dehm group developed a method to deform thin films in TEM (Legros et al., 2001; Dehm, 2009). By using the difference in the thermal expansion coefficients between the film and the substrate, the film can experience tensile or compressive stress when they are heated by a conventional heating holder. However, the strain induced by the substrate on the thin film is very limited. Another problem is that the film-substrate diffusion or chemical reaction may exist during the heating process which will complicate the experimental results. Han group developed a bimetallic extensor equipped with a TEM heating holder (Wang et al., 2010). This bimetallic extensor is made of two pieces of bimetallic strips and a copper ring with a diameter of 3 mm. The bimetallic strips fixed on the copper ring are placed into the heating holder of TEM. The bimetallic strips will be induced to bend by heating. In order to achieve a significant deflection at lower operational temperature ($< 100\,°C$), the two materials must have a large mismatch in thermal expansion coefficients. This bimetallic technique is successfully used to deform NWs, nanofilms in high-resolution TEM. With this device, the strain rate is also controllable in the range from 10^{-2} to 10^{-5} s^{-1} by controlling the increasing rate of temperature.

Using these novel methods and techniques, many unusual deformation behaviors have been revealed for different nanostructured materials, including semiconductor NWs, metallic NWs and films, amorphous-structured nano-spheres and NWs. These unusual deformation behaviors are different from their bulk counterparts involving elastic-plastic transition, full-partial dislocation transition, crystalline-amorphous transition, face-centered cubic to body-centered tetragonal (fcc-bct) structure transition etc. Deformation mechanisms can be well understood through these atomic-scale structural evolution processes.

7.1.3 Elastic-plastic transition and its atomic mechanism of semiconductor NWs

Semiconductor bulk materials usually exhibit brittleness at room temperature due to their high stress for activating the dislocation nucleation/motion. However, when the size of materials goes down to a small scale, the defect-free structure normally makes the small-size materials survive with high fracture stress, which could eventually allow the materials to have chance to overcome the critical stress to achieve elastic-plastic transition. Recently, more and more experimental results show this transition in different semiconductor materials, such as, Si (Östlund et al., 2009; Tang et al., 2012; Stauffer et al., 2012), SiC (Zhang et al., 2007), Ge (Smith et al., 2010), GaAs (Michler et al., 2007; Östlund et al., 2011) etc. Fig. 7.1.1 gives an example revealing a plastic event of a single Si NW during its bending process. Fig. 7.1.1(a) and (b) show low magnified TEM images captured at a time interval about 400 s. According to the same reference (the yellow dashed line) shown in Fig. 7.1.1(a) and (b), it was revealed that the average bent strain increased from 2.13%

to be about 2.29%. From the enlarged images (Fig. 7.1.1(e), (f)), which were taken from the white framed regions of Fig. 7.1.1(c), (d), it can be clearly seen that the surface steps increased from six atomic layers to seven.

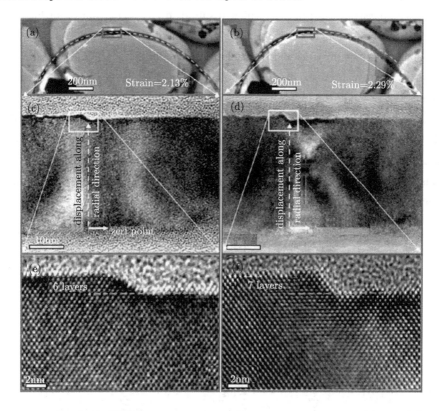

Fig. 7.1.1 Single Si NW bent process was shown at atomic scale. (a) and (b) Low-magnified TEM images captured at a time interval about 400 s with the bent strain increased from 2.13 to 2.29%. (c) and (d) Enlarged images taken from the green framed regions of (a) and (b), respectively. (e) and (f) HRTEM images taken from the white framed regions of (c) and (d), respectively. The HRTEM images show the surfaces steps increased from six atomic layers to seven.

How do these "brittle" materials achieve brittle-ductile transition? What are the main plastic events? The HRTEM images can provide us with direct atomic information. Si NWs fabricated by a thermal evaporation technique are almost free from dislocations. However, a large number of dislocations would emerge during the in situ tensile or bending processes. That is to say, the dislocations can nucleate for NWs. Meantime, motion of these nucleated dislocations can also be observed during the deformation process, which resulted in the formation of surface steps. While at the beginning of elastic-plastic stage, the observed dislocations are all not dissociated perfect 60° dislocations belonging to shuffle sets (Wang et al., 2000). However, with the increase of applied strain, the partial dislocations were captured in bent Si NW when the maximum bent strain surpassed 6.9%. Partial dislocations belong to the glide sets (Bulatov et al., 2001; Mitchell et al., 2003) with higher activation energy compared to the shuffle sets (Kaxiras and Duesbery, 1993). Fig. 7.1.2 provides such HRTEM images. Fig. 7.1.2(a) (no partials) and Fig. 7.1.2(b) (emission of the partial from the same area) provide two typical atomic-scale images showing the partial dislocation emission process at the atomic scale directly.

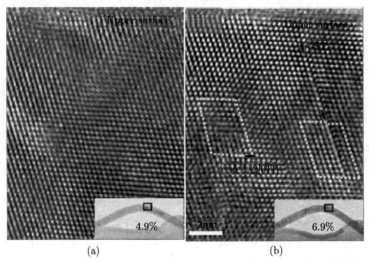

(a)   (b)

Fig. 7.1.2 In situ observation of partial dislocation nucleation in the Si NWs when the bent strain surpasses 6.9%. (a) The Si NW with 4.9% bent strain and (b) the HRTEM image of the same region when the bent strain has been increased to approximately 6.9%.

For semiconductor nanomaterial, dislocations are the main plastic events, including nucleation and motion. Due to the low dimensionality and defect-free structure, high strain and high stress can be applied, which may supply enough energy to activate the dislocations. Thus, the elastic-plastic transitions could take place. For diamond-structure materials, such as Si, SiC, the full dislocations belonging to shuffle set are the main plastic events, while with the increase of strain, the partial dislocations belonging to the glide sets can be activated.

7.1.4 Crystalline-amorphous transition of semiconductor NWs

It is well known that the crystalline materials can be transformed to amorphous state when they are under huge pressure. Nanoindentation is an effective technique to provide a local high pressure on materials and can make a transition from crystal to amorphous structure. In the past, many studies have reported this phenomenon and tried to understand this process. These ex situ TEM experiments provide some details of the microstructure of strained region. But it is difficult to fully understand the transition process because of the post-observations of indentation. Due to the development of in situ HRTEM techniques, it makes it possible to achieve direct atomic structural evolution of crystalline-amorphous transition.

For the first time, Han et al. (2007) provided a direct and in situ atomic-level imaging of crystalline-amorphous transition process for a tensile Si NW. They performed the tensile tests for single crystalline Si NWs in TEM by using their developed technique. They found the Si NWs can exhibit large-strain plasticity at near room temperature and achieve crystalline-amorphous transition at necked region before fracture. Fig. 7.1.3 shows the deformed process and atomic-scale details of a single Si NW. Fig. 7.1.3(a) and (b) show low-magnified images of a Si NW being extended at the initial elastic–plastic transition and in the final broken state, respectively. Fig. 7.1.3(c) is a HRTEM image of the axially extended Si NW at the initial elastic–plastic transition. Fig. 7.1.3(d) is an HRTEM image of the extended Si NW at a later time. Fig. 7.1.3(e) is an enlarged HRTEM image taken from the framed region indicated in Fig. 7.1.3(c) showing the dislocation structure. Fig. 7.1.3(f) shows the two-dimensional (2D) projection of the diamond-type crystalline structure of the Si lattice along [110]. The two types of well-known {111} shuffle and glide planes are indicated in Fig. 7.1.3(f). Three fast Fourier transform (FFT) diffraction patterns are shown in Fig. 7.1.3(g)–(i), corresponding to the areas A–C (Fig. 7.1.3(d)) of the NW, respectively. From the earlier HRTEM image (Fig. 7.1.3(c)), the crystalline structure can be clearly seen, though there are many dislocations at this stage. While with the increase of strain, this NW necked and the necking place changed to discorded structure. This also can be verified by

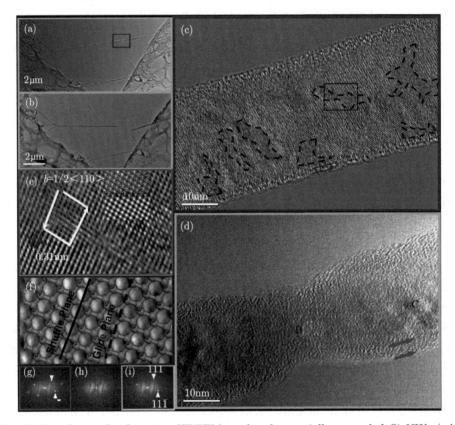

Fig. 7.1.3 Atomic-level in situ HRTEM study of an axially extended Si NW. (a,b) Low magnification images prior to and after the Si NW is plastically broken. c) HRTEM image of the Si NW in a status of elastic-plastic transition and prior to severe plastic deformation. (d) The same Si NW after large-strain plastic deformation with a neck in the middle of the NW. (e) Enlarged HRTEM image taken from the framed region in (c), which indicates the initiation features of plasticity with dislocations. (f) Schematic atomic structure of Si atomic lattice projected along the [110] orientation; the shuffle and glide planes are indicated. (g)–(i) The FFT electron diffraction patterns taken from areas A–C in (c), respectively.

the corresponding FFT diffraction pattern (Fig. 7.1.3(h)).

However, it is regretful that this tensile experiment did not provide more details about how the crystalline structure changed to amorphous structure. This is due to the difficulty to capture the HRTEM image for such a long suspending nanowire. Vibration is almost inevitable. This transition process was revealed during the bending case. According to the mentioned above, full dislocation activities were prevalent in the deformed Si NWs. These full dislocations were observed to be mobile and, therefore, have chances to interact. Fig. 7.1.4 shows the obvious movement of dislocations and the in situ observation of Lomer lock dislocation formed by a dislocation reaction at the atomic scale. In Fig. 7.1.4a, a dislocation with a

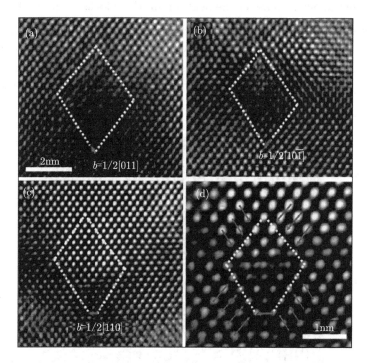

Fig. 7.1.4 The obvious movement of dislocations and the in situ observation of a Lomer lock dislocation formed by a dislocation reaction at atomic scale. (a) A dislocation with a Burgers vector of a/2[011] (b) after the former dislocation escaped, another dislocation (b = a/2 [10-1]) was nucleated; (c) the Lomer lock dislocation formed by a dislocation reaction; (d) the enlarged HRTEM image showing the Lomer lock structure.

Burgers vector of a/2[10-1] was nucleated at the bottom surface of the bent Si NW and then escaped. Fig. 7.1.4b was recorded from the same area of Fig. 7.1.4(a) and shows another dislocation ($b = 1/2$ [10-1]) that was nucleated and then formed a new structural dislocation by reaction (as shown in Fig. 7.1.4(c)). Local Burgers circuits were drawn to identify the Burgers vector b. The enlarged HRTEM image is given in Fig. 7.1.4(d). Extra planes are seen for both the (-111) and the (1-11) planes. This configuration represents a Lomer dislocation exhibiting the Burgers vector of a/2[110]. The Lomer dislocation was formed by the interaction of two full dislocations with Burgers vectors a/2[10-1] and a/2[011], respectively, moving under applied strain on two intersecting slip planes, (-111) and (1-11).

With continuous increase of the bent strain, these locks could not be unzipped, and the locked lattices became disordered, leading to amorphization. Fig. 7.1.5(a)–(d) are taken from a same area and show the in situ observation of Lomer dislocation formation and the subsequent crystalline to amorphous transition process at atomic scale. Through Fig. 7.1.5(a)–(d), the corresponding maximum bent strains were 10.3, 12.8, 13.3, and 14.3%, respectively. Fig. 7.1.5(a) and (b) indicate the dis-

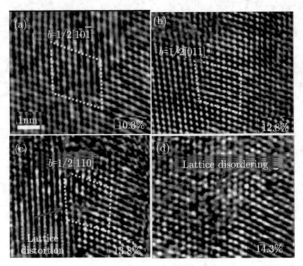

Fig. 7.1.5 The enlarged HRTEM images taken from the same region. The corresponding strains are 10.3%, 12.8%, 13.8%, and 14.3% for (a)–(d), respectively. The c-a transition process was directly observed at atomic scale.

location nucleation and motion. These activated dislocations were driven to meet under high strain/stress and then formed a Lomer dislocation by reaction in Fig. 7.1.5c. The Lomer dislocation junction was expected to be sessile because it can glide in neither of the slip planes containing the reactant dislocations. High stresses are needed to break the Lomer lock, and its destruction process has only been observed in metals (Madec et al., 2003; Rodney and Phillips, 1999). This sessile dislocation will make the local stress to be very high as the strain increases. As it is shown in Fig. 7.1.5d, unlike the destruction process of Lomer dislocation locks in metals, the lattices of the Lomer dislocation lock became highly distorted and finally completely disordered and transformed to be the amorphous state as the strain rate increased. This is the atomic mechanism of crystalline-amorphous transition for strained Si NWs which can be used to understand the same structured materials.

7.1.5 Atomic-scale mechanisms of unusual plasticity of nanocrystalline metals

Structural evolution, deformation behaviors of metals have always been attracting much attention due to the importance of the properties. In recent years, nanocrystalline materials have been found that they have ultra-high strength which implies different deformation behaviors from their bulk counterparts. With the development of in situ HRTEM techniques, more and more studies have provided atomic-scale information and unusual deformation behaviors. They enrich our knowledge largely and make us better understand the deformation mechanisms at atomic scale.

1. Size effect on the types of dislocations

Full dislocations or partial dislocations are dominated during the deformation which is not only depended on materials but also the size. For example, plastic deformation of bulk Cu is usually mediated by the slip of full dislocations. Yue et al. (2012) quantitatively revealed an obvious effect of the sample dimensions on the plasticity mechanisms of tensile Cu single crystalline NWs with diameters between 70–1000 nm by in situ HRTEM. As the diameter reduces to about 150 nm, the normal full dislocation slip is taken over by partial dislocation mediated plasticity. Figure presents an example of Cu NW with a diameter of 166 nm. The loading axis is approximately parallel to $\langle 111 \rangle$. From the HRTEM images, twin boundary (TB), stacking faults (SF) can be clearly seen. They are formed by partial dislocations emitted from TBs. This phenomenon has been also observed in nanocrystalline Au (Zheng et al., 2010).

2. Deformation behaviors of ultra-small nanograins film

The generation, motion, interaction, storage, and annihilation of dislocations are well understood in plastic deformation of conventional polycrystalline metals. However, when the grain size (diameter d) is refined to nanometer scale ($d < 10$ nm), how the dislocations behave becomes an unsettled issue. It has been proposed that in this d regime the strength of metal would exhibit an inverse Hall-Petch d dependence (Chokshi et al., 1989) and during plastic deformation dislocation activities subside, giving way entirely to the sliding of grain boundaries (GBs). The $d < 10$ nm regime is rarely accessed in experiments, it is the realm of molecular dynamics (MD) simulations (Van Swygenhoven, 2002; Yip, 1998; Schiotz and Jacobson, 2003), which did predict persistent dislocation activities, as well as their interactions and storage. Recently, Wang et al. (2010) investigated the dislocation behaviors at atomic scale for a nanograined Pt ultrathin film with an average d of 9 nm by using bimetallic extensor in a TEM. They monitored dislocation dynamics and observed Lomer lock formation through the reaction of full dislocations and their destruction processes at atomic scale. They also provided direct and statistical evidences that the full dislocations behavior can be transformed to partial dislocations behavior as d goes below a critical value. These experimental evidences demonstrate clearly that dislocations behaviors are highly active even in such tiny grains ($d < 10$ nm) which is important to elaborate the deformation mechanisms in nanocrystalline metals.

3. Deformation behaviors of twin-structured crystalline metals

Hall-Petch relation and inverse Hall-Petch relation are usually used to describe the relationship between strength and the size of grains. The Hall-Petch relation predicts that the yield strength increases with the decrease of grain size. However, if the grains reach a size small enough (< 100 nm), the yield strength would either remains constant or decreases with decreasing grains size (Conrad and Narayan, 2000). This phenomenon has been termed the inverse Hall-Petch effect. This similar trend was also revealed between twin thickness and strength in nanotwinned copper. Lu et al. (2009) found that the strength increases with decreasing twin thickness, reaching a maximum at 15 nm. They proposed a mechanism based on post-mortem observation: The strongest twin thickness originates from a transition in the yielding mechanism from the slip transfer across twin boundaries to the activity of preexisting easy dislocation sources. Following the above report, Li et al. (2010) provided an understanding of the atomic mechanism based on molecular dynamics simulation. There is a critical twin thickness for a given grain size, above it, partial dislocations intersecting with the twin boundary dominate the plastic deformation, which results in strengthening; below it, partial dislocations glide along the twin boundary, which results in softening. Yue et al. (2012) directly observed the deformation processes for Cu film with different twin thickness by in HRTEM and convinced this deformation mechanism from experiment.

4. Direct atomic-scale observation of continuous and reversible lattice deformation far beyond the elastic limit in Ni NW

As the material dimensions decrease to nanometer scales, large elastic strains of ~8%, close to the theoretical limit, have been observed in many regimes (Yue et al., 2011; Wu et al., 2005; Wei et al., 2012). However, if inelastic events associated with dislocations (Greer and Nix, 2006; Ah et al., 2009), cracking (Deng et al., 2011), twinning (Li et al., 2010) and first-order phase transitions (Diao et al., 2003) can be largely suppressed under confining loading conditions in small regions, a continuous and reversible lattice strain beyond 8% may become possible. The upper limit of lattice strain is of obvious interest in materials science and physics. Theorists have attempted to compute it from atomistic models since the 1920s (Polanyi, 1921; Frenkel, 1926). Such estimates predicted that the upper limit of lattice deformation strain in metals can surpass 30% (Luo et al., 2002; Clatterbuck et al., 2003), but it has never been observed in experiments. Recently, Wang et al. (2013) performed in situ bending experiments of Ni NWs in a HRTEM and observed ultralarge elastic lattice deformation up to 34.6% induced by continuous lattice shear from 70.5° to 109°. These atomic-scale images also provide with the methods how the large shear strains accomplish the continuous lattice changes from the original FCC through orthogonal path, to the tetragonal lattice and finally, to a re-oriented FCC. Fig. 7.1.6 provides such an atomic scale insights. Fig. 7.1.6(a) is a typical atomic-scale image of the initial atomic structure of the Ni NW showing the inter-planar angle (α) 70.5° of the {111} planes. Fig. 7.1.6(b) is an atomic image showing that the dislocation-free lattice shear gives rise to the gradual increase in α from ~70.5° to ~90° and then to ~109°, which corresponds to the continuous lattice changes from the original FCC through orthogonal path, to the tetragonal lattice and finally, to a re-oriented FCC. Interestingly, the ultra-large lattice shear is fully reversible. When further straining was imposed, the continuous lattice shearing was terminated by dislocations/small angle boundaries. Fig. 7.1.6(d) is a schematic drawing of the projected atomic configurations corresponding to the HRTEM images of Fig. 7.1.2(b). Fig. 7.1.6(e)–(g) illustrate the 3D structures and side views of the FCC lattice with shear strains of 0, 17.3% and 34.6%, respectively. This phenomenon

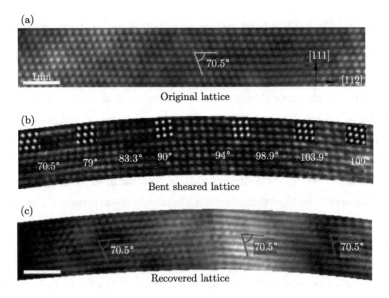

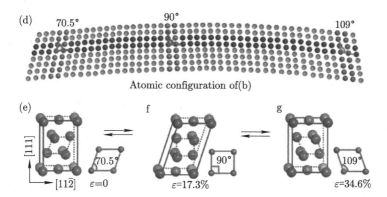

Fig. 7.1.6 Continuous and reversible lattice strain far beyond the elastic limit. (a) HRTEM image of a Ni nanowire captured when the bent strain is ∼1.9%. (b) Enlarged atomic-scale image shows the continuous increase in α from 71° to approximately 109° is clearly demonstrated. The simulated HRTEM images are inserted accordingly. (c) HRTEM image captured after the strain released. No square lattice and other type sheared lattices were observed, indicated the 34.6% lattice shear strain was fully recovered. (d) Atomic configurations that corresponding to (b) to show the ultra-large continuous lattice shear from 0 to 34.6% more clearly. (e)–(g) Schematic 3D structures and corresponding side views show the shear and recover process of the FCC lattice. The scale bars are for 1 nm.

is different from the traditional phase transformation.

7.1.6 Large plasticity of nanoscale glass

Glasses, including oxide glass, metallic glass are known to be brittle and fracture upon any mechanical deformation for shape change at room temperature. They are usually shaped at above their glass-transition temperature. However, these brittle objects have been found that they can exhibit large plastic strain under applied stress when their size is down to nanoscale. Fig. 7.1.7 shows a typical deformation response of amorphous silica nanoparticles in compression tests inside a TEM (Zheng et al., 2010). Fig. 7.1.7(a) displays the centered dark-field image of a silica particle with a diameter of 510 nm before the compression test. After imaging for the positioning of the sample and the diamond flat punch, the beam was blocked off with the condenser lens aperture, and the silica particle was compressed with the Hysitron Picoindenter. The particle is plastically deformable apparently (Fig. 7.1.7(b)). Then continued to compress this particle with the electron beam-on, it was shaped into a pancake (Fig. 7.1.7(c)). This large plastic behavior was also found in tensile tests for single amorphous silica NWs. Elongation exceeding 215% was exhibited during the tensile process for a NW with diameter of 36 nm. HRTEM images show that there are no cracks, shear bands and crystallization. The similar phenomenon has also been demonstrated for small-volume metallic glass by Guo et al. (2007). However, due to the difficulty to provide more details for amorphous structure by HRTEM, atomic-scale deformation mechanisms are dependent on molecular dynamics simulations. More powerful experimental methods or techniques should be developed.

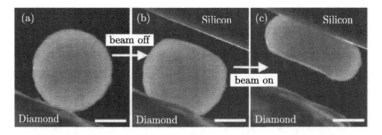

Fig. 7.1.7 Nanocompression of amorphous silica particle inside a TEM. Two consecutive compression runs were performed with the electron beam being off and on, respectively. Panels a–c show the centered dark-field images. The scale bars are for 200 nm.

7.1.7 Summary and future prospects

For nano-mechanics, in situ HRTEM experiments provide the possibility to directly observe the deformation processes and dynamic observations, which are essential for understanding the deformation mechanisms of nanomaterials at atomic scale. In the past decades, significant progress has been made regarding the deformation dynamics of NMs using new developed methods and techniques. More and more new phenomena and mechanisms have been provided. However, open questions remain regarding the dynamics of dislocation interactions in small volumes. The quantification of strain-stress output companying with the atomic-scale images simultaneously is still a difficult technique which should be made a more progress.

For the following decades, a golden age for conducting investigations of the experimental mechanics of materials at atomic scale can be expected to benefit from the huge development of two aspects. One is the wide application of Cscorrector. This makes resolution of TEM enter sub-angstrom age. The huge progress in spatial resolution can make probing ability great improve. The other is the improvement of CCD. For in situ experiments, the time resolution is very important to obtain experimental details. Nowadays, experiments on picosecond, or even femtosecond time scales, have been carried out by applying pulsed-electron packets, or electron pulses. So it is excited to expect the development of in situ real-time high-resolution TEM.

7.2 In situ TEM investigations on Ga/In filled nanotubes

Yihua Gao and Yoshio bando

7.2.1 Introduction

1. Carbon nanotubes

Carbon nanotubes (CNTs) or CNT-like materials may have existed for a long time since their growth complies with the energy minimizing principle, as suggested in the literature (Gao et al., 2008; Tibbetts, 1984; Charlier, 1997). In the early 1960s, CNT-like materials were reported by Bacon (1960) but were thought to be graphite whiskers. In 1991, it was realized that CNT (Iijima, 1991) was a new member of the carbon family, which had expanded just 6 years earlier when fullerene (Kroto et al., 1985) joined the three former members of graphite, diamond and amorphous carbon. The bonding in CNT is sp^2 hybridized and CNT can be considered as graphitic sheets with a hexagonal lattice wrapped up into a seamless cylinder. Dramatically, graphene (Novoselov et al., 2004) having a single graphitic sheet structure was added to this family 13 years after the discovery of CNT. Due to their unique morphology, nanosized scale, and novel physic-chemical properties,

CNTs have attracted much attention among scientists in different fields, including physics, chemistry, materials and mechanics. In the past two decades, great effort has been made to identify and develop new applications of CNTs which do not exist today but for which there may be great demand tomorrow. For example, four years after the discovery of CNTs, working on field emission characteristics (de Heer et al., 2004) demonstrated that CNTs could be used in field-effect devices, and two commercial products, namely high-brightness luminescent elements and an X-ray tube, have hit the market. In addition to these applications based on the field-emitting effect, more than 50 other applications have been developed (Baughman et al., 2002; Xiong et al., 2011). These applications can be classified into several fields: 1) electronics; 2) mechanics; 3) electro-mechanics; 4) hydrogen storage; 5) sensors and detectors for materials, etc. However, there was no applications of CNTs in the thermal field until the invention of the carbon nanothermometer (Gao and Bando, 2002).

2. Carbon nanotube filling

Due to the reductive characteristics of carbon and one-dimensional (1D) shape of CNTs, CNTs can be used as a template for the growth of other 1D material (Dai et al., 1995; Han et al., 1997; 1998). Pederson's simulation work (1992) on the energy of CNT central emptiness predicted that the emptiness could be filled with materials. To date, many materials, including a number of single elements, have been used to fill CNTs, e.g. S, Cr, Mn, Fe, Co, Ni, Cu, Zn, Ge, Se, Ru, Pd, Ag, Sn, Sb, I, Cs, Sm, Gd, Dy, Yb, Re, Au, Pb, and Bi. Even so, our filling with Ga into CNTs was the first application in thermology, and other applications of element-filled CNTs have since been reported (Pan et al., 2007; Lipert et al., 2010; Shi and Cong, 2008; Soldano et al., 2010).

The carbon nanothermometer can be used for various temperature measurements in microenvironments, e.g. working temperature in fuel cells, electronic circuits, and localized micro-zone laser temperature, etc. In our invention, we filled Ga (in liquid state over a large temperature range of $29.78 - 2403\,°C$) into CNTs and confirmed that the linear volume expansion of a nanoscale Ga column (diameter 75 nm) has the same volumetric coefficient ($\alpha_{Ga} = 0.95 \pm 0.06 \times 10^{-4}$

per °C) in the range of 50−500 °C as that of macroscale Ga state ($\alpha_0 = 1.015 \times 10^{-4}$ per °C). We explained the changeless expansion coefficient using a thermodynamic analysis (Gao and Bando, 2002). Another investigation showed that In, an element in the Ga family, can also be filled in CNTs (Gao et al., 2002), but its behavior is not as good as that of Ga. Usually, in situ microenvironment temperature measurement is very difficult to achieve in a transmission electron microscope (TEM) or scanning electron microscope (SEM). Thus, temperature recording is important and we solve this problem by using the oxidization of a Ga tip surface in the open end of a CNT (Gao et al., 2003). To measure low temperatures below the melting point of ice, the unusual freezing and melting of Ga encapsulated in CNTs was investigated (Liu et al., 2004). To avoid the possible degradation of the wrapping carbon layer of a carbon nanothermometer in high-temperature measurement, Ga-filled MgO (Li et al., 2003; Su et al., 2012), In-filled In_2O_3, (Li et al., 2003) and Ga, In-filled silica nanotube thermometers were invented. The Ga-filled MgO nanotube has also a higher sensitivity and a very interesting cross-sectional shape: not round but square. For nanothermometers, Ga/In-filled nanotubes can also act as metal-semiconductor nanowire hetero-junctions (Zhan et al., 2005) and metal vapor absorbents (Liu et al., 2002). By using GaN and Ga_2O_3 as precursor materials for the bulk synthesis of liquid Ga-filled CNTs (Zhan et al., 2005; Dorozhkin et al., 2005), transforming temperature data to electrical resistance data, the carbon nanothermometer paved the way for actual application in microenvironments. We have also made efforts to put this type structure into real use by using In-filled MgO nanotubes as sensor with a protective temperature ∼150 °C (Fu et al., 2012). Interestingly, we found a novel effect named electric-hydraulic expansion effect in a silica shelled gallium microball-nanotube structure and the effect may introduce nanotransmission: electric-hydraulic drive, which will be a new working mechanism for micro-nano-electromechanical system (Gao et al., 2011). We studied electrically driven gallium movement in carbon nanotubes with the hope to design electrically driven nanomass delivery and nanoswitches (Sun and Gao, 2012). Finally, we discussed the problems hindering real-world application for nanothermometers, nanotransmission, nanosensors and nanoswitchs using a metal filled CNT.

3. Comparison of carbon nanothermometer with Hg thermometer

Fig. 7.2.1 schematically compares an ordinary Hg thermometer and the created nanothermometer containing Ga. In an ordinary thermometer, the filling material Hg, with a volume expansion coefficient of 1.8×10^{-4} per °C (Gray et al., 1972), has a short measuring range because of its shorter liquid range, $-38.87 - 356.58$ °C. In a nanothermometer, the relative error ($< 3\%$, $\pi \alpha_C/\alpha_{Ga}$) due to expansion of carbon can be neglected because of its very small linear expansion $\alpha_C \sim \pm 1.0 \times 10^{-6}$ per °C in the base plane. Ga has a long liquid range and a very low vapor pressure [$\log(p/\text{Pa}) = 11.76 - 13,984 T^{-1} - 0.3413 \log T$, T in K] (Lide, 1990), e.g. $5.4689 \times 10^{-6} - 3.9086 \times 10^{-5}$ Pa in the range of $600 - 650$ °C. The significant differences between a nanothermometer and an Hg thermometer are: (i) the nanothermometer is suitable for microenvironments because of its tiny size, and (ii) the nanothermometer has a broader measurement range.

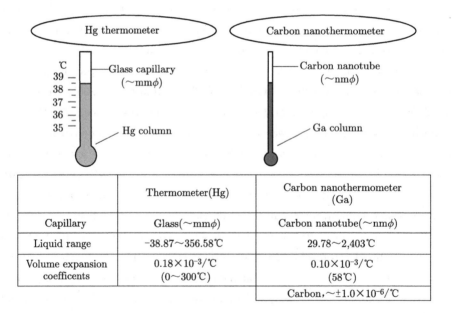

	Thermometer(Hg)	Carbon nanothermometer (Ga)
Capillary	Glass($\sim$mmϕ)	Carbon nanotube($\sim$nmϕ)
Liquid range	$-38.87 \sim 356.58$°C	$29.78 \sim 2,403$°C
Volume expansion coefficents	0.18×10^{-3}/°C ($0 \sim 300$°C)	0.10×10^{-3}/°C (58°C) Carbon, $\sim \pm 1.0 \times 10^{-6}$/°C

Fig. 7.2.1 Comparison between an ordinary Hg thermometer and the Ga-filled carbon nanothermometer.

7.2.2 In situ investigations on metal filled nanotubes

1. Experiment for synthesizing and studying carbon nanothermometers

The carbon nanothermometer was produced in a vertical radio-frequency (RF) furnace as shown in Fig. 7.2.2(a) and (b). The furnace consists of a clear fused-quartz tube 50 cm in length, 12 cm in diameter and 0.25 cm in thickness. This tube contains an inductively heated cylinder of high-purity graphite, which is 7 cm in length, 4.5 cm in outer diameter and 3.5 cm in inner diameter. An uncovered graphite or BN crucible, which was ∼2 cm in diameter and ∼2 cm in height, was put in the cylinder. In the C/BN crucible, the reactant was a homogeneous mixture of Ga_2O_3 and pure amorphous active carbon in a weight ratio of 7.8:1. The C cylinder had one graphite fiber coat and one outlet C pipe on its top. Pure N_2 gas flow was introduced into the furnace. The reactant was heated at 1360 °C for 1 − 2 h. After the process, the reactant in the C crucible disappeared, while some powder materials were deposited on the inner surface of the outlet pipe, where the temperature was estimated to be lower than ∼810 °C. Ga/In-filled oxide nanotubes

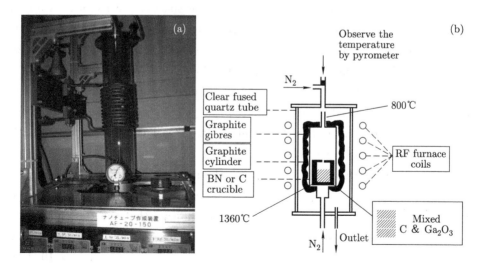

Fig. 7.2.2 (a) Image of RF furnace for carbon nanothermometer synthesis; (b) Schematic of the furnace.

have also been fabricated in the RF furnace by using a similar method and different starting materials.

The carbon/oxide thermometers were collected and analyzed by two types of high resolution TEM (HRTEM), JEM-3000F with an X-ray energy dispersive spectrometer (EDS), and JEM-3100 FEF with Ω energy filter system for electron energy loss spectrum (EELS) and its mapping, and X-ray energy dispersive spectrometer (EDS) system for line scanning and mapping. The behavior of the nanothermometer in the TEM was controlled by holding it in a Gatan heating holder and its twin system (Hot stage power supply, Model 628-0500). Using a Gatan model 900 cold stage, the freezing and melting of Ga encapsulated in CNTs were directly observed by the JEM-3000F TEM. The cold stage used liquid nitrogen for cooling and its cooling rate could be easily adjusted to as low as 0.5 °C per min prior to freezing and melting by changing the heating current.

2. Carbon nanothermometer: nanotube thermometer containing gallium

Fig. 7.2.3 shows a CNT thermometer. A round tip, a longer Ga column, a hollow spaces, a shorter Ga column, and one more hollow spaces are consecutively assembled within the CNT from left to right. The length and outer diameter $2r_0$ of the CNT are 9,180 nm and 85 nm, respectively, whereas the length and diameter $2r_1$ of the longer Ga column are 7560 nm (equals $H_{p_0 t_0}$) and 75 nm, respectively; Fig. 7.2.4(a)–(h) shows that the Ga level rises or falls consistently when the column temperature is increased or decreased in the range 18 – 500 °C. The variation with temperature of the height of the Ga meniscus in its CNT is plotted in Fig. 7.2.3(i), using the level at 58 °C as a reference.

The volumetric change of liquid Ga in the macroscopic state upon heating is described by $v_t = v_0(1 + \alpha_{Ga}\Delta t)$, where v_t and v_0 are the volumes at temperatures t and t_0 respectively, $\Delta t = t - t_0$. Our results indicate that this description also is applied to nanoquantities of 1D liquid Ga: calculations, on the basis that its change in volume with temperature gives a value of $\alpha_{Ga} = 0.95 \pm 0.06 \times 10^{-4}$ per °C. In this respect, the expansion coefficient differs from another basic thermal property, the melting point, which is greatly influenced by the surface effect. For

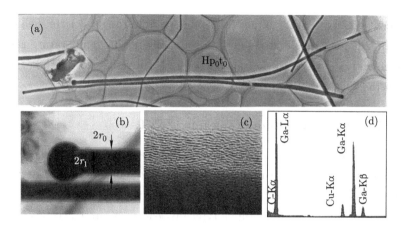

Fig. 7.2.3 (a) Morphology of a Ga-filled CNT. (b) The C-sealed round tip; (c) HRTEM image of the C wrapping layer. The d spacing between the wave-shaped fringes is $\sim$0.34 nm; (d) EDS spectrum of the region in (c). The peaks of C-Kα (0.28 keV), Ga-Lα (1.10 keV), Ga-Kα (9.24 keV) and Ga-Kβ (10.26 keV) are labeled. The Cu-Kα peak was generated from the TEM grid.

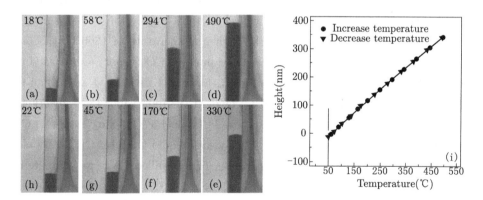

Fig. 7.2.4 Level of the Ga meniscus changing with temperature increase (a)–(d) and decrease (d)–(h). (i) Height of Ga meniscus plotted against temperature, measured in steps of $30-50\,^\circ$C.

such a nanothermometer, the temperature can be measured as $t = 58 + \Delta H/0.753$, where ΔH (in nm) is the height change of the Ga column from t to 58 °C. Our experiment confirmed that 50 – 500 °C is a reliable range: (i) the height of the liquid Ga column varies linearly and reproducibly in this temperature range, and (ii) we have tried to extend the range to 600 °C, but failed due to the evaporation of Ga in the TEM starting from 550 °C.

3. Possible growth diagram of Ga filled CNTs

Generally, there are two ways to produce material-filled CNTs. The first entail filling pre-existing CNTs with a molten media, followed by wet chemistry solution methods. The second is to simultaneously produce the nanotubes and their fillings during arc-discharge. These methods typically lead to discrete crystals; a continuous crystal length does not exceed ~1 µm. Our approach relies on a substantially different method and produces CNTs filled with a long 1D nanoscale liquid Ga column (~10 µm), making it readily available for actual applications. The growth of the present nanothermometer may be stimulated by Ga drops at the bottom (or top) of the CNT (see Fig. 7.2.5. (a)–(e)). Ga drops provide discrete nucleation spots for vapor-vapor (VV) reaction between Ga_2O and CO vapors and continuous growth of Ga-filled CNTs. The growth of the Ga-filled CNTs is assumed to include two chemical reactions. At the higher temperature of 1360 °C, the reaction Ga_2O_3 (solid) + 2C (solid) → Ga_2O (vapor) + 2CO (vapor) occurs as Gibbs energy (Barin et al., 1989) falls at −140 kJ, forming 1 mol Ga_2O and 2 mol CO vapor. The lower temperature (~810 °C) favors the vapor–vapor (VV) reaction, n Ga_2O (vapor) + $(n+2)$ CO (vapor) → 2n Ga (liquid) + C (solid) + $(n+1)$ CO_2 (vapor), leading to the formation of 1D nanoscale liquid Ga-filled CNTs. By measuring the outer and inner diameters $2r_0$, $2r_1$ of a Ga-filled CNT, we estimate the value of $n = m_c r_{Ga} r_1^2 / [2m_{Ga} r_c (r_0^2 - r_1^2)]$, where m_c = 12 g/mol, r_{Ga} = 6.10 g/cm³, m_{Ga} = 69.72 g/mol, and r_c~2.00 g/cm³. For the Ga-filled CNT in Fig. 7.2.5(a), n is ~1. A series of calculations on the Gibbs energy change for the VV reaction at n~1 illustrates that it can only occur at a temperature lower than 818 °C, which is fully consistent with the fact that the Ga-filled CNTs were obtained in the lower temperature zone (~810 °C). Our carbon nanothermometer is quite different from S. Boyer's ordinary Ga-filled temperature responsive device (Boyer, 1931), both in fabrication procedure and measurement environment. S. Boyer's device is used for

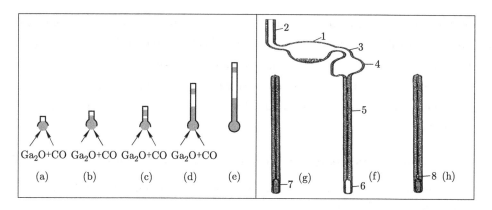

Fig. 7.2.5 (a)–(e) Schematic of fabricating a carbon nanothermometer; (f)–(h) Procedure of fabricating S. Boyer's Ga temperature device for macroscopic environments.

macroenvironments and is made by pouring liquid Ga into a glass tube, pumping and heating impurity gas, and sealing off the capillary glass, etc. (see Fig. 7.2.5(g)-(h)). First, the Ga metal or Ga-containing alloy is placed in a treatment bulb 1, (f), which is connected by conduit 2 to a vacuum pump, and by a conduit 3 to a reservoir 4. This reservoir in turn is connected to the stem 5 and bulb 6 of a thermometer which preferably is made of fused quartz. After thorough exhausting, during which the quartz container is highly heated to drive off water vapor and gases, the Ga in the bulb 1 is heated to volatilize the skin of gallium chloride and possible traces of oxide, these impurities being deposited in the upper part of the bulb 1. The metallic Ga is left in the bulb 1 in a clean and bright condition. The Ga is heated to redness to free the metal from hydrogen and other gases, the operation of the vacuum pump being continued to remove the liberated gas. Preferably, the metal is cooled and reheated a number of times and transferred to the bulb 4 by tilting the apparatus. The conduit 3 is finally sealed off by fusion. A required amount of metal is caused to run into the bore of the stem 5 and to enter the bulb 6. The thermometer is then placed in an oven leaving the bulb 4 projecting and is heated several times to a temperature of about 800 °C or higher, to completely eliminate gas from the metal. Gas bubbles are caused to enter the chamber 4 by tilting and tapping the thermometer. The stem of the thermometer finally is sealed off from the bulb 4, leaving the completed thermometer as shown in (g) containing a quantity 7 suitable for service. Most of the gas eliminated during the last stage of manufacture is contained in the sealed-off bulb; the small amount of gas remaining in the thermometer stem does no harm. When plastic metal is alloyed with Ga, the thermometer tube preferably is provided with an expansion chamber, as indicated at 8 in (h).

4. Thermodynamic analysis of surface effect on expansion of Ga in CNT

It is found that the expansion characteristics of the Ga column in the CNT virtually match those of Ga in the macroscopic state (Gao and Bando, 2002). To explain this, we derived the equation

$$\alpha_{Ga} = \frac{\alpha_0}{1 + \kappa p_{t_0,\text{inner}}} + \frac{\kappa(p_{t,\text{inner}} - p_{t_0,\text{inner}})}{(1 + \kappa p_{t_0,\text{inner}})(t - t_0)} \qquad (7.2.1)$$

for the measured expansion coefficient α_{Ga} at temperature t via a series of quantitative thermodynamic analyses (Gao and Bando, 2002) after considering inner pressure $p_{t,\text{inner}}$ at temperature t and $p_{t_0,\text{inner}}$ at t_0 (Koster et al., 1970), and κ the volume compression coefficient. When the thermometer shown in Fig. 7.2.3(a) was heated in the microscope, the tip-level contact angle θ of the longer Ga column vs temperature was also measured in the range $50-500\,°C$, as shown in Fig. 7.2.6(a). Fig. 7.2.6(b) shows the estimated inner pressure $p_{t,\text{inner}} \sim t$. For a Ga column of 37.5, 5 and 0.94 nm radii at 230 °C, the coefficient α_{Ga} was calculated as 0.997, 0.975 and $0.868\alpha_0$, respectively, as shown in Fig. 7.2.6(c). Therefore, our measured expansion coefficient α_{Ga} of the 1D nanoscale Ga (corresponding to 37.5 nm radius), $0.95 \pm 0.06 \times 10^{-4}$ per °C (virtually equal to α_0, 1.015×10^{-4} per °C) can be clearly understood. The above quantitative analysis revealed that the measured coefficient α_{Ga} is in fact affected by the column radius. However, it is almost equal to α_0, which can be directly taken as the expansion coefficient for calibration of a nanothermometer when Ga diameter $\geqslant 10$ nm.

5. Temperature recording in a Ga-filled CNT thermometer

Fig. 7.2.7 shows the temperature recording method by showing the morphology of one nanothermometer. At first, Ga-filled CNTs were identified in the JEM-3000F at

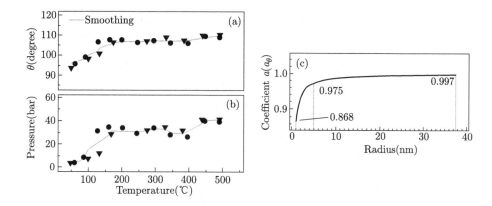

Fig. 7.2.6 (a) Correlations between the contact angle θ of a Ga tip vs temperature t; (b) The inner pressure $p_{t,\text{inner}}$ vs temperature t. The red and blue markers correspond to temperature increase and decrease, respectively, whereas the green line corresponds to the smoothed experiment data; (c) Coefficient α vs radius r_1 at 230 °C.

room temperature (21 °C), as shown in Fig. 7.2.7(a), (b) and (c). The CNT contains a continuous Ga column with a diameter ranging from 48 to 145 nm, a length H_0 of 12090 nm, and a volume V_0 of 9.586×10^7 nm. We then put them in an ordinary natural air-occupied resistance furnace, at a temperature previously controlled by a calibrated thermocouple. After 20 min, we extracted the nanothermometer, cooled it to room temperature, and inserted it back into the TEM, shown in Fig. 7.2.7(d), (e) and (f). The Ga-tip in the closed tube end did not change (Fig.7.2.7(e) and (b)). In contrast, there was a notable change in Ga-column height $\Delta H = 170$ nm, corresponding to a volume change ΔV of 2.810×10^6 nm^3 (Fig.7.2.7(f) and (c)). Our investigation showed that the remaining height ΔH at room temperature is due to the gallium oxide tip (Gao et al., 2003), which forms during measurement and sticks to the C wall tightly even after cooling. We calculated that the temperature experienced by the nanothermometer inside the furnace was 341 °C; there is a small deviation ~5% between it and the controlled temperature of 358 °C. Reproducibility of the temperature measurements using the present technique was additionally verified by means of another Ga-filled CNT heated in air to 190 °C. This method demonstrates that Ga-filled CNT thermometers are fully capable of recording the temperature in an oxygen-containing microenvironment. It should be noted that the nanothermometer should be sealed at one end, open and half-full at the other end, to provide a channel for the entrance of O_2 from outside and space for expansion, respectively.

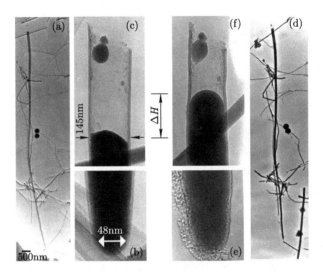

Fig. 7.2.7 (a)–(f) TEM images of a carbon nanothermometer containing a continuous Ga column, taken at 21 °C. (a), (b), and (c) before placing the thermometer into an air-occupied furnace; (d), (e), and (f) after extracting the thermometer from the furnace, i.e. after measuring its temperature.

6. Melting, expansion and electric transport behaviors of indium in CNTs and MgO nanotubes

The single-phase In-filled CNTs and MgO nanotubes (In, $a = 0.325$ nm and $c = 0.494$ nm; JCPDs: 5-642, melting point: 156.6 °C) were synthesized in the vertical RF furnace. Fig.7.2.8 shows a typical CNT. By comparing Fig.7.2.8(a) with Fig. 7.2.3(a), it is found that In-filled CNTs usually lack a round tip with a size larger than the diameter of the main body, different from Ga-filled CNTs. This phenomenon may be due to two factors: 1) The surface tension of In on C ($\sim$500 mN/m) is smaller than that of Ga on C ($\sim$700 mN/m); 2) The density of In is larger

($\sim$6.5 g/cm^3) than that of Ga ($\sim$5.6 g/cm^3) at near 810 °C. During the growth, the larger surface tension of Ga can support a round Ga droplet with a larger volume/weight ratio, while the lower surface tension of In can only support an In droplet with a smaller volume/weight ratio when the metal particles act as catalyst on the top. Fig. 7.2.8(b) shows the morphology of the In-filled CNT after heating to 377 °C and then cooling to 21 °C. In Fig. 7.2.8(c)–(f), the shape and height (stage I in Fig. 7.2.8(k)) of the In meniscus inside the CNT change even if the temperature is lower than the melting point of In (156.6 °C). This phenomenon is due to an interfacial effect between In and C-wall, and surface effects in nanoscale materials (Goldstein et al., 1992) when the thermal effect of an electron beam (Chabala, 1992) is negligible by minimizing the beam current. The equation

$$\lambda(2r_1 - \lambda) = \frac{2\Delta\sigma r_1 \theta T_0}{\Delta h_f (T_0 - T)} \qquad (7.2.2)$$

was derived (Awschalom and Warnock, 1987) to calculate the liquid out-shell thickness λ during the melting process in a column at a given temperature T below T^* (the melting point T^* of an element in the nanoscale state is lower than its standard melting point T_0), where r_1 is the column radii, $\Delta\sigma$ is the difference between the solid-tube wall interfacial energy and the liquid-tube wall interfacial energy, θ is the molar volume of a liquid at melting point T^*, and Δh_f is the heat of fusion. According to this equation, the phenomenon whose height increases with temperature below the melting point can be explained qualitatively as: for In-filled CNTs, the higher the temperature ($T \uparrow$), the more liquid In that forms ($\lambda \uparrow$), and the larger the volume of the column because the density of liquid In is smaller than that of solid In, hence the greater the change and higher level of meniscus shape. Fig. 7.2.8(g)–(j) display the In meniscus at temperatures above melting point. In Fig. 7.2.8(k), process I corresponds to melting. In stage II of Fig. 7.2.8(k), the column expands with coefficient 3.25×10^{-4} per °C, nearly three times the coefficient α_{In} (1.16×10^{-4} per °C) of liquid indium in the macroscopic state. Process III corresponds to the decrease in meniscus height when 377 °C is kept for 5 min. Process IV corresponds to the decrease in temperature. Process V means that the meniscus remains unchanged even if the temperature decreases. The phenomenon described above may be due to some N_2 bubbles dissolving in indium during the material synthesis stage, which then expand rapidly as the temperature increases. However, any existing N_2 may escape when the temperature is kept at 377 °C for 5 min, leading to a rapid decrease in height. In the temperature decreasing stage IV of Fig.7.2.8(k), the expansion coefficient α_{In} is 1.00×10^{-4} per °C, virtually matching the coefficient in the macroscopic state. However, comparing Fig. 7.2.8(a) and (b), when the temperature decreases back to 21 °C, the end-tip of the In column cannot recover to the initial position before heating. This may be due to the lack of a round tip, which may provide a high tension drawing back the column during the temperature decreases. Our findings suggest that two problems need to be addressed if In-filled CNTs are to be used as a nanothermometer: (i) post-heating of the as-deposited indium-filled CNTs to avoid gas atoms becoming occupied in indium; and (ii) how to restore the initial position of the In meniscus during the temperature decrease stage.

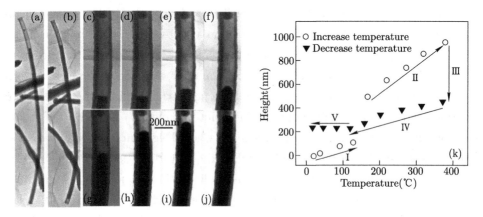

Fig. 7.2.8 (a) Morphology of an In-filled CNT; (b) Morphology of the In-filled CNT after heating to 377 °C and then cooling to 21 °C; (c)–(f) Shape and height of the In meniscus inside the CNT at 21 °C, 35 °C, 89 °C, and 126 °C; (g)–(j) Shape and height of the In meniscus at 170 °C, 270 °C, 322 °C, and 377 °C, respectively; (k) Meniscus height as a function of temperature.

Fig. 7.2.9 shows the melting, expansion behavior and electric transport of In-filled MgO nanotubes. The melting and expansion behaviors are very similar to the behaviors of In-filled CNTs, as shown in Fig. 7.2.9(a)–(j) and described by

Y. Fu et al. (2012). Fig. 7.2.9(k) shows that the central empty column length

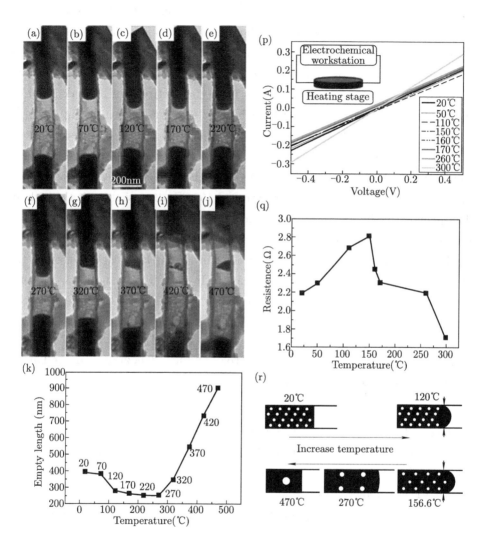

Fig. 7.2.9 Morphology of the In inside the MgO nanotube at 20 °C (a), 70 °C (b), 120 °C (c), 170 °C (d), 220 °C (e), 270 °C (f), 320 °C (g), 370 °C (h), 420 °C (i), and 470 °C (j). (k) The central empty column length versus the temperature. (p) I–V characteristic curves of the disk at different temperatures. (q) Resistance of the disk as a function of temperature. (r) Diagram the In Column with gas nanopores changing with temperature.

changed with the temperature. This above experiment suggests that the In-filled MgO nanotube is not suitable to act as a nanothermometer. However, it seems that massive In-filled MgO nanotubes can be fabricated as temperature protective sensors. Massive In-filled MgO nanotubes were collected and pressed into a piece of disk with a diameter of 13 mm and a thickness of 1.8 mm. The I-V characteristic curves of this disk were investigated by using an electrochemical workstation at different temperatures controlled by a heating stage, as shown in Fig. 7.2.9(p). According to these I-V characteristics, we calculated the resistances of disk at different temperatures, as shown in Fig. 7.2.9(q). It reveals that the resistance of sample increases first during 20 – 150 ℃ and then dramatically decreased upon increasing temperature above 150 ℃. Around 150 ℃, the resistance is the highest. The phenomena in Fig. 7.2.9(p, q) can be explained by the existence of many insulator N_2 gas nanopores and In nanoparticles separated by the gas nanopores in the In filling, which are generated during the growth of In filled MgO nanotubes (Fu et al., 2012). It is obvious that these massive In-filled MgO nanotubes can serve as the temperature protective sensor for some specials devices with a protective temperature ~150 ℃, such as a Nd-Fe-B magnet device.

7. Low-temperature nanothermometer: unusual freezing and melting of Ga encapsulated in CNTs

Fig. 7.2.10(a)–(e) shows a Ga-filled CNT at various temperature stages during cooling and heating. To minimize electron beam heating, the smallest condenser aperture (4th, 20 μm) was used, resulting in an electron beam intensity nearly 56 times weaker than normal imaging conditions (1st condenser aperture, 150 μm). The CNT was cooled down from room temperature (21 ℃) to −90 ℃, then reheated to 90 ℃. From (a)–(c), and the Ga column is liquid and varies linearly with temperature (see Fig. 7.2.10(f)). Freezing (solidification) occurred at −80 ℃. This was evidenced by a sharp decrease in Ga volume at this temperature (see both Fig. 7.2.10(c) and (f)), which was different from the 3.1% expansion on the solidification of α-Ga phase at 30 ℃. In Fig. 7.2.10(f), the length variations are relative changes in length comparing with the initial length at 21 ℃ before cooling. The linear portions of the two curves which correspond to the liquid Ga state do not overlap, because the Ga tip surface did not return to the same position upon melt-

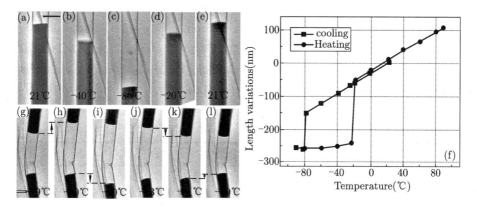

Fig. 7.2.10 TEM micrographs showing Ga volume contraction and expansion inside two CNTs upon cooling and heating. (a)–(e) Scale bar, 100 nm. (a) At 21 °C before cooling; (b) At −40 °C; (c) At −80 °C, solidification occurred; (d) The crystallized Ga melted at −20 °C; (e) Reheated to room temperature (21 °C); (f) The Ga column length variations plotted against temperature. The blue curve corresponds to cooling and the red curve corresponds to heating. (g)–(l) TEM micrographs showing solidification and melting of both β-Ga and γ-Ga in another CNT. Scale bar, 100 nm. (g) At −69 °C, both Ga columns were liquid; (h) The upper Ga column crystallized in γ phase at −70 °C; (i) The lower Ga column crystallized in β phase at −80 °C; (j) At −23 °C, both Ga columns were in the solid state; (k) γ-Ga (upper column) melted at −21 °C; (l) β-Ga (lower column) melted at −20 °C.

ing and reheating to the former temperature. We found that this irreversible effect was caused by a slight change in shape of the CNT upon freezing of Ga. It seemed that the CNT diameter decreased slightly upon Ga solidification. After 10 cycles of freezing and melting treatment, the measured diameter of the CNT is only about 96% of its original value. It is likely that the solidification involves the expulsion of some carbon atoms from the shell network. The behavior and electron diffraction analysis (Liu et al., 2004) show that the Ga is β-phase with the melting point of −20 °C (Awschalom and Warnock, 1987) rather than α-phase. γ-Ga encapsulated in CNTs has also been found for both β-Ga and γ-Ga encapsulated in the same

CNT (see Fig.7.2.10(g)–(l)), which shows their melting and freezing behavior. The arrows and the dashed lines indicate the sudden large volume contractions in (h) upper column, γ-Ga, $-70\,°C$ and (i) lower column, β-Ga, $-80\,°C$, and sudden large expansions in (k) upper column, γ-Ga, $-21\,°C$ and (l) lower column, β-Ga, $-20\,°C$, only $1\,°C$ apart upon solidification and melting. For comparison, crystallography data of α-, β-, and γ-Ga phases are listed in Table 7.2.1.

Table 7.2.1 Crystallography data of α-, β-, and γ-Ga.

	α-Ga	β-Ga	γ-Ga
Symmetry	Orthorhombic	Monoclinic	Orthorhombic
Space group	$Cmca$	$C2/c$	$Cmcm$
Lattice parameters (nm)	$a = 0.4519$ $b = 0.7660$ $c = 0.4525$	$a = 0.2766$ $b = 0.8053$ $c = 0.3332$ $\beta = 92°$	$a = 1.0593$ $b = 1.3523$ $c = 0.5203$
No. of atoms/cell	8	4	40
Cell density	0.43	0.45	0.45
Melting point	$29.8\,°C$	$-16.3\,°C$	$-35.6\,°C$

The presently observed deviation in Ga melting point from its bulk form could be due to a combination of several factors: 1) The interaction between Ga and the CNT wall being prominent at low temperature; 2) Uncontrollable impurities; 3) The high vacuum in the TEM column. A series of observations on CNTs with different diameters ranging from 100 to 200 nm showed no dependence of melting point on CNT diameter; in all cases, both β-Ga and γ-Ga melted at around $-20\,°C$. If the encapsulated Ga is cooled down but is not frozen, its tip surface will return to its initial position when reheated to the former temperature, and the characteristic curve of Ga volumetric expansion is a single straight line. Therefore, a Ga-filled CNT serves as a perfect nanothermometer during freezing and can be used in a far broader temperature range ($-69 - 500\,°C$) than its initial range ($50 - 500\,°C$). We derived a value of α_{Ga} at 1.008×10^{-4} per $°C$ in the temperature range of

−69 – 90 °C, which is exceptionally consistent with that of bulk Ga.

8. Ga-filled silica nanotubes and single-crystal MgO nanotubes with high sensitivity and wide measuring range

A carbon nanothermometer may not withstand temperatures higher than 600 °C (Golberg et al., 2001) and a substitute material should be found. MgO is such a material, and is widely used as protective cases for thermocouples and capillary tubes, and as high-temperature furnace linings and crucibles for fusion of certain metals such as aluminum, copper, and silver. On the other hand, single-crystal MgO is desired for building up nanostructure because nanostructure needs to be strong enough. Therefore, single-crystal MgO nanotubes filled with Ga were fabricated and demonstrated as nanothermometers having a wide temperature range.

X-ray diffraction and TEM observations revealed that the nanostructures were Ga-filled MgO nanotubes, while MgO is cubic ($a = 0.420$ nm, JCPDs: 45-0946). The nanotubes were several micrometers long and approximately 30 – 100 nm in outer dimensions. The hollow cavity of the nanotubes exhibited uniform dimensions, 20 – 60 nm wide, along the tube axes. Most of the nanotubes were closed at both ends and filled with Ga in the hollow cavities.

It is interesting that the present MgO nanotubes exhibit square cross-sections and are well-crystallized, as shown in Fig. 7.2.11 (a) and (b). The growth direction of the nanotubes is the [100] direction of MgO. The Ga-filled MgO nanotubes may grow via the reaction: $Mg + Ga_2O_3 \rightarrow MgO + Ga$. The same growth rate in the cross sections along the two normal directions peculiar to the cubic structure of MgO may result in the formation of the final square-like cross-section morphology. The liquid Ga in the center of the growing nanotube promoted the growth of tubular nanostructures. TEM images of a Ga-filled nanotube were recorded in the temperature range of 30 – 694 °C, as shown in Fig. 7.2.11 (c). The two Ga-filled fragments, with an entire length of 1940 nm, are sealed in the nanotube and separated by a distance of 109 nm at 30 °C, leaving the central tube part unfilled. The distance between the tips of the two parts decreased during the heating process, and increased during the cooling process. The dependence of the distance on temperature is depicted in Fig. 7.2.11(d), which displays a linear relationship. This naturally makes the Ga-filled MgO nanotubes the perfect wide-temperature range nanothermometer. Using the distance between the two fragments of Ga at 30 °C as a reference, a temperature value can be easily obtained by routine measurement of the distance d (nm), expressed by the formula $T = T_0 + (d_0 - d)/(0.719 \times 10^{-4} \times L_0)$, where L_0 and d_0 are the entire length of a Ga filling and the distance between the two fragments at T_0, respectively, and d is the distance at temperature T. The measured expansion coefficient α_{Ga}, 0.719×10^{-4} per °C, is much smaller than α_0, 1.015×10^{-4} per °C, which can be explained as follows: If we consider the linear thermal expansion of the cavity of the MgO nanotube during heating and adopt the linear expansion coefficient ($\alpha_{MgO} = 0.14 \times 10^{-4}$ per °C) of MgO, a volume expansion coefficient of 0.999×10^{-4} per °C may be calculated for the liquid Ga filling ($\alpha_{Ga} + 2\alpha_{MgO}$), the same value as 1.015×10^{-4} per °C (Gray et al., 1972). The working range for each particular Ga-filled MgO thermometer solely depends on whole length L_0 and space d_0 between two Ga columns. For instance, in the thermometer we observed, the effective range of use is up to 694 °C, which is higher

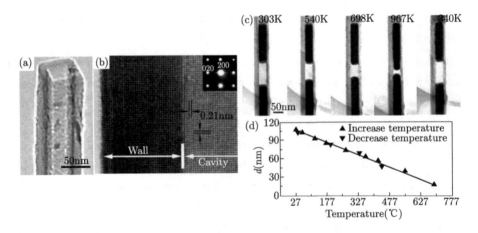

Fig. 7.2.11 (a) TEM image of an open-ended MgO nanotube; (b) Representative high-resolution TEM image of an MgO nanotube without filling and its corresponding electron diffraction pattern; (c) Consecutive TEM images of a Ga-filled MgO nanotube taken at different temperatures; (d) Distance between the tips of the two separate Ga fragments in (c) as a function of temperature.

than 500 °C of carbon nanothermometer. The broader range may be due to the well-crystallized MgO wall, which is stronger and denser than the poor-crystallized C wall in CNT at high temperature and prevents Ga vapor from escaping from the nanotubes.

In-filled MgO, In_2O_3 and silica NTs were also fabricated, studied and confirmed as wide-temperature range nanothermometers. The above investigations on Ga-filled nanothermometers indicate that they may have a wide measurement range of $-69 - 700$ °C, very different from the micro cryogenic resistance thermometer for turbulence measurements, with a range of -269 to -193 °C. Till now, it has been demonstrated that the highest temperature up to 900 °C can be measured by using a Ga-filled silica NT. In this work, we have also got a higher sensitivity of 6.02 nm/ °C, 8 times of that of the first Ga-filled carbon nanothermometer.

9. A liquid-Ga-filled carbon nanotube: a miniature temperature sensor and electrical switch

Our previous works have resolved the basic problems for potential applications of carbon nanothermometers. However, the above temperature measurement observations in TEM are time-consuming, delicate, and rather expensive. We initiated a search for possible integration of individual Ga-filled CNTs into a stand-alone device with straight-forward temperature to transform temperature data to electrical resistance data. Alternatively, we propose electrical control of a Ga-filled CNT nanothermometer/nanosensor as another approach for measuring the dependence of the resistance of Ga-filled CNT on temperature. We have solved the problems to enable such control: (i) The NT shell contacts the inner Ga liquid column well (< 0.2 kΩ µm) in a nearly perfect ohmic state. (ii) No noticeable Schottky barrier exists on the NT inner shells or Ga filling. (iii) The unfilled regions of the NT are diffusive conductors with a semi metallic resistance per length of $\rho \approx 25$kΩ µm^{-1}. (iv) The Ga-filled regions, on the contrary, are perfect metallic conductors with a resistance per length of 0.35kΩ µm^{-1}, corresponding to the inner Ga liquid column. In other words, in the filled NT regions the current should flow through the inner gallium rather than through the outer carbon sheath. Such properties could make Ga-filled NTs applicable as NT electrical switches and/or temperature sensors, provided that the Ga in those NTs is liquid at room temperature and can thus be manipulated by temperature, an electron beam, and/or mechanical influence.

These findings provide the basis for creating the first electrically controlled NT temperature sensor. The schematics of the proposed device and its working principle are shown in Fig. 7.2.12. An individual Ga-filled NT is placed across two metal electrodes and is connected into the resistance measurement circuit (Fig. 7.2.12 (a)). A gap of the required length is produced in the Ga column at room temperature, T_{room}, for example, by an AFM tip (Fig. 7.2.12(b)). Subsequent heating/cooling leads to an increase/decrease of the gap length. In Fig. 7.2.12(c), $R(T)$ decreases linearly with temperature T as the gap in the Ga filling decreases linearly with T. Temperature points 1-4, marked on the curve, correspond to those in (b). As soon as the gap in the Ga filling disappears, $R(T)$ shows a switching action:

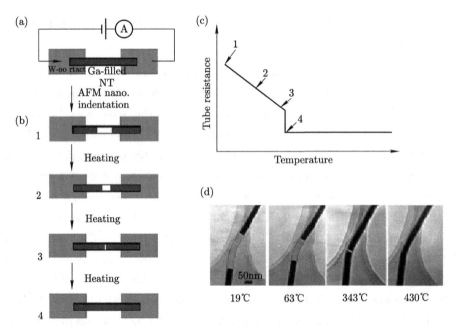

Fig. 7.2.12 (a) Individual Ga-filled CNT is placed across two metal electrodes; (b) A gap of length l is produced in the CNT Ga-filling by AFM nano-indentation; (c) Expected temperature dependence of the tube resistance; (d) TEM verification of temperature-induced shrinkage in a Ga-filled gap.

a sharp drop occurs due to the value of contact resistance between Ga and CNT shells. Fig. 7.2.12(d) shows the TEM verification of temperature-induced shrinkage in a Ga-filled gap. The experiment on resistance also confirmed the expected temperature dependence of the tube resistance. This device can be used as either a thermometer or an electrical switch.

10. Electron-beam induced electric-hydraulic expansion in a silica-shelled gallium microball-nanotube structure

Silica-shelled liquid Ga-filled microball-nanotube (SGBT) was fabricated using a radio frequency furnace. Fig. 7.2.13(a)–(c) displays such a SGBT with a diameter ratio ~ 7 of microball to nanotube. The SGBTs were placed on a C-film-covered Cu-grid for in situ TEM study. When the EB was converged from Fig.7.2.13(a) to (c), the Ga microball inside the cavity quickly expanded to the stable position along the silica nanotube channel. This expansion process is too short (< 100 ms) to be resolved by naked eyes and is remarkably faster than the liquid Ga expansion during Gatan heating stage experiment reported previously. Once the EB was diverged, the Ga tip fell to its original position within a similar short time. The approximate linear correlations of a Ga-column height H (and a volume change $\Delta V/V_0$) vs J is outlined in Fig. 7.2.13(d), round dots show the experimental values. Fig. 7.2.13 (a) reflects the transition from solid to liquid, at which the critical EB current density is 548.9 A/m^2 and the EB current I received by the microball ($= J \times$ the area receiving EB) is 1.973 nA. Under such conditions Ga microball began to flow into the silica nanotube (Fig. 7.2.13(a)) and its temperature is actually the melting temperature of Ga, 29.78 °C. It means that the EB irradiation of 548.9 A/m^2 increases the microball temperature by 9.78 °C from room temperature T_{RT}, 20 °C.

In Fig. 7.2.13(b) and (c), where J are 853.0, 1728.0 A/m^2, and the receiving current I are 3.066, 4.093 nA, respectively, the temperature increments are analyzed as following. Since the Ga microball has a maximum thickness of ~ 2140 nm, and it is covered with low Z-materials (O and Si, which can minimize the electron backscattering) (Williams and Carter, 1996), and the mean free path of 200 kV EB in Ga is only ~ 40 nm, all EB electrons should be scattered at least dozens of times and stopped by the Ga microball. Therefore, nearly all received EB energy Q is transferred to the heat and an equilibrium temperature is reached. According to the Fourier heat transfer law and the boundary conditions, i.e. TEM Cu-grid has the room temperature 20 °C, it is calculated that the EB heating effect raises the microball temperature to 35.2 and 40.3 °C ($\Delta T = T - T_{RT} = 15.2, 20.3$ °C) in Fig. 7.2.13(b) and (c), respectively. The temperature increments 9.78, 15.2 and 20.3 °C for Fig. 7.2.13(a)–(c) have the same linear rule of ΔT vs current I as that in Hobbs' theoretical calculation for silica. However, our calculated increments are almost 10 times of Hobbs's temperature change. Even so, the observed overall expansion ratios of 2.47×10^{-3} and 8.74×10^{-3} in Fig. 7.2.13(b) and (c) are 4.50, 8.25 times of the volume change ratios, 5.48×10^{-4} and 1.06×10^{-3}, corresponding to the two temperature changes of 15.2 and 20.3 °C, respectively (Fig. 7.2.13(e)).

As it is well known, a volume of the matter is determined by pressure and temperature. Here, since EB heating effect could not explain the observed abnormal large expansion, a huge inner pressure p_{in} must be considered during the EB irradiation, similar to the situation in the solid lattice under strong EB irradiation ($p_{in} \sim 400000$ bar at $J \sim 6000000$ A/m^2). The expansions from pressure p_{in} are represented as square dots in Fig. 7.2.13 (d) after subtraction the temperature change part represented by the triangles. The huge inner pressure p_{in} can be calculated according to Eq. (7.2.3)

$$\Delta \rho / \rho_0 = -\Delta V / V_0 = 2.202 \times 10^{-6} p_{in} (p_{in} \text{ in bar}) \qquad (7.2.3)$$

where ρ_0 (~ 6115 kg/m^3) and V_0 are the density and volume of the Ga microball at 29.78 °C, respectively. The dependence of p_{in} on current density J is shown in Fig.7.2.13(f). The huge p_{in} does not originate from the mass bombardment of 200 kV electrons because its contribution is negligible, e.g. 3.153 Pa at 1728.0 A/m^2 (Fig. 7.2.13(c)). This can only expand the Ga liquid by $\Delta V / V_0 = 6.943 \times 10^{-11}$ according to Eq. (7.2.3). The huge p_{in} is originated from the electrostatic repelling force between positive Ga ions generated under EB irradiation. Considering all the regarded processes, namely, EB irradiation, Ga ions' appearances and their repellings, huge inner pressure generation and volume expansion, we proposed that a process named by us as EB induced Electric-Hydraulic Expansion (EHE) effect takes place.

The inner pressure p_{in} vs EB density J can be plotted in Fig. 7.2.13(f), and the overall expansion height in each of Fig. 7.2.13(a)–(c) contains two parts: H_T (smaller) from the EB heating effect and H_p (larger) from the EB induced EHE. The SGBT under the regarded EHE behaves like a micro-nano liquid semi-spring (change height $H_p \geqslant 0$, $J \sim$ const $\times H_p$, as the right part of Fig. 7.2.13(g)). At the equilibrium state, p_{in} equals the sum of pressures p_{st} and p_{vt}, stemming from the surface tension γ and the resistance against the volume change, respectively. The p_{st} can be described by the Laplace equation $p_{st} = -2\gamma \cos\theta/r$, where θ is the contact angle (Fig. 7.2.13(g)) For the state depicted in Fig. 7.2.13(c) at $\sim$1728.0 A/m^2, p_{st} is calculated as 25.852 bar, thus being only 0.742% of the inner pressure p_{in} (3478 bar). To explain Fig.7.2.13(g) well, we use Fig. 7.2.13(h) as an example to address the changes of p_{in}, p_{vt}, $p_{in} - p_{vt}$ with the moving height H' (before reaching the equilibrium height H) at the state marked in Fig. 7.2.13(c). Using electrostatics and Eq. (7.2.3), we can derive $p_{in} = 8574(1 + H'/1070)^{-2}$ and $p_{vt} = 6.4888(H' - 74)$, where p_{in} and p_{vt} are in bar, H' is in the range of 74–610 nm and 74 nm is the height only due to the heating effect in Fig. 7.2.13(c). It means that: 1) for the arrowed equilibrium state, i.e. $p_{in} = p_{vt}$, the stable height H is 610 nm. 2) When $p_{in} - p_{vt} > 0$, the inner pressure p_{in} will continue to decrease and push the Ga tip to move along the arrowed direction until p_{vt} increases to the same value as p_{in}, i.e. the arrowed equilibrium state. The analysis in Fig.7.2.13(f)–(h), combining Fig.7.2.13(d), reveals that the driving equilibrium distance H_p is linearly controlled by the current density J via the inner pressure p_{in}. Similar to a hydraulic-driving macrosystem based on the Pascal principle, the present silica-shelled Ga microball-nanotube can also experience the hydraulic driving based on the linear EHE effect.

The above analysis shows the necessary existence of huge pressure p_{in} and Ga ions. Now we discuss the processes of generation and accumulation of Ga ions, and the relation between p_{in} and the surface density of Ga ions. Generally, for EB-irradiated materials, the electrons in the uppermost states have a chance to receive a rather high kinetic energy and to emit off (Cazaux, 1995). For a material with good conductance, the electrically neutral equilibrium can be quickly restored by the electrons coming from surroundings. However, if the material has a poor conductance, this equilibrium is difficult to be restored because the electrons move from outside slowly and positive charging persists. In our case, the conductive Ga microball is wrapped by an insulating silica shield. The uppermost electrons of Ga atoms bombarded by EB may receive rather high kinetic energy, and be emitted from Ga and in the end, out of the silica shell. The electrical neutrality of Ga atoms cannot be restored due to the poor conductance of silica. Some Ga atoms in the microball become positive Ga ions after losing electrons. The Ga ions repel each other, induce an inner pressure p_{in} and drag Ga atoms along the nanotube. Therefore, the important parameters, i.e. surface positive charge density σ and voltage U can be calculated (Fig.7.2.13(i)) based on a classic electrostatics analysis.

$$\sigma = (1 + H/r_0)\sqrt{\varepsilon_0 p_{in}} \qquad (7.2.4)$$

$$U = (r_0 + H)\sqrt{p_{in}/\varepsilon_0} \qquad (7.2.5)$$

where $\varepsilon_0 = 8.85 \times 10^{-12}$ C^2/N $\cdot$ m^2. For example, at the state displayed in Fig.7.2.13(c), $J = 1728.0$ A/m^2, σ and U values are calculated as 0.087 C/m^2 and 10.547 kV, respectively. 0.087 C/m^2 means that only 1.931% of the total Ga surface atoms (or 14.50 ppm of total Ga atoms in the microball) lost 2 electrons per atom, where the surface density of Ga atoms, $\sigma_{atom} = (\rho_0/m_{Ga})^{2/3} \sim 1.408 \times 10^{19}$/m^2 is derived by O.G. Shpyrko et al. (2006), and $m_{Ga} = 11.58 \times 10^{-26}$ kg. Fig. 7.2.13(i) clearly shows that the surface charge density σ increases with EB density J, which should be a characteristic of collision cascade.

Due to the regarded positive charging the potential exists on the Ga microball surface. In addition, the Ga microball is grounded through the silica shell, the carbon film on TEM grid, the TEM holder and the TEM column (Fig.7.2.13(j) and (k)), and the transport of electrons under EB irradiation on the SGBT can be considered as an electric circuit, as modeled in Fig.7.2.13(l). When 200 kV electrons strike the Ga microball, they are stopped by it, so the microball can be considered as a resistor with a high resistance R. Meanwhile, EB irradiation turned some Ga atoms into positively charged Ga ions, which can be considered as a battery with a voltage u'. The voltage u' is determined by the EB current density J and the resistance $R_C (= R')$ of the wrapping silica. In the end, the electrons stopped by the microball will flow outside through the bottom silica shell. The Ga ions will stay on the surface of the Ga microball due to the resistance of the bottom silica shell and the microball capacitance C. Therefore, an electric circuit depicted in Fig.7.2.13(l) is established.

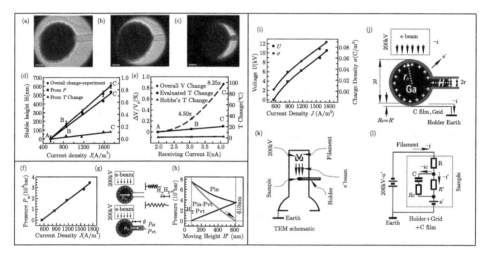

Fig. 7.2.13 (a)–(c) Three TEM images of Ga microball expansion at various J (548.9, 853.0 and 1728.0 A/m^2, scale bar: 500 nm). (d) The dependences of the stable height H and volume change $\Delta V/V_0$ of the Ga microball on J values. (e) Diagram comparing the calculated linear temperature changes with the linear Hobbs's temperature change at the same receiving currents. (f) The plot of pin vs the density J. (g) The schematics for the pressure equilibrium state. (h) The plots of $p_{in}, p_{vt}, p_{in} - p_{vt}$ vs the moving height H' for the state "C" before reaching the equilibrium. (i) The σ and U dependences of J. (j) Ga ions are generated under EB irradiation. (k) The TEM gun, Ga microball, silica wrapping layer and the Earth form an electric circuit (l). (l) An electric circuit model to describe the accumulation of Ga ions during EB irradiation.

Based on a series of equations, the positive charge q of Ga ions at time t can be derived as

$$q = \frac{CU_0 R'}{R + R'}(1 - e^{-\beta t}) \qquad (7.2.6)$$

where $U_0 = 200$ kV, $\beta = (R + R')/C(RR' + RR_c + R'R_c)$, $C = 4\pi\varepsilon_0 r_0 = 1.19 \times$

10^{-16} F. For Fig.7.2.13(c), β is calculated as 3098 s^{-1} and 99.9999812% of the ultimately stable Ga ions can be charged into the Ga microball capacitor within 5 ms. Because this short charging process coincides with the Ga tip expanding process, we conclude that our analysis is consistent with the experiment, where the expansion is indeed very fast ($<$ 100 ms).

In summary, under in situ EB irradiation on a SGBT, an abnormally large and fast Ga liquid expansion was observed. Our analysis indicated that it is related to an EHE effect due to a huge inner pressure p_{in} induced by the repelling Coulomb force between positively charged Ga ions on the Ga microball surface. Based on the expansion ratio and the classical electrostatics, the p_{in} and the surface charge density were calculated. A circuit model is proposed to calculate the accumulation of the Ga ions and fast expansion of the Ga microball along the tube channel. The described EHE effect may pave a way toward development of a novel interesting micro-nano hydraulic-driving system.

11. Electrically driven gallium movement in CNTs

Metal filled CNTs are broadly studied for its nano-mass transformation under electric field. However, the mechanism of two-terminal electrical circuit bias-assisted migration under different operating conditions is still under debate, where electromigration and thermomigration forces are the two competitive driving forces.

Gallium is a good electrical and thermal conductor, and has the highest liquid range (29.78 – 2403 °C) among all metals. CNTs have been encapsulated with Ga for two significant nanoscale applications: nanothermometers and temperature nanosensors–nanoswitches. Herein, we use Ga-filled CNTs to study electrically driven mass delivery at a higher current ($\sim$15 mA) and resistance changes at a lower current ($\sim$2 mA) in two different Ga-filled CNTs, respectively, in order to exploit the possible application mechanism based on the observed phenomena.

Then the experiments were performed for the double-electrode scheme, a Ga-filled nanotube was put in a tight contact with the gold and tungsten electrodes. When a positive bias +1.5 V is applied on the left gold tip (served as the anode), a current of high density flows through the nanotube. The gallium filling modifications under the current flow are displayed in Fig. 7.2.14(a)–(f). The tip of the Ga column is sharpened and becomes like a needle in Fig. 7.2.14(c)–(e). Gallium mass is gradually diminished and finally becomes a nearly empty nanotube which is seen Fig. 7.2.14(f). During the whole process, gallium in the CNTs does not show any trends for thermal expansion, however, the filling is transported from right to left *via* the "wind force" which is directly proportional to the current density j. The gallium movement in the nanotube is driven by the following force F as Eq. (7.2.7).

$$F = F_w + F_d = e\rho(Z_w + Z_d)j = m^*\frac{dv}{dt} \qquad (7.2.7)$$

where the force F is the sum of an electron mediated "wind force" (Sorbello, 1998), F_w and a direct electrostatic force, F_d. ρ is electrical resistivity, m^* is the transport mass and variable quantity, v is a velocity, Z_w and Z_d are the effective valences for the wind and the direct force mediated processes, respectively. For self-diffusion in semiconductors (Kandel adn Kaxiras, 1996), Z_d may dominate over Z_w, while, for metals, Z_d is low and the wind force F_w is considered to dominate (Rous and Bly, 2000). In our experiment, gallium acted as an electron donor, $Z_d > 0$, so the orientation of electrostatic force, F_d, is along the electric field to the right. However, the migrated direction of gallium is opposing the electric field, which implies that the main force driving the gallium migration is the wind force F_w from negatively charged carriers (electrons), i.e., $Z_w < 0$. Under electromigration gallium mass is gradually transported through the CNT to the anode. The relevancy between gallium transport mass and time is addressed in Fig. 7.2.14(g). Error bars come from the uncertainty in the transport Ga mass m^*. From the states "a" to "d", the transport mass speed is rather low of ca. 1.328 fg/s as it is determined from the slope of the curve, however, it is still twice faster than copper movement (0.5 fg/s) across the CNT (Golberg et al., 2007). After the state "e", the gallium mass decreases more quickly and the transport is accelerated to a speed of ∼10.345 fg/s and then it totally and suddenly disappeared at state "f". The smaller Ga mass in the CNT has the faster transport speed under the electromigration, because the smaller remaining gallium can play a smaller shielding role in resisting the transport and the transport speed increases proportional to time until state "e". After that, the transport speed has a higher acceleration, as shown from state "e" to "f". In order to measure the electrical transport properties of CNTs with different amounts of gallium, the STM tip was electrically biased to 0.5 V in a contact mode, and the corresponding current of several mA was passed through the nanotube. A set of nearly linear $I-V$ curves (Fig 7.2.14(h)) shows good Ohmic contact between the testing nanotube and both electrodes. Due to the effect of the higher current, gallium column was gradually separated from the internal surface

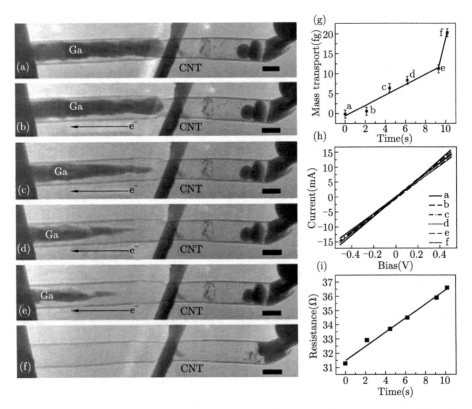

Fig. 7.2.14 Six continuous TEM images showing the induced movement of gallium in a CNT. (Scale bar: 100 nm) (a) A gold tip (on the left) and a tungsten tip (on the right) are in electrical contact with Ga-filled CNT. (b)–(e) At successive time, gallium morphological change under higher current (~15 mA) electromigration. (f) Gallium was transported completely to the anode, leaving empty and clean CNT. (g) Corresponding Gallium mass transport speed plots with time. (h) Serials of representative $I-V$ curves with a decrease of gallium mass transport in the CNT under the influence of electromigration. (i) A nominal increase in resistance from curve "a" (R = 31.5Ω) to "f" (R = 36.7Ω).

of CNT. Therefore, from state "a" to "f", the system resistance only marginally varies, and the conductivity decreases slightly under reduction of the gallium mass. The I-V curve "f" clarifies the measured electrical resistance of the whole gold–CNT–tungsten circuit. Following the simple relation, $R = V/I$, a nominal increase in resistance from curve "a" ($R = 31.5\Omega$) to "f" ($R = 36.7\Omega$) can be determined, as shown in Fig. 7.2.14(i). The reason for the resistance increasing is the shorter electron path through metal Ga and the longer electron path through CNT with a poorer conductivity comparing to Ga.

As shown in Fig. 7.2.15(a), we then choose a short tube with a tight contact between Ga filling and a W electrode on one side and with a CNT-gold contact on the other side. A positive bias +1.5 V was applied to the gold tip, and a current was established through the nanotube. When a lower current passes through CNT, joined forces of the thermomigration, electromigration and binding should be taken into account (Zhao et al., 2010), and gallium begins to migrate slowly along the direction of the electron flow, that is, towards the anode side (Fig. 7.2.15(b)–(d)). Fig. 7.2.15(c) showed a deformation of the CNT because of the increased temperature. However, there is no obvious expansion from Fig. 7.2.15(a)–(c) for the filling Ga, which means that the temperature increase of Ga is negligible. Dramatically, gallium migrates quickly to the anode as shown in Fig. 7.2.15(d). In this case, the gallium melts and an electron mediated "wind force" should be considered as the main force to drive the gallium to the Au anode, the same driving direction as shown in Fig. 7.2.15. The difference is that the Ga is pressed to the Au anode in Fig. 7.2.14, while Ga is pulled to Au anode in Fig. 7.2.15. Since both the wall of nanotube and Ga column are conducive to carrying the current, the electrical resistance of a Ga-filled CNT is influenced not only by its length and cross-sectional area, but also mainly affected by the Ga column length. With an increase of the gallium column under current-driven, a series of consecutively I-V curves (Fig. 7.2.15(e)) have been examined. The initial state in curve a has an electrical resistance of 2.564 kΩ, then become 640 Ω and 510Ω in curves b and c, respectively. This values are obvious larger than the values in Fig. 7.2.14 ($\sim$34Ω). The possible reason is that only a portion of the tip in Fig. 7.2.15 sticks to the Au anode, leading to bigger contact resistance. Comparing to the linearity of I-V in

Fig. 7.2.14(h), the nonlinearity of curves a, b and c in Fig. 7.2.15(e) may originate from the continuous decreases of contact resistance during current flow. With Ga mass migration, the electrical resistance of the nanotube dramatically drops, as shown by the curves b and c. When Ga has finally been transported to the anode, the circuit was completely unblocked. The inset in Fig. 7.2.15(e) of the curve d in detail illustrates the tiny electrical resistance is approximate to 0.4 Ω. Therefore, under the low current driving, the mass migration of gallium can critically affect the electrical transport property of a Ga-filled CNT and the Ga-filled CNT may serve as an electrically-driven switch and becomes electrically-driven delivery at once the current exceeds a certain value due to the good fluidity of liquid Ga. Here, because the Ga-filled CNT is open and the Ga filling has contacted to the Au anode under the electric driving, no reversible effect has been observed. However, we can consider the switch is a disposable switch, or a closed Ga-filled CNT can be a reversible switch under electric-driving, similar to the reversible switch under heat-driving.

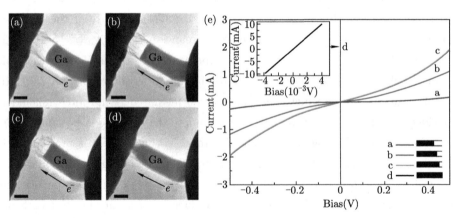

Fig. 7.2.15 (a)–(d) Sequential TEM images with gallium morphological change in a CNT under the electric driving inside the STM-TEM setup. (Scale bar: 100 nm) (e) Related electrical performance demonstrated by the I-V curves with the movement of the gallium in the nanotube.

In summary, electric-driven gallium movement in carbon nanotubes has been discussed in detail. A higher current (∼15 mA) makes the gallium migrate sharply toward the anode, which increases its mass transport speed with time, in the range of 0 to more than 10.345 fg/s. By contrast, a lower current (~2 mA) only drives gallium to contact the anode, which decreases the resistance of the nanocomposite sharply, from 2.564 kΩ to 0.4 Ω. These results are valuable for designing electric-driven nanomass delivery and nanoswitches.

7.2.3 Conclusion and outlook

Using Metal Ga/In filled CNTs and oxide nanotubes, several applications are exploited as nanothermometers (Mathur, 2002), nanotransmission, nanosensor and nanoswitch based on thermal effect, electric-hydraulic effect, conductor-insulator difference and "electric wind force", respectively. However, two problems remain: 1) How to obtain a high yield of half-full pure Ga/In-filled CNTs or oxide nanotubes. A half-full nanotube means that there is a space for the expansion of Ga; 2) How to manipulate and integrate a nanothermometer with an electric circuit. If we can solve the two problems, we can put this Ga/In filled CNTs and oxide nanotubes into a near future use.

7.3 In situ TEM electrical measurements

Shize Yang, Xuedong Bai and Enge Wang

7.3.1 STM-TEM joint technique

Scanning tunneling microscope invented by Gerd Binnig and Heinrich Rohrer in 1981, has gained great success with resolution down to 0.1 nm and manipulation ability of a single atom. But STM has its limitations. The shape of STM tip is usually unknown and may change during operation. The distance between STM tip and sample is hard to be accurately measured. TEM can directly image both crystalline structure by high resolution TEM and electron diffraction patterns to determine crystalline structure of materials. Modern analytical TEM is usually equipped with energy dispersive X-ray spectroscopy (EDS) and electron energy loss spectroscopy (EELS). Those have made TEM a powerful and comprehensive tool in characterization of materials structure and chemical composition. In traditional study, the structure and physical property of materials are measured during separate processes, thus complicating direct correlation. The central idea of in situ measurement is that structure and crystal structure are measured at the same time. Building an STM inside TEM combines the capability of the two and offers a direct correlation between structure and physical property.

The challenge to incorporate STM into TEM is that the space is quite limited between the pole-piece of a TEM where TEM sample is placed. In most TEM, the sample is put on the tip of a specimen holder which is 20–40 cm long. The specimen holder is mounted to the TEM through a side entry vacuum port. The sample is located between the two pole pieces of objective lens. The gap is usually 4 – 5 mm between the two pole pieces. A traditional STM is too large to be directly placed into a TEM. There are two ways to combine them together. One is to incorporate an STM into a TEM specimen holder. The other is to redesign TEM. Most researchers choose the first as it is easy to realize.

Gerber and co-workers reported the first combination of STM with SEM in 1986, several years after the invention of STM. T. Ichinokawa et al. also combined STM with SEM to improve the resolution of SEM at the time. The first STM-TEM system was reported by Spence and colleagues. They used the combined system to study the mechanism of STM operation and imaging. Takayanagi et al. reported similar system combing an STM with TEM operating in a reflection electron microscopy mode. Several other groups over the world reported their design since then. Zhonglin Wang (2000) at Georgia Institute of Technology reported their design in a review article published in 2000. H. Olin and his co-workers (2003) published a design as shown in Fig. 7.3.1 and that is the basis of commercialized Nanofactory TEM-STM system. Xuedong Bai at Institute of Physics, Chinese Academy of Sciences made their TEM-STM system in 2004 as shown in Fig. 7.3.2.

One technical problem is how to increase the movement range of STM probe. Usually STM probe can be precisely positioned through a piezoceramic. Although piezoceramic provides resolution down to several angstroms, it can only move the probe in a range of several micrometers. It is not easy to place the probe so near that probe can move to the intended position. Several designs were adopted to move the probe by a larger distance of several millimeters. Some chose an inch

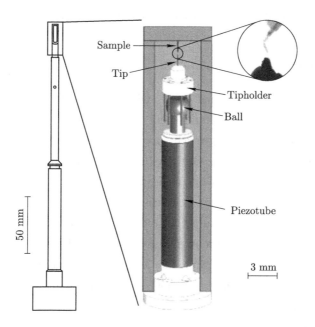

Fig. 7.3.1 Scheme of Nanofactory STM-TEM holder.

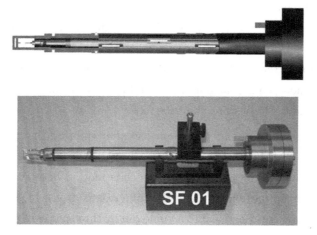

Fig. 7.3.2 STM-TEM holder developed by SF01 group at Institute of Physics, CAS.

worm driver. Some other people used a micrometer screw or a stepper motor.

1. System design

The STM-TEM holder designed by H. Olin et al. is shown in Fig. 7.3.1. The design is quite delicate and simple. This design provides full three dimensional coarse movements while saving much space so that it can be easily mounted to a TEM specimen holder. It utilizes a three-dimensional inertial slider mechanism for coarsely moving the STM tip while using the same piezotube to precisely position the tip. The tip holder with six legs catches the sapphire ball tightly. By applying different voltages to the piezotube, the piezotube will move the sapphire in three directions. If the sapphire ball moves slowly, the tip holder will move together. The friction between the legs of the tip holder and the ball is large enough. So the STM tip can be precisely positioned and moved by slowly changing the voltage. If the sapphire ball moves quickly enough, the tip holder will overcome the friction and slide against the ball. By applying a saw tooth voltage pulse to the piezotube, the sapphire ball will be rapidly moved to a direction and the tip holder will slide a step against the ball and be contrary to the ball movement. By applying different voltage pulses tip holder can be moved in three directions. Steps ranging from 0.5 μm to 30 μm can be easily made perpendicular to the tube axis. Along the tube axis steps range from 0.05 to 1.5 μm. According to the datasheet of Nanofactory STM-TEM holder, the dynamic range is 2 mm and the resolution is 0.2 Å perpendicular to tube axis and 0.025 Å along the tube axis.

In this design the tip holder has four legs that embrace the sapphire ball. The coarse movement of the tip in the direction perpendicular to the piezotube axis is similar to the above design which uses the slide motion between the ball and tip holder legs. The tip movement along the tube axis is driven by a special designed carriage. The carriage is fastened to a guide rail by three pieces of piezoceramics. By applying a voltage to the three pieces of piezoceramics the carriage will be moved by a step. By sequentially retracting the voltage on the piezoceramics quickly one by one, the carriage will slide against the piezoceramic by one step. In this way the STM tip can be moved by more than 10 millimeters.

2. Specimen preparation

The specimen preparation process for STM-TEM is different from that for traditional TEM as shown in Fig. 7.3.3. In a traditional TEM you place the sample on a TEM grid which is penetrated by electron beam. In STM-TEM both the sample and the STM tip needs to be placed perpendicular to the electron beam. So the sample is usually mounted on a special support which also serves as an electrode. The support in most cases is a thin gold or platinum wire about 3 mm in diameter. And it is placed in an area where e-beam can cover the area under study. The STM tip is placed on the opposite side.

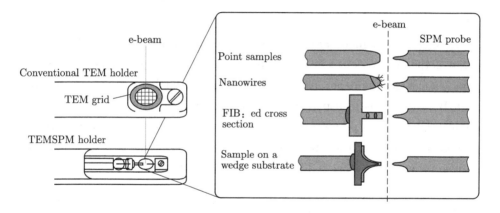

Fig. 7.3.3 Specimen mounting design for in situ TEM (Alexandra Nafari et al, 2011)

Nanowires and nanotubes are common sample studied using STM-TEM. Their sample is relative easy to prepare. Samples can be mounted to either the supporting wire or directly to the STM probe tip. Electrochemically sharpened tungsten/gold tip is widely used as the STM probe. The tip radius can be as small as several nanometers. The supporting wire is cut at the end to be sharp. Sharp tip will increase the chance of finding individual isolated object. And also the sample on the tip is more likely not to be shadowed. Usually, to ensure good electrical contact, the wire support or STM probe is firstly dipped into electrically conducting glue, which usually is silver glue or carbon paste. And then the wire or STM tip is dipped into the powder of sample gently. But sometimes, the wire or STM tip is directly dipped into the samples and the surface adhesion force will leave some sample on the tip. Growing the sample directly on the supporting wire is the best choice. This will ensure good mechanical and electrical contacts.

However, by this "dipping" way the chances of observing a single object is low. To obtain a single individual object, L. de Knoop (2005) used an FIB equipped with nanomanipulator to find a single nanotube and welded the nanotube to a gold wire. Liu et al. (2008) used a dielectrophoresis method to attach individual ZnO nanowires to the tungsten tip. If the nanowires are grown directly on a silicon substrate, the substrate should be cut carefully into small slabs. And to preserve the nanowires on the edge, the substrate needs to be cut from the side and lets the silicon break.

Heterostructure material can also be studied by in situ TEM. The layered substrate is the first to be cut into small pieces. And then the specimen is thinned by grinding against sandpaper. After that the sample is further thinned by ion milling or FIB so that it is transparent to electrons. But the implanted ion in FIB will change the electrical property of the material. In case that the electrical property is to be determined, the sample should be thinned by grinding and ion milling. The sample is mounted on half a TEM grid at the end of TEM holder.

3. Possible drawbacks and problems

The issue of irradiation effects must be addressed. The electron beam in a TEM has a high energy in a range of 100 – 300 keV. High energy electron could damage the specimen. This might happen when taking TEM images. But by changing the electron energy, electron density on the sample and exposure time, the damage can be controlled. For carbon nanotubes, accelerating voltage below 120 kV is recommended because this is the threshold for knock on damage. But even at low voltages, the damage will be made with a prolonged exposure time. Sometimes this can help to modify the sample in certain ways. Single atom wide gold wire and carbon wire have been achieved in this way. Electron beam irradiation will induce amorphous carbon deposition. And this is often used to improve electrical contacts between one-dimensional sample and metallic support.

7.3.2 Electrical manipulation inside TEM

1. Electrical migration

Inside TEM, movements of particles and clusters can clearly been seen. Using STM-TEM design, voltage can be applied across nanomaterials. In an early work, mass transport on the surface of carbon nanotubes was demonstrated as shown in Fig. 7.3.4. B. C. Regan et al. (2004) showed their work that used carbon nanotube to convey indium particles. They firstly coated the arc-grown multiwall nanotubes with indium by thermal evaporation. The nanotubes were then loaded to the TEM stage and approached by a freshly etched tungsten tip. By applying a voltage across individual nanotubes, indium particles melt during joule heating. The melted particles had a round shape which indicated that the particle was molten. By recording a movie, they observed the movements of the molten particles and quantitative

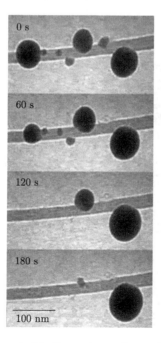

Fig. 7.3.4 Movement of indium atoms observed by in situ TEM.

analysis, which showed that the mass lost on one side tallied the mass added on the other side. They then reversed the voltage direction and found that the mass move was reversed too, thus indicating that voltage gradient might dictate the mass transport. By adjusting the voltage form, the mass transport could be controlled delicately. They developed a quantitative model to describe the phenomenon. In the model, indium atoms were treated as two-dimensional gases. Indium atoms diffused due to temperature gradient but would be biased by electric field. Electric field could tune the thermodynamic equilibrium point of indium surface concentration, thereby controlling growing or shrinking of particles at a specific site.

Mass transport inside single nanotube was first demonstrated by K. Svensson et al. (2004). MWNTs were filled with iron and placed on a metal wire by conductive paste as shown in Fig. 7.3.5. A gold tip was used to approach the MWCNT and a contact was made. When a voltage was applied cross, a current would pass through the CNT and the CNT would be heated to a high temperature due to joule heating. The iron core was broken into small clusters and then moved toward the anode. Various sized nanotubes were measured and a threshold in current density around 7×10^6 A/cm^2 was found. The force for electromigration is determined by:

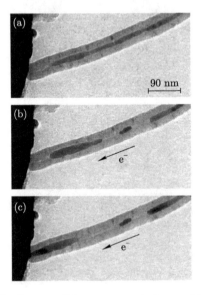

Fig. 7.3.5 Iron particles movement observed by TEM.

$$F = F_d + F_W = eZ_d E + eZ_W E \qquad (7.3.1)$$

where E is the electric field, and Z_d and Z_w are the effective valences for the direct and the wind force respectively. The direct force is due to charged particles or accumulation of charge due to scattering of current carriers. The wind force is from momentum transfer caused by scattering of current carriers against defects. An individual MWCNT is demonstrated to serve as a pipette to retrieve and deposit an iron particle on a nanotube purposely. And this might apply to other electrically conducting and low adhesion materials.

Many other experiments have been done since the early work. Dmitri Golberg et al. (2007) showed that copper filled amorphous CNTs could behave like nanoscale switches or rheostats. The resistance could be varied by tuning the copper filling ratio inside the CNTs. And copper can even flow from one CNT to another by contacting. The copper mass flow rate was estimated to be about 0.5 fg/s. Copper was found to move towards the cathode side opposite to the direction predicted by electromigration theory. Dong et al. (2007) observed copper inside MWNTs moved along the electron flow direction and a flow rate of 120 ag/s was reached. They demonstrated that copper-filled CNT could be soldered to another CNT. Dong et al. (2009) showed their results that metal particles could move through walls from the cap of one CNT into another CNT. When the voltage was reversed, the metal particle moved in the opposite direction. Molecular dynamics simulation was performed to understand the phenomenon. And the simulation revealed that repulsive charges enlarged ions separation and resulted in mass transport. Jiong Zhao et al. (2010) systematically investigated mass transport in CNTs under voltage bias. They considered three mass transport mechanisms. The first was thermal evaporation which caused a particle to change its size. The second was thermal diffusion which was the temperature gradient driven movement. The third was electromigration which was the current driven movement. They studied the Fe nanoparticles inside MWCNT experimentally. It was found that thermal gradient force and electromigration force depended on size of the particle, length of the nanotube, position of the particle and magnitude of current density. Thermal evaporation took place when the nanoparticle was heated enough. At higher temperature, thermomigration or electromigraion occurred. They suggest that the loading configuration of nanoparticles be controlled and a cooling system be used to minimize thermal migration. In that sense nanoparticle movement can be controlled precisely by electromigraion.

2. Joule heating induced phenomenon

When a current passes through a resistor, heat will be generated at a power of $P = I^2R$. More heat will be generated if the resistance is larger. And this consequence has many applications.

John Cumings et al. (2000) showed that a single MWCNT could be peeled and sharpened layer by layer with the outer layer removed first. Yuan-Chih Chang et al. (2008) studied MWCNT under electrical current. By applying a voltage of 4 V, the carbon nanotube broke in the middle forming a telescopic tip. The inner layer protruded. The outermost layer of the telescopic tip could be contacted and removed by electrical current. By contacting the protruding inner layer of the telescopic MWCNT and applying a moderate current, the inner layer of the MWCNT could be pulled out. With high precision piezodriven manipulator they were able to reinsert the inner CNT back.

J.Y. Huang et al. (2005) studied dynamic behaviors of carbon nanotubes under in situ Joule heating systematically. In one experiment, they used MWNT grown on a carbon fiber by CVD which was fixed on to a gold wire directly. The MWNT was contacted by another gold tip. When a current of 240 µA, broken-down first appeared at the midpoint of the nanotube. Under a constant of 3 V, instant drops of current could be observed corresponding to the breakdown of each wall. The sequential wall-by-wall breakdown was recorded by TEM images. Breakdown starting from both inner wall and outer wall were observed. The observations imply that transport in MWNTs was diffusive rather than ballistic.

In another experiment, Huang and others (2006) observed kink motion in carbon nanotubes by in situ TEM as shown in Fig. 7.3.6. Under voltage bias, carbon nanotubes could be heated up to 2000 ℃. And at such a high temperature, kink motion was observed by applying tensile stress. The kinks originated from vacancy was formed and aggregated, and moved longitudinally or spirally along the carbon nanotubes. Vacancies and interstitials are active at elevated temperatures and they tend to aggregate to form a kink. Once a kink is originated, it tends to migrate under external stress and at high temperatures. As the atoms beside the kink are stressed and are likely to evaporate at elevated temperature. It is proven that the high temperatures and tensile stress es are the origin of kink motion.

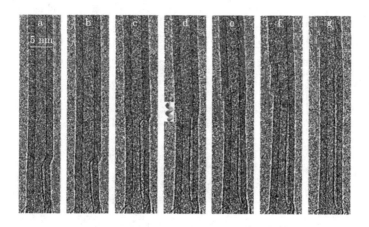

Fig. 7.3.6   Movement of a kink defect recorded by in situ TEM.

In another work, Huang et al. (2007) discovered enhanced ductile behavior of double-wall and triple-wall carbon nanotubes. Tensile elongation reached 190% and diameter reduction reached 90%. Point defects such as vacancies and interstitials were resolved from TEM images. Detailed analysis of high resolution TEM imaging revealed that the super elongation was attributed to a high temperature creep deformation.

Huang et al. (2007) also studied the dynamic behavior of giant fullerenes as shown in Fig. 7.3.7. At high temperatures, the volume of giant fullerene could reduce more than 100 in fold, while the shell of fullerene remained intact. Initially, MWNTs were contacted by two gold electrodes. Giant fullerenes were formed by electrical breakdown of the original MWNTs. At high temperatures, giant fullerenes shrank continuously until they reached critical dimensions. The carbon atoms losing rate was constant with time. Theoretical calculations suggest that the pentagon sites are preferred for carbon atoms to evaporate. The pentagons remained constantly 12 and tallied with the constant evaporation rate.

In another work, Huang et al. (2008) studied dislocation dynamics in MWCNTs. Gliding, climbing, and glide-climb interactions of a $1/2\langle 0001\rangle$ sessile dislocation

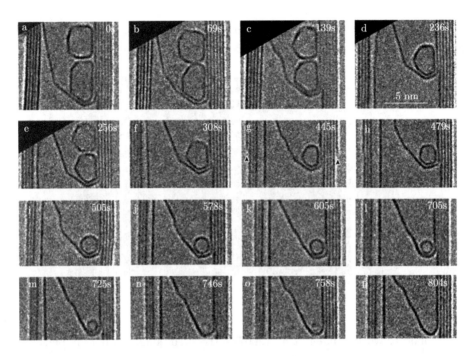

Fig. 7.3.7 Dynamic behaviors of giant fullerenes recorded by in situ TEM.

were in situ recorded. The dislocation gliding led to the cross-linking of different shells of the CNT. Cracks were formed with dislocation climb. Dislocation glide-climb interactions created kinks. It was found that dislocation loops could act as the channels for mass transport.

Chuanhong Jin et al. (2008) studied the metal atom catalyzed enlargement of fullerenes inside TEM as shown in Fig. 7.3.8. Individual MWNT was contacted by platinum and tungsten probe. Fullerene molecules could be formed by thermal shrinkage of the inner shells. Tungsten particles were introduced into the MWCNT via vacuum arc induced sputtering. The tungsten cluster was found to migrate continuously on the fullerene cage. Local deformation was found on the tungsten cluster adsorption site. Regions around the tungsten cluster grew first. Further growth was initiated when the tungsten cluster moved to another site on the same cage. Throughout the experiment, the cage grew radially without elongation. This confirmed that round shape of fullerene cage was energetically preferred. The diameter of the fullerene reached 1.1 nm finally. The crucial role of tungsten cluster was validated by two aspects: 1) smaller fullerene molecule never grew; 2) larger fullerene molecule started to grow only after the adsorption of tungsten metal.

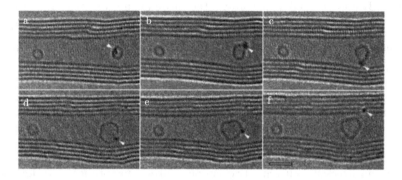

Fig. 7.3.8 Dynamic behaviors of fullerenes. Iron particle is indicated by a white arrow.

In another work, Jin et al. (2008) monitored the vacancy activities with real-time in situ TEM. Large vacancies were observed to migrate and coalesce. Experiment was carried out in a field emission TEM (JEOL-2010F) operated at 120kV close to the knock-on threshold of carbon. Few walled carbon nanotubes were fabricated in situ using a layer-by-layer peeling process.

3. In situ welding and cutting

Ming-Sheng Wang et al. (2008) studied field evaporation of CNTs. It was found that evaporation occurred where it protruded sharply as local field was the strongest. The region became blunt with the removal of atoms. Voltage had to be raised to maintain evaporation. And hence, more protruding regions would be evaporated. The end of CNT would be smoother and smoother. Finally the end became very regular and circular. It was found that electron irradiation could promote the evaporation process. By irradiating electrons on specific sites, the CNT end surface direction could be controlled. The evaporation rate in this method can be controlled down to 0.1 nm·s^{-1}, which offers a very precise way to control CNT length.

Xianlong Wei et al. (2007) presented a method to cut and sharpen nanotubes using a "nanoknife". The nanoknife was produced inside SEM. Individual CNT was contacted by tungsten tip and platinum electrode respectively. A segment of CNT was left attached to the tungsten tip by in situ breakdown and was used as a nanoknife. The nanoknife was contacted to a nanotube to be cut at a selected position. By applying a 5 V dc voltage, the carbon nanotube was cut at the contact point. It was found that length of knife should be smaller than nanotubes to be cut. Diameter had little effect on the cutting. To sharpen a CNT, the nanoknife contacted the CNT tip. Under a bias, the CNT lost atoms at the contact area and became thinner.

Atsuko Nagataki et al. (2009) studied head-to-head coalescence process of two capped carbon nanotubes. In the experiment, Carbon nanotubes were fixed on a Pt coated Si substrate and further approached by a Pt-coated Si cantilever. By applying an excessive current, two capped carbon nanotubes were formed. The achieved carbon nanotubes could be thinned layer by layer by bridging them and applying an excessive current. Finally, both the carbon nanotubes were cut from the innermost shell and had capped tips. Contact force could be determined. As the spring constant of the Si cantilever was 0.03 N/m and bending could be determined from direct TEM image. Results showed that contact force was related to the "current for coalescence" and was divided into three categories. Constant temperature tight-binding molecular dynamics simulations were performed to understand the observed phenomena. The three categories might correspond to single sp^3-like bond, single sp^2-like bond and double sp^3-like bond respectively.

7.3.3 Electrical transport measurement

1. Field emission

Field emission is often referred to electrons emission due to electrostatic field. The emission current can be affected by many factors. Under TEM, it is possible to relate the field emission property of individual nanoobjects with its structure.

Zhonglin Wang et al. (2002) observed field emission of individual carbon nanotubes. Structural damage is recorded. The damage is found to happen in two ways. One is pilling of graphitic layers by pieces and segments, the other is concentrically stripping.

Xu Zhi el at. (2005) studied field emission of a carbon nanotube in mechanical resonance. Field emission was induced by a dc voltage applied longitudinally and mechanical resonance was stimulated by applying a transverse ac field. It is found that the frequency of the field emission current is twice that of the mechanical resonance.

Chuanhong Jin et al. (2005) studied field emission of single carbon nanotubes inside TEM. Emission geometry and vacuum contaminates was studied respectively. It was found that protruding tip geometry is good for field emission. Generally the worse the vacuum levels the higher the emission current. Large emission at several tens of µA for a long time would damage the structure and cause emission failure.

Wei Wei et al. (2007) studied the failure mechanism of field emitting carbon nanotubes both theoretically and experimentally. The tip cooling effect due to evaporation of electrons was considered. Based on their model they propose that the highest temperature point is located interior to the CNT tip, thus indicating that failure occurs segment by segment.

Ming-Sheng Wang et al. (2008) used in situ TEM to fabricate CNTs with conical tips, as shown in Fig. 7.3.9(a). Field emission curves of different tip formations were measured and quantitatively analyzed using F-N equation as shown in Fig. 7.3.9(b) and (c). Three regions can be found in the F-N plots. In the A region where Joule heating is negligible F-N theory agrees with experiment very well. In the B region, the field emission current is larger than expected from F-N theory. This might be due to current heating which promotes thermionic emission. In region C, emission current is unstable. Structural change occurs due to thermally assisted field evaporation. It was also found that 1 nm is a critical radius for F-N theory. The maximum stable field emission current is determined by cap structure.

In another work, Ming-Sheng Wang et al. (2010) capped a tungsten tip with carbon "onion". The tungsten tip was first melted into a ball to remove contamination and improve conductance. Then the tungsten tip approached a carbon onion that was placed on the edge of the gold electrode. Firstly, a constant small current less than 1 µA was applied. Then the current was raised to 1 mA. And half of the carbon onion was fixed onto the tungsten tip suddenly. This fusion reaction was due to the sharp change of current and joule heating effect. The carbon onion

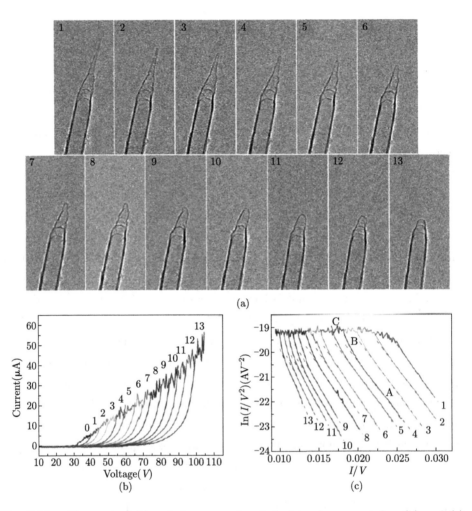

Fig. 7.3.9 Change in field emission property of single carbon nanotubes (b) and (c) with different tip formations (a) as recorded by real-time TEM recording and in situ measurement.

capped tungsten tip showed enhanced field emission and low turn-on voltage. The carbon onion emitter could sustain an emission current of 40 − 100 μA. The maximum emission current was found to be related with the diameter of the carbon onion and tungsten tip, the contact area, and the distance between anode and cathode. As carbon, tungsten, tungsten carbide had high melting temperature and the calculated temperature before failure was below melting points. The possible mechanism for failure was a conduction channel which led to discharge. Fluctuation of emission current for 1 hour was found to be 21.6%. By connecting a 6 MΩ resistor in series, 1 hour fluctuation reduced to 5.7%. The final state of carbon onion after 1 hour emission at a few μA was good, indicating long term endurance. By connecting a 25 MΩ resistor and at an emission current of 3 μA, 30 minutes current fluctuation reduced to 1.9%.

Xianlong Wei et al. (2012) studied field emission of individual graphene nanoribbons (GNR) by in situ TEM. Graphene nanoribbon was contacted to gold electrode and tungsten probe "1" respectively. Another Tungsten tip "2" was employed to collect electrons. Gold electrode is grounded. A pump voltage (V_{pump}) was applied to tungsten tip 1 and a collect voltage (V_{collect}) was applied to tungsten tip 2. It was found that emission current I_{collect} increased V_{pump} at fixed V_{collect}. It indicated that a large emission current was measured with a lower electric field. This contradicted with field emission theory. It was suggested that internal drive by electric field or heat might account for the measured electron emission. Electrons were found to emit under less than 3V bias corresponding to an electric field of 10^6 V/m which is 2 orders of magnitude than common field emission. Graphene nanoribbons can be treated as two dimensional solids. Therefore, electron emission can occur in two ways. One is parallel emission from one dimensional edge which can be explained by phonon-assisted electron emission. The other is perpendicular emission from two dimensional surfaces which is the conventional field emission and thermionic emission. Thermionic emission requires high temperature of 3000 K which will damage GNR. Assuming the thermionic emission model, the experimental data cannot be well fitted. These two reasons nullified thermionic emission. PAEE model was adopted to validate phonon-assisted field emission. The result fitted the experiment very well. Electrons can be emitted from all surfaces with the same probability. So two-dimensional perpendicular emission dominates the emission current. Emission current can reach 4.6 nA or 12.7 A/cm^2, which is about 3 or 4 orders of magnitude higher than graphene films.

2. Electrical transport property

Based on nanomanipulator, two terminal transport can be easily measured. Various materials have been studied with this method including nanotubes and nanowires. The effect of material structure, electron irradiation, as well as in situ deformation can be related to transport directly.

Stefan Frank et al. (1998) found the quantized conductance of multiwall carbon nanotubes. The MWNTs under study were produced by arc and mounted to a gold wire fixed on the STM probe. The counter electrode was a heatable copper reservoir containing mercury. By dipping the protruding MWNTs into the mercury, nanotubes were cleansed. Using STM probe, electrical contact was made. The nanotube was then driven in and out of the liquid metal with electrical conductance measured. Fig. 7.3.10(a) shows the relation between conductance and time. The tip was moving at a speed of ± 5 μm/s. The conductance of contacted nanotube was $1G_0$ and remained for a distance of 2 μm. Sequence of $1G_0$ steps were observed as shown in Fig. 7.3.10(c). This is because other tubes come into contact. $0.5G_0$ steps appeared before $1G_0$ steps as shown in Fig. 7.3.10(c). This is related to the tip

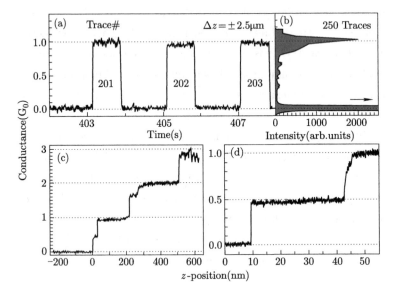

Fig. 7.3.10 Quantum conductance of carbon nanotubes recorded by in situ TEM.

structure of the nanotubes. It was found that about 30% of the nanotubes had tapered tips which could be twice the diameter of the respective tubes. The reduction in conductance can be explained by tip-to-shaft interface, which elastically scatters electrons thus reducing the transmission coefficient. Similar experiments were carried out with molten gallium and Cerrolow-117. The similar results obtained indicated that metal does not affect the transport. The nanotubes endured voltages as high as 6 V for a long time, with a current density larger than 10^7 Acm^{-2}. Supposing the power dissipated uniformly, the maximum temperature would be 20000 K. This is impossible as nanotubes burn at about 700 °C. This indicated a ballistic transport. The measured conductance was lower than the predicted value of $2G_0$ for independent SWNT and not scale with layers as predicted for MWCT. It is suggested that only the outer layer contributed to transport as resistivity is high perpendicular to axis.

Hideaki Ohnishi et al. (1998) studied transport property of quantum gold atom wires as shown in Fig. 7.3.11. The experiment was conducted inside an ultrahigh-vacuum HRTEM at 10^{-8} Pa. A gold tip was mechanically sharpened and mounted to a piezoelectric manipulator. The tip was then dipped into a gold island deposited

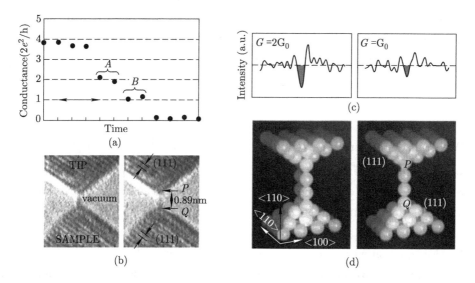

Fig. 7.3.11 Quantum conductance of gold chains.

on a copper wire by the manipulator. After that, the tip was retracted at a constant speed. During the process video images were recorded at 33-ms per frame and conductance was measured at the same time. Steps in conductance along with thinning of gold bridge were observed. The tip and gold substrate are along the [110] direction and the (200) lattice planes was observed. Repeated experiments validates the linear strands of gold atoms have a unit conductance of G_0. Fig. 7.3.11(a) shows the change in conductance during retracing the tip. The conductance has steps of G_0. TEM image for steps A and B are shown in Fig. 7.3.11(b) left and right panels respectively. Intensity profile in Fig. 7.3.11(c) shows that profile peak of the left strand is twice that of the right. This suggests that the right strand is a single row of gold atom and the left has two rows.

D. Erts and his co-workers (2000) investigated Maxwell and Sharvin conductance of gold point contacts using TEM-STM technique. For the diffusive transport regime, the electron mean free path l is smaller than the contact radius. The conductance is given by Maxwell,

$$G_M = \frac{2a}{\rho} \quad (7.3.2)$$

where ρ is the resistivity of the metal which is given by $\rho = \frac{mv_F}{ne^2l}$, where m is the electron mass, v_F the Fermi velocity, n the electron density, and e the electron charge. For point contacts ($l \gg a$), the ballistic transport dominates. The conductance is given by Sharvin to be,

$$G_S = \frac{3\pi a^2}{4\rho l} \quad (7.3.3)$$

Wexler made the expression, $G_W = G_S \left[1 + \frac{3\pi}{8}\Gamma(K)\frac{a}{l}\right]^{-1}$, where $\Gamma(K)$ is a slowly varying function of the order unity and $K = l/a$ is the Knudsen number. The resistance of point contacts with radius less than 0.4 nm had step like behavior. The relation of point contacts conductance with square radius was shown. Only for radius below 1 nm, a good fit to Sharvin conductance was reached. The data fitted Wexler equation with a mean free path of $l = 3.8$ nm. This was much lower than the bulk value of 37 nm at room temperature. The low value could be accounted by a high density of scattering centers generated during the point contact formation.

Temperature effect was also taken into consideration. A voltage up to 140 mV corresponding to a voltage of 540 K was applied. Linear I/V curves observed indicated that scattering was mainly due to scattering centers.

Z. Y. Zhang et al. (2006) showed a method to retrieve parameters of semiconducting nanowires by simple two point I-V measurement. The structure is described by metal-semiconductor-metal (M-S-M) model as shown in Fig. 7.3.12. The I-V curve at large bias appeared to be a straight line and the slope was related to the resistance of the nanowire. The M-S-M structure was treated as two Schottky barriers connected back to back, connected in series with a semiconductor resistor. The I-V curve measured can be either symmetric or rectifying. The current couldn't be explained by thermionic emission. Tunneling current played a key role under reverse bias. At a reverse voltage V at normal temperature, the tunneling current was mainly due to thermionic field emission. Classic thermionic emission applied under forward bias. If a positive voltage V is applied to the right metal electrode, Schottky barrier 1 on the left is reversed while Schottky barrier on the right is forward biased. The voltage distribution can be divided into three parts, voltage on the barrier $1V_1$, voltage on the nanowire V_2 and voltage on the right barrier $2V_3$,

$$V = V_1 + V_2 + V_3 \qquad (7.3.4)$$

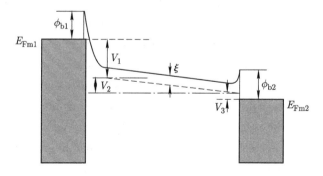

Fig. 7.3.12   M-S-M model.

At low bias, the current is small and the voltage is mainly on the Schottky barrier. At higher bias larger than 0.5 V, the voltage on the reversed Schottky barrier V_1 increases rapidly and comes to domination. At 5 V, the current becomes notable and V_2 becomes important while V_3 remains small. V_1 begins to saturate at total V of 10 V. At total voltage larger than 17 V, the voltage V_2 on the nanowire increases linearly. The resistance of the nanowire can be found, $R = \dfrac{dV_2}{dI} \approx \dfrac{dV}{dI}$. At intermediate bias, the total current I is given by

$$\ln I = (\ln)(SJ) = \ln S + V\left(\dfrac{q}{kT} - \dfrac{1}{E_0}\right) + \ln J_S, \qquad (7.3.5)$$

where J is the current density through the Schottky barrier, S is the contact area, E_0 is a parameter connected with carrier density given by $E_0 = E_{00}\coth(E_{00}/kT)$, where $E_{00} = (n/m*\varepsilon)^{1/2}\dfrac{\hbar q}{2}$ and J_S is a slowly varying function of bias. So $q/(kT) - 1/E_0$ is given by the slope of $\ln I$-V curves. Electron concentration n can be determined by E_0 and electron mobility is given by the formula $\mu = 1/nq\rho$.

Xuedong Bai et al. (2007) studied the electrical transport of individual boron nitride nanotubes under in situ deformation. The experiments were carried out inside a 300 kV field emission high resolution TEM. The nanotube was first attached onto an STM tip or the opposite electrode using a sticky graphite paste. Electrical contact can be made by manipulating boron nitride nanotube to approach the opposite electrode. To make a tight physical contact, several dozens of nA was applied. A local high temperature can be generated which formed a perfect contact. Atomically sharp STM tip can be inserted into the hollow channels of nanotube to make better physical contact. After the electrical contact was made, bending curvature of the nanotube was gradually increased by piezomanipulator. During the process, I-V curves were recorded simultaneously as shown in Fig. 7.3.13. Without bending, electrical current can hardly be detected. Under slight bending, a current of several nA at a few volts was recorded. With increasing in the bending curvature and at large bias, the resistance decreased to 480 MΩ. With further increasing in bending, resistance decreased to 260 MΩ. When the bending was released, the I-V characteristics also recovered. At moderate bending, the current values suggest that the nanotube behave like semiconductor. Electrode of gold, platinum and tungsten were tested. The observed result is the same. Theoretically, the band gap of a single-walled boron nitride nanotube could be changed within a large range from 2 eV to

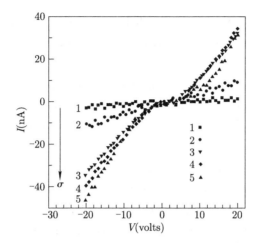

Fig. 7.3.13 I-V curves measured by in situ TEM.

5 eV by cross-section flattening. In the study, the nanotube became flattened while bent. The spacing between the two opposite sides of the inner wall became very close to interlayer basal plane spacing in the standard layered BN system. Another possible reason for the deformation-driven transport may be the bending induced in-shell defect formation, which should also modify the band structure. The defects of voids, vacancies and antisite atoms may appear on bending and disappear after a load release. Local electronic structure was probed using EELS. The fine structure of a σ^* peak and the intensity ratio of π^* and σ^* peaks obviously changed. I-V hysteresis was found by first sweeping from a negative to a positive bias and then sweeping from positive to negative. This result suggests a piezoelectric effect. A simple mode was introduced. Boron nitride nanotubes are noncentrosymmetric and are naturally polar thus suggestive of piezoelectricity. In a bent nanotube, there is a spontaneous polarization along the nanotube. The grounded electrode A has a potential of nearly zero, another electrode B has a negative potential and the polarization field is formed from A to B. When a negative voltage is applied to B, the applied field coincides with the polarization field. While a positive voltage is applied to B, the applied field in the reversed direction of the polarization field. Then the current under positive bias might be larger than that under negative bias. When the voltage sweeping pathway was reversed, from positive to negative, at an

applied voltage of 60 V, the applied field and polarization field effects balanced out and the current become zero. Hereafter, the polarization field dominated the carrier transport at low bias.

Kaihui Liu et al. (2008) studied the transport property of individual ZnO nanowire. It was found that with increasing bending curvature, the electrical conductance dropped significantly. Y. G. Wang et al. (2011) studied single ZnSe nanowire transport using in situ TEM. It was found that in situ introduced strain improved conductance remarkably. According to theoretical calculation, band gap is narrowed upon tensile strain. The electron affinity increases with bandgap narrowing thus reducing Schottky barrier height. This accounts for the reduction in threshold voltage. It is suggested that the strain induced band modification and piezopotential contributes to the enhanced conductance.

In another work, Kaihui Liu et al. (2009) studied chirality resolved transport of double-walled nanotubes (DWNT) based on in situ field effect transistor. The specimen holder was specially designed to load SiO_2/Si substrates for FET operation. The substrates had micro fabricated narrow slits with width of 5 – 20 μm for electron beam to pass through. Electron diffraction pattern of nanotubes spanning the slits can be imaged. Ultra long individual double-walled nanotubes with length more than 10 μm were used. Palladium metal was used to form the top contacts. According to electron diffraction patterns of the DWNT, chirality indices could be identified. DWNT can be categorized into four types by outer/inner shells: metallic (M)/semiconducting(S), M/M, S/S, and S/M. As shown in Fig. 7.3.14, the transport property of four different types of DWNT was measured. The data showed that gate voltage had no modulation effect for M/M and M/S nanotubes, while the S/S type behaved like p-type semiconductor with on/off ratio as high as 10^4. For S/M type, the gate voltage had a rather small modulation with on/off current ratio less than 20.

Yang Liu et al. (2011) showed a method to retrieve thermal distribution across a single nanowire based on M-S-M structure. Thermal conductivity and ZT coefficient of ZnO nanowire were obtained. In the experiment, a single ZnO nanowire was contacted by tungsten tip and platinum electrode on the two ends. At low bias, the electrical parameters could be extracted. While at higher bias, negative differential

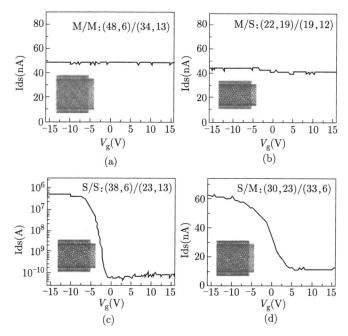

Fig. 7.3.14 Chiral indexes related transport of double-walled nanotubes.

conductance was observed due to elevated temperature. At this region the thermal conductivity could be calculated using a self-consistent method. For each I-V point, one dimensional thermal conduction equation is

$$A\nabla(\kappa\nabla T) + \frac{I^2 R_{\text{NW}}}{L} = 0 \tag{7.3.6}$$

where A is the cross-sectional area of the nanowire, κ is the thermal conductivity, T is the absolute temperature, I is the current, R_{NW} is the resistance and L is the nanowire length between the electrodes. Black body radiation is ignored as it is estimated to be 3 orders of magnitude less than nanowire dissipation. Resistance of the nanowire can be written as

$$R_{\text{NW}} = \int_0^L \frac{dx}{en(x)\mu(x)A} \tag{7.3.7}$$

where $n(x)$ and $\mu(x)$ are the carrier concentration and mobility respectively at point x along the nanowire, e is the charge of electron. $n(T)$ and $\mu(T)$ should be obtained to calculate $R_{\text{NW}}(T)$. Considering thermal activation of carriers and making some simplification, $n(T)$ is:

$$n(T) = n(300\text{K}) \exp\left[E_a\left(\frac{1}{300k_\text{B}} - \frac{1}{k_\text{B}T}\right)\right] \quad (7.3.8)$$

Considering scattering mechanism, carrier mobility is:

$$\frac{1}{\mu} = \frac{1}{\mu_\text{L}} + \frac{1}{\mu_\text{I}} + \frac{1}{\mu_\text{SR}} \quad (7.3.9)$$

where μ_L is the lattice vibration scattering mobility, μ_I is the ionized impurity scattering mobility, and μ_SR is the surface roughness scattering mobility. At high temperature, electron-phonon scattering dominates and μ is limited mainly by lattice vibration. At 300 K, the ZnO nanowire is a weak degenerate semiconductor and it is assumed that the lattice vibration scattering mobility is inversely proportional to temperature. This gives

$$\mu(T) = \mu(300\text{K})\frac{300\text{K}}{T} \quad (7.3.10)$$

For each I-V data point, a starting thermal conductivity and resistance is assumed. And then the temperature profile along the nanowire can be calculated using finite element method. The total resistance can be recalculated using the above equations. If the calculated current V_NW/R_NW is larger or smaller than the measured value, thermal conductivity is decreased or increased. The procedure continues until the calculated current agrees with the experiment value. (κ, T) pair for each I-V point is calculated in the same way. Temperature distribution along the nanowire and the effects of applied voltage is shown in Fig. 7.3.15. According to extracted thermal conductivity versus temperature, a relationship of $\kappa = 1/(\alpha T + \beta T^2)$ is found. This is suggestive of four phonon scattering.

3. Resistance switching effect

Peng Gao et al. (2010) studied cerium oxide film using in situ TEM and found reversible electro-driven redox process. CeO_2 films were grown on conductive Nb doped $SrTiO_3$ substrates by pulsed laser deposition. Conventional mechanical polishing and argon ion milling were used to make cross-sectional specimen. An electrochemically sharpened gold tip was fixed on the piezomanipulator to make electrical contact with the CeO_2 films. In the absence of an electrical field, the HRTEM

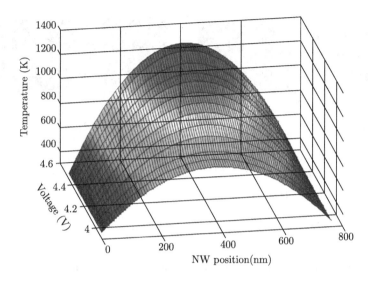

Fig. 7.3.15 Temperature distribution on the nanowire by calculation.

image and ED pattern along the ⟨110⟩ zone axis of CeO_2 was recorded. When a bias was applied to the sample, structural changes took place as shown in Fig. 7.3.16. Extra diffraction spots were also observed under the applied electric field. The superlattice reflections indicate that oxygen anions have been removed from the CeO_2 film and the introduced oxygen vacancies were structurally ordered as Ce_2O_3. The process is as follows. Under electrical field, oxygen anions at the surface will be attracted toward the positive electrode and driven out of the surface. Anions inside the films will be attracted toward the surface. This means that oxygen vacancies will be formed and will diffuse in a reversed direction and propagate into the interior of the crystal. The oxygen vacancies will cause the corrugation of two adjacent cation packing layers as the original oxygen anions leave the lattice. This will attract other negatively charged oxygen anions toward the vacancies and will lead to a series of migrations as the local positive charges at the vacancy sites are compensated for by adjacent oxygen anions. Thus the oxygen migration process can be viewed as a continuous diffusion of the anions. EELS spectra were also recorded and the result corroborated the Ce^{4+} to Ce^{3+} transition upon electric field.

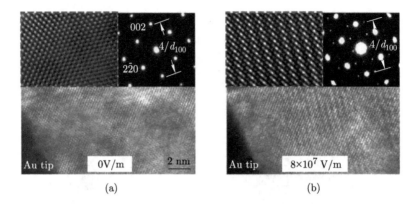

Fig. 7.3.16 Electrically induced phase change of CeO_2.

Upon electrical field the oxygen anions are activated to migrate because of the lowered barrier. The oxygen vacancies that formed at the surface migrated to the cathode and then assembled. So CeO_2 near the cathode was first reduced to Ce_2O_3 and the superstructure strips emerged near the cathode. When more and more oxygen anions were left, the superstructure strips propagated to the anode. At the beginning, higher oxide of cerium is put into an equilibrium state with oxygen molecules in the gas phase. The oxygen chemical potential (μ_O) is closely related to the equilibrium stability, described as a function of temperature (T) and oxygen partial pressure (p),

$$\mu_O = \mu_O^0 + RT\ln(p/p^0) \qquad (7.3.11)$$

where μ_O^0 is the standard chemical potential of oxygen, R is the gas constant, and p^0 is the standard pressure of 1 atm. In the presence of an electrical field, the equilibrium is determined by both the oxygen chemical potential and the electrical potential. The electrochemical potential is

$$\mu_O = \mu_O^0 + RT\ln(p/p^0) + ZF\phi \qquad (7.3.12)$$

where Z is 1 for an electron and 2 for an oxygen anion, F is the Faraday's constant, and ϕ is the electrostatic potential. The electrochemical potential determines the migration of anions. During this process, the electrons are supplied by one electrode and transform Ce^{4+} into Ce^{3+} and simultaneously take an electron from oxygen anion O^{2-} to oxidize it into an O atom with other electrode. At room temperature and 2×10^{-5} Pa, the critical potential is about 2.8 V. In the experiments, by increasing bias, phase transition can be much more easily observed. When the electric field was removed, the thermodynamic equilibrium state was broken and the oxygen content of the cerium oxide went back to the case before the electrical field was applied. This indicates that electrostatic potential plays a key role in the phase transformation of cerium oxides.

Zhi Xu et al. (2010) studied the switching behavior of $Ag/Ag_2S/W$ sandwich structure inside TEM as shown in Fig. 7.3.17 (a). In the experiment tungsten tip was grounded. And positive or negative voltage was applied to the Ag electrode. By sweeping voltage from 0 mV $-$200 mV $-$0 mV $-$300 mV $-$0 mV, a clear bipolar switching I-V curve was measured as shown in Fig. 7.3.17(d). At the beginning, the current is very low. By sweeping to positive voltages, the conductance suddenly increases at 130 mV, and the nanodevice transformed to the on-state. Subsequent sweeping of the bias back to negative values suddenly decreased the conductance at -145 mV, and the device went into the off-state. The on-off current ratio was larger than 5 orders of magnitude. During voltage scanning, the corresponding morphology changes were found in the Ag_2S portion as shown in Fig. 7.3.17(b) and (c). When the current suddenly increased (off-to-on transition), a piece of some new nanocrystal was seen to grow. The crystal kept its shape until the current suddenly dropped (on-to-off transition), then it shrank back immediately. This morphology change repeatedly took place in parallel with the on/off switching cycles. When the switch was turned on, both argentite and pure Ag fringes appeared suggesting that pure Ag and argentite Ag_2S phase together had formed the regarded conducting channel. Comparing ON state images with OFF state ones, it was found that a part of the acanthite phase transformed into the argentite phase, the argentite phase grew bigger and excessive Ag accumulated and formed a Ag nanocrystal. These composed conduction channels. Based on the experimental result, the switching mechanism of Ag_2S MIM structure was elucidated. Initially, Ag_2S is in the acanthite nonconductive phase. When a bias is applied, the Ag cations start to migrate, at the same time the acanthite phase transforms into the argentite one. The silver cations can be reduced anywhere along their migration way, but their reduction

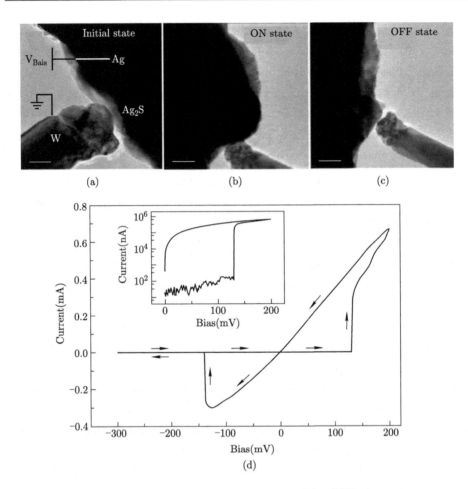

Fig. 7.3.17 Resistance switching property of Ag_2S/W structure.

occurs mostly at the interface between the Ag_2S electrolyte and the cathode. The Ag atoms in the silver electrode are ionized and dissolved into Ag_2S to supply Ag cations. Due to Ag mass transport, the volume of Ag_2S electrolyte within the conducting channel increases, and the mixture of Ag and argentite grows on the original acanthite surface. The pure Ag crystal starts to grow, while more and more acanthite Ag_2S transforms into argentite Ag_2S. A sudden resistance drop occurs when the conducting nanochannel entirely connects two electrodes. After applying a negative bias, the Ag crystal starts to be ionized and Ag cations move toward the anode. When the Ag-rich environment disappears, the argentite phase transforms back to the acanthite phase. Once the argentite $Ag_2S + Ag$ conducting channel becomes trespassed by the nonconductive acanthite phase, the channel breaks, the resistance dramatically increases, and the whole device is turned off. When a positive bias is applied, the conducting channel starts to rebuild itself. In the argentite phase, the S anions form a rigid body-centered cubic lattice with Ag cations randomly occupying 1/3 of the 12d sites in the lattice. The observed result that the switching occurs at a bias of 130 meV is consistent with calculated minimum barrier of 89 meV.

Nelson et al. (2011) studied ferroelectric switching using in situ TEM. The domain nucleation and evolution of $BiFeO_3$ thin film was studied. They used (110) oriented $TbScO_3$ as substrates. Then they used 20-nm $La_{0.7}Sr_{0.3}MnO_3$ as the buffer electrode and 100-nm (001) polarized $BiFeO_3$ was deposited as shown in Fig. 7.3.18(a). Cross-sectional sample of 60 nm in thickness was made. Electrically etched tungsten tip was used as counter electrode. Inhomogeneous out-of-plane electrical field was introduced by applying a bias, thus causing transitions among the eight possible $\langle 111 \rangle$ polarization directions. Three switching types are feasible, denoted by the angular rotation of polarization vector of 71°, 109° and 180°. They found that 71° rotations are preferred. Bias was applied to the tungsten tip with $La_{0.7}Sr_{0.3}MnO_3$ bottom electrode grounded. Ferroelectric domains started to nucleate upon a critical electrical field which is determined by local defects. TEM

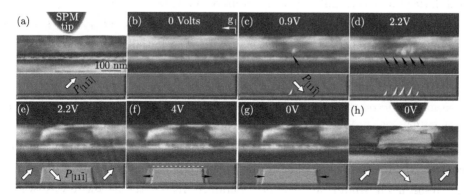

Fig. 7.3.18   Domain dynamics under electric field observed by in situ TEM.

images showed that nucleation started from the bottom interface as shown in Fig. 7.3.18(b)–(h). It showed that along the $La_{0.7}Sr_{0.3}MnO_3$ interface there was broad band with a strong negative field favoring $[11\bar{1}]$ polarization nucleation. This agreed with TEM images. The rectangular domains formed at the $La_{0.7}Sr_{0.3}MnO_3$ interface were metastable. Different domains grew or shrank under bias. Upon a critical voltage of 2.2 V, the domains assembled to form a single domain. Nuclei moved 200 nm laterally across the interface, more than three times the contact area of the tip. This deviated from the kinetically limited models describing ferroelectric switching.

Yuchao Yang et al. (2012) observed the conducting filament in nanoscale resistive memories. They deposited amorphous Si directly to a tungsten tip and form a W/a-Si/Ag sandwich structure. Voltage was applied to the Ag electrode with respect to the tungsten tip. *I-V* curves were recorded. At 490 s, the resistance sharply dropped. TEM images recorded the growth process of filament in real time. It was found that filament started to grow from the active electrode and the filament has a conical shape with the broad base at Ag electrode side. In situ erasing process showed that filament dissolution started from the filament/inert electrode interface.

4. Lithium ion transport

Jianyu Huang et al. (2010) studied the electrochemical lithiation of a single SnO_2 nanowire electrode inside TEM as shown in Fig. 7.3.19(a). A nanoscale electrochemical device was successfully constructed inside TEM using a single SnO_2 nanowire as an anode, an ionic liquid electrolyte, and a cathode of $LiCoO_2$ particles. A bias of -3.5 V was applied to the SnO_2 nanowire. The initial SnO_2 nanowire had a length of 16 μm and a diameter of 188 nm. Electrochemical reaction was observed to initiate from the point of contact between the SnO_2 nanowire and electrolyte. The solid state reaction front propagated along the nanowire away from the electrolyte. The diameter and length of the SnO2 nanowire expanded as the reaction front propagated as shown in Fig. 7.3.19(b). Real time TEM observations showed contrast change from crystalline to amorphous. At 625 s, the nanowire began to flex resulting in bending and coil of a spiral. At 1860 s, the nanowire was twisted suggesting extensive plastic deformation and microstructural changes. The charg-

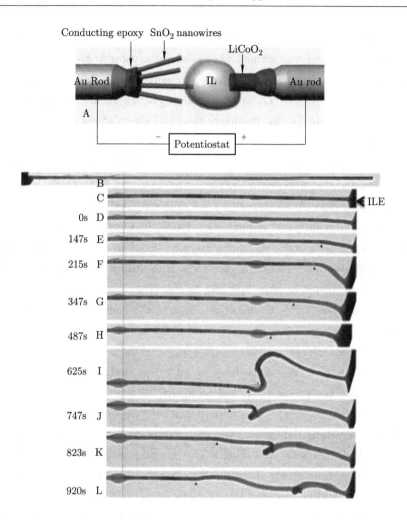

Fig. 7.3.19 Scheme of experimental design and in situ lithiation of single SnO₂ nanowire.

ing process took about half an hour. The nanowire was extended by ~60%, its diameter expanded ~45% and total volume expanded ~240%. Detailed investigation on the propagation front was performed. Based on EELS result, it was found that the nanowire after charging was composed of Li_xSn and Sn particles dispersed in an amorphous Li_2O matrix. The reacting interface had a high density of dislocations and was named "Medusa zone" due to dislocations "snaking" away from the interface. The dislocation cores would form effective Li transport channels and expedited Li insertion into the crystalline interior. On the other hand the amorphous phase observed was formed via the direct crystal-to-glass transition. After discharging, Li_xSn alloy nanoparticles were changed back to Sn and the diameter of the nanowire decreased. The overall volume change during discharge was much smaller than charging. During discharge the Li_2O glass did not contribute and only the Li_xSn precipitates were active.

Yang Liu and Jianyu Huang et al. (2011) studied the lithiation-delithiation process of individual aluminum nanowires by in situ TEM. Al NWs with naturally oxidized surface Al_2O_3 layer ($\sim 4-5$ nm) was used. The electrochemical cell was composed of a single Al nanowire electrode, Li_2O solid electrolyte and a bulk Li metal counter electrode. Al nanowires were fixed to an Al rod by conductive epoxy, functioning as the working electrode. Bulk lithium metal was scratched by a tungsten wire inside a glove box and in less than 2 seconds the tungsten wire was loaded into the TEM. Li metal attached to the tungsten wire worked as the counter and reference electrodes. The naturally grown Li_2O layer served as the solid electrolyte. Different voltages were applied to the working electrode for lithiation and delithiation. When a voltage of -2 V was applied, the lithiation started. After lithiation, the pristine straight nanowire became curved and the diameter increased. The initial single crystalline nanowire was converted to polycrystalline LiAl alloy. The volume expansion ratio after lithiation was assumed to be 100%. After delithiation, nanovoids were found. The size and number of nanovoids increase with time. After four cycles of delithiation, nanovoids could be found everywhere in the nanowire and nanocrystalline Al particles could be indentified from electron diffraction points. The void dynamics were caused by the dealloying of Li from LiAl. Combining elemental mapping EELS, it was found that the surface Al_2O_3 layer first transformed into Li-Al-O glass. This glass state prevented rupture during lithiation and acted as a solid electrolyte.

In another work, Li Zhong et al. (2011) studied lithiation mechanism of individual SnO_2 nanowires in a flooding geometry. In was found that lithiation started by multiple stripes parallel to (020) plane. These stripes served as reaction fronts afterwards. A high density of dislocations and enlarged interplanar spacing was found in the stripes. Those provided effective path for Li ion transport. The stripes increased after further lithiation and finally merged into one stripe. Large elongation, volume expansion and amorphization were observed.

Xiaohua Liu et al. (2011) used Si nanowire as anode and studied the electrochemical lithiation process using in situ TEM. Potential of −4 V and 0 V which were used for charging and discharging respectively, were applied to the anode with respect to the cathode at room temperature. Four types of Si nanowires were used: intrinsic, C-coated, heavily P-doped and C-coated plus heavily P-doped. It was found that ultrafast and full electrochemical lithiation was achieved for C-coated, P-doped and C-coated plus P-doped nanowires. In contrast, lithiation of intrinsic Si nanowires were slow and incomplete, forming amorphous Li_xSi alloy. The other types of Si nanowires showed a charging rate 1-2 orders of magnitude faster and formed crystalline $Li_{15}Si_4$ phase. Rupture was not observed despite high volume expansion of ∼300%.

Yang Liu and Huang et al. (2011) studied in situ lithiation of MWCNTs. An individual MWCNT was used as anode. A layer of Li_2O served as solid electrolyte and bulk Li metal was cathode. In situ observation revealed that a uniform layer of Li_2O several nanometers thick formed on the surface of MWCNT. The layer remained after delithiation resulting in a loss of capacity. The lithiated MWCNT was pulled and pushed by piezomanipulator. Sharp fracture edge was observed indicating brittle fracture of lithiated MWCNT. Molecular orbital theoretical calculations were performed. It was suggested that interlayer Li caused mechanical weakening via a "point-force" effect. Charge transfer from lithium to carbon contributed to chemical weakening.

7.4 Several advanced electron microscopy methods and their applications on materials science

Renchao Che and Qi Cao

During the past several decades, the field of TEM has been characterized by technological breakthrough in microscopy design. These developments aim mainly at an improvement of the resolution in order to simplify the direct visual interpretation of the images. Nowadays, a resolution of the order of 50 pm can be achieved by state-of-the-art instrumentation (Dahmen et al., 2009). With the development of aberration correctors (Rose, 1990), the resolution of a TEM image is no longer limited by the quality of the lenses but is restricted by the "width" of the atom itself which is determined by the electrostatic potential and the thermal motion of the atom (Van Dyck et al., 2003). Whenever that limit is reached, the images that contain all the information can be obtained using electrons.

It must be stressed that nowadays, we can totally achieve spatial resolution at the atomic scale not only in transmission electron microscopy (TEM) mode or in scanning transmission electron microscopy (STEM) mode, but also in electron energy loss spectroscopy (EELS) mode which furthermore yields chemical and subsequently electronic information at that same level (Van Tendeloo et al., 2012). By utilizing EELS, we can obtain the distribution information of electron energy losses, and consequently identify the phases, composition, elemental mapping as well as valence state of different elements from details of the energy loss spectra. Moreover, based on the energy loss near-edge structure (ELNES) of EELS spectra, electron energy loss magnetic chiral dichroism (EMCD) technique was proposed by Hébert C. and Schattschneider P. (2003) and then experimentally developed in 2006 (Schattschneider et al.). Momentum transfer in EMCD is considered to be equivalent to the polarization vector in XMCD, allowing the measurement of the magnetic circular dichroism without spin polarization of an electron beam. These two techniques endow modern TEM capability to get abundant information of electron structures of samples. On the other hand, another technique, referred as electron holography (HOLO), equipped TEM competence to obtained electrostatic potential information either inside or around sample particles. These advanced techniques, though vary from each other, all do help to break the limitation attributed to "width" of the atom itself and obtain higher resolution of TEM and more chemical as well as structural information of advanced materials.

In this chapter, brief introductions including their ultimate principles of the three advanced techniques mentioned above, typically, the EELS, EMCD and HOLO, are involved. Also, practical application examples of the three techniques are given in order to provide a comprehensive understanding of what specific information on earth can these techniques obtain and how they actually do.

7.4.1 Electron energy loss spectroscopy (EELS)

1. Brief introduction

Modern electron microscopy reaches its ultimate potential when combined with analytical techniques that provide chemical and electronic structure information. One of the most important, and also the most commonly used techniques to obtain such chemical information are based on the inelastic scattering of fast electrons which interacted with atoms in a material. During this scattering process, an atom is excited to an unoccupied state with an energy ΔE above the ground state. The fast electron, as is shown in Fig. 7.4.1(a) and (b), has provided this energy and loses ΔE. This loss can be determined using an electron energy loss spectrum (EELS) spectrometer.

The EELS mode of TEM was developed accompanied with the origination of electron energy filter system which is shown in Fig. 7.4.1(c) and (d). In EELS mode, energy loss (ΔE) information is selected to require useful image contrast. Typically, zero-loss imaging can be improved by eliminating chromatic blurring, and thus the loss imaging could provide compositional contrast and subsequently elemental mapping. In addition, more phases/composition and elemental valence state information can be indentified from detailed EELS spectra and by measure or resolve the distribution of electron energy losses of specific samples.

2. Characterization of LaOFeP superconductor: an example

The superconductivity of LaOFeP was first discovered in 2006 (Kamihara et al., 2006). Recently, La$[O_xF_{1-x}]$FeAs was found to be a superconductor at T_c about 26 K. From a structural point of view, higher superconducting T_c in La$[O_xF_{1-x}]$FeAs could be achieved by replacing P with As and partial substitution of O^{2-} by F^- in LaOFeP (Kamihara et al., 2008). One of the arresting characteristics of this type of materials is occurrence of superconductivity at about 4 K with the presence of Fe since ferromagnetism and superconductivity are considered to compete in conventional superconductors although, in principle, any metal might become a superconductor in its nonmagnetic state at a sufficiently low temperature (Shimizu et al.,

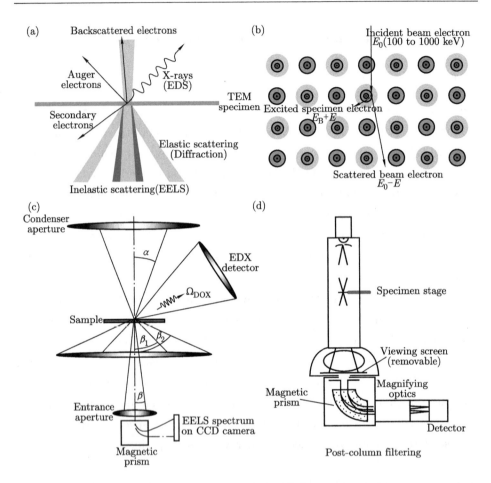

Fig. 7.4.1 Sketch for (a) the interaction and signal generating between the electron beam and specimen; (b) an atomic-scale view of electron energy loss in TEM; (c) a STEM setup including EELS, EDX and HAADF detection and (d) a typical post-column electron energy filter system.

2001). Iron has $3d$ transition electrons and exhibits ferromagnetic character. Therefore, LaOFeP differs possibly from conventional superconductors. The number of iron based superconductors identified so far is still limited. Three examples are iron under extreme pressures (15 − 30 GPa), filled skutterudite $LnFe_4P_{12}$ (Ln = Y, La) that has been prepared at high temperatures and high pressures and $[La_{1-x}Sr_x]$

OFeAs (Wen et al., 2008). The discovery of superconductivity in LaOFeP provides a new case for researching a superconductor containing a ferromagnetic element. Actually, the strong electron correlation, coexistence, and competition between superconductivity and ferromagnetism have been extensively discussed as a critical issue for many years (Coronado et al., 2000; Pfleiderer et al., 2001). The understanding of the electronic structure is important to obtain insight into the superconductivity mechanism. Correlation between the electronic structure and superconductivity has attracted intensive interest. LaOFeP is an iron-containing superconductor and is prepared through a solid state reaction method or a high-temperature process together with an arc melting process (Kamihara et al., 2006). Lebègue studied the electronic structure of LaOFeP by means of ab initio calculations using density functional theory (DFT) (Lebegue, 2007). His research focused on the charge transfer in this layered material and suggested that the La-O and Fe-P intralayer bonding present a significant covalency, whereas the interlayer bonding is ionic. A significant two dimensional character of this system is found. And further, as is introduced next, Li et al. developed a facile two-step solid reaction synthesis for the high quality LaOFeP superconducting materials. They investigated the electronic band structure and magnetic moment of LaOFeP by using first-principles method. Also, the effect of the atomic position of P on the magnetic moment is also discussed. They have observed certain EELS features from LaOFeP that can be attributed to the collective plasmon excitations, interband transition, or core-level transitions. Orientation dependence for typical edges in LaOFeP is also examined in comparison with the theoretical simulations. Moreover, annealing experiments at 1200 °C confirm that low oxygen pressure is a key factor for the occurrence of superconductivity in LaOFeP. Since the F-doped LaOFeP exhibits higher superconductivity T_c, the influence of F-doping on electronic structure and Fermi surface is briefly discussed.

Experimental and Instrumentation Details: LaOFeP sample used in this study was synthesized by mixing the starting materials with appropriate composition and pressing them into pellets. The pellets were enclosed inside an evacuated quartz tube. The vacuum of the tube was better than 10^{-5} Pa. High purity argon was filled to prevent implosion of the silica tubes. The first calcinations were done at 1200 °C for 12 h (sample without annealing). Then, the pellet was ground into powder and pressed into a pellet again for the second period of calcination (sample with annealing). High purity Ar gas and about 0.1 g La_2O_3 powder was added into the silica tube. The sintering temperature and time were the same as the first time and sample B was obtained. The purpose of the second sintering was to clarify the effect of low oxygen annealing on the superconductivity improvement of LaOFeP.

An FEI Tecnai F20 TEM equipped with a post-column Gatan imaging filter (GIF) was used for the EELS measurement and TEM imaging. The energy resolution was about 0.85 eV, determined by the full-width at half-maximum (FWHM) of the zero-loss peak. To avoid electron-channeling effects, the selected grain was tilted slightly off the zone axis by $1-2°$. The convergence angle was about 0.7 mrad and the collection angle was about 3 mrad. The crystal structure of LaOFeP for the band structure calculation is based on the tetragonal structure with lattice parameters $a = b = 0.3964$ nm and $c = 0.8512$ nm and space group of $P4/nmm$. The self-consistency was carried out on 4000 total k-points, which corresponded to a $20 \times 20 \times 9$ k-mesh in the irreducible Brillouin zone (BZ). The Fermi surfaces (FS) for LaOFeP and F-doped LaOFeP were calculated on a $27 \times 27 \times 12$ mesh containing 10000 k-points in total. Theoretical data of core-loss EELS and low-energy spectra were simulated within the WIEN 2k code. For the F-doped LaOFeP system, the F ion was considered to randomly occupy the O position so the virtual crystal approximation was used to simulate the effect of F-doping. In the F-doped system, 5000 k-points were used in the self-consistency to obtain the convergence of total energy.

Results and Discussion: Three kinds of magnetic states were calculated including the nonmagnetic (NM), ferromagnetic (FM) and antiferromagnetic (AFM) state, as is shown in Fig. 7.4.2. In the AFM state, we constructed an AFM ordering by flipping the spin orientation of one Fe atom in the cell, as illustrated in Fig. 7.4.1(c), in which the space group was changed into $P\bar{4}m2$. The heights of La and P were, respectively, relaxed in each model. The calculated results for the three kinds of magnetic states indicate that the ground state of LaOFeP is a weakly magnetic state. The magnetic moment of an iron atom calculated in the FM state is $0.14\mu_B$/atom. The calculation of the AFM state almost converges into an NM state with a magnetic moment of $0.006\mu_B$ for each Fe atom. Although the calculated total energy of the three states is quite similar, the FM state has the lowest total energy of the three.

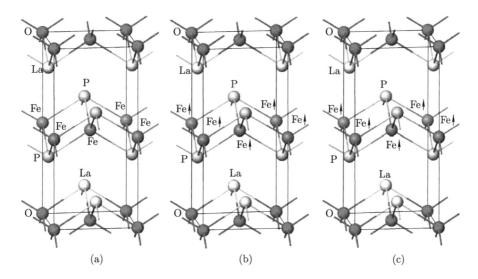

Fig. 7.4.2 The calculation model for (a) NM state, (b) FM state and (c) AFM state. The arrows indicate the spin orientation of Fe atoms in each model.

Fig. 7.4.3 shows the calculated density of states (DOS) near the FM level. It can be seen that the main contribution arises from the Fe-P layer (Lebegue, 2007). It should also be noted that the profiles of spin up DOS and spin down DOS for the Fe atom are almost symmetric, revealing that the Fe atom does not form a long-range magnetic ordering in this system.

Fig. 7.4.4 shows the electronic band structural features of LaOFeP. Fig. 7.4.4(a) is a brief energy level scheme for LaOFeP, qualitatively illustrating the bonding states and anti-bonding states from -20 eV below E_F to 3.0 eV above E_F. Within deeper binding energy regions (-16 — -20 eV and -14 — -16 eV), the electronic states are mainly composed of LaO entity. Fig. 7.4.4(b) and 7.4.4(c) are the spin-up and spin-down band structures for LaOFeP from -6.0 to 3.0 eV. Fundamental conductivity property is governed by the electronic states near the Fermi level. In the case of LaOFeP, hybridization states composed of Fe $3d$ with P $3p$ have large

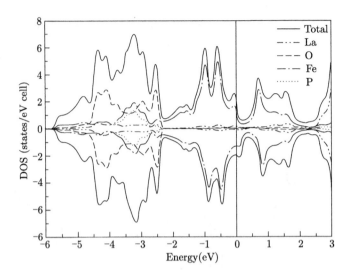

Fig. 7.4.3 The calculated density of states (DOS) of LaOFeP.

weights and represent the main contribution to the DOS near Fermi energy.

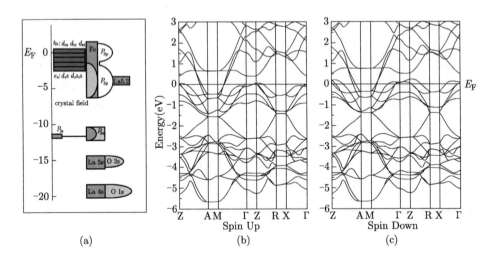

Fig. 7.4.4 (a) Schematic of LaOFeP energy level scheme from -20 eV below E_F to 3.0 eV above E_F; (b) The calculated spin up band structure and (c) spin down band structure of LaOFeP. 0 eV denotes the Fermi level.

Referring back to the DOS in Fig. 7.4.3, we can identify that the ten bands near the E_F ranging from -2.5 to 2.5 eV are mainly composed of Fe $3d$ states. Detailed analysis of the atomic orbital decomposition from wave functions is performed according to the crystal field theory. One Fe atom and four P atoms form a FeP4 tetrahedron. Two P atoms are located upward from the Fe atom layer and make a P-Fe-P bonding angle of $120.18°$. The other two P atoms are located downward from the Fe atom layer and form a P-Fe-P bonding angle of $104.39°$. Hence, this distorted FeP4 tetrahedron is suppressed along the z axis of LaOFeP. The Fe $3d$ bands are crystal field split in the tetrahedral P environment. Triple degenerate t_{2g} states (d_{xy}, d_{xz}, d_{yz}) are located above the doubly degenerate e_g states (d_{z^2}, $d_{x^2-y^2}$). If we set the c axis of LaOFeP as the z direction of the FeP4 tetrahedron, disposal calculation of the Fe $3d$ orbital shows that the dz^2 orbital extends along the c axis of LaOFeP and both the $d_{x^2-y^2}$ and d_{xy} orbitals disperse within the $a-b$ planes. Sub-bands of Fe $3d$ and P $3p$ intersect with the Fermi level. These partially filled bands might play an important role in the superconductivity and provide an opportunity for the improvement of superconductivity induced by hole-doping or electron-doping. It can be identified from Fig. 7.4.4 that there are two e bonding orbitals (dz^2 and dx^2-y^2) about -2.5 to -1.0 eV below the E_F and two e anti-bonding orbitals (also dz^2 and dx^2-y^2) about 0 to 0.8 eV above E_F. Above these two orbitals, in the range from -1.0 eV to 0 eV and from 0.4 eV to 2.5 eV, the dominating part is the bonding and anti-bonding orbitals of the three t_2 orbitals (d_{xy}, d_{yz}, and d_{xz}). Below the E_F, five Fe $3d$ bonding orbitals are completely occupied, one Fe d^2z anti-bonding orbital and two t_{2g} anti-bonding orbitals are partially occupied, so there are nearly six electrons filling in the Fe $3d$ orbitals. The ionic charge state of Fe in this compound is nearly Fe^{2+}. The $3d$ electrons in two equivalent Fe atoms occupy these orbitals by 50% probabilities, respectively, resulting in symmetric spin up DOS and spin down DOS, so the whole system shows very weak magnetic moment.

The calculations also revealed that the height of the P atom inside the LaOFeP lattice influences the magnetic moment of the Fe atom notably. For example, when using the experimental height of the P atom in the calculations, the length of the Fe-P bond is 0.2286 nm and the calculated magnetic moment of Fe is 0.08 μ_B/atom.

And when we use an optimal height of P atom in the calculations, the length of the Fe-P bond changes to 0.2243 nm, a decrease of about 2% compared to the experimental value, but the magnetic moment of Fe increases to 0.14 μ_B/atom. By further adjusting the height of the P atom, the magnetic moment of Fe might completely disappear. Also, the influence of the height of a La atom on the magnetic moment of this system was investigated, but no distinct effect was observed. This fact indicates that it is the specific structure of Fe-P tetrahedron that induces the low magnetic moment of Fe in this material.

Electron states around the Fermi surface affect the superconductivity measurably, especially in the case of hole-doping or cation-doping high-temperature cuprate superconductors (HTS) (Dessau et al., 1993; Tokura and Nagaosa, 2000; Lang et al., 2002). The Fermi surfaces in LaOFeP were simulated to explore the origination of its superconductivity. According to the band structure of LaOFeP in Fig. 7.4.4(b)–(c), there are five energy levels across the E_F in both spin up bands and spin down bands, which correspond to five Fermi surfaces in spin up state and spin down state, respectively, as shown in Fig. 7.4.5. Two cylindrical FSs centered along the A-M direction (Fig. 7.4.4(c) and Fig. 7.4.5(c)) exist in both spin up and spin down orientations. According to the band structure of LaOFeP in Fig. 7.4.4, these two Fermi surfaces are attributed to three t_{2g} orbitals. That is to say, they mainly have the characteristics of Fe d_{xy}, d_{yz} and d_{xz}. The other three FSs are centered by the $Z-\Gamma$ direction in which the larger one is attributed to the d_{z^2} orbital, while the two smaller FSs (as shown in Fig. 7.4.4(c) and Fig. 7.4.5(c)) have the characteristics of the $d_{xz} + d_{yz}$ orbital. This is different from the high-T_c cuprates in which the electrons near the E_F are mostly occupied by Cu $d_{x^2-y^2}$ states. This kind of difference is mainly caused by the different coordination condition between Fe-P and Cu-O. The former is tetrahedral, while the latter is octahedral. Comparing the Fermi surfaces in the spin up and spin down orientations, it should be noted that the weak magnetic moment of the Fe atom originates from the electrons in the d_z^2 orbital. Not only are the Fermi surfaces corresponding to the d_z^2 orbital (as shown in Fig. 7.4.4(a) and Fig. 7.4.5(a)) obviously different in their spin up and spin down orientations, but also the position of their energy levels containing d_z^2 characteristics near the Z and Γ points are different: the spin down d_z^2 orbital is about 0.2 eV higher than the spin up d_z^2 orbital.

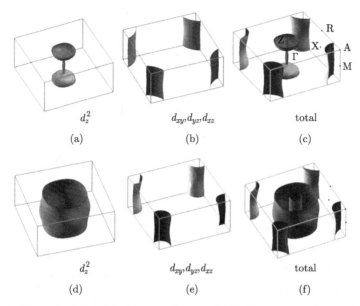

Fig. 7.4.5 The calculated (a)–(c) spin down and (d)–(f) Fermi surfaces of LaOFeP.

Next, the anisotropic properties of the electronic states in and the correlation with the electronic band structure of LaOFeP are established. Fig. 7.4.6(a) shows the crystal structure model of LaOFeP. A two-dimensional feature of LaOFeP can also be confirmed from the SEM observation and HRTEM imaging, as shown in Fig. 7.4.6(c)–(d). Fig. 7.4.6(b) shows the EELS spectra in the low energy range recorded from the LaOFeP superconductor with the incident electron beam parallel with and perpendicular to the c axis, respectively. The main difference between the two spectra is peak c at around 43.5 eV. Peak c shows a slight bump feature. It is expected that the spectrum measured with the momentum transfer $q//c$ should reveal more interlayer bonding contribution than the spectra recorded with $q \perp c$.

In the $q \perp c$ spectra, the data should show a major contribution from the intralayer bonding feature such as La-O bonding and Fe-P bonding. Interlayer bonding is composed of the covalent bonding between the $(La^{3+}O^{2-})$ layer and the $(Fe^{2+}P^{3-})$ layer. The interaction between the two layers is mainly formed by weak hybridization of La and P atoms. Therefore, in the case of the incident beam per-

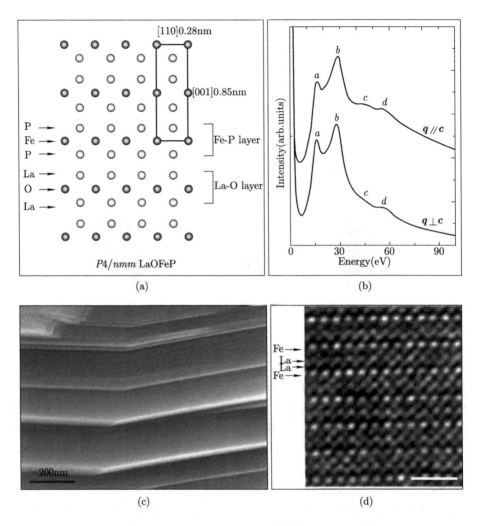

Fig. 7.4.6 (a) Atomic model of LaOFeP along the $[1\bar{1}0]$ direction; (b) The low loss EELS spectra with momentum transfer $q//c$ and $q\perp c$ for LaOFeP, respectively; (c) SEM image of a LaOFeP crystal showing its layered structure; (d) HRTEM image of LaOFeP, scale bar=1 nm.

pendicular to the c axis (for instance, $q\perp c$), such La-P bonding is not the major contribution, and excitation from $a-b$ planes is the major contribution. Via careful analysis on the data of DOS and the corresponding transition matrix elements, peak c can be directly interpreted as the transition from La 5s to La 4p states hybridized with P 3p states. This orientation dependence of EELS results basically from the anisotropic feature of the unoccupied electronic structures because the LaOFeP crystal belongs to a type of two-dimensional layered structure. The peak at 16.8 eV, as shown in Fig. 7.4.6(b), could be assigned to the extended O 2s state to the O 2p-La 4p vacant hybridization states. The 28.5 eV peak shows a bump bulk feature. Accordingly, peak b can be identified to be a plasmon peak. And then, Peak d at 55.3 eV is assigned to be Fe $M_{2,3}$ edges, corresponding to the transition from the Fe $3p3/2$ and $3p1/2$ states to the empty continuum states.

Compared with other techniques, EELS significantly extends the energy range and is a useful technique for analysis of the dielectric properties on a microstructure level. Fig. 7.4.7 shows the dielectric functions obtained by Kramer-Kronig analysis

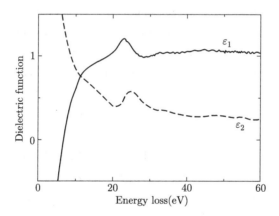

Fig. 7.4.7   Real part and imaginary part of dielectric function of LaOFeP.

(KKA) of the spectrum in the low loss energy region. To make the KKA reliable, the contribution from the elastic peak was subtracted, and the multiple scattering effects were removed by performing a Fourier-log method. Hence, a single scattering distribution was obtained via this deconvolution process. The imaginary part ε_2 of the dielectric function was obtained by

$$\varepsilon_{2ii} = \frac{4\pi^2 e^2}{m^2(\omega - \Delta_c/\hbar)V} \sum_{v,c,k} |\langle ck|p_i|v|k\rangle|^2 \times \delta(E_{ck} - E_{vk} - \hbar\omega)$$

where E_{ck} and E_{vk} are the quasi-particle energies approximated to the eigenvalues, $|ck\rangle$ and $|vk\rangle$ are the Bloch functions of the conduction and valence bands, V is the volume of the unit cell, and p_i is the momentum operator with $I = x$, y or z direction corresponding to the Cartesian axes. Then, the KKA was performed to calculate the real part ε_1 of dielectric function. The minimum value, maximum value and the intersecting points between the x axis and ε_1 curves include the information of interband transition. Peaks in the imaginary part ε_1 and ε_2 are mainly associated with the interband transitions, intraband transitions and collective plasmon excitations.

Fig. 7.4.8(a) displays the magnetic susceptibility curves for the LaOFeP sample before and after annealing. This annealing was done together with about 0.1 g La_2O_3 powder sealed in a silica tube to supply an oxygen source. The strong diamagnetic signal demonstrates the bulk superconducting character in the annealed LaOFeP (sample B). In contrast, no diamagnetic signal could be detected from the LaOFeP product before annealing (sample A) although a decrease in susceptibility value along with decreasing temperature was observed. Fig. 7.4.8(b) shows La $L_{2,3}$ edges core-loss EELS spectra for samples A and B. Simulated data with energy broadening 0.1 eV are used for comparison. Before comparison, the background of the experimental spectrum was subtracted and multi-scattering effects were deconvoluted. Peak a and peak c derive from the spin-orbit splitting Fe $2p_{3/2}$ and $2p_{1/2}$ edges of LaOFeP, respectively. The ratio of L_3 to L_2 is estimated to be about 4.4 − 4.6, giving a Fe valence of around 2.0 for LaOFeP for both annealed and unannealed samples. This Fe valence estimation is consistent with the above calculation.

The most remarkable phenomenon revealed in this spectrum is peak b, to the right side of peak a. Peak b can be observed from the annealed LaOFeP sample and further confirmed by DFT simulations. However, this peak does not show up in the data from the LaOFeP sample without annealing (the top curve) although the X-ray diffraction pattern of both specimens does not show any obvious difference. Detailed simulation analysis on the EELS data and DOS data reveal that peak a corresponds to the transition from Fe $2p$ to Fe $3d_{z^2}$-P $3p$ hybridized vacant states and peak b corresponds to the transition from Fe $2p$ to Fe $3d_{x^2+y^2}$-P $3p$ hybridized vacant states.

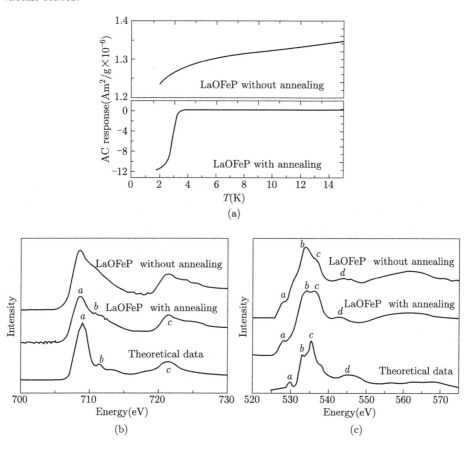

Fig. 7.4.8 (a) Magnetic susceptibility curves of LaOFeP before and after annealing under low O_2 concentration. EELS spectra of LaOFeP in the energy regions of (b) iron $L_{2,3}$ ionization edges and (c) oxygen K ionization edges. The calculated data for LaOFeP are shown for comparison. The spectra have been offset in the vertical direction for clarity.

To make the analysis reliable, multiple scattering effects have been deconvoluted from the original spectrum in order to obtain the single scattering spectrum. Accordingly, this single scattering spectrum can be reasonably compared to the electronic DOS data. The basic difference between the LaOFeP sample before and after annealing is that superconductivity can be found at 4.2 K in annealed LaOFeP, but superconductivity cannot be found in LaOFeP without annealing. The occurrence of superconductivity induced by annealing is similar to such phenomena found in high-T_c superconductors such as $YBa_2Cu_3O_{7-\delta}$ and $La_{2-x}Sr_xCuO_4$. (Loram et al., 1993; Millis and Monien, 1993)

Fig. 7.4.8(c) shows the oxygen K-edge core-loss electron energy loss spectra for LaOFeP before and after annealing. On each spectrum, four evident peaks (a)–(d) can be observed. Peak a corresponds to the transition from O $1s$ states toward La $4d_{z^2}$-O $2p$ joint vacant states. The b–d excitations arise chiefly from O $1s$ states to La $4d_{xy}$-O $2p$, La $4d_{xz+yz}$-O $2p$ and La $4d_{x^2+y^2}$-O $2p$ hybridized states. It should be noted that peak c's contribution becomes more dominant in the superconducting sample with annealing. The effect of annealing on the superconductivity mechanism of LaOFeP is a complicated issue. For example, it is possible that the exact chemical formula of LaOFeP after this annealing might be transformed to $LaO_{1-\delta}FeP$ and the carrier density might change. This issue will be addressed in future work.

Fig. 7.4.9 shows the EELS core-loss edges of La. White lines can be observed, representing La $M_{4,5}$ and the relative intensity ratio between them is similar to the theoretical calculation data ($I_{L_3} > I_{L_2}$, which is different from the La_2O_3 case). Dipole-forbidden M_2 and M_3 edges were also found at the $1100-1250$ eV range,

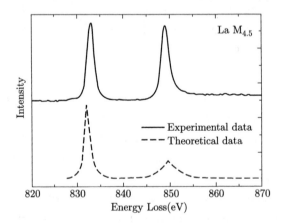

Fig. 7.4.9 Experimental and simulated EELS spectra of LaOFeP in the region of La $M_{4,5}$ ionization edges.

which is not shown here.

The data discussed above reveal the two-dimensional character of LaOFeP revealed by orientation-dependent EELS results. Anisotropic EELS spectra depend on momentum transfer along different orthogonal directions of the selected LaOFeP crystallite sheet. In the $q//c$ spectra, the momentum transfer resulting from interlayer bonding is easily detected. However, in the $q \perp c$ spectra, the main momentum transfer is in the La-O layer or the Fe-P layer. Therefore, we observe that the edge at about 55 eV is more intense for momentum transfer q perpendicular to c than parallel to c. By comparison, the layers of Fe-P in the a-b planes of LaOFeP are believed to play a role similar to the CuO_2 planes in the high temperature superconducting oxides. From the carrier density point of view, our annealing experiment shows that a low O_2 atmosphere is important to the superconductivity of LaOFeP.

Conclusions: The electronic band structure and magnetic moment of LaOFeP were investigated with first-principles calculations. The calculations for NM, FM, and AFM states of LaOFeP indicate that the Fe atom does not form a long-range magnetic ordering. The Fe atom in this system shows very weak magnetic moment, which is a necessary condition for the appearance of superconductivity in this material. The Fe magnetic moment is sensitive to the height of the P atom. Analysis of the band structure shows that the charge state of Fe ion in this compound is nearly Fe^{2+}. The Fermi surface analysis indicates that there are five FSs corresponding to the Fe $3d$ orbital characteristics, respectively. Experimental results of EELS fundamentally agree with the DFT band calculation for the LaOFeP superconductor. Besides, an interpretation of experimental EELS spectra for the low-loss spectra and core-loss spectra of LaOFeP was presented. Low-energy EELS spectra show that an obvious peak is found at about 44 eV measured with the momentum transfer $q//c$ and not found with the $q \perp c$ spectra. EELS spectra with the direction of momentum transfer implies that this state results from La-P hybridization. Annealing experiments indicate that low oxygen pressure induces the occurrence of superconductivity of LaOFeP, which is evidenced by the O K and Fe $L_{2,3}$ edges. It is supposed that our results might give some reference to the understanding of the superconductivity of LaOFeP.

7.4.2 Electron magnetic chiral dichroism (EMCD)

1. Brief introduction

Circular dichroism (CD) is caused by the existence in the medium of some physical properties related to handedness and electron spin. The EMCD technique is based on EELS detection, which can provide chemical information at sub-nanometer scale, such as type, coordination and valence states of the constituent chemical elements. Meanwhile, Lorentz microscopy is able to determine the orientation of a domain wall within a thin film, but not normal to the film plane. The latter orientation can be determined with EMCD because its detection depends on the orientation and strength of magnetization along the beam direction of a transmission electron microscope (TEM). The limitations of EMCD include incompatibility with automatic beam scanning due to the requirement of two-beam diffraction condition, as well as difficulty of operation, weak signals and dependence of the signals on the sample thickness. The magnetic field of the objective lens in a TEM would align the magnetic moments in the sample parallel to the optical axis (i.e. out of the sample's plane). Therefore, an EMCD experiment has to be carried out in Lorentz mode, with the objective lens switched off.

The EMCD technique, which is based on electron energy loss near-edge structure (ELNES), is quite similar to the X-ray magnetic circular dichroism (XMCD). Momentum transfer in EMCD could be regarded equivalent to the polarization vector of XMCD, which fundamentally allows the measurement of the magnetic CD escaping from influencing by spin polarization of an electron beam. Chiral electronic transitions can be detected with a standard TEM equipped with an EELS system or an energy filter system. And owing to its ultrahigh spatial resolution, EMCD technique is gradually becoming a key microscopic characterization method for spintronics and nanomagnetism (Zhang et al., 2009).

2. Characterization of magnetic domain walls: an example

Spin-dependent transport of charge carriers has been extensively studied in tunneling magnetoresistance systems owing to their immense potential for magnetoelectric applications and a variety of interesting physical properties (Parkin et al., 2008). Thin ferromagnetic NiFe layers are usually used as electrodes in such systems. The magnetic domain walls inside these ferromagnetic layers are crucial for determining the tunneling spin polarizations and are particularly associated with the magnetic long-range order in a 3D Heisenberg system (Wu et al., 2004). Undoubtedly, information on the wall types in magnetic thin films is important for various magnetic devices. Efficient techniques are still lacking for characterizing the magnetic properties of ultrathin electrode films with high spatial resolution. Only a few direct studies have been reported on the spin states of individual domain walls in thin ferromagnetic layers, and there is still no consensus on how to reliably determine the type of an individual wall in a magnetic domain. Most methods used for this purpose, such as Bitter powder technique, magneto-optical Kerr effect, Lorentz microscopy, electron holography, magnetic force microscopy and X-ray topography, have insufficient selectivity or spatial resolution to directly relate the microstructure and magnetic character of individual domain walls (Tonomura et al., 1980).

Experimental Details: A thin permalloy film was chosen as a ferromagnetic specimen. Permalloy has many advantages such as high magnetic permeability, low coercivity and magnetostriction and significantly anisotropic magnetoresistance. Permalloys typically have a face-centered cubic crystal structure with a lattice constant of approximately 0.355 nm for a nickel concentration of about 80%. A two-beam diffraction condition required for the EMCD experiment is relatively easy to achieve in such material. TEM samples were prepared by mechanical polishing and dimpling. An ion milling system (Gatan-691) with a liquid-nitrogen-cooled stage was used for thinning the samples. The in situ Lorentz microscopy examinations were carried out using a JEM-2100F TEM equipped with a field emission gun, with the objective lens switched off and the objective mini-lens on. A post column Gatan imaging filter (GIF-Tridium) was used for EELS measurements. The energy resolution was about 0.70 eV, as evaluated using the full width at half maximum of the zero-loss peak. To avoid electron channeling effects, the selected grain was slightly tilted off the zone axis (by $2-4°$). The convergence angle was about 0.7 mrad ($q = 0.4$ nm^{-1}) and the collection angle was about 3 mrad ($q = 1.7$ nm^{-1}).

Results and Discussion: The structure of cross-tie walls was established by Hebert and Schattschneider in 1958 using Bitter pattern methods. The Neel wall is spaced at regular intervals by short cross Bloch lines, denoted as "cross ties", each terminating in a free end as shown in Fig. 7.4.10(a). Domain wall creeping of the circle-tie walls is connected with circular Bloch line movements and a variation of the main wall curvature. Fig. 7.4.10(b) shows an under-focal Lorentz microscopy image of a cross-tie wall. The dark line corresponds to a main domain wall: a 180° Neel-type wall. The two shorter white lines are cross ties and therefore the intersecting points are cross Bloch lines, as marked by the arrows. The dark points between the cross ties are circular Bloch lines. Fig. 7.4.10(c) presents another Lorentz image from the same specimen, showing a circular Bloch line with a stronger contrast.

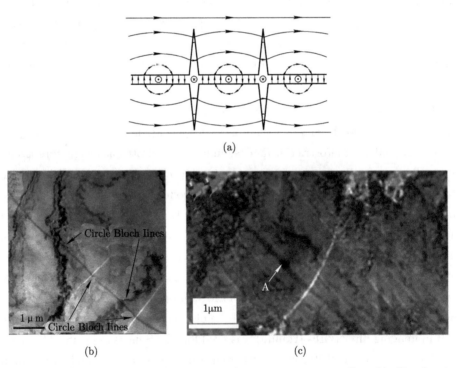

Fig. 7.4.10 (a) A diagram of the cross-tie wall structure in a permalloy thin film showing flux closures within the film plane and a Lorentz TEM image (b) of cross-tie walls and (c) showing a circular Bloch line marked by A.

To confirm that the dark and bright lines are magnetic domain walls, two sequential images were recorded with an under-focus of about -25 µm and an over-focus of $+25$ µm. The transmitted electrons are deflected by the Lorentz force, the vector of which is tuned by the defocus values. Hence, "convergent" or "divergent" wall images were formed as shown in Fig. 7.4.11.

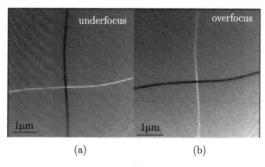

Fig. 7.4.11 Lorentz images of magnetic domain walls: two sequential images were recorded with (a) an under-focal value of -25 µm and (b) an over-focal value of $+25$ µm.

EELS examinations were performed to detect magnetic dichroism at several locations on the permalloy film. Since EMCD signals are usually an order of magnitude weaker than XMCD signals, thus condensed beam of convergent-beam electron diffraction (CBED) mode was used in experiments. The angular convergence and the focal values of the incident beam were controlled by adjusting both the condenser lens and the prefield of the objective minilens in Lorentz mode. In two-beam CBED, an aperture selects only the central (000) beam and diffracted (111) beam G in the back focal plane, as shown in the inset of Fig. 7.4.12. A Thales circle was made up of the coherent transmission and diffraction beams and was drawn with a diameter G to intersect the 0 and G disks. The strongest dichroic signals can be detected at the top and bottom points on the Thales circle, which, respectively, correspond to q^+ and q^- wave vectors. The EELS detector was positioned at the two moment transfer points. From the diffraction dynamic point of view, the imaginary part of the mixed dynamic form factor has maxima at the top and bottom positions where pseudovectors are also of the maximum (Schattschneider et al., 2006).

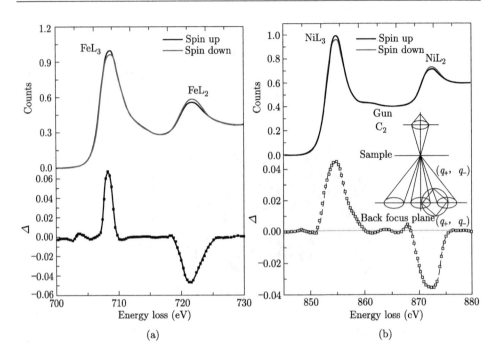

Fig. 7.4.12 Typical EMCD spectra near the (a) Fe-$L_{2,3}$ and (b) Ni-$L_{2,3}$ edges recorded at a cross Bloch line. The inset outlines the diffraction geometry where "+" and "−" correspond to the superposition of two groups of momentum transfer vectors.

Fig. 7.4.12 shows the EMCD energy spectra around the Fe $L_{2,3}$ and Ni $L_{2,3}$ edges measured from the cross Bloch line of permalloy film. The existence of an EMCD difference signal confirmed that the Bloch line carries a magnetic flux parallel to the electron beam. The EMCD signals (normalized spin-up/down difference $\Delta > 0.06$) are strong enough to be visible without magnification. Excitations detected at the spin–orbit split $2p_{3/2}$ and $2p_{1/2}$ levels generated two peaks at about 708 and 721 eV, respectively. This dichroic spectroscopy at Fe-L_3 and Fe-L_2 edges results from spin population imbalance around the Fe-3d Fermi surface. There are more spin-up electrons than spin-down electrons when detected at the q^+ position in momentum space. Combining the analysis of Fig. 7.4.10 and Fig. 7.4.12, the orientation of the cross Bloch line was determined as parallel to the electron beam. The signal from electrons at spin-up states is higher than that from electron at spin-down states. In contrast, at the main domain wall, the magnetic flux is within the sample plane, which makes an orthogonal orientation relationship with the incident beam direction and therefore contributes no dichroic signal to EMCD, as shown in Fig. 7.4.13.

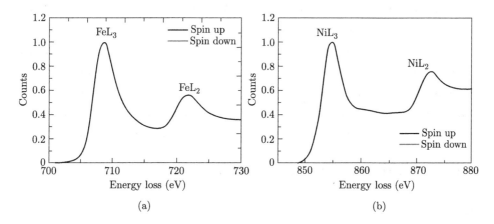

Fig. 7.4.13 Typical EMCD spectra near the (a) Fe-$L_{2,3}$ and (b) Ni-$L_{2,3}$ edges recorded at a circular Bloch line.

The most remarkable phenomenon, revealed from the EMCD difference signal in Fig. 7.4.13, is that the spin orientation at the circular Bloch line is anti-parallel to that at the cross Bloch line. The signal from electrons at spin-down states is stronger than that from electrons at spin-up states. However, the EMCD difference is weaker for the circular Bloch line than cross Bloch line ($\Delta \approx 0.02$), which is discussed as follows. The magnetization orientation in each domain is determined by the competition of magnetostatic energy, magnetoelastic energy, demagnetizing field energy, exchange interaction, magnetocrystalline anisotropy, magnetic dipolar interaction and domain wall energy. The circular Bloch line has a more complicated spatial distribution, which is similar to a three-dimensional spiral contour map, than the cross Bloch line. While the EMCD signal from the circular Bloch line is an average over the circle region, this is not the case for the cross Bloch line.

The spin-up signal is stronger than the spin-down signal at the Fe-L_3 edge at 708 eV of energy loss. Similarly, the Ni-L_3 signal is also higher at the "+" position (spin-up) than at the "−" position (spin-down). Thus, a ferromagnetic order at both the cross Bloch line and circular Bloch line was deduced on the basis of the same spin orientation of Fe and Ni signals. This is because of metallic bonding between Fe and Ni atoms in permalloys, with a parallel spin orientation. Such Fe-Ni metallic bonding is composed of Fe $3d$ electrons with spin-up (spin-down) orientation and Ni $3d$ electrons with spin-up (spin-down) orientation.

Conclusions: Convergent-beam ED and EMCD techniques are combined in Lorentz mode to determine the nature of individual magnetic domain walls using a transmission electron microscope. This technique allows distinguishing of the cross Bloch line, circular Bloch line and Neel wall. Notably, the high spatial resolution enables the study of spin-dependent transport across magnetic domain walls.

7.4.3 Electron holography (HOLO)

1. Brief introduction

Electron holography (HOLO) was originally proposed (Gabor, 1949) as a means of correcting for EM lens aberrations, substantially before the advent of the laser and the use of holography in light optics. The technique is based on the formation of an interference pattern or hologram in the TEM. Its development followed from earlier experiments in electron interferometry, many of which took place at the University of Tübingen, and relied on the development and availability of high-brightness electron sources. The technique overcomes the important limitation of most TEM imaging modes, namely that only spatial distributions of image intensity are recorded. All information about the phase shift of the high-energy electron wave that passes through the specimen is then lost.

By contrast, electron holography allows the phase shift of the electron wave to be recovered. As the phase shift is sensitive to local variations in magnetic and electrostatic potential, the technique can be used to obtain quantitative information about magnetic and electric fields in materials and devices with a spatial resolution that can approach the nanometer scale. This capability is of great importance for the study of a wide variety of material properties, such as the characterization of magnetic domain walls in spintronic devices (Junginger et al., 2007) and the factors that affect the coercive fields of individual magnetic nanostructures (Bromwich et al., 2006).

The original work (Gabor, 1949) described the reconstruction of an image by illuminating an "in-line" electron hologram with a parallel beam of light and using a spherical-aberration-correcting plate and an astigmatism corrector, but the image reconstructed in this way is disturbed y a ghost or conjugate twin image. The mode of electron holography that is most often used for tackling problems in materials science is instead the off-axis, or sideband, mode, which is available n many modern electron microscopes and has been applied to the characterization of materials as diverse as quantum well structures, magnetoresistive films, nanowires and semi conductor devices.

The EM geometry for the TEM mode of off-axis HOLO is shown schematically in Fig. 7.4.14. A field-emission electron gun (FEG) is used to provide a highly coherent source of electrons. In reality, the source is never perfectly coherent, but the degree of coherence must be such that an interference fringe pattern of sufficient quality can be recorded within a reasonable acquisition time, during which specimen and/or beam drift must be negligible. Although electron holograms have historically been recorded on photographic film, digital acquisition using charge-coupled-device cameras is now common practice. To obtain an off-axis electron hologram, the specimen is positioned so that it covers approximately half the field of view. A voltage is then applied to an electrostatic biprism, which is usually located in place of one of the conventional selected-area apertures in the microscope. The biprism is analogous to a glass prism in light optics, but takes the form of a fine (< 1 μm of diameter) wire that is often made from gold-coated quartz. The voltage applied to the biprism acts to tilt a "reference" electron wave that passes through vacuum with respect to the electron wave that passes through the specimen. The two waves are allowed to overlap and interfere. If the electron source is sufficiently coherent then, in addition to a bright-field image of the specimen, an interference fringe pattern is formed on the detector in the overlap region. Just as in a textbook "double-slit experiment", electrons are emitted one by one from the field-emission electron gun in the microscope. After being deflected by the biprism, they reach the detector and are detected individually as particles. When large numbers of electrons have accumulated, their wave-like properties become apparent and an interference fringe pattern is built up.

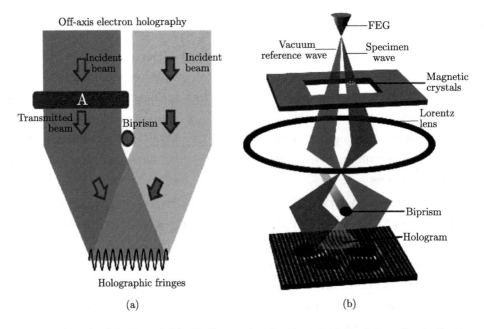

Fig. 7.4.14 The (a) 2D and (b) 3D illustration for the principle of the off-axis electron holography technique. Zone A in (a) refers to an unstrained reference region of crystalline samples.

The amplitude and the phase shift of the electron wave that leaves the specimen are recorded in the intensities and the positions of the interference fringes in the hologram, respectively. The phase shift is sensitive to the in-plane component of the magnetic induction and the electrostatic potential in the specimen. Although the technique can be used to record phase information at atomic resolution (to improve the interpretable resolution beyond the point resolution of the microscope) and to measure variations in specimen thickness and mean inner potential, here we concentrate on the medium-resolution application of the technique to characterize longer-range electromagnetic fields. For studies of magnetic materials, a Lorentz lens (a high-strength minilens) allows the microscope to be operated at high magnification with the conventional objective lens switched off and the specimen either in magnetic-field-free conditions or in a chosen externally applied magnetic field. The typical field of view of such measurements is between a few hundred nanometers and a few micrometers, and the spatial resolution of the recorded electrostatic or magnetic fields can approach or exceed 5 nm.

2. Characterization of FePt alloy nanorods: an example

Iron-platinum alloys are important both scientifically and technologically. Since a FePt alloy possesses high uniaxial magnetocrystalline anisotropy ($K_u \approx 7 \times 10^6$ J/m^3) and good chemical stability, they have potential applications both to high performance permanent magnets and to ultrahigh density magnetic recording media (Boyen et al., 2005; Elkins et al., 2003). FePt nanomaterials have stimulated great interest in recent years. Previous work about the synthesis of FePt nanoscale materials has mainly focused on wet chemical processing such as the chemical solution-phase reaction, seed-mediated growth and electrodeposition (Che et al., 2005). The control of the shape and size of magnetic nanostructures has been becoming a more important issue because magnetic properties of nanostructures strongly depend on their shape and size, so that many researchers have explored fabrication methods for size- and shape-controlled magnetic nanostructures (Smith et al., 2000). Electron beam-induced deposition (EBID) is one of the most promising methods for size and position controllable fabrication and used to fabricate nanostructures with various materials including magnetic materials. EBID has many advantages, including perfect flexibility in substrates selectivity, easily achieving 3D structures at nanometer scale and fine fabrication resolution (Takeguchi et al., 2005). Thus, a fabrication method to construct magnetic FePt alloy nanorods and the quantitatively HOLO characterization for their residual magnetic flux density are presented as follows.

Experimental Details: Self-standing nanorods were grown at an edge of molybdenum substrate by EBID inside the chamber of a FEG-SEM (JEOL JSM-7800UHV). The working voltage of JEOL JSM-7800UHV was 30 kV and the beam current used to make deposition was 8×10^{-10} A. Both iron pentacarbonyl and a cyclopentadienylplatinum (IV)-trimethyl precursor were simultaneously introduced into the SEM chamber through a nozzle with a 0.2 mm inner diameter. The ratio between the two gases was maintained to be 1:1 in order to satisfy the stoichiometric ratio of the FePt phase. The partial pressure inside the microscope column was controlled at 1.2×10^{-4} Pa. A computer-controlled scan generator was used to maintain a constant beam scanning speed of about 3 nm/s. Then, an in situ annealing at 600 °C under about 2×10^{-6} Pa vacuum was performed inside the SEM chamber, so that the as-deposited nanorods could be crystallized. The 3D morphology of the nanorods was observed from about 90° different directions and it was found that the cross section of nanorods was an ellipse with a long/short axes ratio about 5:1. The line-width was directly measured from SEM and TEM images, and the thickness was determined by an amplitude image reconstructed from an off-axis electron hologram, as mentioned later. A series of analysis methods, including HRTEM, ED, HOLO and EDX characterizations were performed using a JEOL JEM-3000F FEG-TEM operated at 300 kV. Electron holograms were acquired by a Gatan-Ultrascan CCD camera with an acquisitive image size of 1024×1024 pixels.

Results and Discussion: Fig. 7.4.15(a) is a low-magnification TEM image showing an array of nanorods perpendicularly formed on a Mo substrate edge by an EBID technique. The line-width of the nanorods is uniform of about 26 nm and the length of the nanorods is around 360 nm. The line-width and length could be controlled accurately by changing the electron beam current and beam scanning speed for deposition (Takeguchi et al., 2005). The composition of the nanorods was examined by EDX analysis. EDX analysis was carried out at different areas of the nanorods, indicating that the compositional ratio of Fe to Pt was approximately 1:1 and distributed homogeneously in the nanorods. This might benefit from the constant scanning speed of the electron beam. A typical EDX spectrum from one nanorod was shown in Fig. 7.4.15(b), indicating that there are iron, platinum and carbon elements inside the nanorod. The Mo signal at 2.29 keV comes from the substrate. Fe, Pt and C were considered to be decomposed from the $Fe(CO)_5$ and $(C_5H_5)Pt(CH_3)_3$ molecules absorbed at the surface of Mo substrate. It should be noted that platinum and iron elements are dominant and carbon is considerably little. This means that the electron beam dissociated precursor molecules by cutting Fe-$(CO)_5$, C-H, C-C and Pt-CH_3 bonds, and consequently nonvolatile metal parts were deposited along the beam scanning line, and therefore nanorods were formed. Most gaseous organic ligands were pumped out.

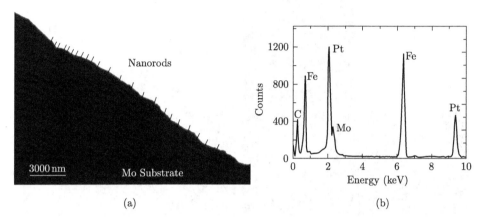

Fig. 7.4.15 Morphology and composition of the nanorods. (a) Low magnification TEM image showing nanorods grown at the edge of the Mo substrate; (b) Representative EDX spectrum taken from the nanorods.

Fig. 7.4.16(a) shows a bright-field TEM image of one nanorod after annealing at 600 °C for 2 h. It was found that the FePt nanorod was composed of a chain of metallic nanoparticles encapsulated within a thin layer of carbon-containing sheath. It is supposed that the outside carbon-containing sheath has effects both on assembling this chain of particles to be a 1D nanorod and on protecting the inside metallic phase from oxidation corrosion. Fig. 7.4.16(b)–(d) are three microdiffraction patterns recorded from different areas of the nanorod using a 10 nm electron probe. Fig. 7.4.16(b)–(d) could be indexed using the tetragonal $L1_0$-FePt structure with lattice parameters: $a = 0.385$ nm and $c = 0.371$ nm, and the zone axes were identified as [010], [$\bar{3}$10] and [$\bar{1}$00] projections, respectively. Fig. 7.4.16(e) shows a high-resolution TEM image taken from this nanorod and the inset is an atomic model of $L1_0$-FePt. It was found that the spacing of the two largest lattices were both of 0.22 nm, and they intersected with an angle of 69.5°. These lattices correspond to the (1$\bar{1}$1) and (111) crystalline planes of the tetragonal $L1_0$-FePt crystal (JCPDS 43-1359). Thus conclusions can be made that the nanorods are composed of crystalline $L1_0$-FePt nanoparticles.

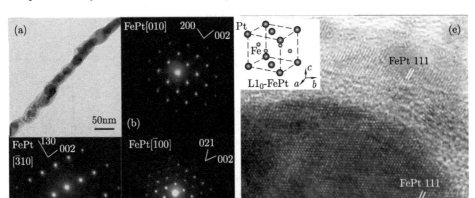

Fig. 7.4.16 Structural determination via ED and HRTEM imaging. (a) Low magnification TEM image of one individual nanorod; ED patterns from this nanorod along the (b) [010], (c) [$\bar{3}$10] and (d) [$\bar{1}$00] zone axis; (e) a HRTEM image of this nanorod viewing along FePt [10$\bar{1}$] orientation and the inset is an atomic structure model of $L1_0$-FePt.

The residual magnetic flux density (B_r) of one individual nanorod was evaluated by an off-axis electron holography method. When the electron beam passes through a local magnetic field, the interaction between the incident electron wave and the magnetic field from the FePt nanorod introduces a local differential phase shift to the exit electron wave. Off-axis electron holography provides an access to the phase shift information of the electron wave after it has traversed through a nanorod. Fig. 7.4.17(a)–(b) show a low magnification TEM image and an electron hologram taken from a FePt nanorod, respectively. Fig. 7.4.17(c) shows the phase image reconstructed from the electron hologram, with an amplification factor of 4. The spacing between adjacent equiphase lines indicates a phase shift value of $\pi/2$. Closed flux lines distributed outside the nanorod represented the magnetic flux lines and clearly indicated the ferromagnetic nature of the nanorod. The magnetization direction was shown to be parallel to the axis orientation of this nanorod. The phase shift value across the nanorod was calculated from the phase profile measured in the area, marked by the black rectangle in Fig. 7.4.17(c). The phase shift value shown in Fig. 7.4.17(d) was averaged over ten pixels perpendicular to the scanning line to improve counting statistics. This phase shift is 7.3 rad measured from Fig. 7.4.17(d).

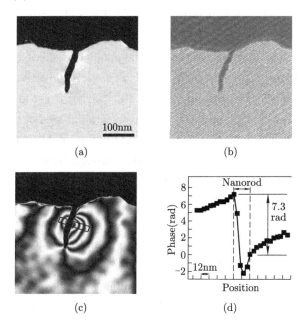

Fig. 7.4.17 (a) Low-magnification TEM image showing one FePt nanorod with Mo substrate; (b) off-axis hologram obtained from this nanorod; (c) phase profile in a direction perpendicular to the nanorod, reconstructed from electron hologram. The black rectangle marks the area used for an averaged line scanning; (d) line profile of the phase distribution of the FePt nanorod indicated by the rectangle.

The magnetic flux density B_r of the FePt nanorod from this phase shift value was calculated as follows. At first, we derived the thickness t of the nanorod by manipulating the amplitude image from an off-axis electron hologram (Gajdardziska-Josifovska and McCartney, 1994). The normalized thickness (t/λ_i) could be approximately obtained from the following expression: $t/\lambda_i = -2\ln(A_0/A_r)$, where i is the mean-free path for inelastic scattering of high-energy electrons (considering nanorod composition and acceleration voltage of microscope), A_0 and A_r stand for the energy-filtered (zero-loss) amplitude of the objective and reference waves, respectively. Thus, the thickness at the center of this nanorod was calculated to be about 152 nm. The line-width of this nanorod was about 26 nm, as mentioned in Fig. 7.4.15(a). Thus, the cross-sectional area S of this nanorod was computed to be $S = (\pi \times a \times b/4) = (\pi \times 26 \times 152 \times 10^{-18})/4 = 3.11 \times 10^{-15}$ m^2, assuming an elliptical geometry.

Finally, the phase shift due to the internal magnetic field of a nanorod could be described as follows: $\Delta \varphi = 2\pi \cdot (e/h) \cdot B \cdot S$, where e is the electron charge (1.60×10^{-19} C), h is the Plank's constant (6.62×10^{-34} J/S), B is the internal magnetic flux density in the nanorod and S is the cross-sectional area of the nanorod. So, a phase shift of 2π radian is caused by a magnetic flux of 4.1×10^{-15} Wb. As the phase difference of the electron wave passing through both sides of this nanorod was about 7.3 rad, which was measured from Fig. 7.4.17(d), the corresponding magnetic flux was calculated to be 4.76×10^{-15} Wb. Hence, the residual magnetic flux density B_r was computed to be about 1.53 T. This value is two to three times larger than that of the Fe-containing nanorods and α-Fe formed by EBID in a similar manner.

Conclusions: Successful fabrication of magnetic alloyed $L1_0$-FePt nanorods was carried out by EBID inside an ultrahigh-vacuum SEM. A mixture gas composed of Fe(CO)$_5$ and (CH$_3$)$_3$ (C$_5$H$_5$)Pt was introduced into the main chamber to make deposition. Crystalline FePt nanorods were obtained via this dual-gas deposition technique with heat treatment and characterized by EDX, SAED and HRTEM. This fabrication method is comparatively easy to be extended to other metallic elements. Further, off-axis electron holography was applied to measure the residual magnetic flux density B_r of the FePt nanorods. The B_r value of the as-prepared FePt nanorod was about 1.53 T, revealing its hard magnetic nature.

References

Agrait N., Rodrigo J. G. and Vieira S., Conductance steps and Quantization in Atomic Size Contacts, Phys. Rev. B, 1993, 47, 12345–12348.

Awschalom D.D., Warnock J., Super cooled liquids and solids in porous-glass. Phys.

Rev. B, 1987, **35**, 6779–6785.

Bacon R., Growth, Structure, and Properties of Graphite Whiskers, J. Appl. Phys., 1960, **31**, 283–290.

Bai, X. D., D. Golberg, Y. Bando, C. Y. Zhi, C. C. Tang, M. Mitome and K. Kurashima. (2007) Deformation-driven electrical transport of individual boron nitride nanotubes, Nano Letters, **7**, 632–637.

Barin I., Sauert F., Ernst S.R., Wang S.S., Thermochemical Data of Pure Substances, VCH Verlagsgesellschaft mbH: D–6940 Weinheim, 1989.

Baughman R.H., Zakhidov A.A., de Heer W.A., Carbon Nanotubes-The Route Toward Applications. Science, 2002, **297**, 787–792.

Boyen H. G., Fauth K., Stahl B., Ziemann P., Kastle G., Weigl F., Banhart F., Hessler M., Schutz G., Gajbhiye N. S., Ellrich J., Hahn H., Buttner M., Garnier M. G. and Oelhafen P. Electronic and magnetic properties of ligand-free FePt nanoparticles. Adv. Mater., 2005, **17** (5), 574–578.

Boyer S., Temperature Responsive Device. US Patent 1,793,303, Feb. 17, 1931.

Bromwich T. J., Kasama T., Chong R. K. K., Dunin-Borkowski R. E., Petford-Long A. K., Heinonen O. G. and Ross C. A.. Remanent magnetic states and interactions in nano-pillars. Nanotechnology, 2006, **17** (17), 4367–4373.

Bulatov V. V., Justo J. F., Cai W., Yip S., Argon A. S., Lenosky T., de Koning M., de la Rubia T. D., Parameter-Free Modelling of Dislocation Motion: the Case of Silicon. Philos. Mag. A, 2001, **81**, 1257–1281.

Cazaux J., Correlations between ionization radiation damage and charging effects in transmission electron microscopy. Ultramicroscopy, 1995, **60**, 411.

Chabala J.M., Oxide-growth kinetics and fractal-like patterning across liquid gallium surfaces, Phys. Rev. B, 1992, **46**, 11346–11357.

Chang, Y.-C., Y.-H. Liaw, Y.-S. Huang, T. Hsu, C.-S. Chang and T.-T. Tsong. (2008) In Situ Tailoring and Manipulation of Carbon Nanotubes, Small, **4**, 2195–2198.

Charlier J.C., Vita D.A., Blase X., Car R., Microscopic Growth Mechanisms for Carbon Nanotubes. Science, 1997, **275**, 647–649.

Che R. C., Takeguchi M., Shimojo M., Zhang W. and Furuya K., Fabrication and electron holography characterization of FePt alloy nanorods. Appl. Phys. Lett.,

2005, **87** (22), 223109.

Chokshi A. H., Rosen A., Karch J., and Gleiter H., On the Validity of the Hall-Petch Relationship in Nanocrystalline Materials. Scr. Metall., 1989, **23**, 1679–1684.

Clatterbuck D. M., Krenn C. R., Cohen M. L. and Morris J.W., Phonon Instabilities and the Ideal Strength of Aluminum. Phys. Rev. Lett. 2003, **91**, 135501.

Conrad H. and Narayan J., On the Grain Size Softening in Nanocrystalline Materials. Scripta Mater, 2000, **42**, 1025–1030.

Coronado E., Galan-Mascaros J. R., Gomez-Garcia C. J. and Laukhin V., Coexistence of ferromagnetism and metallic conductivity in a molecule-based layered compound. Nature, 2000, **408** (6811), 447–449.

Cumings, J., P. G. Collins and A. Zettl. (2000) Materials: Peeling and sharpening multiwall nanotubes, Nature, **406**, 586–586.

Dahmen U., Erni R., Radmilovic V., Kisielowski C., Rossell M. D. and Denes P., Philos. Background, status and future of the Transmission Electron Aberration Corrected Microscope project. Trans. R. Soc. A, 2009, **367** (1903), 3795–3808.

Dai H.J., Wong E.W., Lu Y.Z., Fan S.S., Lieber C.M., Synthesis and Characterization of Carbide Nanorods. Nature, 1995, **375**, 769–772.

Dehm G., Miniaturized Single-Crystalline FCC Metals Deformed in Tension: New Insights in Size-Dependent Plasticity. Prog. Mater. Sci., 2009, **54**, 664–688.

Deng Q. S., et al. Uniform Tensile Elongation of a Metallic Glass Prepared by FIB in the Limit of Suppressed Shear Banding. Acta Mater., 2011, **65**, 6511–6518.

Dessau D. S., Shen Z. X., King D. M., Marshall D. S., Lombardo L. W., Dickinson P. H., Loeser A. G., DiCarlo J., Park C. H., Kapitulnik A. and Spicer W. E., Key features in the measured band-structure of $Bi_2Sr_2CaCu_2O_{8+\delta}$: Flat bands at EF and Fermi-surface nesting. Phys. Rev. Lett., 1993, **71** (17), 2781–2784.

Diao J. K., Gall K. and Dunn M. L., Surface-Stress-Induced Phase Transformation in Metal Nanowires. Nat. Mater. 2003, **2**, 656–660.

Dong L., X. Tao, M. Hamdi, L. Zhang, X. Zhang, A. Ferreira and B. J. Nelson. (2009) Nanotube Fluidic Junctions: Internanotube Attogram Mass Transport through Walls, Nano Letters, **9**, 210–214.

Dong, Tao, L. Zhang, Zhang and B. J. Nelson. (2007) Nanorobotic Spot Welding: Controlled Metal Deposition with Attogram Precision from Copper-Filled

Carbon Nanotubes, Nano Letters, **7**, 58–63.

Dorozhkin P., Tovstong S., Golberg D., Zhan J.H., Ishikawa Y., Shiozawa M., Nakanishi H., Nakata K., Bando Y., A liquid-Ga-fitted carbon nanotube: A miniaturized temperature sensor and electrical switch. Small, 2005, **1**, 1088–1093.

Elkins K. E., Vedantam T. S., Liu J. P., Zeng H., Sun S. H., Ding Y. and Wang Z. L., Ultrafine FePt nanoparticles prepared by the chemical reduction method. Nano Lett., 2003, **3** (12), 1647–1649.

Erts, D., Olin, H., Ryen, L., Olsson, E., Th, ouml, eacute and Thölén, A. (2000) Maxwell and Sharvin conductance in gold point contacts investigated using TEM-STM, Physical Review B, **61**, 12725.

Frank, S., P. Poncharal, Z. L. Wang and W. A. de Heer. (1998) Carbon nanotube quantum resistors, Science, **280**, 1744–1746.

Frenkel J., Zur Theorie der Elastizit. Atsgrenze und der Festigkeit Kristallinischer Körper, Z. Phys. 1926, **37**, 572–609.

Fu Y., Sun M., Tian W. W., Wang J. B., Gao Y. H., Melting, Expansion Behavior and Electric Transport of In-Filling in MgO Nanotubes. J. Nanosci. Nanotechnol., 2012, **12**, 2718–2721.

Gabor D., Microscopy by reconstructed wave-fronts. Proc. R. Soc. Lond. A, 1949, **197** (1051), 454–487.

Gajdardziska-Josifovska M. and McCartney M. R., Elimination of thickness dependence from medium resolution electron holograms. Ultramicroscopy, 1994, **53** (3), 291–296.

Gao Y. H., Bando Y., Golberg D., Melting and expansion behavior of indium in carbon nanotubes. Appl. Phys. Lett., 2002, **81**, 4133–4135.

Gao Y. H., Bando Y., Liu Z.W., Golberg D., Nakanishi H., Temperature measurement using gallium-filled carbon nanotube nanothermometer, Appl. Phys. Lett., 2003, **83**, 2913–2915.

Gao Y. H., Bando Y., Nanothermodynamic analysis of surface effect on expansion characteristics of Ga in carbon nanotubes. Appl. Phys. Lett., 2002, **81**, 3966–3968.

Gao Y. H., Sun M., Su J., Zhi C.Y., Golberg D., Bando Y., Duan X.F., Liquid Ga–

filled carbon nanotube: miniaturized temperature sensor and electrical switch. Appl. Phys. Lett., 2011, **99**, 083112.

Gao Y. H., Zhang H. Y., Wang Y. J., Zhang Q. F., Han X. Y., Li Y. B., Liu Z. W., Zhan J. H., Golberg D., Dorozhkin P., Tovstong S., Huang D. X., Bando Y., Nanotube Thermometer. J. Chin. Electr. Microsc. Soc., 2008, 2752–166.

Gao Y.H., Bando Y., Carbon nanothermometer containing gallium. Nature, 2002, **415**, 599.

Gao, P., Z. C. Kang, W. Y. Fu, W. L. Wang, X. D. Bai and E. G. Wang. (2010) Electrically Driven Redox Process in Cerium Oxides, Journal of the American Chemical Society, **132**, 4197–4201.

Golberg D. et al., Nanomaterial Engineering and Property Studies in a Transmission Electron Microscopy, Adv. Mater., 2012, **24**, 177–194.

Golberg D., Bando Y., Kurashima K., Sato T., Synthesis and characterization of ropes made of BN multiwalled nanotubes, Scr. Mater., 2001, **44**, 1561–1565.

Golberg D., Costa P. M. F. J., Mitome M., Hampel S., Haase D., Mueller C., Leonhardt A., Bando Y., Copper-filled carbon nanotubes: Rheostatlike Behavior and femtogram copper mass transport. Adv Mater., 2007, **19**, 1937.

Goldstein A.N., Echer C.M., Alivisatos A.P., Melting in semiconductor nanocrystals. Science, 1992, **256**, 1425–1427.

Gray D.E., Billings B.H., Bleil D.F., Cook R.K., Crosswhite H.M., Frederiskse H.P.R., Lindsay R.B., Marion J.B., Zemansky M.W., Heat, 4–122 & Thermal expansion, 4–141 American Institute of Physics Handbook, 3rd ed. McGraw-Hill: New York, 1972.

Greer J. R. and Nix W. D., Nanoscale Gold Pillars Strengthened through Dislocation Starvation. Phys. Rev. B, 2006, **73**, 245410.

Guo H., Chen K., Oh Y. et al., Mechanics and Dynamics of the strain-Induced M1-M2 Structural Phase Transition in Individual VO_2 Nanowires. Nano Lett., 2011, **11**, 3207–3213.

Guo H., Yan P. F., Wang Y. B., Tan J., Zhang Z. F., Sui M. L. and Ma E., Tensile Ductility and Necking of Metallic Glass. Nat. Mater., 2007, **6**, 735–739.

Han W.Q., Bando Y., Kurashima K., Sato T., Synthesis of boron nitride nanotubes from carbon nanotubes by a substitution reaction. Appl. Phys. Lett., 1998, **73**,

3085–3087.

Han W.Q., Fan S.S., Li Q.Q., Hu Y.D., Synthesis of gallium nitride nanorods through a carbon nanotube-confined reaction. Science, 1997, **277**, 1287–1289.

Han X. D., Zhang Y. F., Zheng K., Zhang X. N., Zhang Z., Hao Y. J., Guo X. Y., Yuan J. and Wang Z. L., Low-Temperature In Situ Large Strain Plasticity of Ceramic SiC Nanowires and Its atomic-Scale Mechanism, Nano Lett., 2007, **7**, 452–457.

Han X. D., Zhang Z. and Wang Z. L., Experimental Nanomechanics of One-Dimensional Nanomaterials by In Situ Microscopy. Nano: Brief Reports and Reviews, 2007, **2**, 1–23.

Han X. D., Zheng K., Zhang Y. F., Zhang X. N., Zhang Z. and Wang Z. L., Low-Temperature In Situ Large-Strain Plasticity of Silicon Nanowires, Adv. Mater., 2007, **19**, 2112–2118.

Haque M. A. and Saif M. T. A., In-situ Tensile Testing of Nano-scale Specimens in SEM and TEM. Exp. Mech., 2002, **42**, 123–128.

Hébert C. and Schattschneider P., A proposal for dichroic experiments in the electron microscope. Ultramicroscopy, 2003, **96** (3–4), 463–468.

Hirsch P., Howie A., Nicholson R., Pashley D. W. and Whelan M. J., Electron Microscopy of Thin Crystals, Krieger Publishing Company, Malabar, 1967.

Howe J. M., Mori H. and Wang Z. L., In Situ High Resolution Transmission Electron Microscopy in the Study of Nanomaterials and Properties, MRS Bulletin, 2008, **33**, 115–121.

Huang J. Y., Chen S., Jo S. H., Wang Z., Han D. X., Chen G., Dresselhaus M. S. and Ren Z. F., Atomic-Scale Imaging of Wall-by-Wall Breakdown and Concurrent Transport Measurements in Multiwall Carbon Nanotubes, Phys. Rev. Lett., 2005, **94**, 236802.

Huang, J. Y., F. Ding and B. I. Yakobson. (2008) Dislocation Dynamics in Multi-walled Carbon Nanotubes at High Temperatures, Physical Review Letters, **100**, 035503.

Huang, J. Y., F. Ding, K. Jiao and B. I. Yakobson. (2007) Real Time Microscopy, Kinetics, and Mechanism of Giant Fullerene Evaporation, Physical Review Letters, **99**, 175503.

Huang, J. Y., L. Zhong, C. M. Wang, J. P. Sullivan, W. Xu, L. Q. Zhang, S. X. Mao, N. S. Hudak, X. H. Liu, A. Subramanian, H. Y. Fan, L. A. Qi, A. Kushima and J. Li. (2010) In Situ Observation of the Electrochemical Lithiation of a Single SnO2 Nanowire Electrode, Science, **330**, 1515–1520.

Huang, J. Y., S. Chen, Z. F. Ren, Z. Q. Wang, D. Z. Wang, M. Vaziri, Z. Suo, G. Chen and M. S. Dresselhaus. (2006) Kink Formation and Motion in Carbon Nanotubes at High Temperatures, Physical Review Letters, **97**, 075501.

Huang, J. Y., S. Chen, Z. F. Ren, Z. Wang, K. Kempa, M. J. Naughton, G. Chen and M. S. Dresselhaus. (2007) Enhanced Ductile Behavior of Tensile-Elongated Individual Double-Walled and Triple-Walled Carbon Nanotubes at High Temperatures, Physical Review Letters, **98**, 185501.

Iijima S., Helical Microtubules of Graphitic Carbon. Nature, 1991, **354**, 56–58.

Jin, C. H., J. Y. Wang, M. S. Wang, J. Su and L. M. Peng. (2005) In-situ studies of electron field emission of single carbon nanotubes inside the TEM, Carbon, **43**, 1026–1031.

Jin, C., H. Lan, K. Suenaga, L. Peng and S. Iijima. (2008) Metal Atom Catalyzed Enlargement of Fullerenes, Physical Review Letters, **101**, 176102.

Jin, C., K. Suenaga and S. Iijima. (2008) Vacancy Migrations in Carbon Nanotubes, Nano Letters, **8**, 1127–1130.

Junginger F., Klaui M., Backes D., Rudiger U., Kasama T., Dunin-Borkowski R. E., Heyderman L. J., Vaz C. A. F. and Bland J. A. C., Spin torque and heating effects in current-induced domain wall motion probed by transmission electron microscopy. Appl. Phys. Lett., 2007, **90** (13), 132506.

Kamihara Y., Hiramatsu H., Hirano M., Kawamura R., Yanagi H., Kamiya T. and Hosono H., Iron-based layered superconductor: LaOFeP. J. Am. Chem. Soc., 2006, **128** (31), 10012–10013.

Kamihara Y., Watanabe T., Hirano M. and Hosono H., Iron-based layered superconductor La[O_{1-x}Fx]FeAs ($x = 0.05 - 0.12$) with $Tc = 26$ K. J. Am. Chem. Soc., 2008, **130** (11), 3296–3297.

Kandel D., Kaxiras E., Microscopic Theory of Electromigration on Semiconductor Surfaces. Phys. Rev. Lett., 1996, **76**, 1114.

Kaxiras E. and Duesbery M. S., Free Energies of Generalized Stacking Faults in Si and Implications for the Brittle-Ductile Transition. Phys. Rev. Lett., 1993, **70**, 3752–3755.

Kiener D. and Minor A. M., Source Truncation and Exhaustion: Insights from Quantitative In Situ TEM Tensile Testing, Nano Lett., 2011, **11**, 3816–3820.

Kiener D., Hosemann P., Maloy S. A. & Minor A. M., In Situ Nanocompression Testing of Irradiated Copper, Nat. Mater., 2011, **10**, 608–613.

Koster V.H., Hensel F., Franck E.U., Kompressibilitat und thermische Ausdehnung des flussigen Galliums bis 600 °C und 2500 bar. Ber. Bunsenges. Phys. Chem., 1970, **74**, 43–46.

Kroto H.W., Heath J.R., O'Brien S.C., Curl R.F., Smalley R.E., C-60 Buckminsterfullerene. Nature, 1985, **318**, 162–163.

L. de Knoop. (2005) Investigation of iron filled multiwalled carbon nanotubes, Chalmers University of Technology (Dept. of Applied Physics)

Lang K. M., Madhavan V., Hoffman J. E., Hudson E. W., Eisaki H., Uchida S. and Davis. Imaging the granular structure of high-Tc superconductivity in underdoped $Bi_2Sr_2CaCu_2O_{8+\delta}$. Nature, 2002, **415** (6870), 412–416.

Lebegue S. Electronic structure and properties of the Fermi surface of the superconductor LaOFeP. Phys. Rev. B, 2007, **75** (3), 035110.

Legros M., Dehm G., Keller-Flaig R. M., Arzt E., Hemker K. J. and Suresh S., Dynamic Observation of Al Thin Films Plastically Strained in a TEM. Mater. Sci. Eng., 2001, **A15**, 463–467.

Legros M., Gianola D. S. and Hemker K. J., In Situ TEM Observations of Fast Grain-Boundary Motion in Stressed Nanocrystalline Aluminum Films, Acta Mater., 2008, **56**, 3380–3393.

Li S. Z., et al. High-Efficiency Mechanical Energy Storage and Retrieval Using Interfaces in Nanowires. Nano Lett., 2010, **10**, 1774–1779.

Li X. Y., Wei Y. J., Lu L., Lu K. and Gao H. J., Dislocation Nucleation Governed Softening and Maximum Strength in Nano-twinned Metals. Nature, 2010, **464**, 877–880.

Li Y. B., Bando Y., Golberg D., Liu Z.W., Ga-filled single-crystalline MgO nanotube: Wide-temperature range nanothermometer. Appl. Phys. Lett., 2003, **83**, 999–1001.

Li Y. B., Bando Y., Golberg D., Single-Crystalline In2O3 Nanotubes Filled with In. Adv. Mater., 2003, **15**, 581–585.

Lide D.R., Density of liquid elements surface tension of liquid elements vapor pressure, CRC Handbook of Chemistry and Physics, 71st ed. CRC: Ohio 1990–1991.

Lin L. T., Cui T. R., Qin L. C. and Washburn S., Direct Measurement of the Friction Between and Shear Moduli of Shells of Carbon Nanotubes, Phys. Rev. Lett., 2011, **107**, 206101.

Lipert K., Bahr S., Wolny F., Atkinson P., Weißker U., Mühl T., Schmidt O. G., Büchner B., Klingeler R., An individual iron nanowire-filled carbon nanotube probed by micro-Hall magnetometry. Appl. Phys. Lett., 2010, **97**, 212503.

Liu Z.W., Bando Y., Mitome M., Zhan J. H., Unusual Freezing and Melting of Gallium Encapsulated in Carbon Nanotubes. Phys. Rev. Lett., 2004, **93**, 095504.

Liu Z.W., Gao Y.H., Bando Y., Highly effective metal vapor absorbents based on carbon nanotubes. Appl. Phys. Lett., 2002, **81**, 4844–4846.

Liu, K. H., P. Gao, Z. Xu, X. D. Bai and E. G. Wang. (2008) In situ probing electrical response on bending of ZnO nanowires inside transmission electron microscope, Applied Physics Letters, **92**, 213105–213103.

Liu, K. H., W. L. Wang, Z. Xu, X. D. Bai, E. G. Wang, Y. G. Yao, J. Zhang and Z. F. Liu. (2009) Chirality-Dependent Transport Properties of Double-Walled Nanotubes Measured in Situ on Their Field-Effect Transistors, Journal of the American Chemical Society, **131**, 62–63.

Liu, X. H., L. Q. Zhang, L. Zhong, Y. Liu, H. Zheng, J. W. Wang, J. H. Cho, S. A. Dayeh, S. T. Picraux, J. P. Sullivan, S. X. Mao, Z. Z. Ye and J. Y. Huang. (2011) Ultrafast Electrochemical Lithiation of Individual Si Nanowire Anodes, Nano Letters, **11**, 2251–2258.

Liu, Y., H. Zheng, X. H. Liu, S. Huang, T. Zhu, J. W. Wang, A. Kushima, N. S. Hudak, X. Huang, S. L. Zhang, S. X. Mao, X. F. Qian, J. Li and J. Y. Huang. (2011) Lithiation-Induced Embrittlement of Multiwalled Carbon Nanotubes, ACS Nano, **5**, 7245–7253.

Liu, Y., N. S. Hudak, D. L. Huber, S. J. Limmer, J. P. Sullivan and J. Y. Huang. (2011) In Situ Transmission Electron Microscopy Observation of Pulverization of Aluminum Nanowires and Evolution of the Thin Surface Al_2O_3 Layers during

Lithiation-Delithiation Cycles, Nano Letters, **11**, 4188–4194.

Liu, Y., Z. Y. Zhang, X. L. Wei, Q. Li and L. M. Peng. (2011) Simultaneous Electrical and Thermoelectric Parameter Retrieval via Two Terminal Current-Voltage Measurements on Individual ZnO Nanowires, Advanced Functional Materials, **21**, 3900–3906.

Loram J. W., Mirza K. A., Cooper J. R. and Liang W. Y. Electronic specific heat of YBa2Cu3O6+x from 1.8 to 300 K. Phys. Rev. Lett., 1993, **71** (11), 1740–1743.

Lu L., Chen X., Huang X. and Lu K., Revealing the Maximum Strength in Nanotwinned Copper. Science, 2009, **323**, 607–610.

Luo W. D., Roundy D., Cohen M. L. and Morris J. W., Ideal Strength of BCC Molybdenum and Niobium. Phys. Rev. B, 2002, **66**, 094110.

Madec R., Devincre B., Kubin L., Hoc T., Rodney D., The Role of Collinear Interaction in Dislocation-Induced Hardening. Science, 2003, **301**, 1879–1882.

Mathur N., Nanotechnology-Beyond the silicon roadmap, Nature, 2002, **41**, 9573–574.

Michler J., Wasmer K., Meier S., Östlund F. and Leifer K., Plastic Deformation of Gallium Arsenide Micropillars under Uniaxial Compression at Room Temperature. Appl. Phys. Lett., 2007, **90**, 043123.

Millis A. J. and Monien H. Spin gaps and spin dynamics in $La_{2-x}Sr_xCuO_4$ and $YBa_2Cu_3O_{7-\delta}$. Phys. Rev. Lett., 1993, **70** (18), 2810–2813.

Minor A. M., Asif S. A., Shan Z., Stach E. A., Cyrankowski E., Wyrobek T. J. and Warren O. L., A New View of the Onset of Plasticity During the Nanoindentation of Aluminium, Nat. Mater., 2006, **5**, 697–702.

Mitchell T. E., Anderson P. M., Baskes M. I., Chen S. P., Hoagland R. G. and Misra A., Nucleation of Kink Pairs on Partial Dislocations: a New Model for Solution Hardening and Softening, Philos. Mag., 2003, **83**, 1329–1346.

Mompioua F., Legrosa M., Sedlmayrb A., Gianolac D. S., Caillarda D. and Kraft O., Source-Based Strengthening of Sub-Micrometer Al Fibers, Acta Mater., 2006, **60**, 977–983.

Nafari, A., J. Angenete, K. Svensson, A. Sanz-Velasco and H. Olin (2011). Combining Scanning Probe Microscopy and Transmission Electron Microscopy Scanning Probe Microscopy in Nanoscience and Nanotechnology 2. B. Bhushan,

Springer Berlin Heidelberg: 59-99.

Nagataki, A., T. Kawai, Y. Miyamoto, O. Suekane and Y. Nakayama. (2009) Controlling Atomic Joints between Carbon Nanotubes by Electric Current, Physical Review Letters, **102**, 176808.

Nelson, C. T., P. Gao, J. R. Jokisaari, C. Heikes, C. Adamo, A. Melville, S. H. Baek, C. M. Folkman, B. Winchester, Y. J. Gu, Y. M. Liu, K. Zhang, E. G. Wang, J. Y. Li, L. Q. Chen, C. B. Eom, D. G. Schlom and X. Q. Pan. (2011) Domain Dynamics During Ferroelectric Switching, Science, **334**, 968–971.

Novoselov K.S., Geim A.K., Morozov S.V., Jiang D., Zhang Y., Dubonos S.V., Grigorieva I.V., Firsov A.A., Electric field effect in atomically thin carbon films. Science, 2004, **306**, 666–669.

Ohnishi, H., Y. Kondo and K. Takayanagi. (1998) Quantized conductance through individual rows of suspended gold atoms, Nature, **395**, 780–783.

Östlund F., Howie P. R., Ghisleni R., Korte S., Leifer K., Clegg W. J. and Michler J., Ductile-Brittle Transition in Micropillar Compression of GaAs at Room Temperature. Phil. Mag., 2011, **91**, 1190–1199.

Östlund F., Malyska K. R., Leifer K., Hale L. M., Tang Y. Y., Ballarini R., Gerberich W. W., and Michler J., Brittle-to-Ductile Transition in Uniaxial Compression of Silicon Pillars at Room Temperature, Adv. Funct. Mater., 2009, **19**, 2439–2444.

Pan X.L., Fan Z.L., Chen W., Ding Y.J., Luo H.Y., Bao X.H., Enhanced ethanol production inside carbon-nanotube reactors containing catalytic particles. Nature Mater., 2007, **6**, 507–511.

Parkin S. S. P., Hayashi M., and Thomas L.. Magnetic domain-wall racetrack memory. Science, 2008, **320** (5873), 190–194.

Pederson M.R., Broghton J.Q., Nanocapillarity in fullerence tubules. Phys. Rev. Lett., 1992, **69**, 2689–2692.

Peng B., Locascio M., Zapol P. et al., Measurements of Near-Ultimate Strength for Multiwalled Carbon Nanotubes and Irradiation-Induced crosslinking Improvements, Nat. Nanotech., 2008, **3**, 626–631.

Petkov N., In Situ Real-Time TEM Reveals Growth, Transformation and Function in One-Dimensional Nanoscale Materials: From a Nanotechnology Perspective,

Nanotechnology, 2013, **2013**, 1–21.

Pfleiderer C., Uhlarz M., Hayden S. M., Vollmer R., von Lohneysen H., Bernhoeft N. R. and Lonzarich G. G., Coexistence of superconductivity and ferromagnetism in the d-band metal ZrZn2. Nature, 2001, **412** (6842), 58–61.

Polanyi M., Über die Natur des Zerreißvorganges. Z. Phys., 1921, **7**, 323–327.

Regan, B. C., S. Aloni, R. O. Ritchie, U. Dahmen and A. Zettl. (2004) Carbon nanotubes as nanoscale mass conveyors, Nature, **428**, 924–927.

Rodney D. and Phillips R., Structure and Strength of Dislocation Junctions: an Atomic Level Analysis. Phys. Rev. Lett., 1999, **82**, 1704–1707.

Rose H., Outline of a spherically corrected semiaplanatic medium-voltage transmission electron-microscope. Optik, 1990, **85** (1), 19–24.

Rous P. J., Bly D. N., Wind force for adatom electromigration on hetero- geneous surfaces. Phys. Rev. B., 2000, **62**, 8478.

Schattschneider P., Rubino S., Hébert C., Rusz J., Kune J., Novák P., Carlino E., Fabrizioli M., Panaccione G. and Rossi G. Detection of magnetic circular dichroism using a transmission electron microscope. Nature, 2006, **441** (7092), 486–488.

Schiotz J. and Jacobson K.W., A Maximum in the Strength of Nanocrystalline Copper. Science, 2003, **301**, 1357–1359.

Shan Z. W., Mishra R. K., Asif S. A. S., Warren O. L. and Minor A. M., Mechanical Annealing and Source-Limited Deformation in Submicrometre-Diameter Ni Crystals, Nat. Mater., 2008, **7**, 115–119.

Shechtman D., Blech I., D.Gratias and Cahn J., Metallic Phase with Long-Range Orientational Order and No Translational Symmetry. Phys. Rev. Lett., 1984, **53**, 1951–1953.

Shi C. X., Cong H. T., Tuning the coercivity of Fe-filled carbon-nanotube arrays by changing the shape anisotropy of the encapsulated Fe nanoparticles. J. Appl. Phys., 2008, **104**, 034307.

Shimizu K., Kimura T., Furomoto S., Takeda K., Kontani K., Onuki Y. and Amaya K., Superconductivity in the nonmagnetic state of iron under pressure. Nature, 2001, **412** (6844), 316–318.

Shpyrko O.G., Streitel R., Balagurusamy V. S. K., Grigoriev A.Y., Deutsch M.,

Ocko B. M., Meron M., Lin B.H., and Pershan P. S., Surface Crystallization in a Liquid AuSi Alloy, Science, 2006, **31**, 377.

Smith D. A., Holmberg V. C. and Korgel B. A., Flexible Germanium Nanowires: Ideal Strength, Room Temperature Plasticity, and Bendable Semiconductor Fabric. ACS Nano, 2010, **4**, 2356–2362.

Smith D. J., Dunin-Borkowski R. E., McCartney M. R., Kardynal B. and Scheinfein M. R., Interlayer coupling within individual submicron magnetic elements. J. Appl. Phys., 2000, **87** (10), 7400–7404.

Soldano C., Rossella F., Bellani V., Giudicatti S., Kar S., Cobalt Nanocluster-Filled Carbon Nanotube Arrays: Engineered Photonic Bandgap and Optical Reflectivity, ACS Nano, 2010, **4**, 6573–6578.

Sorbello R. S., Theory of electromigration. Solid State Phys., 1998, **51**, 159.

Stauffer D. D., Beaber A., Wagner A., Ugurlu O., Nowak J. L., Mkhoyan K. A., Girshick S. and Gerberich W. W., Strain-Hardening in Submicron Silicon Pillars and Spheres, Acta Mater., 2012, **60**, 2471–2478.

Su J., Sun M., Zhang X. H., Huang Y. L., Gao Y. H., Ga Filled Nanothermometers with High Sensitivity and Wide Measuring Range. J. Nanosci. Nanotechnol., 2012, **12**, 6397–6400.

Sun M. and Gao Y.H., Electrically driven gallium movement in carbon nanotubes. Nanotechnology, 2012, **23**, 065704.

Svensson, K., H. Olin and E. Olsson. (2004) Nanopipettes for Metal Transport, Physical Review Letters, **93**, 145901.

Takeguchi M., Shimojo M., Furuya K. Fabrication of magnetic nanostructures using electron beam induced chemical vapour deposition. Nanotechnology, 2005, **16** (8), 1321–1325.

Tang D. M., Ren C. L., Wang M. S., Wei X. L., Kawamoto N., Liu C., Bando Y., Mitome M., Fukata N. and Golberg D., Mechanical Properties of Si Nanowires as Revealed by In Situ Transmission Electron Microscopy and Molecular Dynamics Simulations, Nano Lett., 2012, **12**, 1898–1904.

Tang D. M., Yin L. C., Li F., Liu C., Yu W. J., Ho P. X.u, Wu B., Lee Y. H., Ma X. L. and Cheng H. M., Carbon Nanotube-Clamped Metal Atomic Chain, Proc. Natl. Acad. Sci. USA, 2010, **107**, 9055–9059.

Tibbetts G.G., Why are carbon filaments tubular? J. Cryst. Growth, 1984, **66**, 632–638.

Tokura Y. and Nagaosa N., Orbital physics in transition-metal oxides. Science, 2000, **288** (5465), 462–468.

Tonomura A., Matsuda T. and Endo J., Direct observation of fine structure of magnetic domain walls by electron holography. Phys. Rev. Lett., 1980, **44** (21), 1430–1433.

Van Dyck D., Van Aert S., den Dekker A. J. and van den Bos A., Is atomic resolution transmission electron microscopy able to resolve and refine amorphous structures. Ultramicroscopy, 2003, **98** (1), 27–42.

Van Swygenhoven H., Grain Boundaries and Dislocations. Science, 2002, **296**, 66–67.

Van Tendeloo G., Bals S., Van Aert S., Verbeeck J. and Van Dyck D., Advanced electron microscopy for advanced materials. Adv. Mater., 2012, **24** (42), 5655–5675.

Wang D., Li F. H. and Zou J., Distinguishing Glide and Shuffle Types for 60° Dislocation in Semicoductors by Field-emission HREM Image Processing. Ultramicroscopy, 2000, **85**, 131–139.

Wang L. H., Han X. D., Liu P., Yue Y. H., Zhang Z. and Ma E., In Situ Observation of Dislocation Behavior in Nanometer Grains, Phys. Rev. Lett., 2010, **105**, 135501.

Wang L. H., Zhang Z. and Han X. D., In Situ Experimental Mechanics of Nanomaterials at the Atomic Scale, NPG Asia Materials, 2013, **5**, 40.

Wang L.H., Liu P., Guan P.F., Yang M.J., Sun J.L., Cheng Y.Q., Hirata A., Zhang Z., Ma E., Chen M.W. and Han X.D., In Situ Atomic-Scale Observation of Continuous and Reversible Lattice Deformation beyond the Elastic Limit, Nat. Commun., 2013, **4**, 2413.

Wang, M. S., D. Golberg and Y. Bando. (2010) Carbon "Onions" as Point Electron Sources, ACS Nano, **4**, 4396–4402.

Wang, M. S., Q. Chen and L. M. Peng. (2008) Field-Emission Characteristics of Individual Carbon Nanotubes with a Conical Tip: The Validity of the Fowler-Nordheim Theory and Maximum Emission Current, Small, **4**, 1907–1912.

Wang, M. S., Q. Chen and L. M. Peng. (2008) Grinding a Nanotube, Advanced Materials, **20**, 724–728.

Wang, Z. L., P. Poncharal and W. A. de Heer. (2000) Nanomeasurements in Transmission Electron Microscopy, Microscopy and Microanalysis, **6**, 224–230.

Wang, Z. L., R. P. Gao, W. A. de Heer and P. Poncharal. (2002) In situ imaging of field emission from individual carbon nanotubes and their structural damage, Applied Physics Letters, **80**, 856–858.

Wei B., Zheng K., Ji Y., Zhang Y. F., Zhang Z. and Han X. D., Size-Dependent Bandgap Modulation of ZnO Nanowires by Tensile Strain, Nano Lett., 2012, **12**, 4595–4599.

Wei X. L., Wang M. S., Bando Y., Golberg D., Tensile Tests on Individual Multi-Walled Boron Nitride Nanotubes, Adv. Mater., 2010, **22**, 4895–4899.

Wei, W., Y. Liu, Y. Wei, K. L. Jiang, L. M. Peng and S. S. Fan. (2007) Tip cooling effect and failure mechanism of field-emitting carbon nanotubes, Nano Letters, **7**, 64–68.

Wei, X. L., Y. Bando and D. Golberg. (2012) Electron Emission from Individual Graphene Nanoribbons Driven by Internal Electric Field, ACS Nano, **6**, 705–711.

Wen H. H., Mu G., Fang L., Yang H. and Zhu X. Y., Superconductivity at 25K in hole-doped (La(1-x)Sr4(x))OFeAs. Europhys. Lett., 2008, **82** (1), 17009.

Williams D. B., and Carter C. B., Transmission Electron Microscopy 2nd edn, Plenum, New York, 1996, 112.

Wu B., Heidelberg A. and Boland J. J., Mechanical Properties of Ultrahigh-Strength Gold Nanowires. Nat. Mater., 2005, **4**, 525–529.

Wu Y. Z., Won C., Scholl A., Doran A., Zhao H. W., Jin X. F. and Qiu Z. Q., Magnetic stripe domains in coupled magnetic sandwiches. Phys. Rev. Lett., 2004, **93** (11), 117205.

XianLong, W., C. Qing, L. Yang and P. LianMao. (2007) Cutting and sharpening carbon nanotubes using a carbon nanotube 'nanoknife', Nanotechnology, **18**, 185503.

Xiong F., Liao A.D., Estrada D., Pop E., Low-Power Switching of Phase-Change Materials with Carbon Nanotube Electrodes. Science, 2011, **332**, 568–570.

Xu, Z., X. D. Bai, E. G. Wang and Z. L. Wang. (2005) Dynamic in situ field emission of a nanotube at electromechanical resonance, Journal of Physics-Condensed Matter, **17**, L507–L512.

Xu Z., Y. Bando, W. L. Wang, X. D. Bai and D. Golberg (2010) Real-Time In Situ HRTEM-Resolved Resistance Switching of Ag_2S Nanoscale Ionic Conductor, ACS Nano, **4**, 2515–2522.

Yang, Y., P. Gao, S. Gaba, T. Chang, X. Pan and W. Lu. (2012) Observation of conducting filament growth in nanoscale resistive memories, Nat Commun, **3**, 732.

Yip S., Nanocrystals: the Strongest Size. Nature, 1998, **391**, 532–533.

Yue Y. H., Liu P., Deng Q. S., Ma E., Zhang Z. and Han X. D., Quantitative Evidence of Crossover toward Partial Dislocation Mediated Plasticity in Copper Single Crystalline Nanowires, Nano Lett., 2012, **12** (8), 4045–4049.

Yue Y. H., Liu P., Zhang Z., Han X. D. and Ma E., Approaching the Theoretical Elastic Limit in Copper Nanowires. Nano Lett., 2011, **11**, 3151–3155.

Yue Y. H., Wang L. H., Zhang Z. and Han X. D., Cross-over of the Plasticity Mechanism in Nanocrystalline Cu. Chin. Phys. Lett. 2012, **29**, 066201.

Zhan J.H., Bando Y., Hu J.Q., Golberg D., Nakanishi H. J., Liquid Gallium Columns Sheathed with Carbon:Bulk Synthesis and Manipulation. J. Phys. Chem. B, 2005, **109**, 11580–11584.

Zhan J.H., Bando Y., Hu J.Q., Liu Z.W., Yin L.W., Golberg D., Fabrication of metal-semiconductor nanowire heterojunctions. Angew. Chem. Int. Ed., 2005, **44**, 2140–2144.

Zhang Y. F., Han X.D., Zheng K., Zhang Z., Direct Observation of Super-Plasticity of Beta-SiC Nanowires at Low Temperature, Adv. Fun. Mater., 2007, **17**, 3435–3440.

Zhang Z. H., Wang X. F., Xu J. B., Muller S., Ronning C. and Li Q., Evidence of intrinsic ferromagnetism in individual dilute magnetic semiconducting nanostructures. Nat. Nanotechnol., 2009, **4** (8), 523–527.

Zhang, Z. Y., C. H. Jin, X. L. Liang, Q. Chen and L. M. Peng. (2006) Current-voltage characteristics and parameter retrieval of semiconducting nanowires, Applied Physics Letters, 89.

Zhao J., Huang J. Q., Wei F., Zhu J., Mass Transportation Mechanism in Electric-Biased Carbon Nanotubes. Nano Lett., 2010, **10**, 4309–4315.

Zheng H., Cao A. J., Weinberger C. R., Huang J. Y., Du K., Wang J. B., Ma Y. Y., Xia Y. N. and Mao S. X., Discrete Plasticity in Sub-10nm-Sized Gold Crystals. Nat. Commun., 2010, **1**, 144.

Zheng K., Wang C. C., Cheng Y. Q., Yue Y. H., Han X. D., Zhang Z., Shan Z. W., Mao S. X., Ye M. M., Yin Y. D. and Ma E., Electron-Beam-Assisted Superplastic Shaping of Nanoscale Amorphous Silica, Nat. Commun., 2010, **1**, 24.

Zhong, L., X. H. Liu, G. F. Wang, S. X. Mao and J. Y. Huang. (2011) Multiple-Stripe Lithiation Mechanism of Individual SnO2 Nanowires in a Flooding Geometry, Physical Review Letters, 106.

Zhu T. and Li J., Ultra-Strength Materials, Prog. Mater. Sci., 2010, **55**, 710–757.

Zhu Y. and Espinosa H. D., An Electromechanical Material Testing System for In Situ Electron Microscopy and Applications. Proc. Natl. Acad. Sci. USA, 2005, **102**, 14503–14508.

8
Helium Ion Microscopy

Daniel Fox, Hongzhou Zhang

8.1 Introduction

Microscopy is a widely used and resilient field which is essential for the analysis of typography, structure and composition right down to the atomic scale. The desired output from any form of microscope is either quantifiable or easily interpretable results, or both. The scanning electron microscope (SEM) provides images which are generally very simple to interpret as they appear similar to images which we are used to interpreting in the real world on a daily basis. A modern field emission SEM typically has a $\sim$1 nm resolution under optimised operating conditions. Within this chapter we will introduce a new type of scanning beam microscope released in 2007 by Carl Zeiss. The SEM is a mature technology which has been pushed near the edge of its fundamental limits. The diffraction limit which makes optical microscopy unsuitable for nanoscale imaging has now become an issue for the SEM. Furthermore, even if the electrons can be focused to a sub-nanometre size probe, the electron beam still suffers severe deflections within the sample. This has the effect of producing signal far from the probe location which leads to a reduction in resolution. The Orion Plus helium ion microscope (HIM) has been developed to overcome these issues. The HIM uses a very well-focused probe of helium ions. These helium ions also remain more collimated within the sample than an electron beam. Ultimately this leads to a better imaging resolution in the HIM. The HIM also provides enhanced surface sensitivity and material contrast compared to an SEM. The HIM has the added benefit of providing functionality unachievable in an SEM. The HIM can image a wider range of samples such as biological materials and polymer based samples due to its efficient charge compensation system. Finally, as it is an ion beam microscope similar to the gallium focused ion beam (FIB) the HIM displays strong grain contrast and can also be used to pattern and modify surfaces at the nanoscale. Of course limitations do exist, as with any microscopy technique, and these will also be discussed throughout the course of this chapter.

The HIM has many of the features which have made SEM analysis so accessible such as straight forward sample preparation and image interpretation. This coupled with its enhanced surface sensitivity and few angstrom resolution give the HIM the potential to become a powerful microscopy technique in its own right.

The helium ion microscope (HIM) requires lenses to control the trajectory of the ions and to focus them to a single point on a given plane. Modern electron microscopes use electromagnetic lenses for this purpose. However, at the acceleration voltages used in the HIM, electromagnetic lenses are less effective at focusing ions than electrostatic lenses. The reason for this is that the force experienced by a charged particle when travelling through an electromagnetic lens is directly proportional to its velocity, $\boldsymbol{F} = q\boldsymbol{v} \times \boldsymbol{B}$. After an electron is accelerated through a 30 kV potential its velocity is more than 85 times greater than that of a helium ion accelerated through the same potential. This is due to the large difference in mass between the two particles. In electrostatic lenses the force experienced by a particle is independent of its velocity, $\boldsymbol{F} = q\boldsymbol{E}$. Electrostatic lenses are therefore more effective at focusing the relatively slow helium ions. A simplified illustration of the two lens system of the HIM is shown in Fig. 8.1.1.

Electrostatic lenses are not without their disadvantages. Electrostatic lenses introduce more severe aberrations than electromagnetic lenses. The focal length is highly dependent on the distance of the particle from the optic axis of the lens, and its angle to the axis. The effects of the strong spherical aberration can be minimised by introducing a small beam limiting aperture to reject off axis particles. This leads to a trade-off between beam current and probe size. Very small beam limiting apertures (5 μm diameter) are commonly used in the helium ion microscope for the highest resolution imaging. Such small aperture sizes provide a problem in the electron microscope due to the effects of diffraction. Helium ions have a larger mass and a shorter de Broglie wavelength, preventing them from being diffracted by apertures of this size.

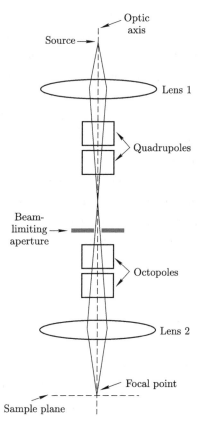

Fig. 8.1.1   An illustration of the principal ion-optics components in the column of the helium ion microscope.

Chromatic aberration is another aberration which limits the ultimate resolution of the microscope. Chromatic aberration is a serious limitation to the probe sizes achievable in conventional ion beam systems. In the FIB the liquid metal ion source (LMIS) generates gallium ions with a relatively large energy spread of $5 \sim 10$ eV (Gianuzzi and Stevie, 2006). As these ions travel at different velocities through the lens, the faster ions spend less time in the electrostatic field and experience less deflection, slower ions are more strongly focused due to their extended interaction time with the field of the lens. Ions of different energies are therefore focused on different planes. This results in the ions being focused on a disc area on the sample, instead of a single point. This chromatic aberration currently limits the resolution of the FIB to $3 \sim 5$ nm at best. However, the source in the HIM, which will be discussed in detail in Section 8.2.1 is a gas field ion source (GFIS). This type of ion source can produce an ion beam with an energy spread of just 0.5 eV, leading to a significant reduction in chromatic aberration over the FIB.

The minimal aberration contributions from Cs and Cc result in an absence of significant aberrations in the system. As long as the octopoles are optimised by the user to minimise the effect of astigmatism of the probe, then sub nanometre imaging is easily achieved with < 0.35 nm routinely demonstrated. This resolution is not only dependent on the aberration minimisation, but also on the size of the source and the size of the region in the sample from which signal is generated. This will be discussed in detail in Section 8.2.2.

8.2 Principles

In this section we will introduce some of the physics which make this microscope unique and allow it to capture highly detailed images at high magnification. We will discuss design features and attachments which enhance the tool's functionality, as well as some of the practical considerations and limitations which must be understood.

8.2.1 The source

The source consists of the emitter, a cryogenically cooled, sharpened tungsten tip, which is held at a positive bias of $25000 \sim 35000$ V during operation. The bias is adjusted outside of this range, from $10000 \sim 45000$ V, during the trimer building process. This process allows the selective removal of atoms from the tip until a stable formation of a pyramid of atoms with a terminating layer of just three atoms is achieved. The voltage is then adjusted within the operating range in order to maximise the electric field strength at this trimer of atoms. Helium gas is then flowed into the region of the emitter. The electric field is only large enough to cause significant ionisation of helium atoms at the location of the trimer atoms; emission is effectively confined to this region. The ionised helium is then accelerated by the electric field through a hole in the extractor plate and continues down the column of the microscope (see Fig. 8.2.1(a)).

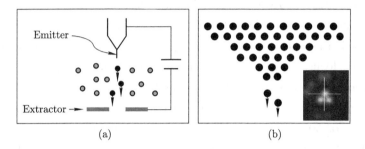

Fig. 8.2.1 The source of the helium ion microscope. (a) The blue circles represent helium gas while the red circles represent helium ions. (b) The structure of the tip of the emitter, the green circles are tungsten and the red circles are helium which has been ionised at the tip of the emitter. Inset is an SFIM image of the trimer at the emitter tip.

The upper column quadropoles can be used to scan the helium ion beam across the beam limiting aperture. When a uniform conductive surface is present on the sample plane, this process produces an emission image showing the regions of the most intense emission on the emitter. Such a scanning field ion microscope (SFIM) image is shown inset in Fig. 8.2.1(b). This SFIM image allows us to observe the trimer, and its intensity, when adjusting the source bias to find the best imaging voltage (BIV). The BIV is the voltage at which the emitter is kept in order to obtain the most beam current from the source. The SFIM image is also used to centre the emission from a single atom onto the optic axis of the microscope, ensuring the signal comes from the smallest sized region possible.

The source unit on the more recent HIM features a floating extractor. This extractor plate can be positively biased, having the effect of reducing the acceleration voltage of the microscope while allowing the tip bias to be kept constant. This feature provides variable voltage operation of the tool. The acceleration voltage can be set as low as 5 kV.

8.2.2 Beam-sample interaction

We have seen that the source of the HIM produces signal from a region almost entirely concentrated to a single atom. We also know that the lens system introduces very little aberration into the beam. The final factor affecting the resolution of the microscope is the volume within the sample from which signal is generated. Even with a probe focused onto a sub-nanometer area, the interaction of this probe with the sample can cause signal to be generated from a region far from the point of incidence of the probe on the sample. In Fig. 8.2.2(a) we can see the simulated distribution of 30 keV electrons after interaction within a silicon sample. This was modelled using the CASINO software package (Hovongton and Drouin, 2013). As well as penetrating deep into the sample, the beam electrons experience large angle deflections and backscattering in the sample. This will lead to electrons being collected by the detector which do not carry information about the sample in the direct vicinity of the probe. In comparison, the interaction of a 30 keV helium ion beam in a silicon sample is shown on the right hand side in Fig. 8.2.2(a). The ion interactions were modelled using the SRIM software package (Ziegler et al., 2008). The helium ion interaction volume is far more localised within the sample.

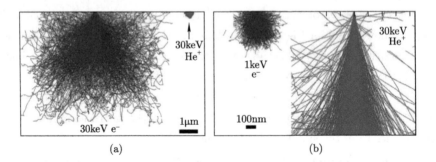

Fig. 8.2.2 (a) 30 keV electron beam interaction volume compared with 30 keV helium ion beam interaction volume in silicon. (b) 1 keV electron beam interaction volume compared with part of the 30 kV helium ion beam interaction volume in silicon.

As well as the helium ions producing secondary electrons from a more localised region in the sample, Petrov and Vyvenko demonstrated that the secondary electrons generated by 30 keV helium ions have a lower energy compared to those generated by 30 keV electrons (Petrov and Vyvenko, 2011). The majority of secondary electrons produced by the ~30 keV helium ion beam have an energy of less than 2 eV. The energy distribution of these secondary electrons is shown in the graph in Fig. 8.2.3. This results in the mean escape depth of the secondary electrons produced by helium ions being limited to typically 2 nm or less (Ramachandra et al., 2009; Rodenburg et al., 2010). Secondary electrons produced in the SEM have a higher average energy of ~5 eV. This is an important fact when we consider the use of low energy electron beams to reduce the size of the interaction volume. In Fig. 8.2.2(b) we see the simulated distribution of 1 keV electrons in silicon. The electrons are confined to a much smaller volume at this reduced energy and do not penetrate as deep as the 30 keV helium ion beam. However, the electrons generated by the helium ion beam are low energy secondary electrons and only escape from within a few nanometres of the surface of the sample. This results in the volume from which secondary electrons are generated by the helium ion beam still being much smaller than that of a low energy electron beam. As well as the fact that a low energy electron beam still cannot produce signal from a smaller region than the helium ion beam, low energy electron beams cannot usually be focused to such confined regions as higher energy electron beams. Imaging with low energy electron beams also generally limits the user to a very short working distance (< 5 mm).

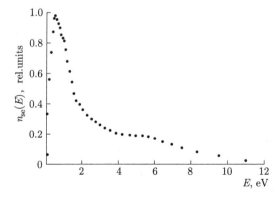

Fig. 8.2.3 Distribution of the energy of secondary electrons produced by a ~30 keV helium ion beam.

8.2.3 Signal generation

The signals generated within the sample by the helium ion beam are different than those generated by the electron beam of an SEM. The electron beam interacts strongly with sample electrons, transferring a large amount of energy to these electrons. These sample electrons are known as SE1 electrons and can then go on to escape from the sample and be detected. Beam electrons can also backscatter from deep within the material; backscattered electron imaging has the benefit of containing material (Z) contrast as the electrons have undergone an elastic collision with an atomic nucleus. However, backscattered electrons can also generate SE2 electrons. SE2 electrons are generated from a region not directly at the beam probe, leading to a reduction in resolution. The backscattered electrons (BSE) can also generate SE3 electrons after leaving the sample, SE3 electrons are generated from surfaces within the chamber such as the pole piece of the final lens. SE3 electrons again lead to a loss of resolution. An illustration of the regions from which these various types of secondary electrons are generated in the SEM is shown in Fig. 8.2.4.

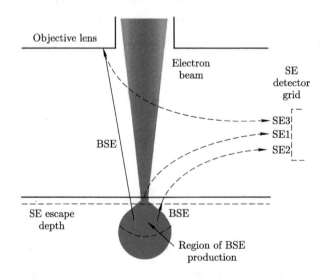

Fig. 8.2.4 Illustration of the different secondary electron types generated in the SEM.

Experiments have been performed by Chen et al. which demonstrate the difference in signals produced in the SEM and the HIM. By using regions of silicon implanted with different doses of gallium, they were able to study the contrast for a range of detectors in the SEM and HIM as a function of doping concentration as shown in Fig. 8.2.5. Fig. 8.2.5(a) shows the images acquired in an SEM and a HIM using different detection modes. The detection configuration and the detectors are the same for HIM-ET and SEM-ET, however, they exhibit different contrast (ET refers to the Everhart-Thornley detector which is a chamber mounted secondary electron detector). The contrast in HIM-ET is quite similar to SEM-Inlens which indicates the doping contrast depends on the detection configuration. The signal detected in the HIM is most likely the same as that detected in SEM using an in-lens detector. This is strong evidence that the signal generated by helium-ion irradiation is quite different from that generated by electron beam irradiation. A quantitative comparison is shown in Fig 8.2.5(b) in terms of the correlation between the contrast and the doping concentration.

The in-lens detector in the SEM is designed to accept primarily SE1 type electrons. This leads to a better imaging resolution than the in-chamber secondary electron detector can provide. The reason the secondary electron detector in the HIM chamber produces contrast similar to that of the SEM in-lens detector is due to the helium ion interaction and signal generation process within the sample.

Within the helium ion microscope the signal produced is predominantly due to the generation of SE1 electrons. The helium ions do have the possibility of backscattering from atomic nuclei within the sample. However, this backscatter probability is low, especially for low Z materials (typically around 1%). The number of SE2 and SE3 electrons is therefore low. The number of SE2 and SE3 electrons will increase when imaging materials with a large atomic mass as the probability of producing backscattered helium ions increases.

The $\sim$30 keV helium ions used to image in the HIM transfer the majority of their energy to electrons within the sample. The large difference in mass between the helium ions and the electrons results in only a small loss of energy from the beam particle. Each helium ion can therefore go on to produce several secondary electrons. As this energy transfer is small the secondary electrons have a lower

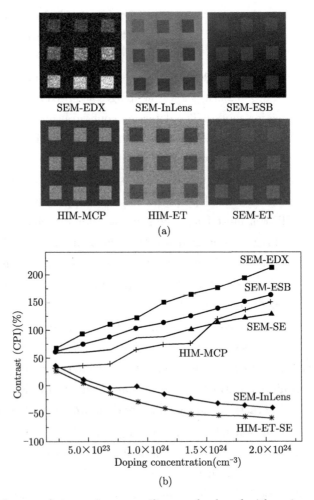

Fig. 8.2.5 (a) Images of nine regions on a silicon wafer doped with various concentrations of gallium. The six different images were acquired using a range of detectors in both SEM and HIM in order to compare contrast. (b) Quantitative comparison between contrast and doping concentration.

energy than those produced within the SEM. This leads to reduced depth from which the secondary electrons can escape. This results in the detected signal in the HIM originating from a very small volume near the surface of the sample.

One area where the HIM is limited compared to the SEM is in the quantative analysis of sample composition. The electron beam in the SEM can displace core shell electrons in the sample. The position left by the ejected core shell electron is filled by an electron from a higher energy level. This electron releases a characteristic energy which can be detected and used to identify the material composition. This energy is released either in the form of an X-ray or an Auger electron. This signal is not produced abundantly in the HIM due to the low X-ray generation cross-section, alternative compositional analysis techniques have been developed and will be discussed in a later section.

The primary beam of helium ions also causes the generation of optical signals in a process similar to cathodoluminescence known as ionoluminescence. This signal can be used to map the optical activity of samples with a very high spatial resolution and is complementary to secondary electron imaging.

8.2.4 Resolution

The resolution in any microscope is determined by a combination of a number of factors. In a scanning microscope the size of the imaging probe is one of the most important factors and is primarily limited by the following parameters. Firstly, the size of the source of the beam; in this case the helium ions are primarily generated within the immediate region of a single atom. This is the smallest source achievable, resulting in a source brightness value of $\sim 5 \times 10^9$ A $\cdot$ cm^{-2} $\cdot$ sr^{-1}, similar to that of a cold field emission electron gun, and approximately three orders of magnitude better than that of other ion beam sources. Secondly, lens aberrations have the effect of spreading the imaging probe into a disc shape. Chromatic aberration is not a large issue due to the low energy spread of the ions and spherical aberration can be largely removed by using a small beam limiting aperture. Thirdly, the effect of diffraction from the beam limiting aperture; this is an issue in with lower energy electron microscopes as the wavelength of the electrons is comparable to the size of aperture diameter. This limits the minimum aperture size that can be used in electron microscopes. This diffraction effect is reduced for beam particles with shorter de Broglie wavelengths. Using the non-relativistic equation:

$$\lambda_{\text{deB}} = \frac{hc}{\sqrt{2E_K m_0 c^2}} \tag{8.2.1}$$

As h (Planck's constant) and c (the speed of light) are constants we can calculate that 30 keV helium ions have a wavelength around 85 times smaller than 30 keV electrons. The final probe size at the specimen plane is found by adding all of the beam broadening contributions in quadrature. Each of these effects on the probe size can be well compensated for within the HIM.

Another factor on which the resolution depends is the volume within the sample from which signal is generated. We have seen that the helium ions generate mainly SE1 type electrons which carry the highest resolution information. These electrons also have a very low energy and therefore a limited escape depth. The ability to form a small probe and also the fact that electrons are generated from a small region near the surface of the sample have led to a demonstrated imaging resolution of 0.24 nm.

8.2.5 Depth of field

The depth of field (DOF) is the largest difference in height between two objects which remain in focus in an image. When imaging large, curved objects it is advantageous to keep the entire object in focus. This requires a large DOF. The depth of field is related to the convergence angle, α, of the beam on the sample. A highly convergent beam will become poorly focused with any small change in distance (z) along the optical axis, as shown in Fig. 8.2.6. Conversely, a more collimated beam with a small convergence angle will remain better focused over a larger range of z. A beam with a small convergence angle can then keep features of different heights in better focus than a highly convergent beam. Very small apertures are used in the HIM compared to the SEM. The HIM is also not limited to such short working distances as the SEM, especially low voltage SEM. Therefore α and z can both be adjusted to give a very large DOF in the HIM.

The images in Fig. 8.2.7 are from coated cancer cells. Fig. 8.2.7(a) and (d) are HIM and SEM images respectively. They were acquired at regions close to the edges of the cells where the height of the sample changes abruptly with a large magnitude. As the cells are about of the same size, we can assume that the curvature at the edge is more or less of the same order of magnitude. If we take two areas with a fixed distance, we can compare the depth-of-fields. Fig. 8.2.7(b) is the focused area

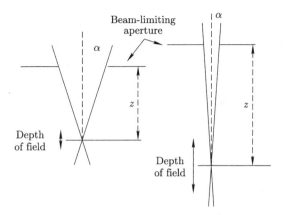

Fig. 8.2.6 Illustration of the effect of convergence angle and working distance on depth of field. On the left is a small DOF due to a large convergence angle and a small working distance (z). On the right is a much larger DOF due to a reduced convergence angle and a larger working distance.

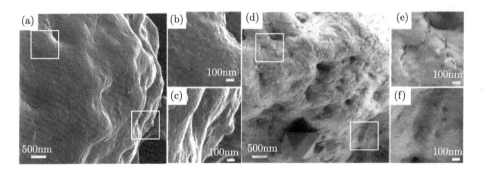

Fig. 8.2.7 (a) HIM image of a highly curved edge of a Caco-2 cell. The regions highlighted in (b) and (c) show no loss of focus for areas at different heights. (d) SEM image of the edge of another Caco-2 cell taken at the same magnification as (a). The highlighted area in (e) is in focus, however the area at a different height shown in (f) is clearly more blurred and out of focus than (e).

during observation and it shows detailed features of the coating film such as the granular features and the nanosized cracks. These details are still clearly seen in Fig. 8.2.7(c). SEM images at focus (Fig. 8.2.7(e)) can also reveal this information;

however in an area at a different height (Fig. 8.2.7(f)) the detail is clearly blurred. The large depth-of-field of the HIM enables the observation of the whole cancer cell without losing important local details with a large range of sample height.

8.2.6 Sample damage

The helium ion microscope shares many similarities with the SEM; it is however still a focused ion beam microscope. The 30 keV helium ions cause atomic displacements within the sample, whereas an SEM cannot. At high doses these displacements lead to amorphisation and removal of surface material by sputtering. The extension of sample damage was investigated by Fox et al. (2012). They prepared a ~100 nm thick silicon lamella by FIB lift-out. The lamella sidewall was then exposed to a high dose scan of helium ions at a glancing angle. This experimental set-up was used in order to observe the damage caused and the extension of the damage at high resolution in a transmission electron microscope (TEM). The silicon lamella after FIB lift-out is shown in Fig. 8.2.8(a). The orientation of the sample to the helium ion beam is illustrated in Fig. 8.2.8(b). The helium ion beam milled a hole through the sample as seen in Fig. 8.2.8(c) and (d). Fig. 8.2.8(c) is a bright-field TEM image, Fig. 8.2.8(d) is a high-angle annular dark field (HAADF) image. Above the hole a wedge of material was produced, marked I in Fig. 8.2.8(d). Below the hole a circular region with extensive damage was produced by the introduction of helium bubbles into the material (II in Fig. 8.2.8(d)). A SRIM simulation of the distribution of 35 keV helium ions in silicon is shown inset in Fig. 8.2.8(d). The simulated distribution of helium ions agrees well with the observed position of bubbles within the sample. Two obvious effects of high dose helium ion irradiation have been highlighted so far, the sputtering of material which can be used to pattern nanoscale features and the introduction of helium bubbles beneath the sample surface. Further to these effects, an extension of damage was observed around these regions. This area is not as badly affected by the beam as it is not at, or directly below, the exposed region. The reason for this damage extension is that the ion beam does experience high angle deflections; these events are less frequent than in an SEM but at high irradiation dose these ions have an observable effect. The amorphous region extends ~100 nm beyond the exposed area. Again this is a reasonable result based on the

SRIM simulation.

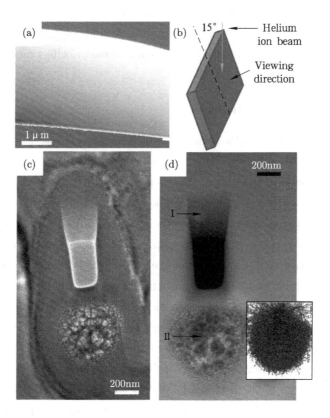

Fig. 8.2.8 (a) SEM image of a silicon lamella after FIB lift-out. (b) Illustration of the geometry of the helium ion beam irradiation. The red arrow represents the helium ions which are incident on the face of the sample at an angle of 15°. (c) Bright field TEM image of the area modified by helium ions. (d) HAADF image of the modified area. I shows the location of the wedge shape and II shows the circular area with bubbles. Inset is a SRIM simulation of 35 keV helium ions in silicon with the same scale as the image.

The effect of high dose irradiation in the HIM is made clear by these results. What we can see is essentially a cross-section of the effect of the irradiation on the sample. The milling effect is demonstrated, the subsurface implantation of ions is also directly observed. And finally the lateral extension of the damage caused by the beam is observed to extend well beyond the region of the beam probe. All of these effects must be considered when exposing a sample to high beam doses. In fact these effects are generally present to some extent and may need to be considered when even minimal beam damage may have a detrimental effect on the final sample required.

8.2.7 Charge compensation

As discussed earlier the HIM generally produces several secondary electrons for each incident ion. In an electrically conductive material this is not an issue as new electrons can flow into the region. However in electrically insulating materials this local charging effect is not naturally compensated for by the sample. This charging effect is also observed in the SEM however it is build-up of negative charges that occurs in the SEM, this is not easily compensated effectively. In the HIM the local build-up of positive charge can be effectively neutralised by a broad beam electron flood gun. The flood gun is pulsed on between scans of the ion beam in order to replenish the area of the sample with electrons. Fig. 8.2.9 (a) below shows the imaging of a polymer surface at high magnification that can be achieved after optimising the flood gun. Fig. 8.2.9(b) is an image from the same location acquired immediately after Fig. 8.2.9(a) but with the flood gun switched off. It is clear that the charging nature of the material leaves the sample depleted of electrons and it becomes difficult to collect any signal without the charge compensation given by

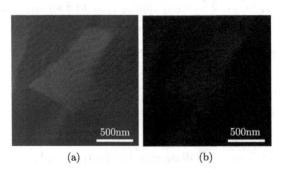

Fig. 8.2.9 (a) A nano-flake embedded in a polymer surface. (b) The same area as (a) imaged immediately after switching off the electron flood gun.

the flood gun.

8.2.8 Contamination

In an SEM the observation of contamination can generally be greatly reduced by working at a high beam energy. This is due to the increased signal generation volume at high beam energies. Less signal is being produced from the near surface area and the SEM is operating with less surface sensitivity. However, in the HIM the signal is generated from within a few nanometres of the surface. Let's then consider the effect of a build-up of a 2 nm thick carbon contamination layer due to hydrocarbons which have been polymerised by the beam. This layer is now the region from which most of the signal will be generated. We are no longer imaging the original sample surface. In order to minimise the effect of hydrocarbon contamination each Orion tool is fitted with a plasma cleaner. This plasma cleaner uses an air plasma to clean samples in the load lock before introducing them into the chamber. The chamber itself is also plasma cleaned, typically on a weekly basis, in order to further minimise contamination. Unfortunately many of the samples which are most interesting to image in the HIM, such as biological samples or polymer samples, cannot be plasma cleaned as this would cause them to be damaged. The effects of contamination are most obvious when working at high magnification. The sample surface may become completely coated in a single high resolution scan. This process is illustrated in Fig. 8.2.10(a), a high magnification scan was acquired and then a second scan was acquired at a lower magnification in order to show the effect of the first scan. Another commonly observed effect on an unclean sample is the "icicles" which can build up around the edge of an image while scanning. This is shown in Fig. 8.2.10(b). Again these are an indication of a build-up of contamination.

It is clearly advantageous to have such inherent surface sensitivity in the HIM for imaging under ideal conditions. However the sensitivity of the tool requires the operator to take serious care when considering what sources of contamination their sample may be exposed to during preparation. Cleaning steps such as plasma cleaning, baking or UV exposure are a necessity when working with the HIM.

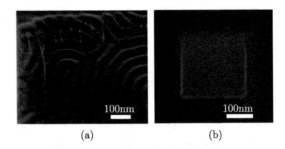

Fig. 8.2.10 (a) High resolution image of a block co-polymer sample. "Icicles" of contamination can be seen growing from the top of the image. (b) Image of a square of contamination deposited by a single high magnification scan of the beam.

8.3 Imaging techniques

In this section we will present alternative imaging techniques to the standard secondary electron imaging mode. These modes of imaging can provide complementary information to the secondary electron images, as well as providing quantitative sample analysis.

8.3.1 Backscattered ion imaging

Although the HIM cannot produce backscattered electrons which contain Z contrast, backscattered helium ions are produced. These ions also scatter from the nucleus of an atom within the sample in a process which is essentially a low energy version of Rutherford backscattered analysis. The backscattered ion yield is therefore roughly proportional to the Z number of the element from which the sample is composed. A graph of backscatter yield versus sample composition is shown in Fig. 8.3.1 (Kostinski and Yao, 2011). The probability of such an event is low for low Z materials, however for heavier metals the probability of a backscatter event can rise to 10%. It is clear from the graph that this probability does not depend linearly on Z, instead the graph is oscillating. This non-linearity means that this technique cannot be used to quantitatively identify materials, but does give an indication of composition for very different Z number materials.

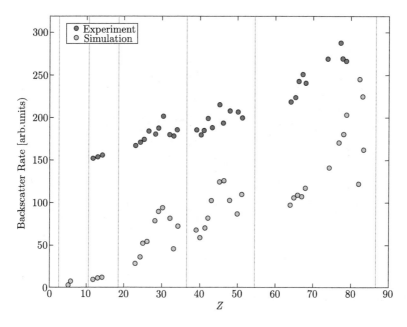

Fig. 8.3.1 The simulated and experimental non-monotonic variation of backscattered helium ion rate as a function of atomic mass (Z).

As this imaging technique is not particularly sensitive to surface topography it can provide material contrast. An interesting phenomenon which can be studied and even exploited with this imaging mode is that of ion channelling. Ions can channel through a material with a reduced probability of a nuclear scattering event when a crystal is aligned on axis with the incident ion beam. This results in the ability to identify different crystal grains. Backscattered ion imaging is achieved with an annular micro channel plate (MCP) detector which is inserted above the sample and below the pole piece of the column. This has the effect of limiting the working distance at which the sample can be placed as the stage must be lowered before the detector can be inserted. Backscattered ion images and secondary electron images can be captured simultaneously and provide complementary information about the sample. Below in Fig. 8.3.2 are images of a tungsten filament captured in the helium ion microscope. Fig. 8.3.2(a) is a secondary electron image showing the surface topography. Fig. 8.3.2(b) is a backscattered helium image showing strong grain contrast.

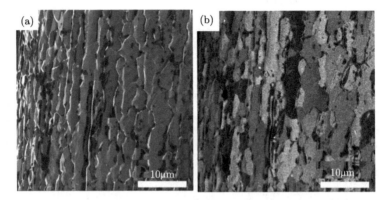

Fig. 8.3.2 A Tungsten filament. (a) Secondary electron image produced by the primary helium ion beam showing surface topography. (b) Backscattered ion image from the same location showing strong grain contrast.

8.3.2 Rutherford backscattering spectrometry

It was discussed in a previous section that the HIM has a low X-ray generation cross-section. This means that we cannot use X-ray detection for compositional analysis. In order to overcome this limitation alternative compositional analysis methods have been investigated. Backscattered ion energies can be detected and analysed in order to determine material composition or thickness. This is a low energy form of Rutherford backscattering spectrometry (RBS). In theory this works well but many practical limitations exist. The detector developed for this analysis is a modified silicon drift detector (SDD) with a resolution of $\sim$4 keV. It is the inherent design of this detector which prevents further gains in energy resolution. Furthermore, we know that as the ion travels through a sample it gradually gives up most of its energy to the sample electrons. This means that if an ion backscatters from deep within the sample it will reach the detector with less energy than an ion which backscatters from near the surface, even if the ions scatter from the same type of nucleus. This will lead to a large background at low energy in the spectrum for bulk materials. This technique therefore finds its niche application in thin film analysis.

These films must have large uniform areas, such as tens of square microns, in order to get sufficient signal to the detector. The acquired spectrum from the tool must then be modelled in the SIMNRA software package in order to fit the peaks and then material composition or thickness may be identified.

8.3.3 Transmission mode

Another detection configuration which may be employed within the HIM is the detection of transmitted ions. This mode is not currently a standard option on the microscope, but a homemade version may be constructed by placing a high SE yield material beneath the sample and angling it towards the detector. Transmission mode is particularly useful when using the helium ion beam to direct write nanopores into thin membranes of material. In this case transmission can be used to confirm when penetration through the membrane has occurred. Transmission mode is not suited to materials below a certain thickness; generally a region less than 10 nm thick and a hole show almost no contrast difference. Shown below in Fig. 8.3.3 is an array of 5 nm diameter nanopores milled into a SiN membrane, the scale bar is 50 nm (Yang et al., 2011).

Fig. 8.3.3 Transmission helium ion image of an array of 5 nm pores in a silicon nitride membrane. Scale bar is 50 nm.

8.4 Applications

This section comprises a review of the systems to which helium ion microscopy has been applied. The applications of the microscope are shown to be highly diverse. There are many record breaking applications which have already been demonstrated and these will be highlighted here.

8.4.1 Biological material imaging

The helium ion microscope provides the unique ability to directly image the surface of biological materials due to its efficient charge compensation system. A well optimised electron flood gun allows the observation of nanostructures at the surface of the material. Fig. 8.4.1 shows a high magnification image of the uncoated surface of a dried human platelet. This type of sample would have to be coated in a thin layer of metal in order to be clearly imaged in an SEM.

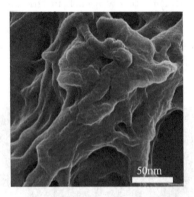

Fig. 8.4.1   High magnification image of an uncoated platelet surface demonstrating the charge compensation ability of the flood gun.

The metal coating typically applied to biological samples, and non-conductive materials in general, can be avoided in the HIM. This allows detailed surface analysis to be done with the HIM. The application of metal coatings obscures such surface detail as seen in Fig. 8.4.2. Fig. 8.4.2 shows two groups of cancer cells. Fig. 8.4.2(a) is a HIM image of cancer cells which have been coated with a layer of gold to compensate for the charging effect. Fig. 8.4.2(b) is a HIM image of a cluster of the same type of cells at the same magnification. However in Fig. 8.4.2(b) the electron flood gun is used for charge compensation. It can clearly be seen that the application of a gold coating layer buries much of the surface information and fills voids on the surface. Any measurements of surface porosity or surface area would be greatly affected by the coating layer.

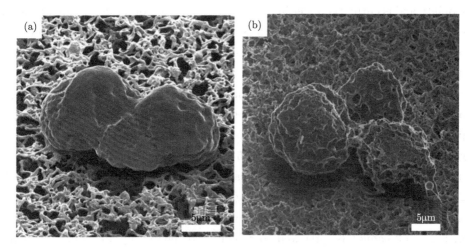

Fig. 8.4.2 (a) HIM image of cancer cells which have been coated with a layer of gold. (b) HIM image of uncoated cancer cells showing unobscured surface detail.

8.4.2 Nanopatterning and nanomodification

The helium ion microscope is a focused ion beam microscope, much like the gallium ion FIB. Although the beam-sample interactions in both microscopes are quite different, the HIM still has the ability to remove sample material by the process of sputtering. The sputtering yield refers to the number of atoms removed per incident ion. In the gallium FIB the sputtering yield for an ion energy of 30 keV in a silicon target is 2.28. In the HIM, again with a 30 keV ion energy, the sputtering yield is 0.0485, approximately 50 times less than the gallium ion sputtering yield. This low yield of sputtered ions makes the HIM ideally suited to the accurate removal of small volumes of material. The FIB cannot mill so accurately at such a nanoscale due to its inferior resolution and high sputtering yield. The FIB is however more suitable for removing relatively large volumes of material, which the HIM cannot. The very well-focused probe of helium ions in the HIM, coupled with the low backscatter yield, allow the HIM to define features as small as 4 nm.

Graphene is a recently isolated material which consists of a single atomic layer of carbon atoms bound in a honeycomb structure. Graphene has a very high carrier mobility and is currently under intense research interest for a range of applications. One limitation of graphene which must be overcome is its lack of an electronic band-gap. This means that a semiconductor device made from graphene could not be switched to an "off" state. However, it has been found that a band-gap can be created in graphene by confining graphene to ribbons with specific edge orientations. The width of the nanoribbon is inversely proportional to the band-gap produced. A band-gap of 0.5 eV at room temperature has been demonstrated by a graphene nanoribbon with a width of 2.5 nm (Tapaszto et al., 2008). The graphene nanoribbons in Fig.8.4.3 were all fabricated with the HIM (Pickard and Scipioni, 2014). The graphene nanoribbon in Fig. 8.4.3(a) has a width of 5 nm. The ten nanoribbons in Fig. 8.4.3(b) are each 10 nm wide. The graphene nanoribbons shown in Fig. 8.4.3 cannot be prepared so rapidly and reliably by any other technique (Zhou and Loh, 2010). These ribbons have exciting potential in next generation electronic devices.

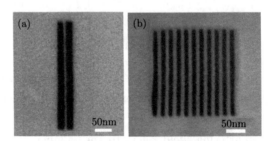

Fig. 8.4.3 (a) A 5 nm wide ribbon of suspended graphene. (b) A high density comb of 10 nm wide ribbons in suspended graphene.

In Fig. 8.3.3 the ability of the HIM to reliably pattern 5 nm nanopores was shown. The same group have demonstrated nanopores with diameters as small as 3.7 nm. Such structures have applications in DNA sequencing (Schneider et al.,

2010) and as gas separation membranes (Jiang et al., 2009).

The HIM has the ability to write complex patterns by uploading a binary image into the software program. An example of a pattern of the Trinity College Dublin logo generated by the tool is shown in Fig. 8.4.4. The integration of an external scan control system is required for the writing of patterns where more control over the beam is required. Parameters which have an effect on the milling which require an external scan control system include the direction of scanning or even the order in which features are milled.

Fig. 8.4.4 Pattern of the Trinity College Dublin logo milled into a thin platinum film. This pattern was produced by uploading a binary image into the basic patterning software on the system interface.

While the HIM has the ability to sputter material from a surface and mill very well defined features, the process does have some limitations. 30 keV helium ions incident on a silicon surface will come to rest at an average depth of ∼280 nm [SRIM]. The dose required for milling is several orders of magnitude higher than that typically required for imaging. At these high doses the large number of helium ions which have built up under the surface produce strain in the sample. The helium build up will go on to produce nano-bubbles like those shown in Fig. 8.4.5(a) (Livengood et al., 2009). Eventually these bubbles join to form large subsurface cavities as shown in Fig. 8.4.5(b). This subsurface void formation is a problem when attempting to pattern bulk materials. The result of this limitation is that nano-milling in the HIM is almost entirely confined to membranes or other thin film materials. When the sample to be milled is sufficiently thin, i.e. less than the average stopping range of the helium ions, then most of the beam particles pass through the sample. The formation of bubbles can then be avoided and milling of larger areas can be performed.

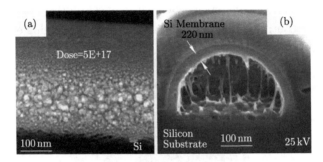

Fig. 8.4.5 (a) TEM cross section image of silicon implanted with a dose of 5×10^{17} helium ions/cm^2. (b) SEM image of a cross section of a bubble produced by implanting the silicon sample with 1.3×10^{18} ions/cm^2.

Fluorine assisted etching can be used to increase the rate of material removal and reduce the issue of sub-surface implantation and void formation. Fluorine can be introduced into the region of the sample surface by a gas injection needle. This feature is not included as standard on the tool. Fluorine has the undesirable effect of continuing to spontaneously etch the sample even with the ion beam blanked. In order to attempt to control this effect the chamber is quickly flushed with an inert gas after milling to remove the fluorine and prevent further etching. This approach to increasing the rate of milling is not ideal for well controlled nano-structuring.

8.4.3 Gas injection system

The HIM can be fitted with an aftermarket gas injection system (GIS). The GIS is used to introduce a gas species into the region of the sample surface through a retractable needle. This gas species is then decomposed by the ion beam, activating the key component of the gas. The gases most commonly used allow deposition of metals such as platinum or tungsten and insulators such as silicon oxide. This deposition process is known as ion beam induced deposition or IBID. Injection of a fluorine compound such as XeF$_2$ can also be achieved. This gas is decomposed by the ion beam and the fluorine gas then facilitates an enhanced material removal rate when attempting to mill a sample. The GIS also uses a purge gas, typically nitrogen, to flush away injected gases when they are no longer required. As discussed earlier, the fluorine assisted etching can help to reduce the sub-surface void formation and subsequent swelling effect which is associated with the high dose irradiation used during milling. This is due to the lower ion dose required to remove a volume of material during fluorine assisted etching. This enhanced material removal rate is evidenced on the TaN film shown in Fig. 8.4.6 (Alkemade and van Veldhoven, 2011). In Fig. 8.4.6(a) fluorine has facilitated the removal of material from the six lines exposed by the ion beam. The pattern has a 12 nm half-pitch. In Fig. 8.4.6(b) the same ion dose has had only a minor effect on the sample and has not resulted in significant milling.

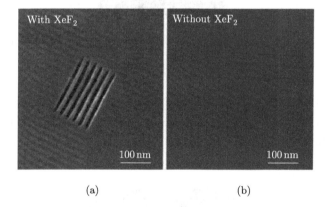

Fig. 8.4.6 (a) A TaN film with six lines exposed to helium ion irradiation in a fluorine gas environment. (b) The same material exposed to the same ion dose, but without the presence of fluorine.

Decomposition of metals by the ion beam has been demonstrated to produce features within close proximity. This indicates that the precursor decomposition is quite well confined to the region of incidence of the ion beam. Scipioni et al. (2011) deposited a series of parallel lines with a decreasing pitch for each pair. The series of deposited platinum lines are shown in Fig. 8.4.7. The lines could be completely resolved down to a pitch of 16 nm.

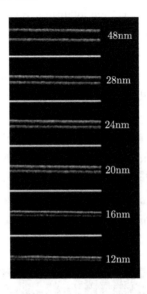

Fig. 8.4.7 A series of pairs of platinum lines deposited by IBID with a decreasing pitch. The two lines can still be fully resolved at 16 nm pitch.

More complex patterns can be deposited with the use of an external beam scan control system. An array of platinum nanopillars with a pitch of 50 nm is shown in Fig. 8.4.8 (Hill et al., 2012). This image clearly shows the high density of features which can be directly written onto a surface by IBID.

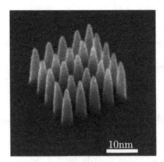

Fig. 8.4.8 Platinum nanopillar array with a pitch of 50 nm written by IBID.

8.4.4 Lithography

Electron beam lithography (EBL) is a well-established method for patterning nanoscale structures and devices. In EBL a thin layer of resist is spun onto a substrate. The resist is then exposed by the electron beam. Depending on the type of resist used (either negative or positive) either the exposed or the unexposed areas can be removed by chemical etching. This technique has a typical pitch resolution limit of about 60 nm. EBL also suffers from the "proximity effect". This effect is due to various beam scattering mechanisms that occur within the resist. When a dense matrix of features is patterned, each feature also has a contribution from the exposure of their neighbours which causes the features to be larger than intended. This contribution is due to the forward scattering of beam electrons in the resist layer and backscattered electrons from the substrate.

Helium ion beam lithography has been investigated. The helium ion beam remains well collimated in the resist layer leading to a reduction in forward scattering. There are also a greatly reduced number of backscatter events in the HIM. These factors result in smaller feature sizes, greater pattern densities and no observable proximity effect. In Fig. 8.4.9(a) a series of lines have been written in a 5 nm thick HSQ resist layer (Sidorkin et al., 2009). These lines have a 6 nm line width and a 15 nm pitch. Fig. 8.4.9(b) and (c) show dot arrays, again written in a 5 nm thick HSQ layer. These arrays have dot sizes of 6 ± 1 nm and pitches of 24 nm and 14 nm respectively.

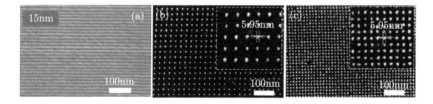

Fig. 8.4.9 Features produced by helium ion beam lithography on 5 nm thick HSQ resist. (a) A series of lines with 6 nm line widths and a pitch of 15 nm. (b) A dot array with 6 nm dot size and 24 nm pitch. (c) A dot array with 6 nm dot size and 14 nm pitch. No proximity effect correction was required.

These patterns demonstrate feature sizes and densities which are beyond the limits of EBL technology. These patterns require no complicated dose correction to account for the proximity effect observed in EBL. The exposure time is also reduced for HIM lithography when compared with EBL at the same beam energy. This is due to two reasons. Firstly the 30 keV helium ions in the HIM yield more secondary electrons per beam particle than the SEM. And secondly the HIM produces secondary electrons with lower energies which increases the resist sensitivity. These effects combined result in a 4.4 times increased exposure rate in the HIM for HSQ resist.

8.4.5 Surface chemistry sensitivity

The helium ion microscope produces lower energy secondary electrons than an SEM. These secondary electrons have a short mean free path within the sample. As well as not being able to travel far within the sample, the electrons need to reach the surface of the sample with a sufficient energy to overcome the work function of the material. Only then can they escape into the vacuum of the chamber and be collected by the detector. The secondary electron yield in the HIM is therefore more sensitive to surface modification than an SEM. These modifications may be in the form of physical implantation or chemical modification. These processes modify the work function of the surface of the sample and alter the number of electrons which can overcome this barrier. An example of this enhanced sensitivity is shown in Fig. 8.4.10 (Scipioni et al., 2008). Fig. 8.4.10(a) is a HIM image of a self-assembled organic monolayer of the chemical NBPT which has been exposed with e-beam lithography. The terminal group of the exposed areas is modified from NO_2 to NH_2. There is no change to the surface topography, the sample remains flat. The exposed circles and ovals can clearly be distinguished from the surrounding, unexposed area. Fig. 8.4.10(b) is an image of the same sample taken in an SEM, no contrast is observed due to the reduced surface sensitivity of SEM. These images clearly show the ability of the HIM to resolve monolayer chemical modifications which are difficult to resolve by other methods.

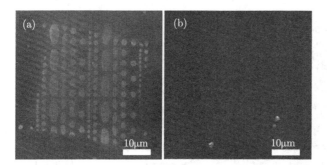

Fig. 8.4.10 Self-assembled monolayer of NBPT exposed with e-beam lithography which modifies the terminal group from NO_2 to NH_2. (a) HIM image showing the monolayer sensitivity of the tool. (b) SEM image shows no contrast.

8.4.6 Dopant contrast analysis

The physical doping of materials is a widespread approach to locally modifying the properties of the material. In the semiconductor industry the dimensions of critical features are now just a hundred atoms in length. Higher resolution quantitative dopant mapping with high sensitivity is critical for the development of future devices. Jepson et al. (Jepson et al., 2009) investigated the ability of the HIM to resolve different doping concentrations. They used boron doped silicon layers with seven different dopant concentrations ranging from 4×10^{15} to 5×10^{19} cm^{-3}. The sample is shown in Fig. 8.4.11(a). The brighter lines on the right are the more highly doped areas. It is clear that the HIM does have the ability to distinguish regions with different dopant concentrations. Due to the enhanced resolution of the HIM it may provide a better solution to SEM dopant analysis. Fig. 8.4.11(b) is a graph of the contrast detected as a function of the doping level. From this graph we can see that the HIM shows more sensitivity to dopant concentration changes than SEM at high doping levels. The SEM is however more sensitive to low dopant concentrations.

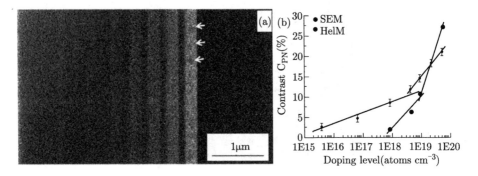

Fig. 8.4.11 (a) HIM image of a silicon surface with seven lines with a range of boron dopant concentrations. The most heavily doped lines are those with the highest intensities on the right. (b) A graph of the image contrast observed in both the SEM and HIM for a range of doping levels.

8.5 Current/Future developments

This section presents some of the potential of the HIM for future applications. These features are not readily available on the standard tool but have been implemented on individual systems for specific applications.

Near-UHV operation

We have shown the helium ion microscope to be a very surface sensitive microscope. While samples can be prepared with the greatest of care in terms of avoiding contamination, the chamber of the microscope is not an ultra-high vacuum (UHV) environment. The chamber of the HIM is at best vacuum $\sim 3 \times 10^{-4}$ Pa. UHV applies to pressures below 10^{-7} Pa, three orders of magnitude better than the HIM chamber pressure. For many surface sensitive applications, such as scanning tunnelling microscopy (STM), a UHV chamber is required in order to reduce the effects of contaminants and surface adsorbates on the sample. The resolution of the HIM has been shown to be as good as 0.24 nm. With this kind of resolution we would expect to be able to resolve the lattice structure of some materials with large lattice parameters. One such material is catalase. Catalase is an enzyme with a crystal structure with lattice spacings of 8.75 nm and 6.85 nm. We would expect to resolve this structure relatively easily in the HIM. However, due to the presence of an adsorbed gas layer on the sample in the non-UHV HIM chamber, a small amount of beam scattering occurs just above the sample surface. This effect is not generally noticeable when imaging, but when pushing the tool to acquire lattice resolution images this effect appears to have a detrimental result on the images acquired. Channelling of the beam through the crystal structure of the catalase is required in order to observe a lattice image. However this channelling effect is very sensitive to surface contaminants. In fact it appears to be so affected by the unavoidable surface adsorbates that no successful lattice resolved images have been demonstrated on the standard HIM tool.

A near-UHV (NUHV) Orion chamber has been produced and is currently in operation in the University of Twente, The Netherlands. The process of building a NUHV chamber requires more than a simple upgrade. It requires a different chamber entirely. A diffusion pump is fitted to the chamber and copper gaskets are used in place of rubber O-rings to achieve the improvement in pressure. The base chamber pressure of the NUHV Orion is $\sim 10^{-6}$ Pa, an improvement of two orders of magnitude over the chamber pressure in the standard Orion system. This NUHV chamber vacuum results in improved channelling contrast, and even allows the observation of the catalase lattice structure. A lattice resolution image of catalase acquired on the NUHV system is shown in Fig. 8.5.1 (Lysse and Carl Zeiss).

Fig. 8.5.1 An image of the resolved lattice structure of a catalase crystal acquired in the customised near-UHV Orion system.

The NUHV chamber also results in a reduced sputtering rate for some samples such as gold nanoparticles. This makes it possible to deliver a higher dose to a small area of the sample when imaging in order to acquire high resolution images with greater contrast. This improvement in contrast and channelling results in images like the one shown in Fig. 8.5.2. Striking channelling contrast from the structural defects in the gold crystal is clearly observed.

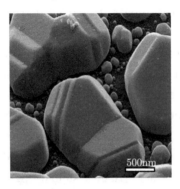

Fig. 8.5.2   An image of a standard gold on carbon sample. This image was acquired in customised near-UHV Orion system. Strong channelling contrast is observed from the various structural defects in the gold nanoparticles.

8.6 Conclusion

This chapter has detailed the fundamental principles behind the helium ion microscope. These have been compared primarily to the principles of SEM. The HIM has demonstrated resolution beyond that of an SEM. The reasons for the improved resolution and enhanced surface sensitivity have been discussed. Some limitations which exist and considerations which must be made have also been presented. The HIM represents the next generation of scanning beam microscopes for high resolution analysis of bulk samples. The modes of operation which can be implemented in the tool to provide alternative forms of sample imaging and analysis were also discussed. The analysis of backscattered helium spectroscopy may provide a useful tool in the field of thin film analysis. Novel imaging techniques are provided by backscattered helium imaging allowing the direct observation of strong material contrast and grain boundaries.

The applications of these enhanced imaging and analysis techniques have been reviewed. The HIM has been shown not only to produce highly detailed high magnification images, but also nanostructures materials which can only be produced in such a way by this tool. The ability of the microscope to selective pattern and modify surfaces and also subsequently image the modification in situ gives the HIM many potential applications which only a focused ion beam microscope can achieve. This new field of helium ion microscopy is still in its infancy but already it has been used to push beyond the boundaries of mature technologies such as electron beam lithography. The work reviewed in this chapter provides sufficient evidence that the HIM is a tool with many applications. As the technology matures many more applications are likely to be discovered, and more records are likely to be broken. Even at this early stage, helium ion microscopy has already proven itself a powerful technique from which any microscopy laboratory would surely find benefit.

References

Alkemade, P. F. A. and van Veldhoven, E. Deposition, milling and etching with a focused helium ion beam, in Nanofabrication: techniques and principles. Springer 2011.

Fox, D., Chen, Y., Faulkner, C. C., and Zhang H., Nano-structuring, surface and bulk modification with a focused helium ion beam. Beilstein Journal of Nanotechnology, 2012, **3**, 579–585.

Hill, R., Notte, J. A., Scipioni, L., Scanning helium ion microscopy, in Advances in imaging and electron physics, 2012, **170**, 1–272.

Hovongton, P. and Drouin, D., monte Carlo SImulation of electron trajectory in solids, 2013, http://www.gel.usherbrooke.ca/casino/index.html.

Jepson, M. A. E., Inkson, B. J., Liu, X., Scipioni, L., Rodenburg, C., Quantitative dopant contrast in the helium ion microscope. Europhysics Letters, 2009, **86**, 26005.

Jiang, D., Cooper, V. R., and Dai, S., Porous graphene as the ultimate membrane for gas separation. Nano Letters, 2009, **9** (12), 4019–4024.

Kostinski, S. and Yao, N., Rutherford backscattering oscillation in scanning helium-ion microscopy. Journal of Applied Physics, 2011, **109**, 064311.

L. Gianuzzi and F. A. Stevie. Introduction to Focused Ion Beams, chapter The Basic FIB Instrument, page 7. Springer, 2006.

Livengood, R., Tan, S., Greenzweig, Y., Notte, J., McVey, S., Subsurface damage from helium ions as a function of dose, beam energy, and dose rate. Journal of Vacuum Science & Technology B: Microelectronics and Nanometer Structures, 2009, **27** (6), 3244–3249.

Lysse, A., Carl Zeiss Inc. Peabody, MA.

Petrov, Y. and Vyvenko, O., Secondary electron emission spectra and energy selective imaging in helium ion microscope. In Society of Photo-Optical Instrumentation Engineers (SPIE) Conference Series, volume 8036 of Society of Photo-Optical Instrumentation Engineers (SPIE) Conference Series, May 2011.

Pickard, D. and Scipioni, L., Graphene nano-ribbon patterning in the orion plus, Carl Zeiss application note, 2014, http://www.tcd.ie/Physics/ultramicroscopy/teaching/PY5019/HIM/AN_Nano-Pore_Milling_with_the_Helium_Ion_Microscope.pdf.

Ramachandra, R., Griffin, B., and Joy, D., A model of secondary electron imaging in the helium ion scanning microscope. Ultramicroscopy, 2009, **109** (6), 748–757.

Rodenburg, C., Liu, X., Jepson, M. A. E., Zhou, Z., Rainforth W. M., and Rodenburg J. M., The role of helium ion microscopy in the characterisation of complex three-dimensional nanostructures. Ultramicroscopy, 2010, **110** (9), 1178–1184.

Schneider, G. F., Kowalczyk, S. W., Calado, V. E., Pandraud G., Zandbergen H. W., Vandersypen L. M. K., and Dekker C., DNA translocation through graphene nanopores. Nano Letters, 2010, **10** (8), 3163–3167.

Scipioni, L., Sanford, C., van Veldhoven, E., Maas, D., A design-of-experiments approach to characterizing beam-induced deposition in the helium ion microscope. Microscopy Today, 2011, **5**, 22–26.

Scipioni, L., Stern, L. A., Notte, J., Sijbrandij, S., Griffin, B., Helium ion microscopy. Advanced materials and processes, 2008, **166** (6), 27–30.

Sidorkin, V., van Veldhoven, E., van der Drift, E., Alkemade, P., Salemink H., Maas D., Sub-10-nm nanolithography with a scanning helium beam. Journal of Vacuum Science & Technology B, 2009, **27**, 18–20.

Tapaszto, L., Dobrik, G., Lambin, P., and Biro, L. P., Tailoring the atomic structure of graphene nanoribbons by scanning tunnelling microscope lithography. Nature Nanotechnology, 2008, **3** (7), 397–401.

Yang, J., Ferranti, D. C., Stern, L. A., Sanford, C. A., Huang, J., Ren, Z., Qin, L.C., and A. R., Hall. Rapid and precise scanning helium ion microscope milling of solid-state nanopores for biomolecule detection. Nanotechnology, 2011, **22** (28): 285–310.

Zhou, Y. and Loh, K. P., Making patterns on graphene. Advanced Materials, 2010, **22** (32), 3615–3620.

Ziegler, J. F., Biersack, J. P., and Ziegler, M. D., Stopping and range of ions in matter. SRIM Co., 2008.

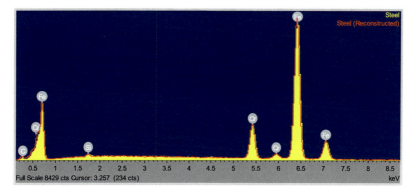

Fig.2.3.3 Comparison between simulated spectrum (red line) and acquired spectrum.

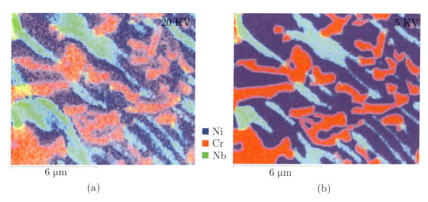

(a)   (b)

Fig.2.3.6 Overlapped Ni, Cr and Nb map showing the better spatial resolution at low kV.

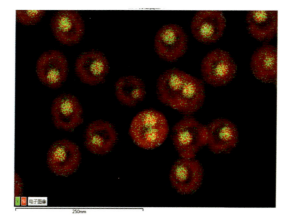

Fig.2.3.7 High spatial resolution element mapping shows the distribution of Si and Y in core/shell nanoballs. The diameter of the balls is about 80 nm. Image courtesy of Wu. W.

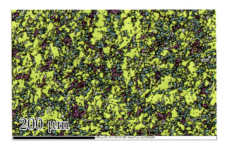

Fig.2.3.18 Texture map of a deformed Cu sheet: 52% cubic fiber (yellow), 24% gamma fiber (teal) and 9% ⟨110⟩ fiber (purple) with 20 degree deviation, image courtesy of Oxford Instruments application notes.

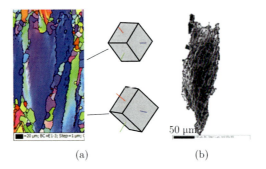

Fig.2.3.20 (a) IPF orientation map of deformed nickel, and (b) local misorientation map showing the sub-grain low angle boundaries.

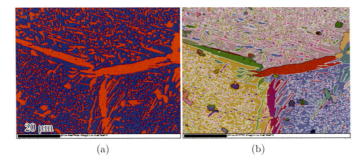

Fig.2.3.21 (a) Phase map and (b) orientation map of the sample.

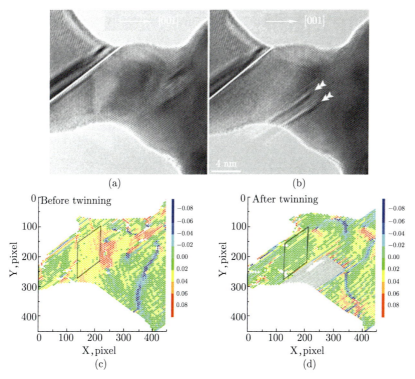

Fig.3.2.2 Deformation twinning with their quantitative strain analysis. (a) and (b) Sequential images captured before and after the twinning partials are emitted; the double arrowheads indicate the location of the deformation twins. (c) and (d) The strain mapping of the HRTEM images of (a) and (b), respectively, showing a strain relaxation immediately after the twin formation. Likewise, the open circles are drawn to represent the twinning partials (Zheng, 2010).

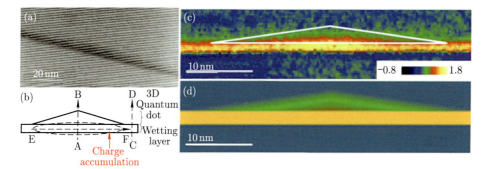

Fig.3.7.9 (a) Electron hologram of individual Ge quantum dot with [110] projection embedded in Si [001] substrate. (b) Sketch of the pyramid shaped dot and wetting layer. (c) Reconstructed phase image of the Ge quantum dot. Phase bar calibrated in radians is shown at bottom right. (d) Simulated phase image of Ge quantum dot based on the pyramid model sketched in (b), the same phase bar as in (c) is applied. (Li et al., 2009 with permission)

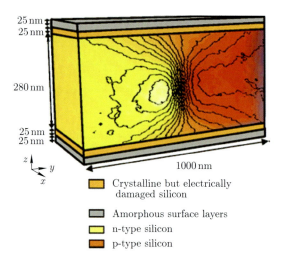

Fig.3.7.10 Tomographic reconstruction of the electrostatic potential arising from a 3-D section of a thin FIB-prepared membrane containing an electrically biased p-n junction.

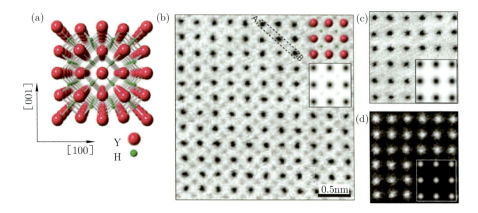

Fig.4.5.4 ABF, BF and ADF–STEM images of the crystalline compound YH_2. (a) Crystal structure of YH_2 viewed from the [010] direction. (b)–(d), ABF, BF and ADF images obtained with the detector ranges 11–22 mrad, 0–22 mrad and 70–150 mrad, respectively. Simulated images are inset in images (b)–(d), and the YH_2 unit-cell projection is overlaid in (b). Ishikawa et al., 2011.

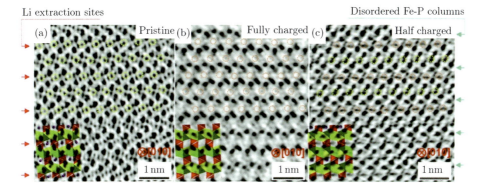

Fig.4.5.5 ABF micrographs showing Li ions of partially delithiated LiFePO$_4$ at every other row. (a) Pristine material with the atomic structure of LiFePO$_4$ shown as inset; (b) fully charged state with the atomic structure of FePO$_4$ shown for comparison; and (c) half charged state showing the Li staging. Note that Li sites are marked by yellow circles and the delithiated sites are marked by orange circles. Gu et al., 2011.

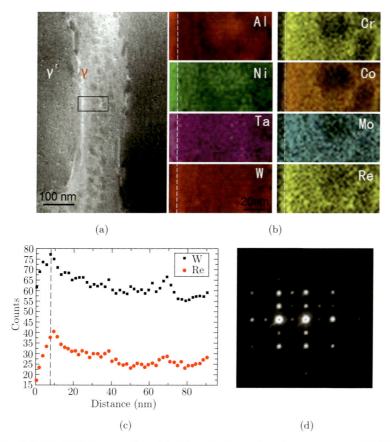

Fig.4.7.1 (a) HAADF image of a nickel-based alloy after the creep test. (b) Element mapping images of major constituents corresponding to areas denoted by a black rectangle in (a). (c) Distribution of element Re and W along the direction perpendicular to the interfaces. Dashed lines schematically indicate the γ/γ' interfaces. (d) Diffraction pattern corresponding to the low-contrast area in γ phases in (a).

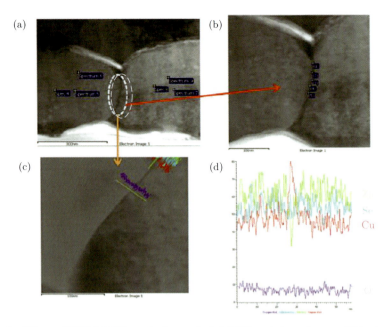

Fig.5.3.6 Ways of TEM/EDS analysis used: (a) EDS point analyses within two neighboring grains, (b) EDS point analyses along the grain boundary, (c) EDS scanning across the boundary, and (d) the magnified view of the resultant profiles in (c).

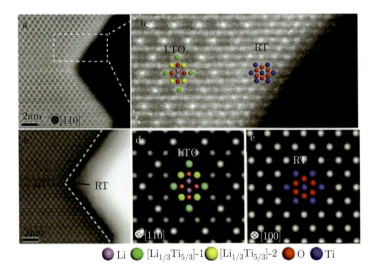

Fig.5.3.9 STEM images of LTO-RT-600: (a, b) HAADF and (c) corresponding ABF images of LTO-RT-600 NSs. The simulated HAADF images of (d) LTO projected from [110] direction and (e) rutile-TiO_2 projected from [100] direction. The insets show the arrangements of atoms.

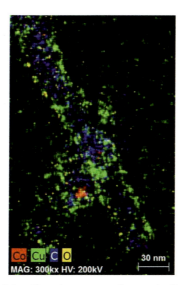

Fig.5.3.10 CuO-coated multi wall carbon nanotubes with Co-catalyst particles analyzed with the XFlash® 6T |30. (Herrmann, S. and Wachtler, T.)

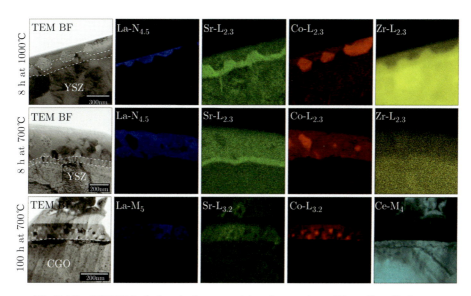

Fig.5.4.5 EFTEM of chemical compositional mapping of $La_{0.5}Sr_{0.5}CoO_{3-\delta}$.

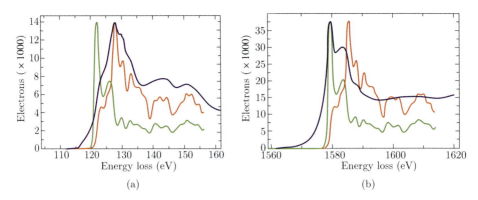

Fig.5.5.7 Comparison of the unoccupied Al p-PDOS with(green) and without(red) a core-hole and ELNES(blue) at the Al-L_1 (left) and Al-K (right) edge in α-Al_2O_3(bulk).

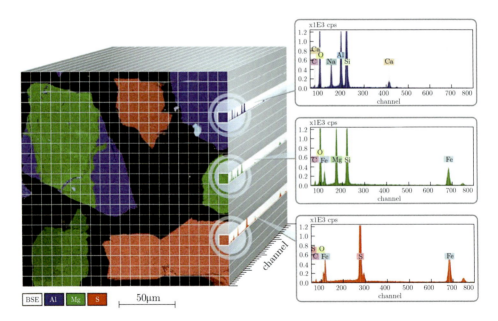

Fig.5.6.2 The hypermap data cube contains a spectrum for each pixel.

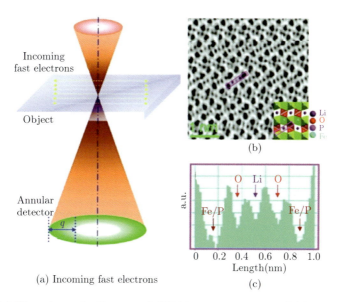

Fig.6.3.1 (a) The schematic diagram of ABF imaging geometry, (b)A demonstration of lithium sites within a LiFePO$_4$ crystal is shown in (b) with the corresponding line profile acquired at the box region shown in (c) to confirm the lithium contrast with respect to oxygen (Gu et al., 2011)

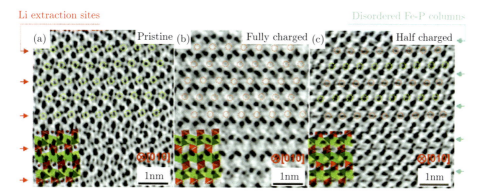

Fig.6.3.2 Initial image in LiFePO$_4$ [010] direction and completed charged and partially charged images are shot through the STEM-ABF technique (Gu et al., 2011).

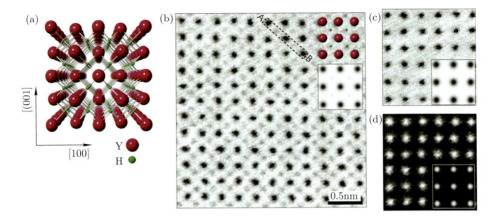

Fig.6.3.3 (a) The structural diagram of crystal YH$_2$ along the [010] direction, (b) ABF image, (c) bright-field image and (d) ADF image (Ishikawa et al., 2011).

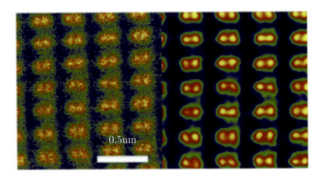

Fig. 6.3.11 Image of ⟨112⟩ Si recorded using a VG microscopes HB603U with Nion aberration corrector operating at 300 kV (left) and image after low-pass filtering and unwarping (right) (Nellist et al., 2004).

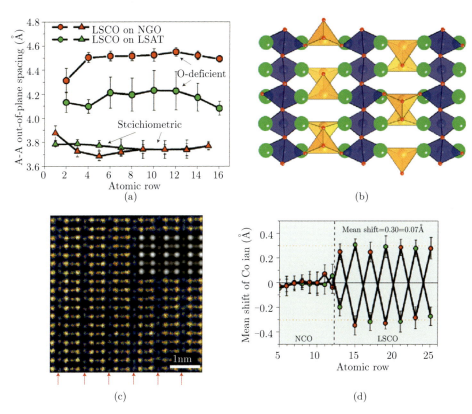

Fig. 6.3.20   Representative lattice spacing change and identified brownmillerite LSCO on NGO.

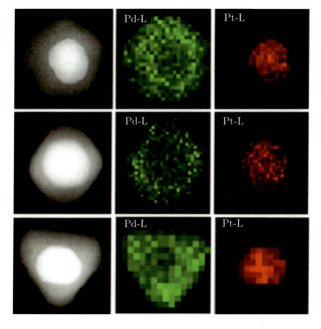

Fig. 6.3.36   EDX mapping of all three polyhedral core-shell nanoparticles (Khanal et al., 2012).

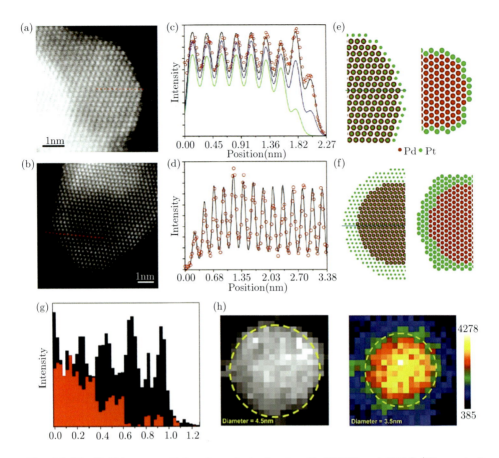

Fig. 6.3.37 Pt-Pd nanoparticles characterized using C_s-STEM and EELS (Wang et al., 2009).

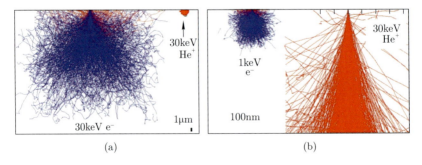

Fig. 8.2.2 (a) 30 keV electron beam interaction volume compared with 30 keV helium ion beam interaction volume in silicon. (b) 1 keV electron beam interaction volume compared with part of the 30 kV helium ion beam interaction volume in silicon.